# 13th International Conference on Development and Modernization of the Manufacturing (RIM 2021)



IOP Conference Series: Materials Science and Engineering
Volume 1208

Sarajevo, Bosnia and Herzegovina
29 September - 1 October 2021

**Editors:**

**Zijah Burzic**
**Budimir Mijovic**
**Fadil Islamovic**
**Atif Hodzic**

ISBN: 978-1-7138-4667-3
ISSN: 1757-8981

**Printed from e-media with permission by:**

Curran Associates, Inc.
57 Morehouse Lane
Red Hook, NY 12571

**Some format issues inherent in the e-media version may also appear in this print version.**



Institute of Physics
Dirac House, Temple Back
Bristol BS1 6BE UK

Phone: 44 1 17 929 7481
Fax: 44 1 17 920 0979

techtracking@iop.org

**Additional copies of this publication are available from:**

Curran Associates, Inc.
57 Morehouse Lane
Red Hook, NY 12571 USA
Phone: 845-758-0400
Fax: 845-758-2633
Email: curran@proceedings.com
Web: www.proceedings.com

# TABLE OF CONTENTS

## Preface

This volume of "IOP Conference Series: Material Science and Engineering" contains the selected papers presented at the 13th International Conference on Development and Modernization of the Manufacturing (RIM 2021), which was held in Sarajevo (Bosnia and Herzegovina) from 29th September to 1st October 2021.

Since 1997, the Faculty of Technical Engineering of the University of Bihać has successfully organized and held 13 international scientific conferences RIM on a bi-annual basis in 4 different countries: Bosnia and Herzegovina, Croatia, Egypt, and Montenegro. The main goal of the Conference is to bring together international scientists, engineers, industrialists, and experts providing them a unique platform for sharing new ideas and the recent developments in various areas of engineering.

The papers of this volume are related to the following thematic areas:

• Mechanical Engineering

• Wood Design and Technologies

• Electrical Engineering and Computer Science

• Technologies and Structures in Civil Engineering

The Organization Committee of the RIM 2021 conference would like to thank all the authors who contributed to this volume as well as to the reviewers, invited speakers, and all conference participants.

### List of Editors:

- Prof. Zijah Burzić, PhD
- Prof. Budimir Mijović, PhD
- Prof. Fadil Islamović, PhD
- Prof. Atif Hodžić, PhD

### List names of Organization Committee:

Fadil Islamović, PhD, president; Atif Hodzić, PhD, vice-president; Aladin Crnkić, PhD, secretary; Damir Hodzić, PhD; Fatka Kulenović, PhD; Edin Mujčić, PhD; Edis Softić, PhD; Šefik Behrem, PhD; Redžo Hasanagić, PhD; Minka Ćehić, PhD; Samir Vojić, PhD; Aida Husetić, PhD; Dženana Gačo, PhD; Bahrudin Hrnjica, PhD; Esad Bajramović, PhD; Ratko Knežević, PhD; MSc Amel Džanić; MSc Amel Toroman; MSc Dejla Ramić; MSc Mustafa Ružnić; MSc Almin Hopovac; MSc Zinaid Kapić; BSc Emir Komić; BSc Sena Handžić; BSc Maida Galić; BSc Nejra Mesić.

### List names of Scientific Committee:

Salah-Eldien Omer, PhD, president (CRO); Fadil Islamović, PhD (BIH); Atif Hodžić, PhD (BIH); Ismet Bišćević, PhD (BIH); Budimir Mijović, PhD (CRO); Thomas Stautmeister, PhD (GER); Milan Jurković, PhD (CRO); Refik Šahinović, PhD (BIH); Fadi Al-Turjman, PhD (TUR); Anatoly Chubinski, PhD (RUS); Necati Agiralioglu, PhD (TUR); Ali Danadeh Mehr, PhD (TUR); Zijah Burzić, PhD (SRB); Milan Šernek, PhD (SLO); Radivoje Topić, PhD (SRB); Branko Božić, PhD (CRO); Hamid Farrokh Ghatte, PhD (TUR); Zoran Trposki, PhD (MKD); Munir Jahic, PhD (BIH); Darko Ujević, PhD (CRO); Slobodan Bojanić, PhD (ESP); Siniša Ilić, PhD (SRB); Smail Klarić, PhD (BIH); Fatka Kulenović, PhD (BIH); Vencislav Grabulov, PhD (SRB); Edin Mujčić, PhD (BIH); Tasko Maneski, PhD (SRB); Sanin Džidić, PhD (BIH); Seid Žapčević, PhD (BIH); Galina Varankina, PhD (RUS); Amir Čaušević, PhD (BIH); Dženana Gačo, PhD (BIH); Mersida

Manjgo, PhD (BIH); Jovan Vučinić, PhD (CRO); Damir Hodžić, PhD (BIH); Endre Romhanji, PhD (SRB); Marko Petrič, PhD (SLO); Šefik Behrem, PhD (BIH); Meri Burzić, PhD (SRB); Zdravko Popović, PhD (SRB); Samir Vojić, PhD (BIH); Enes Hasić, PhD (BIH); Bahrudin Hrnjica, PhD (BIH); Enis Džanić, PhD (BIH); Ramo Šendelj, PhD (MNE); Edis Softić, PhD (BIH); Edin Fatkić, PhD (BIH); Edina Karabegović, PhD (BIH); Goran Zlateski, PhD (MKD); Zlatko Bundalo, PhD (BIH); Blaženka Brlobašić Šajatović, PhD (CRO); Anica Hursa Šajatović, PhD (CRO); Irfan Hošić, PhD (BIH); Ermin Husak , PhD (BIH); Esad Bajramović, PhD (BIH); Aladin Crnkić, PhD (BIH); Redžo Hasanagić, PhD (BIH); Husein Rošić, PhD (BIH); Mehmed Mahmić, PhD (BIH); Minka Ćehić, PhD (BIH); Kemal Gutić, PhD (BIH); Aida Husetić, PhD (BIH); Muamer Peljto, PhD (BIH); Petar Marić, PhD (BIH); Razija Begić, PhD (BIH); Sebastian Dendorfer, PhD (GER); Uwe Reischl, PhD (USA); Vladimir Karanakov, PhD (MKD); Gjorgi Gruevski, PhD (MKD); Dimitar Angelski, PhD (BUG); Damir Kalpić, PhD (CRO); Miha Humar, PhD (SLO); Fuad Hadžikadunić, PhD (BIH); Ibrahim Plančić, PhD (BIH); Nedeljko Vukojević, PhD (BIH); Sabahudin Jašarević, PhD (BIH); Branko Blanuša, PhD (BIH); Sandira Eljsan, PhD (BIH); Tihomir Latinovic, PhD (BIH); Slobodanka Ključanin, PhD (BIH); Mirko Dobrnjac, PhD (BIH); Ranko Babić, PhD (SRB); Mile Petrović, PhD (SRB); Vasilije Petrović, PhD (SRB); Ratko Knežević, PhD (BIH); Nevena Mijajlović, PhD (MNE); Ivana Ognjanović, PhD (MNE); Muharem Kozić, PhD (BiH); Aida Kopljar, PhD (CRO); Jasna Hamzabegović, PhD (BIH); Huska Jukić, PhD (BiH), Albin Muslić, PhD (BIH); Jasmina Bešlagić, PhD (BIH); Belma Bajrić, PhD (BIH); Husein Vilić, PhD (BIH); Halid Makić, PhD (BIH); Azra Bakrač, PhD (BIH); Mirela Kljajić-Dervić, PhD (BIH); Admir Midžić, PhD (BIH); Ifet Šišić, PhD (BIH); Jasmina Ibrahimpašić, PhD (BIH).

**List names of reviewers:**

Galia Marinova, PhD (BGR); Salah-Eldien Omer, PhD (CRO); Edin Mujčić, PhD (BIH); Petar Marić, PhD (BIH); Ahmed El Sayed, PhD (BIH); Tihomir Latinović, PhD (BIH); Selver Softić, PhD (AUT); Muharem Kozić, PhD (BIH); Branko Božić, PhD (CRO); Marko Subotić, PhD (BIH); Darko Motik, PhD (CRO); Goran Zlateski, PhD (MKD); Zoran Trposki, PhD (MKD); Mersida Manjgo, PhD (BIH); Jasna Hamzabegović, PhD (BIH); Siniša Ilić, PhD (SRB); Branko Blanuša, PhD (BIH); Aladin Crnkić, PhD (BIH); Redžo Hasanagić, PhD (BIH); Damir Hodžić, PhD (BIH); Sanel Gredelj, PhD (BIH); Sefik Behrem, PhD (BIH); Fadil Islamovic, PhD (BIH); Gordana Jovanovic Dolecek, PhD (MEX); Samir Vojić, PhD (BIH); Hamid Farrokh Ghatte, PhD (TUR); Fatka Kulenović, PhD (BIH); Esad Bajramović, PhD (BIH); Aida Husetić, PhD (BIH); Miha Humar, PhD (SLO); Mirzet Beganović, PhD (BIH); Minka Ćehić, PhD (BIH); Nebojša Todorović, PhD (SRB); Thomas Stautmeister, PhD (GER); Milanka Điporović-Momčilović, PhD (SRB); Vencislav Grabulov, PhD (SRB); Muamer Peljto, PhD (BIH); Fuad Hadžikadunić, PhD (BIH); Nedim Hodžić, PhD (BIH); Murčo Obućina, PhD (BIH); Mirko Dobrnjac, PhD (BIH); Dragi Tiro, PhD (BIH); Admir Midžić, PhD (BIH); Đulaga Hadžić, PhD (BIH); Džemo Tufekčić, PhD (BIH); Kemal Gutić, PhD (BIH); Sandira Eljšan, PhD (BIH); Ejub Džaferović, PhD (BIH); Alan Topčić, PhD (BIH); Ranko Antunović, PhD (BIH); Dusan Kovacevic, PhD (SRB); Zlatko Bundalo, PhD (BIH); Sebastian Dendofer, PhD (GER); Marko Petrič, PhD (SLO); Milan Šernek, PhD (SLO); Senada Pobrić, PhD (BIH); Budimir Mijović, PhD (CRO); Radivoj M. Topić, PhD (SRB); Dženana Gačo, PhD (BIH); Sanin Džidić, PhD (BIH); Atif Hodžić, PhD (BIH); Zijah Burzić, PhD (SRB); Said Pašalić, PhD (BIH); Bahrudin Hrnjica, PhD (BIH); Seid Žapčević, PhD (BIH); Edis Softić, PhD (BIH); Amir Čaušević, PhD (BIH); Milan Jurković, PhD (CRO).

# Peer review declaration

All papers published in this volume of *IOP Conference Series: Materials Science and Engineering* have been peer reviewed through processes administered by the Editors. Reviews were conducted by expert referees to the professional and scientific standards expected of a proceedings journal published by IOP Publishing.

- **Type of peer review:** Double-blind

- **Conference submission management system:**

Submission of papers and all correspondence between authors, technical secretary, editors, and reviewers were managed with our conference system „Submit System“ (http://rim.tfb.ba/submitsystem/login).

- **Number of submissions received:** 94

- **Number of submissions sent for review:** 89

- **Number of submissions accepted:** 44

- **Acceptance Rate (Number of Submissions Accepted / Number of Submissions Received X 100):** 46,81%

- **Average number of reviews per paper:** 2 to 3

- **Total number of reviewers involved:** 68

- **Any additional info on review process: -**

- **Contact person for queries:**

Aladin Crnkic, PhD, secretary of RIM 2021 conference; aladin.crnkic@unbi.ba; Irfana Ljubijankica bb, 77000 Bihac, Bosnia and Herzegovina

# Towards ergonomics working - machine learning algorithms and musculoskeletal modeling

**Franz Suess, Maximilian Melzner and Sebastian Dendorfer**

Laboratory for Biomechanics, Ostbayerische Technische Hochschule (OTH), Galgenbergstraße 30, 93053 Regensburg, Germany

E-mail: sebastian.dendorfer@oth-r.de

**Abstract**. Ergonomic workplaces lead to fewer work-related musculoskeletal disorders and thus fewer sick days. There are various guidelines to help avoid harmful situations. However, these recommendations are often rather crude and often neglect the complex interaction of biomechanical loading and psychological stress. This study investigates whether machine learning algorithms can be used to predict mechanical and stress-related muscle activity for a standardized motion. For this purpose, experimental data were collected for trunk movement with and without additional psychological stress. Two different algorithms (XGBoost and TensorFlow) were used to model the experimental data. XGBoost in particular predicted the results very well. By combining it with musculoskeletal models, the method shown here can be used for workplace analysis but also for the development of real-time feedback systems in real workplace environments.

## 1. Introduction

Ergonomic workplaces lead to fewer work-related musculoskeletal disorders and thus fewer sick days [1,2]. The design and study of workplaces is the subject of numerous studies, either directly on site or through an experiment in a laboratory where the work environment is replicated. Both methods require time-consuming and expensive preparation with an iterative process for actual optimisation.

Ergonomic workplace assessment tools, such as the Rapid Upper Limb Assessment (RULA) [3], have been developed to assess and minimise risk factors for occupational upper musculoskeletal injury in industrial workers. These include joint positions, the amount of weight-bearing and the amount of weight applied. Furthermore, muscle activity is integrated by describing whether the body position is static or dynamic. However, the assessment of different body postures during an activity is not determined here or only through repeated use. Other guidelines, such as those of the National Institute for Occupational Safety and Health, give values and limits for e.g. intervertebral disc forces during occupational activities [4,5].

Such limits allow for a more accurate analysis of work processes but require tools to determine spinal loading throughout the activity. Musculoskeletal modelling tools (MMT) provide this functionality and have already demonstrated their capabilities in various studies to determine ergonomic and workplace analyses and recommendations. For example, they were able to demonstrate their capabilities in material handling in supermarkets [6]. In another experimental setup, Davis et al. [7] were even able to incorporate mental stress with the help of the EMG-controlled MMT. They were able to show that mental stress increases disc strain and thus provides a tool to incorporate psychological factors. As the analyses of mental stress in real-life situations can be difficult and EMG measurements are prone to errors as they can be influenced by artefacts [8], like movements [9], a tool that simulates the movement in combination with mental stress would overcome these problems. A possible solution could be the use of machine learning tools to estimate muscle activity (even under the influence of mental stress) and provide input data for MMT.

Similar approaches, based purely on mechanical loading, have been used in several studies. For gait analysis Heller et al. [10] used machine learning techniques to reconstructed muscle activity dependent on kinematic data like walking speed as well as joint kinematics. With moderate static loading, Nussbaum et al. [11,12] used experimental data to train and predict lumbar muscle recruitment. Input data of thoracic angle and external load predicted measured activity of erector spinae, rectus abdominus, oblique externus, and latissimus dorsi. Luh Chang et al. [13] and Wang et al. [14] used arm muscle activity in combination with kinematics to train neural network to predict joint torque. All of these studies are purely based on biomechanical loading and neglect the influence of mental stress.

Therefore, the aim of this study was to investigate the feasibility of using machine learning models based on experimental data to predict and incorporate a submaximal subject-specific stress-induced muscle activity (MA) response which might be the basis for a more holistic description of the influence of work place environments on the human body.

## 2. Methods

41 subjects performed 25-degree trunk flexion/extension cycles with and without exposure to cognitive stress, each lasting 120 seconds. The paced auditory serial addition task (PASAT)[15] was used to model cognitive stress (COS) and provides data for the stressed muscle activity. Prior to the stress test, a baseline measurement was taken, which also lasted 120 seconds, providing non-stress data. During the extension, the subject had to push against 5 % maximum load determined by the maximum voluntary contraction test (MVC) beforehand. A dynamometer with an additional TP 500 adapter (CON-TREX® WS; PHYSIOMED AG) set and recorded the motion, moment, and angle (Figure 1). Electromyography (EMG) sensors (Trigno IM sensors from Delsys) recorded six main lower back muscles (m. multifidi, m. erector spinae pars iliocostalis, m. erector spinae pars longissimus - left and right side each).

During the postprocessing a mean cycle for each subject for baseline and stress is computed. Therefore, the data was split at the upper dead centre, aligned to map a full 100 % cycle representing the mean cycle time. The data from each subject was used to compute a mean cycle for torque, speed, and MVC normalized EMG data.

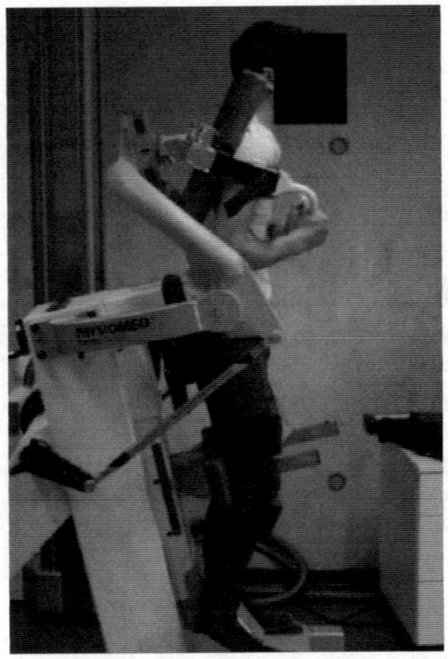

**Figure 1.** CON-TREX® WS dynamometer with TP500 adapter. The subject was attached at the leg, knee, hip and thorax to the device.

The records are divided into baseline and stress data. Furthermore, it is assumed that the subject was maximally stressed under the influence of the COS and absolutely relaxed during the baseline measurement. Based on this assumption, the stress level feature is set to zero for the baseline data and one otherwise. A single training data set contains the kinematic variables: Position, velocity, acceleration, and torque. The subject-specific data weight and height are added to the time series data and provide a sample for training and results in 28297 data points.

### 2.1. *Gradient Boosting*

The library XGBoost (2020d) for Python is the basis for this implementation. First, the data is collected, then stored in a pandas DataFrame and written to disk as a Python pickle object. This approach allows the data to be loaded into memory during the computational steps and saves time for device access and file parsing. The training includes the following parameters and their corresponding values, and only values without defaults are listed here. All other values are documented at the official XGBoost documentation [16]:

- eta: 0.1
- gamma:0
- max_depth: 9
- min_child_weight: 6
- subsample: 0.7

Using these hyperparameters, a machine learning model is trained for each mean EMG channel in the low back as an outcome variable and as a function of subject height, weight, stress level, thoracic flexion angle, thoracic velocity, thoracic acceleration, and applied torque. In the case of the stress level feature, baseline data are coded with a value of zero and one for the stress test. This approach assumes that the subject was in a 100% arousal state and allows the prediction of submaximal muscle activity for a lower stress level.

### 2.2. *Neural Network*

Similar to the gradient tree model, a third-party Python library is used for the neural network (NN) implementation. TensorFlow [17], as an open source multi language library, supports various types of devices and was therefore selected for this implementation. The pre-collected data is used for grid search and Bayesian optimization to perform hyperparameter optimization. The following parameters are part of the optimization: Learning Rate; Epochs, Batch size; Number of neurons in layer 1 - 3. The following table (Table 1) shows the outer boundaries of the search space.

**Table 1.** Parameter boundaries used for hyperparameter grid search. Bayesian optimization uses the optimal parameters as starting parameters.

| Boundary | Learning rate | Epochs | Batch Size | Layer 1 | Layer 2 | Layer 3 |
|---|---|---|---|---|---|---|
| - | 0.01 | 10 | 50 | 14 | 14 | 14 |
| + | 0.001 | 50 | 500 | 28 | 28 | 28 |

The result of the grid search is used as the initial value for the Bayesian optimization package (2020a). The parameter space was further extended to include learning rate; neurons in input layer, number of hidden layers; number of neurons in hidden layer; activation function: batch size; epoch size. Tanh, Relu, and linear function served as possible activation functions for best data fit. Predefined parameters and ranges are adopted by Montavon et al. [18].

During the training process, Bayesian optimization results are used to generate the NN. First, all input variables are mapped to a range between zero and one. This is necessary to avoid any bias due to the different scales of variables. Then, the data is fed into the Tensorflow training routine with a validation sample size of five percent. The mean absolute error (MAE) was chosen to quantify the model quality. The training process was stopped if this value did not change over 10 iterations.

The quality of the two muscle activity regression models is evaluated with the training data of their specific trial using MAE and person correlation. To predict the submaximal load response with the models, the baseline data was used again, but with a load value of 0.5.

## 3. Results

Table 2 and table 3 show the differences between prediction and training data. In general, XGBoost shows better results than the TensorFlow neural network (TNN). This is true for both baseline and stress data. The XGboost models are able to model the data with an MAE below 2.53% m. multifidi muscle activity. In contrast, the TNN error is much larger with up to 68.6% muscle activity in the m. erector spinae pars iliocostalis (Table 2). The similarity analysis with the Pearson correlation coefficient r (Table 3) shows a strong correlation with over 0.78 for the TNN regression model. In this case, the correlation becomes smaller following the increase of MAE. The XGBoost model shows a strong correlation for all muscle activity models above 0.99. In the case of the 50% stress level prediction, both models show different results. In general, the MAE for the XGBoost prediction is lower and closer to the baseline data. Whereas it has its maximum values with 39% and 48% muscle activity in the m. multifidi, the TNN model has its maximum in the m. erector spinae pars iliocostalis with 68% and 113% muscle activity to the baseline and maximum stress. The Pearson correlation coefficient r is smaller for the TNN model, ranging from 0.75 to 0.91 compared to the baseline data and 0.74 to 0.89 for the stress correlation. The XGBoost model varies from 0.85 to 0.95 and 0.77 to 0.92, respectively.

**Table 2.** MAE values for the TNN and XGBoost training evaluation data.

| | Baseline | | Stress | | 50% level to baseline | | 50% level to stress | |
|---|---|---|---|---|---|---|---|---|
| | TNN | XGboost | TNN | XGboost | TNN | XGboost | TNN | XGboost |
| | erector spinae pars longissimus | | | | | | | |
| **right** | 13.55 (4.28) | 1.01 (0.48) | 14.11 (5.99) | 1.08 (0.63) | 7.58 (3.75) | 17.72 (8.44) | 8.69 (5.4) | 26.85 (18.48) |
| **left** | 19.06 (10.2) | 1.12 (0.55) | 19.59 (10.47) | 1.33 (1.37) | 5.97 (2.39) | 21.32 (10.65) | 8.61 (4.9) | 28.22 (15.99) |
| | m. erector spinae pars iliocostalis | | | | | | | |
| **right** | 17.62 (7.7) | 1.00 (0.69) | 20.95 (14.62) | 0.99 (0.69) | 51.87 (39.58) | 21.25 (13.17) | 69.28 (75.20) | 32.78 (33.12) |
| **left** | 49.34 (82.34) | 2.00 (1.81) | 68.60 (176.99) | 1.70 (1.49) | 68.21 (96.73) | 29.59 (24.91) | 113.96 (241.56) | 35.06 (27.8) |
| | m. multifidi | | | | | | | |
| **right** | 35 (16.35) | 2.21 (1.34) | 45.28 (30.28) | 2.53 (2.87) | 43.87 (31.65) | 39 (35.64) | 57.24 (46.31) | 48.08 (41.37) |
| **left** | 34.48 (12.85) | 2.11 (3.97) | 32.10 (14.42) | 1.7 (1.67) | 44.21 (37.57) | 30.42 (52.9) | 56.66 (43.41) | 31.4 (21.62) |
| **mean** | 28.17 | 1.58 | 33.44 | 1.56 | 36.95 | 26.55 | 52.41 | 33.73 |

**Table 3.** Pearson r values for the TNN and XGBoost training evaluation data.

| | Baseline | | Stress | | 50% level to baseline | | 50% level to stress | |
|---|---|---|---|---|---|---|---|---|
| | TNN | XGboost | TNN | XGboost | TNN | XGboost | TNN | XGboost |
| | erector spinae pars longissimus | | | | | | | |
| **right** | 0.97 (0.03) | 0.999 (0.0008) | 0.98 (0.02) | 0.99 (0.001) | 0.84 (0.16) | 0.95 (0.07) | 0.83 (0.16) | 0.92 (0.07) |
| **left** | 0.96 (0.04) | 0.999 (0.0003) | 0.97 (0.02) | 0.999 (0.0004) | 0.91 (0.07) | 0.94 (0.08) | 0.89 (0.08) | 0.92 (0.08) |
| | m. erector spinae pars iliocostalis | | | | | | | |
| **right** | 0.95 (0.04) | 0.999 (0.001) | 0.95 (0.04) | 0.999 (0.001) | 0.79 (0.27) | 0.93 (0.08) | 0.77 (0.25) | 0.88 (0.12) |
| **left** | 0.86 (0.28) | 0.999 (0.002) | 0.87 (0.27) | 0.999 (0.003) | 0.75 (0.27) | 0.93 (0.06) | 0.74 (0.27) | 0.89 (0.08) |
| | m. multifidi | | | | | | | |
| **right** | 0.82 (0.19) | 0.996 (0.004) | 0.78 (0.35) | 0.99 (0.03) | 0.75 (0.29) | 0.85 (0.23) | 0.74 (0.31) | 0.77 (0.27) |
| **left** | 0.86 (0.11) | 0.999 (0.001) | 0.87 (0.14) | 0.999 (0.003) | 0.79 (0.22) | 0.902 (0.17) | 0.78 (0.25) | 0.85 (0.17) |
| **mean** | 0.90 | 0.999 | 0.90 | 0.996 | 0.81 | 0.92 | 0.79 | 0.87 |

The response of the XGBoost for a random subject and a selected muscle (erector spinae pars longissimus) is given in Figure 2. At the 40% position in the cycle, the prediction and target values increase as the subject moves backwards. The curve for the 50 % stress level follows the baseline except in the middle of the backward movement where it moves closer to the stress data. Overall, the predictions are following the experimental data quite well. Results of the TNN model are presented in

Figure 3. The baseline and COS prediction do not follow the target values very closely. The prediction of the load level at 50% reflects this trend.

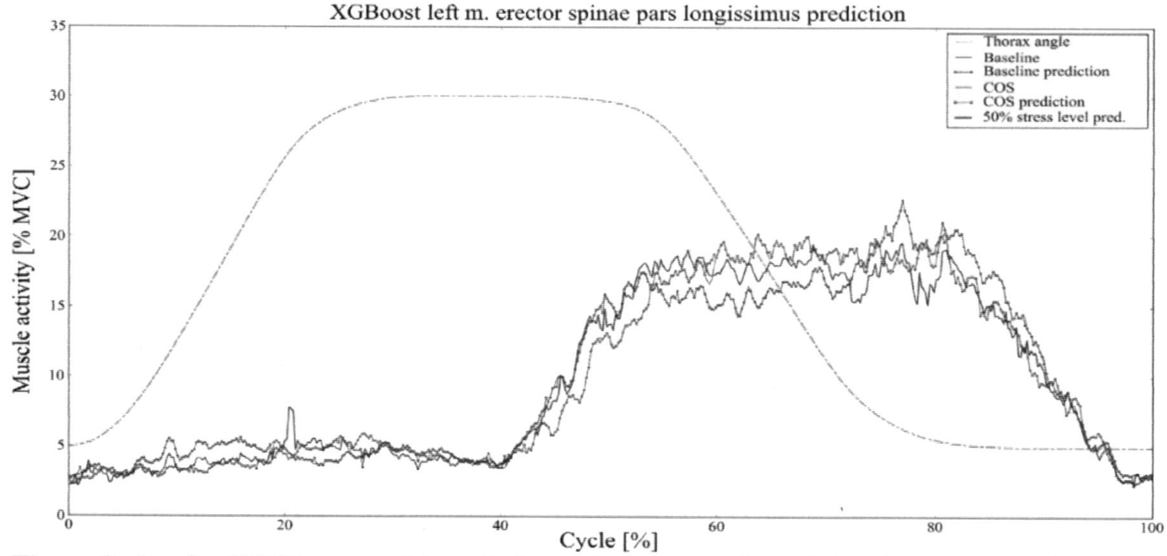


**Figure 2.** Result of XGBoost model prediction and target muscle activity of m. erector spinae pars longissimus. The yellow line indicates the thoracic angle from 5° to 30°.

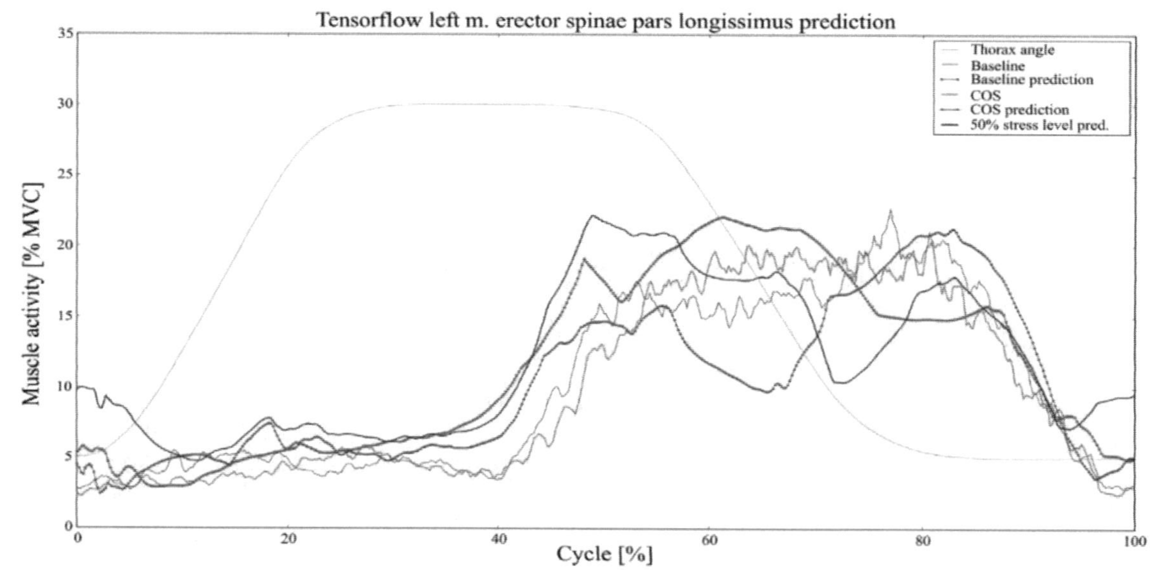


**Figure 3.** TensorFlow model prediction result and target muscle activity of the m. erector spinae pars longissimus. The yellow line indicates the thoracic angle. The yellow line indicates the thoracic angle from 5° to 30.

## 4. Discussion

Machine learning models of lumbar muscle recruitment were developed to predict muscle activity for thoracic movements in response to cognitive stress. Both the XGBoost and TNN models are able to predict the activation pattern as a function of thoracic flexion angle, velocity, torque, and stress level, but with different levels of precision.

In particular, the gradient boost approach shows greater accuracy for the relatively small amount of sample data. For the person correlation as a measure of similarity, this trend is repeated with a higher r-value for the gradient-boost method. The average value of r = 0.9 for TNN found in this study is enhanced by the gradient-boost model which was able to increase the correlation to r = 0.996. With an MAE below 2.53 % in the case of the gradient-boost model, the prediction is quite accurate. Despite a good correlation for the maximum and minimum stress data, this trend is not as accurate for the 50% stress level. In this case, one would expect an approximately equal MAE value when comparing the prediction to the baseline and stress curves, but this is not the case. This situation may be caused by a nonlinear transformation from the baseline to the loading data or by missing features such as lateral torque in the model. In addition, the stress-related kinetic changes are influencing the muscle activity, which may be represented by a non-uniformly distributed MAE value. The accessibility to such models allows numerical studies to identify kinetic parameters causing additional stress-related loads. The example in Figure 3 shows the underestimation and overestimation of muscle activity during backward movement. This leads to a less accurate estimate of submaximals. This trend is smaller in the case of the gradient boost method but is still present.

The relatively small sample size and the unknown actual stress level of the subject directly affect the models and cause the lower accuracy of the TNN model. The use of the data for training and analysis is a direct result of the small sample size. Nevertheless, this approach was taken to establish and identify a procedure if more data were available to train a general model. Prediction of the submaximal stress pattern using kinetic data from the baseline study may not reflect the correct trajectory of motion and may result in a less accurate prediction. In addition, to provide training data for submaximal loads, there is a limitation in accurately determining the stress level. Using an individual scaled value that is independent of personal assessment, submaximal measurements are possible. This option would provide the missing data point and allow for much more accurate training. Therefore, a study design that performs submaximal measurement is necessary to capture a range of stress responses.

The results of this study indicate that the combination of machine learning algorithms and experimental data provides a useful basis for the analysis of stress-induced muscle loads. Combining these methods with musculoskeletal modeling opens up the possibility to study a wide variety of movements and tasks. This could enable more realistic assessments of workplace environments. The future potential of these methods is considerably high. The process of using only very limited input data, e.g., collected by wearables, as control variables for a trained algorithm opens up the possibility of not only evaluating workplaces, but also creating direct feedback for potentially hazardous situations.

## 5. Conclusion

Despite a number of limitations in this study, the results clearly demonstrate that the combined musculoskeletal and mental load can be considered in ergonomic analyses and that the demonstrated methodology could serve as a basis for novel ergonomic workplace designs by applying the technology directly to workers in, for example, wearable-based setups.

IOP Conf. Series: Materials Science and Engineering 1208 (2021) 012001 doi:10.1088/1757-899X/1208/1/012001

**References**

[1] Ross P 1994 *AAOHN Journal* **42** 171.
[2] Heidarimoghadam R, Mohammadfam I, Babamiri M, Soltanian A R, Khotanlou Hand Sohrabi M S 2020 *International journal of occupational safety and ergonomics : JOSE* 1.
[3] McAtamney Land Nigel Corlett E 1993 *Applied Ergonomics* **24** 91.
[4] Nelson GS W H 1981 *Work practices guide for manual lifting* (U.S. Department of Health and Human Services, Public Health Service, Centers for Disease Control, National Institute for Occupational Safety and Health).
[5] Waters T R, Putz-Anderson V, Garg Aand Fine L J 1993 *Ergonomics* **36** 749.
[6] Larsen F G, Svenningsen F P, Andersen M S, Zee M deand Skals S 2020 *Ann Biomed Eng* **48** 805.
[7] Davis K G, Marras W S, Heaney C A, Waters T Rand Gupta P 2002 *Spine* **27** 2645.
[8] Konrad P 2005 *The ABC of EMG. A Practical Introduction to Kinesiological Electromyography* (USA: Noraxon INC.).
[9] Luca C J de, Gilmore L D, Kuznetsov Mand Roy S H 2010 *Journal of Biomechanics* **43** 1573.
[10] Heller B W, Veltink P H, Rijkhoff N J, Rutten W Land Andrews B J 1993 *Biol. Cybern.* **69** 327.
[11] Nussbaum M A, Chaffin D Band Martin B J 1995 *Functional Tissue Engineering* **28** 1015.
[12] Nussbaum M A, Martin B Jand Chaffin D B 1997 *Functional Tissue Engineering* **30** 251..
[13] Luh J-J, Chang G-C, Cheng C-K, Lai J-Sand Kuo T-S 1999 *J Electromyogr Kinesiol* **9** 173.
[14] Wang L, Buchanan T S 2002 *IEEE Trans Neural Syst Rehabil Eng* **10** 30.
[15] Lejuez C W, Kahler C Wand Brown R A 2003 *The Behavior Therapist.*
[16] 2020 *XGBoost Documentation — xgboost 1.2.0-SNAPSHOT documentation.* Available at https://xgboost.readthedocs.io/en/latest/#.
[17] 2020 *TensorFlow.* Available at https://www.tensorflow.org/.
[18] Montavon G, Orr G Band Müller K-R 2012 *Neural networks: tricks of the trade* (Berlin: Springer).

# Energy efficient embedded systems and their application in wireless sensor networks

**Zlatko V Bundalo**

Faculty of Electrical Engineering, University of Banja Luka, Bosnia and Herzegovina

E-mail: zlatko.bundalo@etf.unibl.org

**Abstract.** Embedded systems are specialised electronic systems that perform limited number of fixed operations and are used in many application areas. Such systems are based on using microprocessors for their implementation. Embedded systems are usually part of other systems where they are embedded into some embedding systems. They have to be efficient in electrical energy consumption, size of program code, time of operation, weight and cost. Embedded systems are inexpensive and are used in almost every electronic product or other electronic systems. Many embedded systems are mobile systems supplied by batteries and the available electrical energy must be used efficiently as much as possible. Application areas where embedded systems are used and where minimal consumption of energy is required are battery powered wireless sensor networks. The methods for reduction of energy consumption and for power management in embedded systems are considered and described in this paper. The accent is given on design and application of energy efficient embedded systems in wireless sensor networks and on possibilities to reduce energy consumption in such systems. The methods for energy harvesting, that are very attractive and very useful in wireless sensor networks applications, are also considered and described. One practically implemented battery supplied wireless sensor network for application for environmental data acquisition and monitoring in agriculture is described in the paper.

## 1. Introduction

Development of electronic technologies enabled implementation and application of specialised electronic systems that are embedded into some other systems. Such embedded systems perform limited number of needed operations and are used in many applications [1-9]. Embedded systems are optimised and minimised in size, weight, cost, time of operation and electrical energy consumption. Such systems are used in many electronic product or other electronic systems. Many embedded systems are supplied by batteries and available electrical energy must be used efficiently. Battery powered wireless sensor networks (WSN) are areas where embedded systems are applied and where minimal consumption of energy is required [9-20].

Possibilities and methods used for power management and for reduction of energy consumption in embedded systems are considered and described in this paper. Accent is given on design of energy efficient embedded systems in wireless sensor networks and on possibilities to reduce energy consumption in such systems. Way of design and main characteristics of embedded systems and wireless sensor networks are considered and described first. Then, possibilities and methods for energy consumption reduction in embedded systems are considered and described. Also, methods for energy harvesting, that are very useful and very important in wireless sensor networks applications, are described. Finally, one practically implemented battery powered wireless sensor network is

IOP Conf. Series: Materials Science and Engineering 1208 (2021) 012002 doi:10.1088/1757-899X/1208/1/012002

described. Solar cells are used for energy harvesting and recharging of batteries in sensor nodes. The wireless sensor network is used for environmental data acquisition and monitoring in agriculture.

## 2. Embedded systems

Embedded systems are specialised electronic systems that realize limited number of needed and fixed tasks [1-9]. The tasks do not change during lifetime of the system. Such systems are computer based systems with strong integration of hardware and software and intended for performing of specific functions. They are embedded into some other system, for what provide better functionality and performance. The system is intended to perform a specific task. Designer can to optimize it from the aspect of speed of operation and to reduce size, energy consumption and price of the product. The term embedded indicates that such systems are integral part of some larger system that is known as embedding system. More embedded systems can coexist in one embedding system [1-9].

Embedded systems are inexpensive and used in almost all application areas and almost every electronic product. The basic purposes of a such system are: data collection and storage; data transmission and communication; data processing, monitoring and control; applications specific for users. The main areas of embedded systems application are: consumer electronics, building automation and control, automotive electronics, telecommunication systems, computer peripherals, computer networks, health care and medical systems, industrial automated systems, robotics, military applications [1-9]. Embedded systems are usually real-time systems with more inputs and outputs that respond to more independent events. Also, in many applications, the embedded systems (nodes) are connected in some type of network. Such systems are consisting of many embedded nodes that are communicating over network and perform all needed operations [9-20].

There are two main components in embedded system: hardware and software [1-9]. Hardware performs all necessary operations and should be minimal for needed application. Software defines way of hardware operation for needed application and should also be minimal. Basic subsystems or units of embedded system hardware are: processing unit or processor, memory unit, peripheral units and power supply unit. There are also other hardware components depending on the application. Such components are: reset logic, oscillator (clock generator) circuits, timers logic, interrupt logic, input logic, output logic, communication logic and application specific circuits [1-9]. Figure 1 shows general hardware structure of typical embedded system. In the most situations all this components are not used. Embedded system is always optimized and minimized and uses minimal number of hardware components needed for some concrete application.written in the style of a submission to *J. Phys.: Conf. Ser.*, show the best layout for your paper using Microsoft Word. If you don't wish to use the Word template provided, please use the following page setup measurements.

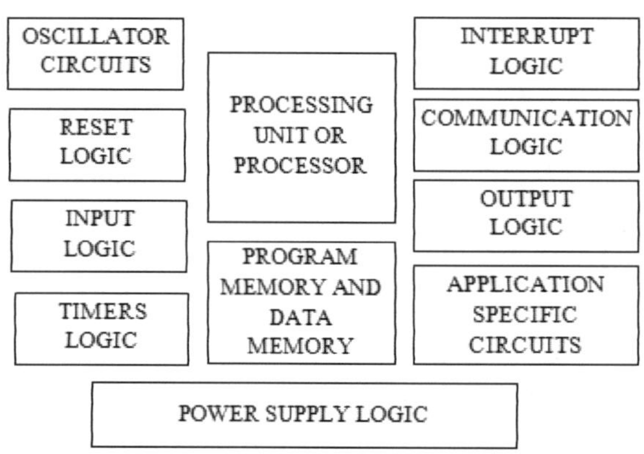


**Figure 1.** General hardware structure of embedded system.

Processing unit or processor is computing component that executes programs and performs all needed operations in the system. It should provide processing power needed to perform the tasks within the system. There are more types of processors that are used in the embedded systems: General

Purpose Processors (GPP) (Microprocessor, Microcontroller, Digital Signal Processor, Multimedia Processor); Application Specific System Processors (ASSP); Application Specific Instruction Processors (ASIP); core(s) of GPP or ASIP processors on Application Specific Integrated Circuit (ASIC) or on Very Large Scale Integration (VLSI) circuit [1-9]. What type of processor is practically used depends of the requirements of the system application. It should be used as simple as possible and as chip as possible processor for concrete application.

Memory unit stores the software (program that the system executes) and the data that is used in the system. It consists of program memory and data memory. As program memory for software storage generally are used Read Only Memory (ROM), Electrically Programmable Read Only Memory (EPROM) and flash type memories. As data memory for data storage generally are used Random Access Memory (RAM) type memories. Such systems do not use disk.

Peripheral units are connected to embedded system and enable communication with the external environment. Input peripheral units are generally connected with the sensors that measure the environment parameters data. The system processes that data and controls the output operations. Output peripheral units are generally connected with the actuators that control the environment parameters. For interconnection of the system with sensors generally are used A/D converters and for interconnection with actuators are used D/A converters.

Software of embedded system is set of programs that define what the system performs and how it performs. The software defines way of data collection, way of interpretation of data from sensors and way of generation of control signals for actuators. The software can consist of program components such as: initialization and configuration, operating system, application software, error handling, debug and maintenance support etc. [1-9]. To be effective as much as possible, the software should be small as much as possible. That is the main reason that many embedded systems use application software only. If the operation system is used then it is Real Time Operation System (RTOS). During design of software for embedded systems it should be taken into consideration the three constraints: available memory, processor speed and need to limit power consumption.

Main characteristics of embedded system are [1-9]:

- Connectable - Connected to physical environment through sensors and actuators,
- Dependable - Reliable, maintainable, available, safe, secure,
- Heterogeneous - Contain analogue and digital parts,
- Application-specific - Application is specific and fixed in advance,
- Dedicated and single-functioned - Usually performs a specialized task repeatedly,
- Tightly constrained - Must be of minimal possible size, operate fast enough to process data in real time, consume minimum power and be inexpensive,
- Reactive - Typically must continually reacts on events and changes from its environment,
- Real time - Usually must complete needed tasks in real time without any delay,
- Microprocessors based - Mainly microprocessor or microcontroller based.
- Memory based - Have memory, since software usually is in ROM memory, but does not need secondary memories.
- Efficient - Small size, low weight, low power consumption, low cost.

Main advantages of embedded system are: easy customizable, low power consumption, low cost, enhanced performances. Main disadvantages of embedded system are: complicated and expensive development process, long time to market [1-9].

Some high demanding real-time applications, such as multimedia and network applications, have large influence on development of embedded systems with using multiprocessor systems-on-chip (MPSoC). It combines the advantages of parallel processing with the high integration levels of systems-on-chip (SoC). There is also trend of networking of embedded systems (nodes). Further development and application of embedded systems will be also positively influenced by development of Artificial Intelligence (AI), Virtual Reality (VR) and Augmented Reality (AR), machine learning, deep learning and the Internet of Things (IoT) or Internet of Everything (IoE) [1-9].

## 3. Wireless sensor networks

Development of wireless communication technologies and embedded electronic systems enabled their application in many areas and different physical environments. Such were created the wireless sensor networks [9-20]. Primary goal of application of such embedded systems is monitoring and control of activities and parameters in different environments. By integration of different embedded systems (nodes) and wireless communication it is possible to implement adequate automated system. Such system performs set of programmed functions for control and monitoring of many values, parameters and performances, by direct or remote access [9-20].

Wireless sensor network (WSN) is larger network of small embedded sensor nodes (SN) capable for data collection, data processing and wireless communication [9-17]. Architecture and organization of WSN is of homogenous or heterogeneous type depending on application and implementation of the network. The goal in every WSN is to establish data collection from source sensor node to destination node. Data flow can be realized by direct communication of two nodes or via intermediary nodes (multihop). Such nodes are directly or indirectly connected with WSN and belong to the WSN network. In source node data are collected from sensors, realized potential processing and transferred to external (sink) node that collects the data. Destination node can be some external device as is computer. In many cases WSN is integrated with other technologies to enable data transfer toward distant data base. It enables centralized monitoring of collected data. Architecture of WSN is analogous to the protocol stack of usual computer network. Layers of WSN protocol stack are: physical layer, data layer and service layer [10]. In every of layers are realized appropriate services.

Each of sensor nodes types has appropriate functionalities and realizes appropriate services. For queries or requests responsible is external node that processes them and transfers toward other nodes in the network. The child node is responsible for providing data, that using data layer, is forwarded back to the source.

In the physical layer can be realized sensor nodes. The sensor nodes can be: external node, cluster node, child node and parent node [9-20]. Cluster nodes have ability of data collection and forwarding (via neighbouring nodes) to the sink node (nodes). Parent nodes are integrated into more than two cluster nodes. For forwarding, processing and generation requests for data toward other nodes responsible is the data layer. Process of data forwarding to group of nodes can be toward WSN or some region of WSN depending on query (request) [10].

WSN can be implemented using direct or indirect communication. Direct communication is realized between source and destination node. In indirect communication more intermediary nodes participate in data forwarding (multihop). What type of the communications should be realized depends on needs, environment and possible constraints. To avoid eventual problems in WSN often is used indirect type of communication [10,11]. WSN use indirect communication if exist obstacles for communication and if is used larger number of sensor nodes. Using such type of WSN decreases energy consumption with realization of efficient data routing through WSN using appropriate routing algorithm.

Complex applications and networks may use very large number of sensor nodes. Such WSN can be implemented in form of cluster organized WSN. Cluster based WSN uses division of WSN into certain number of sensor nodes groups. Every group of sensor nodes has appropriate cluster sensor node. Cluster node has specific role and function of control of neighbouring nodes in the group [10,11]. In such groups all neighbouring nodes of one cluster group exchange data with their cluster node. Cluster nodes forward data to destination node. Data can be then forwarded towards centralized system or data base.

In some WSN it is necessary to enable mobility of sensor nodes. There are applications here is needed that certain types of sensor nodes (source or destination nodes) move. In such networks nodes are moving in the area where the data is collected [9-20].

In large and complex WSN important is also network scalability. Scalability enables maintenance of characteristics and performance in WSN with many sensor nodes using complex protocols for efficient routing. It is important for multihop WSN where exists very large number of sensor nodes [9-17].

To achieve optimal performance of WSN used are more types of WSN topologies. Every of topologies has some advantages and disadvantages. Every topology has wide application depending of application and requirements. Topologies of WSN generally are: star, node-to-node (Peer-to-Peer), mesh and hierarchical or tree topology [9-17]. Forming and using of some topology is based on application of appropriate algorithms.

Main areas of applications of wireless sensor networks are: environmental monitoring, agriculture, civil engineering, military applications, health monitoring.

Wireless sensor node is embedded system that uses several embedded electronic components [9-17]. The components implement tasks of measurement, processing of measured values and transmission of processed information toward other sensor nodes, network coordinator or external (sink) node. Process of measurement detects value of some physical parameter in environment that is then processed. After processing the data is sent by wireless way. Wireless communication enables wireless transfer of information and interaction of sensor nodes [9-17]. Architecture of typical wireless sensor node (SN) is shown in Figure 2. The sensor node has five basic subsystems or units: computing or processing subsystem, communication subsystem, sensors subsystem, actuators subsystem and power supply subsystem. Main components of processing subsystems are microprocessor or microcontroller and memory. Communication subsystem consisting of two main components: short range radio transceiver for wireless communication and antenna. Sensors subsystem connects node to external environment and consists of sensors, sensors logic and A/D converters logic. Actuators subsystem also connects node to external environment and consists of actuators, actuators logic and D/A converters logic. Main components of power supply subsystem are battery, DC-DC converter and energy harvesting logic. Some wireless sensor nodes use two additional subsystems: localization subsystem and mobilizer subsystem. Localization subsystem is used to identify location or position of sensor node in applications where sensor nodes are randomly distributed or are mobile. Mobilizer subsystem is used to provide mobility for sensor node in applications where sensor nodes are mobile. In the most practical implementations all this components are not used. Sensor node is always optimized and minimized and uses minimal number of hardware components for some concrete application.

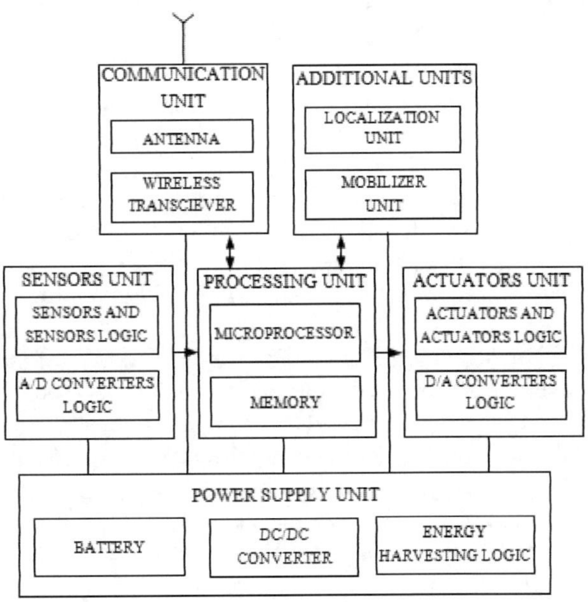


**Figure 2.** Architecture of wireless sensor node.

Wireless communication technologies and wireless standards that are most often used for interconnection of wireless sensor nodes and for implementation of wireless sensor networks are: WiFi, ZigBee, Radio Frequency Identification (RFID), Bluetooth and mobile technologies [9-20].

## 4. Energy efficiency in embedded systems

Embedded systems in large number of applications are realized as battery supplied devices and it is imperative to be implemented as systems with minimal energy consumption [6,21-39]. In such low-power applications available electrical energy must be used very efficiently. Typical low-power applications are: battery operated equipment, mobile communication equipments, wireless communication equipment, consumer electronics, biomedical applications. To reduce power consumption as much as possible in such applications are used CMOS circuits since CMOS is the technology with minimal energy consumption.

### *4.1. Reduction of energy consumption*

The two main sources of total power consumption (dissipation) in digital circuits are static and dynamic power consumption [21-24]:

$$P_{td} = P_{sd} + P_{dd}, \quad (1)$$

where $P_{td}$ is total power consumption, $P_{sd}$ is static consumption and $P_{dd}$ is dynamic power consumption.

Static power consumption in CMOS digital circuits consists of leakage current power dissipation ($P_{lc}$) and static power dissipation ($P_{sp}$):

$$P_{sd} = P_{lc} + P_{sp}. \quad (2)$$

Leakage current power dissipation ($P_{lc}$) is dominant static power consumption. Static power dissipation ($P_{sp}$) is minor static power consumption and practically can be neglected. Leakage current power dissipation ($P_{lc}$) is practically given by:

$$P_{lc} = I_{lc} V_{dd}, \quad (3)$$

were $I_{ls}$ is leakage current and $V_{dd}$ is power supply voltage of the circuit.

Dynamic power consumption in CMOS digital circuits consists of capacitive switching power dissipation ($P_{cs}$) and short circuit power dissipation ($P_{sc}$):

$$P_{dd} = P_{cs} + P_{sc}. \quad (4)$$

Capacitive switching power dissipation ($P_{cs}$) is dominant dynamic power consumption. Short circuit power dissipation ($P_{sc}$) is minor static power consumption but can not be neglected. Average capacitive switching power dissipation ($P_{cs}$) is approximately given by:

$$P_{cs} = p_t C_L V_{dd}^2 f_{clk}, \quad (5)$$

were $C_L$ is capacitive load, $V_{dd}$ is power supply voltage and $f_{clk}$ is clock frequency of the circuit. The product $p_t \cdot C_L$ is called the average switched capacitance. The methods for reducing this capacitance are realized at system, architecture, circuit and technology level.

Average short circuit power dissipation ($P_{sc}$) is approximately given by:

$$P_{sc} = I_{sc} V_{dd}, \quad (6)$$

were $I_{sc}$ is average short circuit current and $V_{dd}$ is power supply voltage of the circuit.

The approximate total average power consumption (dissipation) is then given by:

$$P_{tad} = P_{lc} + P_{cs} + P_{sc} = I_{lc} V_{dd} + p_t C_L V_{dd}^2 f_{clk} + I_{sc} V_{dd}. \quad (7)$$

It can be seen from the last equation (7) that the system power consumption can be reduced by: reduction of power supply voltage $V_{dd}$, reduction of load capacitance $C_L$, reduction of switching activity ($f_{clk}$) of the circuits and reduction of leakage current $I_{lc}$ in the embedded system.

Reduction of power supply voltage $V_{dd}$ has the greatest impact on reduction of total power consumption. It reduces all types of the power consumption. The greatest contribution is in decreasing capacitive switching power dissipation ($P_{cs}$) what is the biggest power consumption in many applications, especially in high speed applications. Reduction of the power supply voltage gives quadratic improvement in decreasing of that type of power consumption. But, reduction of the power supply voltage has one very negative influence what is reduction of speed of operation of the circuits and the system. The value of power supply voltage $V_{dd}$ of every circuit, unit and all system should be

IOP Conf. Series: Materials Science and Engineering 1208 (2021) 012002 doi:10.1088/1757-899X/1208/1/012002

minimized as much as possible for the concrete application of the system. There are more methods for the power supply voltage management that are used in embedded systems.

Reduction of load capacitance $C_L$ decreases capacitive switching power dissipation ($P_{cs}$) what is the biggest power consumption in high speed applications. That reduction gives linear improvement in decreasing of that type of power consumption. Reduction of the load capacitance has also very positive influence on increasing of speed of operation of the circuits and the system. The value of load capacitance $C_L$ of every circuit, unit and all system should be minimized as much as possible for the concrete application of the system. There are more methods for reduction of load capacitance used at technology level, circuits level, architectural level and system level design.

Reduction of switching activity of the circuits in the embedded system reduces dynamic power consumption. Reduction of switching activity is mainly performed by decreasing the clock frequency $f_{clk}$ of the circuits and system. Decreasing of the clock frequency reduces linearly capacitive switching power dissipation ($P_{cs}$). It also reduces average short circuit current $I_{sc}$ and such reduces short circuit power dissipation ($P_{sc}$) of the system. But, reduction of the clock frequency has negative influence in reduction of speed of operation of the circuits and the system. The value of clock frequency $f_{clk}$ of every circuit, unit and all system should be minimized as much as possible for the concrete application of the system. There are more methods for the clock frequency management that are used in embedded systems.

Reduction of leakage current $I_{lc}$ reduces static power consumption. Leakage current is independent of operating frequency. It depends very much on the process technology used for realization of system hardware components. Reduction of leakage current can be mainly obtained by using hardware components realized in technology with smaller leakage currents. Leakage current also decreases with decreasing the power supply voltage value $V_{dd}$ of the circuits and system. But, reduction of the power supply voltage reduces speed of operation of the circuits and the system. There are more methods for reduction of leakage current used at technology level and circuits level design. As the process technology advances the dynamic current decreases, but the static leakage current (for low-voltage submicron technology) tends to increase and such becomes more important.

There are many methods or techniques for power management and reduction of energy consumption or energy saving in embedded systems [21-39]. All the methods can be classified in two types: hardware based methods and software based methods. Hardware based methods are using appropriate design and implementation of hardware elements and complete system hardware that reduce and minimize energy consumption in concrete application. Software based methods are using appropriate design and implementation of software elements and complete system software that reduce and minimize energy consumption in concrete application.

The most often used hardware based methods for reduction of energy consumption in embedded systems are [21-39]:

- Design and implementation using as minimal hardware as possible, where only necessary hardware components should be present, for concrete application,
- Design and implementation using technology that has minimal energy consumption, using CMOS technology,
- Design and implementation using as minimal supply voltage value as possible for concrete application,
- Power gating that involves turning off the supply voltage of hardware blocks whenever is possible or when they are not active, especially blocks with high energy consumption, as are microprocessor and memory,
- Decreasing value of the supply voltage of hardware blocks whenever is possible or when they are not active, especially blocks with high energy consumption, as are microprocessor and memory,
- Using voltage islands if some blocks are slower than others and then is possible to isolate slower blocks, run them at lower frequency and turn down the supply voltage until these blocks meet needed operation speed,

- Using dynamic voltage frequency scaling that is combination of voltage islands and power gating where the designer adjusts the voltage and clock frequency of each block so that it is meeting its operation speed for the current application,
- Design and implementation to have as minimal load capacitances of circuits, blocks and units as possible for concrete application,
- Dividing hardware components into more modules where such modules have smaller load capacitances and smaller number of accesses,
- Reduction of load capacitance of buses reducing their length, using reserved buses instead of multiplexed ones, segmenting buses,
- Design and implementation using as minimal clock frequency value as possible for concrete application,
- Using clock gating technique for reducing dynamic power that shuts off the clock to parts of the system that are inactive and designers use clock gating at the block level to create a standby mode,
- Design and implementation using hardware components realized in technology with as minimal leakage currents as possible for concrete application,
- Design and implementation of the system to be as much as possible in one of inactive states, in power off, power down, idle, sleep or standby mode.

The most often used software based methods for reduction of energy consumption in embedded systems are [21-39]:

- Design and implementation of the software to minimise as much as possible number of operations for concrete application,
- Design and implementation of as short as possible software, with minimal number of instructions, where only necessary software components are present, that also will need minimal program memory capacity for concrete application,
- Using assembler programming language for software development to obtain minimal programs and minimal number of instruction executions,
- Avoiding using program loops, using as much as possible the sequential programs and using unrolling the loop style of programming what decreases number of instruction executions,
- Avoiding as much as possible using program structures as procedures, subroutines, functions and modular programming and using sequential programs what decreases number of instruction executions,
- Avoiding as much as possible using program branches and jumps and using the sequential programs what decreases number of instruction executions,
- Minimising as much as possible using interrupts and if interrupts are used minimising interrupt processing programs what decreases number of instruction executions,
- Avoiding using operating system and using only applicative program what decreases number of instruction executions and if operating system is used use specialised operating systems for embedded systems that have dynamic power management function,
- Design, writting and implementation of the software as one compact and as short as possible program, application program,
- Using such way of data representation and data length that will enable the greatest savings in energy consumption for concrete application,
- Design and implementation of the software to put the system as much as possible in one of inactive states, in power off, power down, idle, sleep or standby mode.

It is common and recommended to use multiple and as much as possible of all these hardware and software techniques for consumption reduction in different parts of embedded system design.

Principle of operation of a typical embedded system is shown in Figure 3

After turning on, configuration and initialization, the system periodically executes the application program and collects input data from environment, processes the data and sends obtained results to environment. Then it waits some needed time or until change of the input data occurs and again repeats all activities in the control loop. Also, usually are used interrupts and interrupt logic for detecting and processing asynchronous events from the environment. The control loop has some duration time $T_c$ and application program execution has some duration time $T_a$, depending on the application. Difference between this two times is the waiting time $T_w$ of the system. Always is:

$$T_c = T_a + T_w. \tag{8}$$

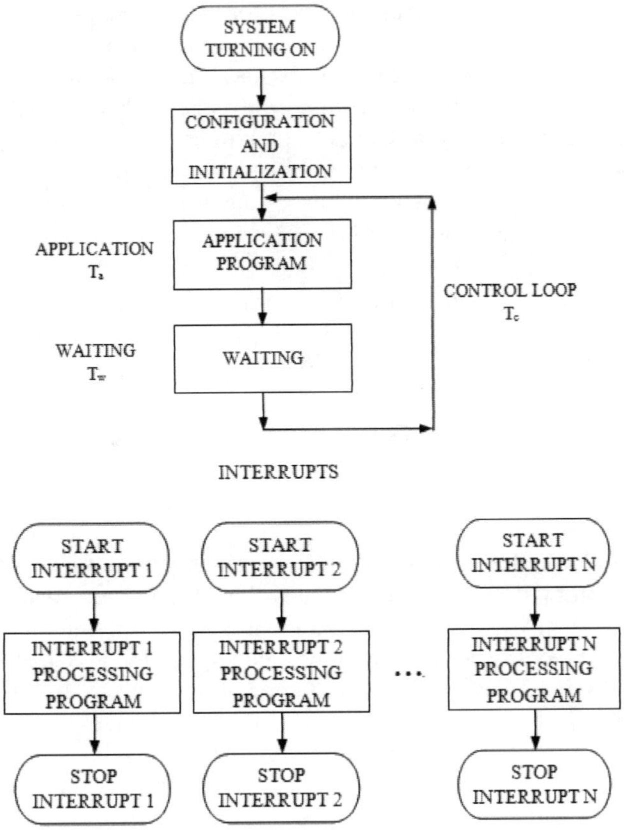


**Figure 3.** Principle of operation of embedded system.

To obtain energy efficient system it is needed to design system such to be control time ($T_c$) as long as possible, application time ($T_a$) as short as possible and waiting time ($T_w$) as long as possible. The application time duration $T_a$ will be minimal if system hardware and software (application program) are optimized and minimal for concrete application. The control loop time duration $T_c$ should be selected to be maximal for concrete application. In such situation the waiting time $T_w$ will be maximal possible. In the waiting time interval should be used all appropriate and most suitable hardware methods for energy consumption reduction. It should be used one of system inactive states (power off, power down, idle, sleep or standby). The best is to use method of power supply disconnection (power off) for all the hardware components if it is possible. It reduces the energy consumption to zero in the waiting time interval.

To increase energy efficiency it should be avoided using interrupts. Interrupts should be used only if it is unavoidable for concrete application of the system. In such situations it should be used minimal number of interrupts. Also, used interrupts should be occurred as rarely as possible, with the lowest possible frequency of occurrence. Programs for interrupt processing (interrupt routines) should also be as much as possible simple and short, with minimal number of instructions and minimal number of branches or jumps.

*4.2. Energy harvesting*

There are many embedded systems that are used in situations and applications where the sources of permanent power energy supply (AC power supply) are not existing or are not available [17,21-24,32-34,39]. Such situation is in application of wireless sensor networks and sensor nodes, where sensor nodes can be randomly placed or be movable and mobile. In such situations it has to be used batteries as sources for power energy supply of the systems. But, batteries have limited lifetime what limits lifetime of the sensors nodes and the systems. Solution for that problem is using rechargeable batteries and extracting energy from ambient sources for the batteries charging. Process of extracting and collecting energy from environmental sources is known as energy harvesting. Such way of embedded systems energy supply enables that the batteries and the system operate continuously and practically without limited lifetime. It also reduces problems of obtaining energy efficiency of embedded systems since it can provide enough energy for system correct and permanent operation [17,24,32].

Main energy environment sources for energy harvesting are:

- Mechanical energy from vibration, pressure and strain,
- Thermal energy from furnaces and other heating sources,
- Solar energy from all light sources, from lighting to the sun,
- Electromagnetic energy captured via inductors, coils and transformers,
- Wind and fluid energy from air and liquid flow,
- Human energy of human movement by foot, human skin and blood,
- Chemical energy from natural chemical processes or biological processes.

By using energy harvesting the embedded system completely alone collects and accumulates energy as it becomes available. In many cases, such energy sources provide energy in small packets. In such situation, the harvesting circuit must stay in active mode permanently, to capture harvestable energy when it becomes available, and to provide output required for the application. The power consumption of the harvester has to be as small as possible. The energy consumed by this circuit must be much smaller than the energy obtained by the ambient sources. Also important characteristic of the harvesting circuit is the capability to store obtained energy as long as possible with minimal loss. Energy harvesting circuits must have very high that capability due to the infrequency of the energy capture activity. The energy harvesting circuit must also economize the stored energy in order to provide correct operation of the supplied system for the concrete application.

Energy harvesting is most suitable for applications demanding small amounts of continuous power or having short periods of high-power consumption, what harvested and stored energy can provide. Wireless sensor nodes are typical devices for such applications. Harvesting ambient energy allows the wireless sensor nodes to operate nearly indefinitely, without their battery failure. Also, energy harvesting devices are usually small. There are sensor nodes that do not use any self-contained energy source. They only harvest energy from the environment. But, the supply of energy may be interrupted some period of time since the power obtained from the harvesting can not be guaranteed all the time.

There are more types of energy harvesting with different degrees of efficiency depending on the application [17,21-24,32-34,39]. Classification of energy harvesting can be based on the type of energy used to harvest the power. Generally, there are three main and most used types of harvesting sources from environment [17,21-24,39]:

- Mechanical vibration,
- Thermoelectric generation,
- Solar cells.

Mechanical vibration energy harvesting is based on the principle that device in movement can generate one of next types of energy: vibration, kinetic or mechanical energy. All types of this energy can be converted into electrical energy and be harvested. For such energy conversion and harvesting can be used piezoelectric, electrostatic and electromagnetic mechanisms and devices.

Piezoelectric materials convert mechanical energy from vibrations, pressure or force into electricity. The advantage of such harvesters is direct generation of needed voltage. They do not need separate voltage source and additional components. But, they have some disadvantages since characteristics of piezoelectric materials are changing with age, pressure and temperature, such materials are breakable and may have leakage of charge.

Electrostatic principle of harvesting is based on changing the capacitance of vibration-dependent varactors. Vibrations separate the planes of initially charged varactor. The mechanical energy is converted into electrical energy. The main advantage of electrostatic converters is ability to integrate into microelectronic devices without added components. Disadvantage of such converters is that they need an additional voltage source for initial charging of the capacitor.

Electromagnetic energy harvesting is based on using electromagnetic induction. The most effective ways of electromagnetic energy harvesting are with using permanent magnets, coil and resonating cantilever beam. Such harvesting has advantage of improved reliability, there is not mechanical contact of any parts and separate voltage source is not needed. The main disadvantage of this type of harvesting is that electromagnetic materials are large in size and complicated to integrate with embedded systems.

Thermoelectric generation energy harvesting use the principle of creating electric potential based on a temperature difference. Voltage is generated when there is temperature difference between two junctions of conducting materials. Thermal energy harvesting uses temperature differences or gradients to generate electricity.

Solar cells or photovoltaic cells energy harvesting is the best known and the most practically used way of harvesting. It uses devices that convert light energy into electrical energy. Typically used is light energy obtained from sunlight. For locations where there is available light and where usage of other types of power supply are not possible or are expensive, usage of solar cells is very convenient solution. In designing sources which harvest solar energy it must have in mind such factors as availability of day light, characteristics of used solar cells and the intensity of the incident light. However, the supply of solar energy is interrupted at some period of time and the power obtained can not be guaranteed all the time.

## 5. Energy efficient embedded wireless sensor network system

As example of application of energy efficient embedded systems here is described implementation of one simpler on ZigBee wireless technology based wireless sensor network. The network was realized for measurement and monitoring of some values from environment. Such systems were described in papers given in references [18-20]. The network is flexible, modular and can be used for different applications. It is possible to realize measurement and monitoring of environment parameters values obtained from sensors in sensor nodes. There are two types of sensor nodes in the network: end or field sensor node and coordinator sensor node. Data read from sensors of end sensor nodes are wirelessly transmitted to coordinator node. Coordinator node is connected with personal computer of PC type. Personal computer is connected with the Internet. Data from coordinator node is transferred to personal computer and stored in appropriate file. Then the data from the file is stored in the local data base. It is also realized Web based application that enables presenting and monitoring of results from sensors in the network, using graphics and diagrams, via Internet and from distance. The measurement, collection and transmission of data from sensors are realized periodically. The network was designed and used for environmental data acquisition and monitoring in agriculture applications. It monitors environment temperature, environment humidity and soil humidity.

The system is based on application of the Arduino Uno boards, sensors and wireless communication modules [40-42]. End sensor node uses Arduino board, wireless module and sensors connected to board that measure temperature and humidity of environment and soil humidity. Coordinator sensor node does not use sensors and does not measure environmental parameters. It coordinates and controls operation of the network. For realization of coordinator sensor node used are Arduino board and wireless module. Hardware components used for implementation of the network are: Arduino Uno boards, wirelless modules Digi Xbee S2 based on ZigBee protocol and sensors for measurement of air temperature, air humidity and soil humidity. For measurement of environmental values used is DHT11 sensor for environment temperature and humidity, and sensor for soil humidity. Sensor DHT11 is of digital type and has integrated components for measurement. Sensor for soil humidity is also of digital type. It gives high logical level on the output if soil is dry and low logical level if soil is wet. The block-scheme of the implemented end sensor nodes is shown in Figure 4.

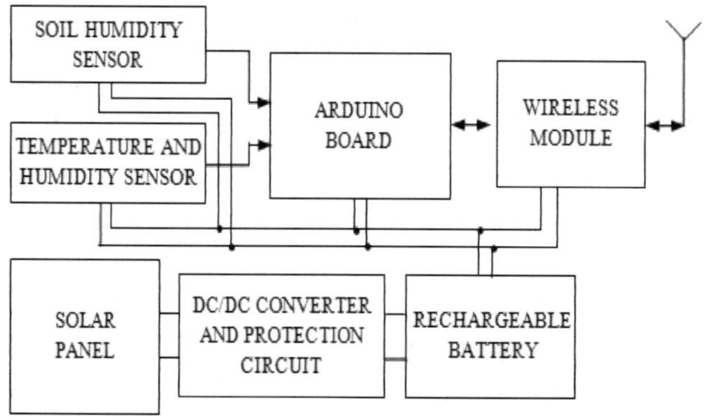


**Figure 4.** Block-scheme of implemented end sensor node.

The end sensor nodes measure and collect measured environment data, aggregate the data and transmit the data to the coordinator node. The coordinator node is connected to computer of PC type that stores the data in the data base, presents the data to the user and statistically processes the data. During the process of measurement and collection of parameters data from sensors it is possible to remotely monitor the data in real time by the Web based application. The block-scheme of the implemented coordinator sensor node is shown in Figure 5.

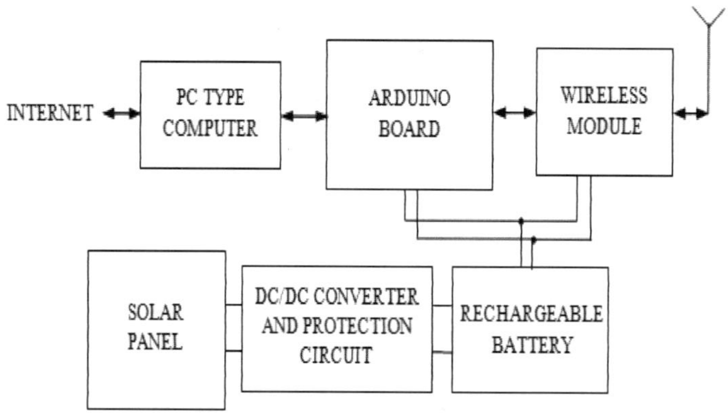


**Figure 5.** Block-scheme of implemented coordinator sensor node.

In order to reduce energy consumption, to increase energy efficiency and to obtain practically unlimited lifetime of the system there were used appropriate methods for decreasing of energy consumption. According to the application the system measures, collects, aggregates and transmits data periodically in appropriate time period. Since are monitored very slow changing processes that time period could be very long, represented in minutes. Also, time of active operation (measurement, collection, aggregation and transmission of data) is very short, less than second. The system is designed and configured such that in active mode is very short time, during collection and transmission of data. All other time the system, including wireless module, is in sleep mode. Also, transmission of data is not occurred after every data collection. Data is collected in the end sensor nodes and transmitted after its aggregation. Period of data transmission is much longer (several minutes) than interval of data collection (less than second). Wireless module is in active mode only during data transmission. All that very much reduces average and total energy consumption of the system. Also, all the sensor nodes use rechargeable battery to power the wireless module, the sensors and the Arduino board. Lithium–ion rechargeable batteries of type 18650 are used. In order to achieve the sensor nodes autonomous operation and extend the lifetime to practically unlimited interval, the batteries are charged using solar cells panel and DC-DC converter with protection circuit.

The designed wireless sensor nodes, the network and all the system are suitable for measurement of ambient temperature, ambient humidity and soil-humidity level for applications in agriculture. For example can be used in some type of vegetable production, grin production or flower production.

Practically obtained results of monitoring by developed and implemented sensor nodes and sensor network placed in environment are shown in Figure 6, Figure 7 and Figure 8. The obtained monitoring results are given in % and °C, for one time interval of monitoring. Figure 6 shows results of ambient temperature monitoring. Figure 7 shows results of ambient humidity monitoring. Results of soil humidity monitoring are shown in Figure 8.

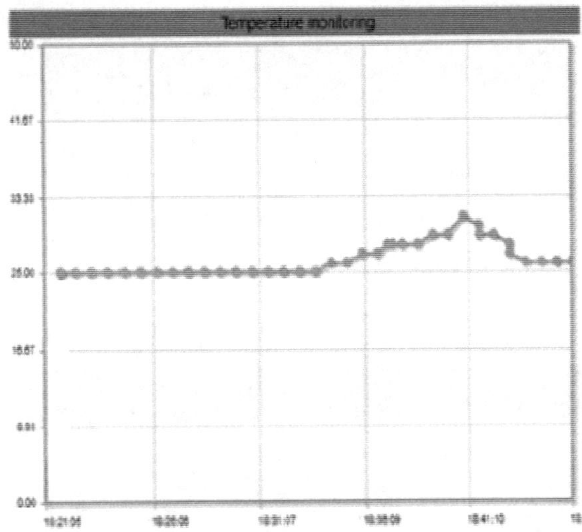


**Figure 6.** Results of environment temperature data monitoring.

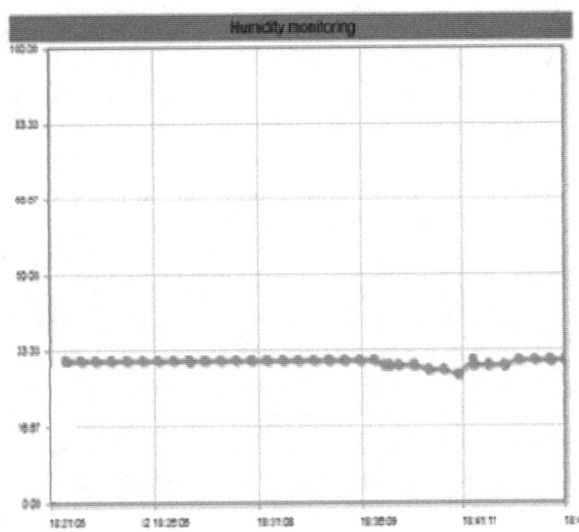


**Figure 7.** Results of environment humidity data monitoring.

Given results were obtained during testing of operation of the sensor nodes, sensor network and the implemented system. It can be seen that the temperature and humidity were constant in the beginning of the testing time interval. Then it was created and generated external heating of the environment of the sensor nodes. That caused increasing of the measured temperature and decreasing of measured humidity. After some specific time the environment heating was stopped and measured temperature and humidity naturally and gradually returned to the initial values. It all can be seed in the Figure 6 and Figure 7. The soil humidity sensor monitors the soil humidity state with the soil humidity threshold. Initially, the humidity level in the soil was low and output state was high (1). After some specific time, it was filled in some amount of water in the soil around the soil humidity sensor. Such, it was achieved appropriate level of soil humidity, greater than the humidity threshold, and sensor output changed to low (0) state. It all can be seed in the Figure 8.

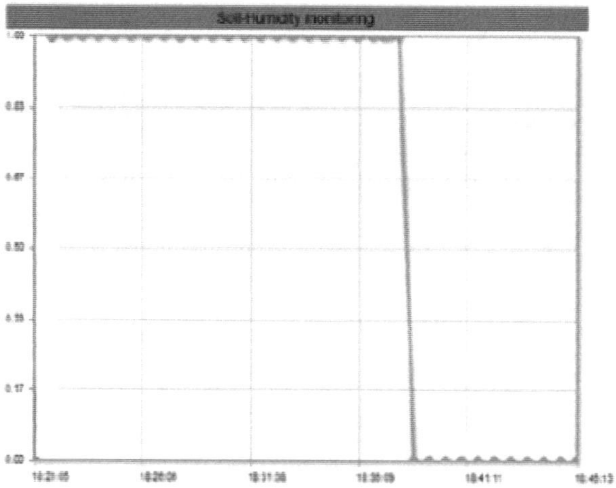

**Figure 8.** Results of soil humidity data monitoring.

## 6. Conclusion

Advancement of electronic technologies enabled application of specialised embedded systems in very many applications. Good characteristics of such systems enable their application in almost all areas where they are embedded in some other embedding systems improving their possibilities and operation. Embedded systems perform only needed operations and are minimised in cost, weight, size, time of operation and energy consumption. Many embedded systems are powered by batteries and must be as much as possible efficient in consumption of electrical energy.

Battery supplied wireless sensor networks are based on application of embedded systems and require minimal consumption of energy. Sensor nodes of such network are embedded systems specialized for application where the network is used and interconnected in appropriate network. Optimization of sensor nodes and optimization of network topology enable realization of very effective wireless sensor networks for almost all possible applications. It is especially interesting for applications in the Internet of Things (IoT) or Internet of Everything (IoE) areas.

There are many methods for reduction of energy consumption in embedded systems. What methods should be used in some situation depends the most on the concrete application of the system. The general and the main recommendation for decreasing energy consumption in embedded systems is to design and implement system with minimal possible hardware and minimal possible software used. All other hardware and software methods for energy consumption reduction contribute to further decreasing of energy consumption of the system. That methods should be selected and applied according to needs and possibilities of the system in the concrete application. For battery supplied embedded systems it is very important to use energy harvesting from the system environment. There are many such possibilities convenient for different applications. One of the most suitable and the most practically used energy harvesting method is using solar energy and solar cells.

Practically designed and implemented battery powered wireless sensor network is based on ZigBee wireless communication technology. It is used for environmental data acquisition and monitoring in agriculture. The ZigBee communication technology is used since it is very suitable wireless technology for such type of applications. Appropriate methods for reduction of energy consumption in the sensor nodes and in the complete wireless sensor network are used. The wireless sensor nodes and the complete sensor network are optimized and maximally adopted to the application. Also, the application enables to collect and transmit sensors data periodically, in long time periods. But, time interval of data collection and transmission is very short. Such, the system is the most of time in waiting state, between that two activities. During the waiting state the system is in the sleep mode. That very much reduces energy consumption of the system. Some other hardware and software methods are also used to reduce energy consumption in the active system state. All that is reducing total energy consumption of the system. Application of rechargeable battery and solar energy harvesting in the sensor nodes enable autonomous operation and practically unlimited the lifetime interval of the nodes and the complete system. The system is simple, modular, very flexible and inexpensive. It is very easy to increase number of sensor nodes to necessary number, to program sensor nodes for use of different sensors, and to program needed data aggregation and processing for concrete application. The system could be very easy, with small modifications, adopted also for other similar applications that require and use simple battery powered wireless sensor nodes and networks.

**References**

[1] Gajski D, Abdi S, Gerstlauer A and Schirner G 2009 Embedded System Design: Modeling, Synthesis and Verification. Springer
[2] Wolf W 2007 High-Performance Embedded Computing: Architectures, Applications, and Methodologies. Morgan Kaufmann
[3] Valvano J 2011 Embedded Microcomputer Systems: Real-Time Interfacing. Cengage Learning
[4] White E 2011 Making Embedded Systems: Design Patterns for Great Software. O'Reilly Media
[5] Stringham G 2010 Hardware/Firmware Interface Design: Best Practices for Improving Embedded Systems Development. Newnes
[6] Schmitz M, Al-Hashimi B and Eles P 2010 System-Level Design Techniques for Energy-Efficient Embedded Systems. Springer
[7] Marwedel P 2020 Embedded System Design: Embedded Systems Foundations of Cyber-Physical Systems, and the Internet of Things. Springer
[8] Hobbs C 2019 Embedded Software Development for Safety-Critical Systems. CRC Press
[9] Agrawal D 2017 Embedded Sensor Systems. Springer
[10] Mahalik N 2007 Sensor Networks and Configuration: Fundamentals, Standards, Platforms and Applications. Springer
[11] Sohraby C, Minoli D and Znati T 2007 Wireless Sensor Networks: Technologies, Protocols and Applications. John Wiley & Sons
[12] Nayak A, Stojmenović I 2010 Wireless Sensor and Actuator Networks: Algorithms and Protocols for Scalable Coordination and Data Communication. John Wiley and Sons
[13] Farahani S 2008 ZigBee Wireless Networks and Transceivers. Elsevier
[14] Chai S, Wang Z, Zhang B, Cui L and Chai R 2020 Wireless Sensor Networks. Springer
[15] Elhoseny M and Hassanien A 2019 Dynamic Wireless Sensor Networks. Springer
[16] Sharma V and Pughat A 2017 Energy-efficient wireless sensor networks. CRC Press
[17] Kanoun O 2019 Energy Harvesting for Wireless Sensor Networks. De Gruyter
[18] Pasalic D, Bundalo Z, Bundalo D and Cvijic B 2015 ZigBee-based data transmission and monitoring wireless smart sensor network integrated with the Internet *Proceedings of Conference MECO2015* pp.240-243.
[19] Pasalic D, Bundalo Z, Bundalo D, Softic F and Cvijic B 2015 Environment data acquisition using energy efficient wireless sensor network *Proceedings of Conference ERK2015* pp. B:31-34
[20] Pasalic D, Bundalo Z, Bundalo D, Softic F and Cvijic B 2016 Environmental data monitoring using wireless sensor networks. *Annals of Faculty Engineering Hunedoara-International Journal of Engineering* vol. 14 no. 2 pp. 99-102
[21] Pistoia F 2008 Battery operated devices and systems: From portable electronics to industrial products. Elsevier
[22] Chedid W and Yu C 2002 Survey on Power Management Techniques for Energy Efficient Computer Systems. Mobile Computing Research Lab. Cleveland State University Cleveland

[23] Zomaya Y and Lee Y 2012 Energy Efficient Distributed Computing Systems. John Wiley and Sons
[24] Stojcev M, Kosanovic M and Golubovic Lj 2009 Power management and energy harvesting techniques for wireless sensor nodes. *Proceedings of Conference TELSIKS2009* pp. 65-72
[25] Wu H, Chen C and Weng K 2021 An Energy-Efficient Strategy for Microcontrollers. *Applied Sciences* 11 2581
[26] Thakkar A, Chaudhari K and Shah M 2020 A Comprehensive Survey on Energy-Efficient Power Management Techniques. *Procedia Computer Science* 167 1189–1199
[27] Salami B, et al. 2020 An Experimental Study of Reduced-Voltage Operation in Modern FPGAs for Neural Network Acceleration. *International Conference on Dependable Systems and Networks (DSN)* 138-149
[28] Papadimitriou G, et al. 2019 Adaptive Voltage/Frequency Scaling and Core Allocation for Balanced Energy and Performance on Multicore CPUs *International Symposium on High-Performance Computer Architecture (HPCA)* 133–146
[29] Rodriguez-Zurrunero R and Araujo A 2021 Adaptive frequency scaling strategy to improve energy efficiency in a tick-less Operating System for resource-constrained embedded device *Future Generation Computer Systems* 124 230–242
[30] Fan K, Cosenza B and Juurlink B 2019 Predictable GPUs frequency scaling for energy and performance. *International Conference on Parallel Processing (ICPP)* pp. 1–10
[31] Rodriguez-Zurrunero R, Araujo A and Lowery M 2021 Methods for lowering the power consumption of OS-based adaptive deep brain stimulation controllers *Sensors* 21 2349
[32] Raghunathan V, Ganerival S and Srivastava M 2006 Emerging techniques for long lived wireless sensor networks *IEEE Communication Magazine* vol. 44 no. 4 pp. 108-114
[33] Alippi C, Anastasi G, Di Francesco M and Roveri M 2009 Energy management in wireless sensor networks with energy-hugry sensors *IEEE Instrumentation and Measurement Magazine* vol. 12 no. 2 pp. 16- 23
[34] Pasalic D, Bundalo Z, Bundalo D, Softic F and Cvijic B 2015 Methods for energy consumption management in wireless sensor networks *Proceedings of Conference RIM2015* pp. 231-236
[35] TouatiY, Ali-Cherif A and Daachi B 2017 Energy Management in Wireless Sensor Networks. Elevier
[36] Lopez-Ardao J, Rodríguez-Rubio R, Suárez-González A, Rodríguez-Pérez M and Sousa-Vieira M 2021 Current Trends on Green Wireless Sensor Networks *Sensors* 21 4281
[37] Chowdhury S and Hossain A 2020 Different Energy Saving Schemes in Wireless Sensor Networks: A Survey *Wireless Personal Communications* 114 2043-2062
[38] Singh J, Kaur R and Singh D 2020 A survey and taxonomy on energy management schemes in wireless sensor networks. *Journal of Systems Architecture* 111
[39] Singh J, Kaur R and Singh D 2021 Energy harvesting in wireless sensor networks: A taxonomic survey. *International Journal of Energy Research* 45 118–140
[40] Atmel Corporation Datasheets. www.atmel.com/products/
[41] Arduino Datasheets. www.arduino.cc/
[42] XBee®/XBee-PRO® ZB RF Modules-ZigBee-v1.x4x 2013 Digi International

# Alternative concept in structural analysis education -AxisVM® X6 FEM software -

**Dušan Kovačević**

Department for Civil Engineering and Geodesy
Laboratory for Testing of Structures
Faculty of Technical Sciences, University of Novi Sad
Trg Dositeja Obradovića 6, 21000 Novi Sad, Serbia
E-mail: dusan@uns.ac.rs

**Abstract**. One short review on the preferable concept in Structural Analysis education by use of finite element method (FEM) software is given. It is generally accepted that good "educationally aided" FEM software includes a pre-processor with intuitive graphically user interface that enable simple and efficient geometrical modelling of structural topology, powerful solver-processor with sophisticated numerical modelling possibilities, post-processor with various types of analysis results presentation, as well as benefit of fast, easy training and satisfaction of use.

Hungarian software company "InterCAD" developed FEM software AxsVM® that satisfies mentioned conditions and much more performances in professional and scientific area. Therefore AxisVM® is embedded in a curriculum of subject "FEM Modelling in Structural Analysis" on Department of Civil Engineering and Geodesy, Faculty of Technical Sciences, University in Novi Sad, Serbia.

## 1. Introduction

New concept of education in structural analysis field and, especially, examination of the knowledge, should be implemented for achievement of complete understanding of the essence of structural behavior. "Old fashion" learning, intended to the "encyclopedic knowledge", by understanding of many "good old" methods, possibly provide wider "quiz-like knowing of lessons" and contribute to some of "technical culture", but it ruins enthusiasm of students in reaching of so-called "functional knowledge". "Functional knowledge" is not only ability of practical use of studying results, but situation what make possible establishing of new knowledge by "full life self-studying". In this regard is one alternative definition of intelligence as "possibility of obtaining of new knowledge by learning". Here is important to make difference between "high school learning" and "college studying" according to need of research on student's level in college, see [1].

Therefore, teaching and studying focused to one dominant method give the results that establish a structural modeling as the most important step in structural analysis and design. In this way, numerical modeling, as a mathematical realization of structural modeling system and FEM as the prevailing concept in the numerical modeling, become "sine qua non" of modern structural analysis education.

The software implementation quality decides how much the advantages of FEM would be realized and presented. Thus, as the key one, is imposed the questions and dilemmas of choice of the FEM software in education.

In connection to this, arises the problem of educational concept. In addition to high knowledge in the primary profession (structural analysis and design) also is needed:

- basic knowledge in FEM software development (numerical mathematics, software design and programming),
- certain knowledge in FEM software distribution (computer technology knowledge at the advanced user level, sense for marketing and public relations) and
- high level knowledge in FEM software competent use (basics of numerical mathematics, advanced use of computer technology).

The key thing is that every technically oriented faculty should make possible this concept in education.

## 2. Actual circumstances in FEM education

Learning of the principles of FEM is initial step in strategy of FEM education according to fact that FEM is widely accepted concept in structural modeling, analysis and design. It is completely justified for education in engineering area, but need one approach that is more practical. For engineers, dilemma what is FEM does not exist: FEM is not the only one of numerical methods for solving of differential equations, but a technology for modeling, analysis and design in many engineering fields, see [2].

Although the dilemma disappeared in the last century (after the Hrenikoff and Courant works), is still going the old joke about how things can be analyzed (different professions differently observe one simple scene), figure 1.

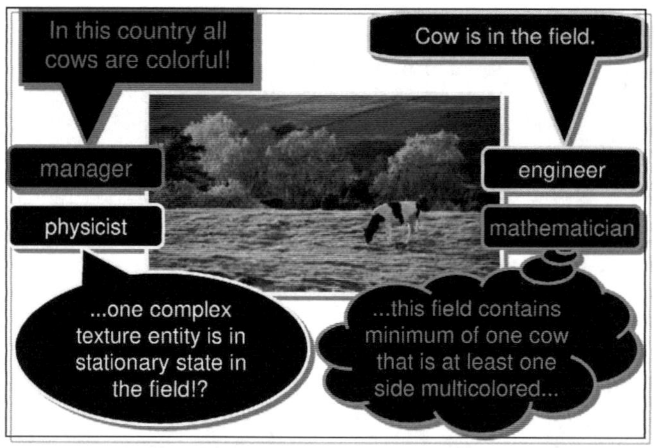


**Figure 1.** Different views about same thing

The old joke illustrates education of FEM that is sometimes overloaded by information that are unnecessary and sometimes hide the essence of FEM. Reason for that is the fact that well educated structural engineers do not practice so-called "dry theoretical approach", but only trainers/educators of other vocation. This circumstance is particularly noticeable in well-established university environments, but, fortunately, things become well, by time, see [3].

## 3. AxisVM® X6 - training tool in desired FEM education concept

AxisVM® ver. X6 is a Microsoft Windows® based FEM software which, because of its extensive analysis capabilities (developed by civil engineers for civil engineers), has verified for successful application in design of projects from large complex structures to small and simple buildings, [4].

AxisVM® performance satisfy needs of everyday design practice, scientific work and education in structural analysis domain. Therefore AxisVM® is embedded in program of following subjects: Theory of plates and Shells, Structural Stability and Dynamics, FEM Modelling in Structural Analysis and

Testing of Structures in our Department for Civil Engineering and Geodesy on Faculty of Technical Sciences, University in Novi Sad, Serbia.

The software performs linear and nonlinear, static and dynamic analysis of beam, membrane, plate, and shell structures. It enables use of an unlimited number of nodes and unlimited number of FEs freely combined in models by use of: plane/space frames/truss FEs, elastically embedded beams, plane stress/strain membranes, bended plates, spatial shells and any combination of mentioned structural parts.

There are three AxisVM® main versions: Professional (fully functional and limitless version with all modelling and computation abilities), Mini (limited version for educational purposes) and Viewer (for review and analysis of results obtained by Professional version). Typical working environment of this software is shown by figure 2.

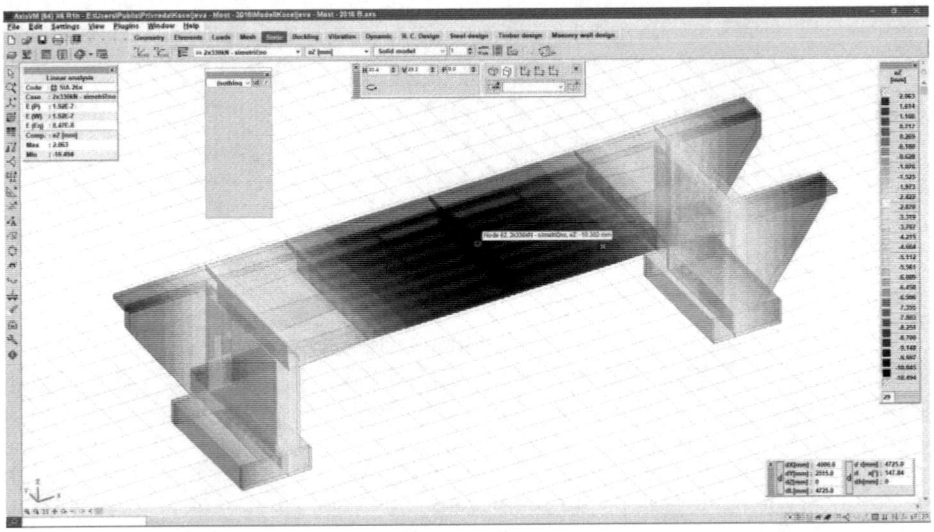

**Figure 2.** Desktop screen of AxisVM® (FEM model of one bridge for test by load)

### 3.1. *AxisVM® X6 Pre-processor (Editor)*

Why pre-processor of FEM software is so important for educational purposes? Probably because of possibility of fast and easy understanding of use of software and taking the information by visual approach. In initial steps of education, students accept the FEM on the best way if the presentation starts from the geometrical topics. Figure 3 shows different method of the geometrical discretization for the same domain. There are examples of good "✓", very good "✓✓", not so good "✓×" or bad "××" FE mesh.

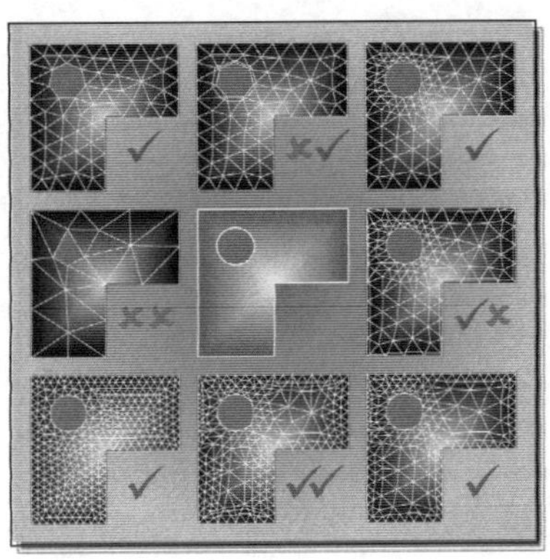

**Figure 3.** Different FEs mesh for same domain

Pre-processor of AxisVM® performs geometrical discretization by which user can to influence on the result in substantial amount. However, it is necessary to draw attention to the advantages and disadvantages of some FE mesh in terms of influencing the result of numerical analysis. Additionally, this pre-processor offers many geometry generation commands (translate, rotate, mirror, scale, with multiple copy or move), powerful selection tools (filtered selection, for example), working with structural parts what allows easy editing of most complex 3D geometries and includes advanced part management. It is included cross-section and material libraries and fully integrated graphical cross-section editor for complex cross-section (all cross-sectional properties are automatically calculated), what is very attractive for educational and design office use, figure 4.

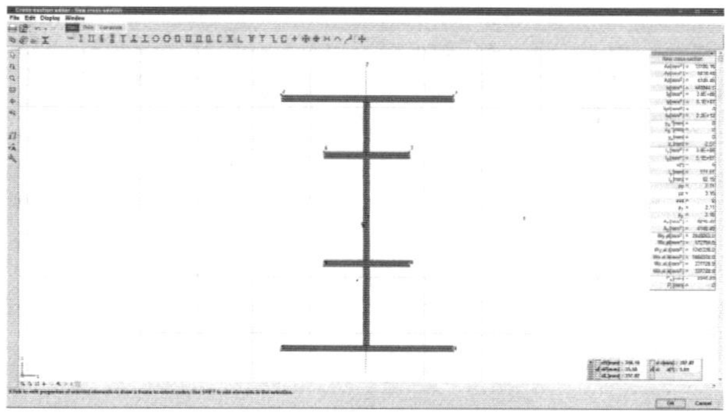

**Figure 4.** Graphically aided cross-section editor

*3.2 AxisVM X6® Processor (Solver)*

AxisVM X6® implements an "object-oriented FE architecture" that inherits more reliability than the classical systems. It provides a variety of FEs for modeling line and surface structures, special elements for modeling boundary conditions and connections and elements with nonlinear capabilities. Line FEs are truss, beam (6- and 7-DOF) and rib. The rib FE is a 3-node isoparametric with quadratic displacement interpolation that can be used taking into account the shear deformations) or in conjunction with surface elements. 7-DOF beam makes possible modeling of warping effects.

Surface FEs are: membrane, plate and shell (these elements are isoparametric quadrilateral (8/9-node) or triangular (6-node) elements, that use quadratic shape functions to interpolate displacements. Plate and shell use Mindlin's plate assumptions in a Heterosis formulation. Example of structrual part with shell FEs are given on figure 5.

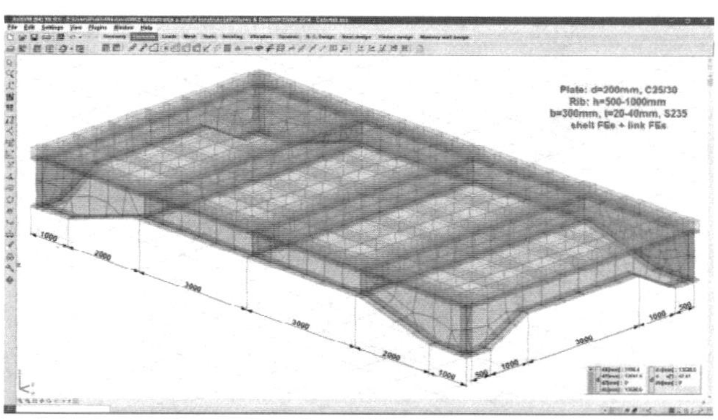


**Figure 5.** AxisVM X6® shell FEs modeling example

Additionally AxisVM® offers Winkler-Pasternak (additionally includes shear layer) type elastic supports for line and surface elements with nonlinear characteristics (tension or compression only, limited resistance) what is excellent for explanation of a real work of foundations for students.

Joint support elements are with arbitrary orientation and stiffness can have nonlinear characteristics (tension or compression only, limited resistance) what gives opportunities for understanding the essence of nonlinear behavior of structures, what is sometimes "mystery" for students and engineers, [5, 6].

There are gap, spring, link and rigid special FEs for modeling of particular phenomena in boundary/interface conditions, see [7].

Various loads can be applied on the nodes and the finite elements, as well as number of load combinations from load cases. Load cases can be classified in load groups for automatic critical internal force calculations.

AxisVM X6® performs most of the analyses, typical in the practical design of civil engineering structures: linear static analysis, buckling analysis with critical force and shape computation, nonlinear static analysis (displacement/force controlled incremental-iterative solution), free vibration analysis, figure 6, dynamic and earthquake analysis (response spectrum, time history and pushover analysis).

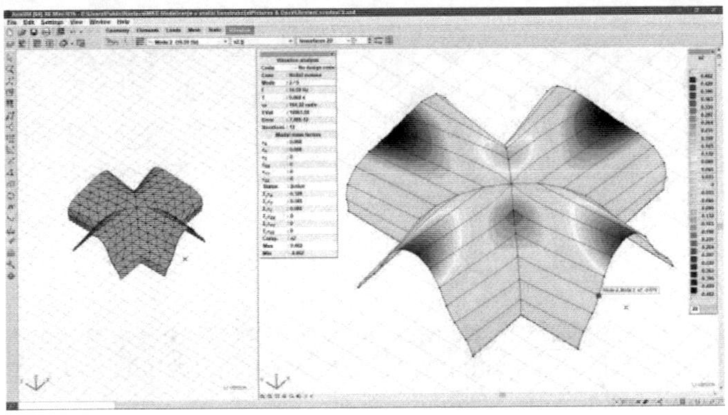

**Figure 6.** AxisVM X6® free vibration analysis

This part of FEM software as well as this area of structural design are subject of many dilemmas, questions and misunderstandings among students and civil engineers.

Here it is necessary to point out that the type and shape are not the same thing. FE type is determined in numerical manner, but a shape is derived from FE geometry. In that sense it is necessary to discuss about some circumstances when type structural systems require dense mesh of FE and discussion about inadequate shapes of FE (left and right side of the figure 7).

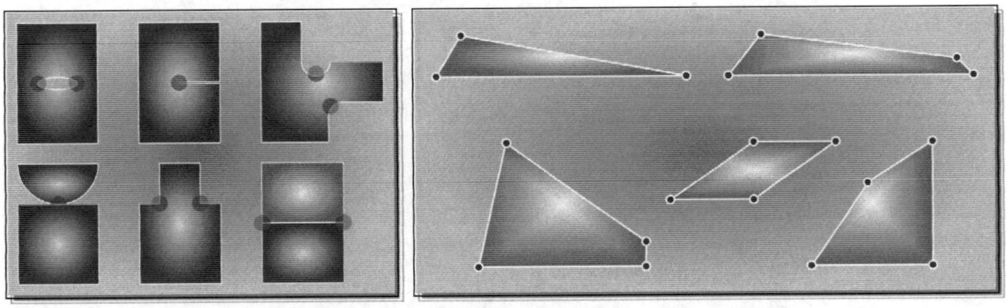

**Figure 7.** Examples that need attention in discretization and examples of "bad" FE shapes

In the first case (areas with the stress concentration) it is necessary to obtain FE with corresponding geometry (triangle) and sufficiently small. In this matter, it is not allowed that shape of FE be distorted over the acceptable limit (too big length/width ratio, sharp angle corner, shape which tend to disappearing of some joints, etc.).

Reducing the size of FE can cause problems not only with the time of computation, but also problems with the convergence and stability of the FE solution, figure 8.

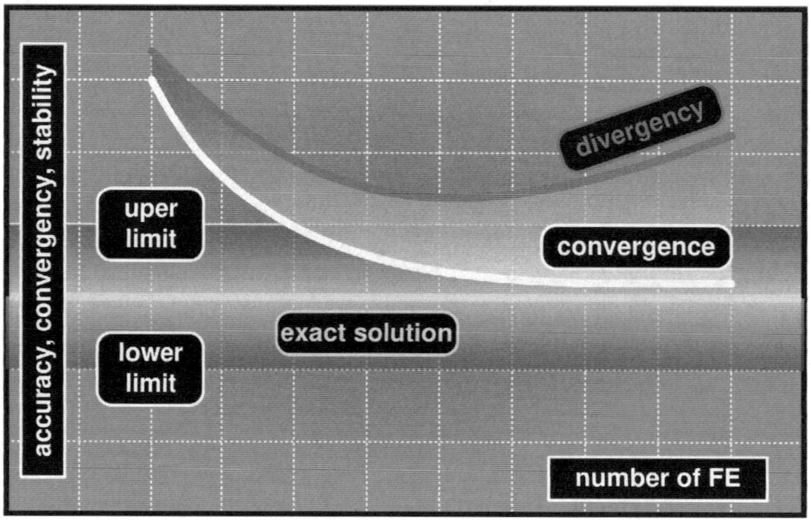


**Figure 8.** Convergence of FE solution

Clearly, in application of FEM software, it is not sufficient to know only the basic problem and the essence of the applied method, but also to know the software performances (in sense of possibilities and limits) and possibilities of numerical methods.

Last, but not least it is important to present various strategies of formation of structural matrices in FEM. Illustrative example is given on figure 9.

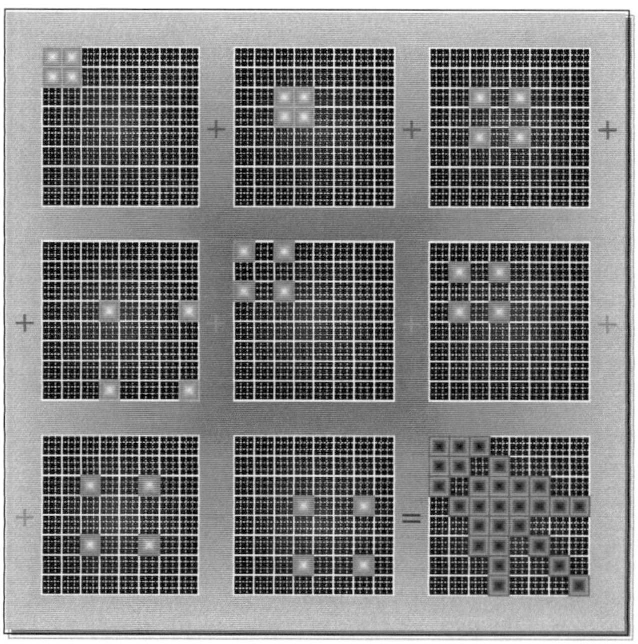

**Figure 9**. Composition-assembling of matrices in FEM

Here is essential to emphasize band shape, symmetricl, diagonally dominant, positive definite and sparse structure of FEM matrices and technology of solving systems of algebraic equations that are strongly developed by expansion of multiprocessor/multithread PC systems and corresponding algorithms for parallel computation.

*3.3 AxisVM X6® Post-processor (Viewer)*

After AxisVM® Pre-processor, as the most attractive for FEM introduction lessons and its Processor, as more educative for understanding of FEM essence, follows a description of Post-processor as most interesting for viewing of results of FEM computation and analysis of possible problems, difficulties, traps and surprises.

AxisVM X6® provides multi-window model analysis results display and design capabilities:

- results display by different types of diagrams with user defined color legend, figure 10,

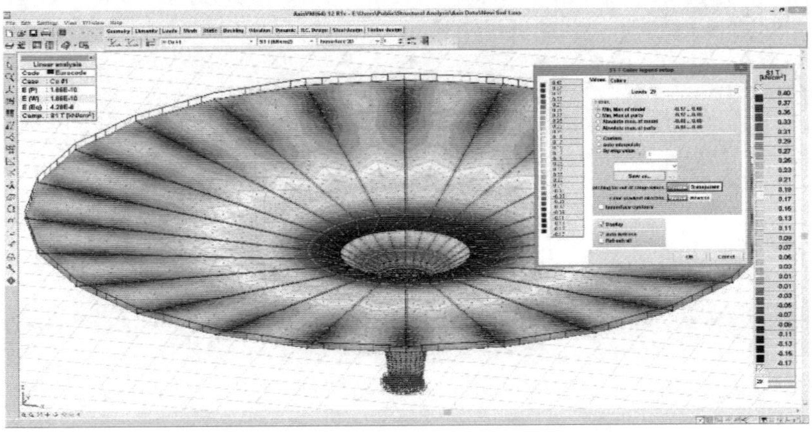

**Figure 10.** AxisVM X6® analysis results display – one space roof structure

- multi window presentation of the different numerical quantities with ability of animation view of results what is very useful in studying of structural behavior,
- integrated Report Generator for create of report templates,
- detailed documentation of strength and stability checks using formulas and substituted values,
- 3D PDF file export of the graphical content, which can be rotated and zoomed in and out within the PDF reader, etc.

## 4. Conclusions

Application of FEM software as specific training tool in the educational process is the area where the professional competence is essential, but together with solid knowledge in the field of computer technology and numerical mathematics. Similar relations are justified in FEM software development and distribution. Fortunately, InterCAD follows that route and therefore AxisVM® is developed "by engineers for engineers" as well as for students and researchers.

Second important thing is the question of technical regulations that should be harmonized regards to strong development of personal computers hardware and FEM software. Progress in FEM education could be motivating in desired trend of innovation of technical rules.

**Acknowledgements**

This paper has been supported by the Ministry of Education, Science and Technological Development, Republic of Serbia, through Project №. 451-03-9/2021-14/200156: "Innovative scientific and artistic research from the FTS activity domain".

IOP Conf. Series: Materials Science and Engineering 1208 (2021) 012003 doi:10.1088/1757-899X/1208/1/012003

## References

[1] Kovačević D 2006 *FEM Modelling in Structural Analysis* (in Serbian) (Belgrade: Građevinska knjiga) p336

[2] Cook R D 1995 *Finite Element Modelling for Stress Analysis* (New York: John Wiley & Sons) p336

[3] Felippa C A 2001 *Advanced Finite Element Methods for Solids, Plates and Shells (ASEN 6367)*, (Department of Aerospace Engineering Sciences, University of Colorado at Boulder)

[4] AxisVM X6® 2021 *User's manual* (Budapest: InterCAD) p650

[5] Kovačević D 2007 *Some aspects of FEM modelling of nonlinear behaviour of civil engineering structures* (invited lecture in Serbian) (Belgrade: Mathematical Institute of Serbian Academy of Science and Arts) p55

[6] Kovačević D Budak I Antić A Nagode A and Kosec B 2013 *FEM Modelling and Analysis in Prevention of the Waterway Dredger's Crane Serviceability Failure* (Engineering Failure Analysis ISSN 1350-6307, No. 1, Vol. 28) pp. 328-339.

[7] Kovačević D 2010 *Theoretical and practical aspects of modelling of boundary and interface conditions of structural systems* (invite lecture in Serbian) (Belgrade: Mathematical Institute of Serbian Academy of Science and Arts) p45

IOP Conf. Series: Materials Science and Engineering 1208 (2021) 012004 doi:10.1088/1757-899X/1208/1/012004

# Melt electrospun fibrous architectures with target geometries

**Budimir Mijović[1], Josip Jelić[1], Petra Brać[1] and Snježana Kirin[2]**

[1]University of Zagreb, Faculty of Textile Technology, Zagreb, Croatia

[2]University of Applied Sciences, Karlovac, Croatia

E-mail: budimir.mijovic@ttf.unizg.hr

**Abstract**. In the melt electrospinning technique, the polymer melt is stretched under high voltage and the cooled to form microfibers structures with a fibre diameter in the tens of micrometres range, although some studies have reported values ranging from hundreds of nanometres to hundreds of micrometres. In this respect, this technique has significance in the biomedical field, where tissue engineering scaffolds with bimodal (nano and micro) fibrous structures are preferred in regard to cell adhesion, spreading and infiltration to final tissue reconstruction. This paper gives a review of recently reported melt electrospinning devices, especially those based on the direct writing principle, and of their comparison with the new melt Spraybase electrospinning device. The Spraybase device provides high precision melt jet deposition into 2D and 3D programmed architectures, with versatile translation speeds of the collector plate in the X-Y and the melt head in the Z direction. The melt spun fibrous architectures are designed depending on the types of tissue cells used in scaffold development.

## 1. Introduction

Electrospinning of polymer melt is a relatively new spinning technology, as seen by the publications in this field that all date after 2011. Literature on polymer melt electrospinning accounts for just 1% of the overall electrospinning literature [1]. Polymer melt electrospinning is an ideal technique for the production of highly porous nano- or microfiber structures that are suitable for biomedical use. In recent decades, melt electrospinning has been recognized as an ecological procedure, as it has eliminated the cytotoxic effects of solvents used in electrospinning [2]. As an ecologically acceptable method of spinning micro- and nanofibers, electrospinning has been attracting the interest of many scientists. There is increased interest also of direct 3D printing of biopolymers with a scaffold design in tissue engineering [3]. Melt electrospinning typically produces microfibers with a diameter ranging from 5-40 µm, though this technology is also capable of producing fibers with a diameter of just 0.5 micrometers [4]. In the field of regenerative medicine, this is highly appliable, as melt electrospinning is capable of producing tissues with a higher resolution than is possible using 3D printing [5]. In recent years, there has been a massive surge in the demand for polymer nanofibers, that are used for a variety of applications, including tissue engineering, protective clothing, filtration and sensor systems [1, 6].

## 2. Materials and methods

The polymer polylactide (PLA) was used for sample preparation. This compound is obtained from natural resources and significantly reduces carbon traces in comparison with other oil-based plastics. The commercial name of PLA is Luminy® L175 (Figure 1). This is a homopolymer with a high melting point and high viscosity that is suitable for fibre spinning. In comparison with standard PLA, this homopolymer polyactide has a higher melting point and accelerated crystallisation rate. The density of the polylactide is 1.24 g/cm$^3$.

**Figure 1.** Polylactide (PLA) [6]

Polylactide electrospinning was performed on the device Spraybase® (AS-1204-000-01; Avectas Ltd., Ireland; Figure 2), at the Department of Fundamental Natural and Engineering Science, Faculty of Textile Technology, University of Zagreb. The polymer melt electrospinning device (Figure 2) consists of the following components: 1) high voltage source (voltage to 20 kV), 2) melt head with polymer container capable of movement along the Z axis, 3) heating system (achieves temperatures to 250°C), 4) air compressor for polymer emission (pressure to 5 bars), 5) metal flat plate collector capable of movement along the X and Y axes, with additional heating, and 6) safety cover.

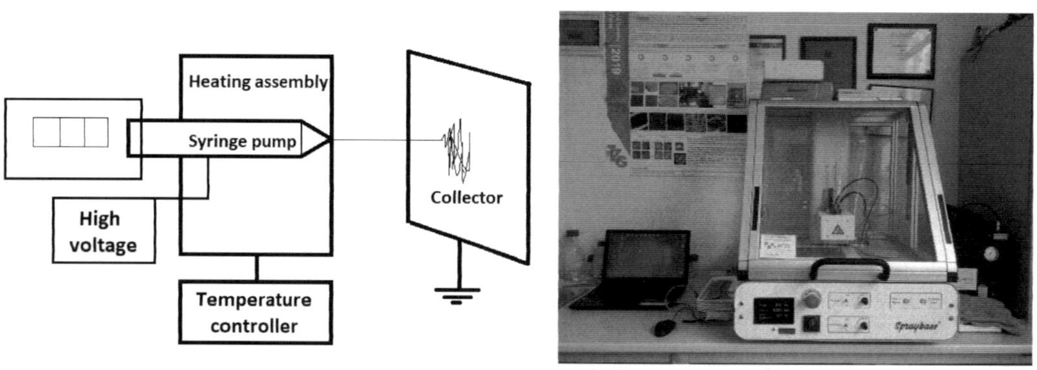


**Figure 2.** Polymer melt electrospinning device Spraybase®.

Working with the polymer melt electrospinning device consists of several steps. The first step is development of the 2D model in the program "SEL program generator". Next, the program and table of coordinates for collector movement is generated in the MSEL control system, which also controls the operation of the device. The desired parameters (voltage, pressure and temperature) are set on the device control panel.

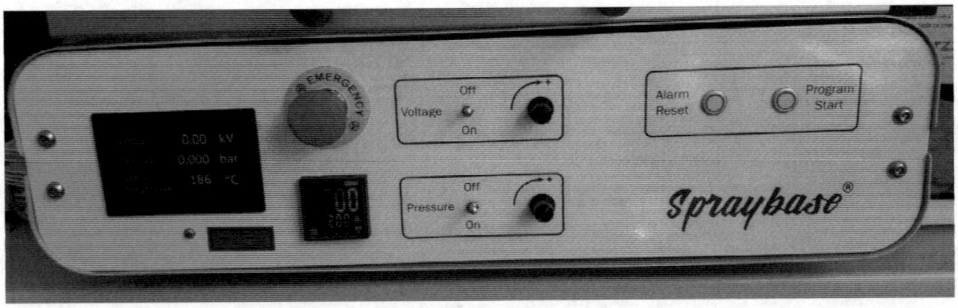


**Figure 3.** Control panel for setting the process parameters.

Optimisation of process parameters pertains to air pressure, melt temperature, voltage and distance of the head from the collector [7]. Optimization involves setting the height of the head from the collector, speed of movement of the collector, number of electrospinning cycles (number of layers) and container filling with polymer granules.

## 3. Results

### *3.1. Sample porosity*

Porosity was calculated for each sample from the first and second group. Porosity was calculated using a digital Dino Lite microscope to photograph every sample, and images were then processed using ImageJ software. After converting images into the appropriate 8-bit format (greyscale), a histogram was compiled, where 0 indicates pores and 255 indicates the polymer. The porosity results of individual samples are given in Table 1.

**Table 1.** Sample porosity

| Group 1 (single layer) | | Group 2 (double layer) | |
|---|---|---|---|
| 1a | 0,70 | 2a | 0,64 |
| 1b | 0,66 | 2b | 0,56 |
| 1c | 0,68 | 2c | 0,58 |
| 1d | 0,71 | 2d | 0,62 |
| 1e | 0,74 | 2e | 0,58 |

The porosity results show a greater porosity in samples of Group 1 (1a), and therefore it can be concluded that adding a layer reduces the sample porosity (Figure 4).

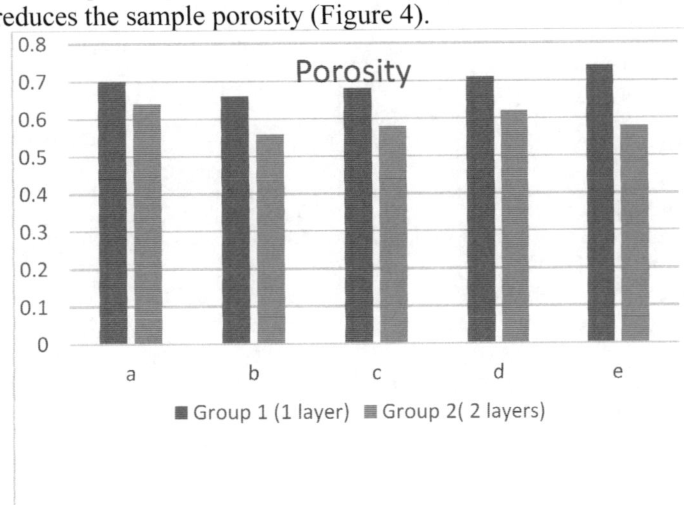


**Figure 4.** Sample porosity

The effective material constants for fiber network can be determined using homogenization procedure concept for fiber network. The strain energy fiber network for representative volume element is equal to strain energy continuum element with effective material constant [17]. The strain energy of the representative volume element under plane stress conditions are

$$U = (1/2)\cdot \langle\varepsilon_{ij}\rangle \cdot C_{ijkl}\, \varepsilon_{kl} \cdot V \qquad (1)$$

where is $V = b\cdot h\cdot 2\cdot r$ - representative volume element, $C_{ijkl}$ are effective elasticity tensor. We assume that microscopic deformation tensor of a fiber segments $\varepsilon_{ij}$ is compatible with effective macroscopic strain $\langle\varepsilon_{ij}\rangle$ of effective continuum (affine transformation). This is bridge relations between fiber segment microstrain $\varepsilon_{ij}$ and macroscopic strain $\langle\varepsilon_{ij}\rangle$ in the effective medium.

*3.2. Air permeability*

Air permeability was calculated in the group of samples with a single layer of PLA polymer created using polymer melt electrospinning. The samples of the first group, all with a single layer, called sample 1a, 1b and 1c were all spun under the same conditions though each sample is unique and therefore was tested separately. Cold air was blown through the sample (10x10 cm) at a rate of 1 m/s and permeability of each sample measured using an anemometer. Samples from Group 2, spun in a double layer, were measured in the same way. The results are given in Table 2.

**Table 2.** Air permeability of samples

| **Sample (1 mm, single layer)** | **1a** | **1b** | **1c** | **1d** | **1e** |
|---|---|---|---|---|---|
| **Thickness (mm)** | 0,4278 | 0,4080 | 0,6198 | 0,4360 | 0,4380 |
| **Air permeability (m/s)** | 5,99 | 5,23 | 4,75 | 5,50 | 5,95 |
| **Temperature (°C)** | 40 | 40,6 | 40,4 | 39 | 40 |
| **Sample (1 mm, single layer)** | **2a** | **2b** | **2c** | **2d** | **2e** |
| **Thickness (mm)** | 0,492 | 0,5956 | 0,7376 | 0,690 | 0,723 |
| **Air permeability (m/s)** | 5,31 | 4,47 | 4,08 | 4,11 | 4,21 |
| **Temperature (°C)** | 39,6 | 39 | 39 | 40 | 39 |

From the results shown in Table 2, Group 1 samples proved to have a higher air permeability, and therefore it can be concluded that by adding more layers, the air permeability is reduced (Figure 5).

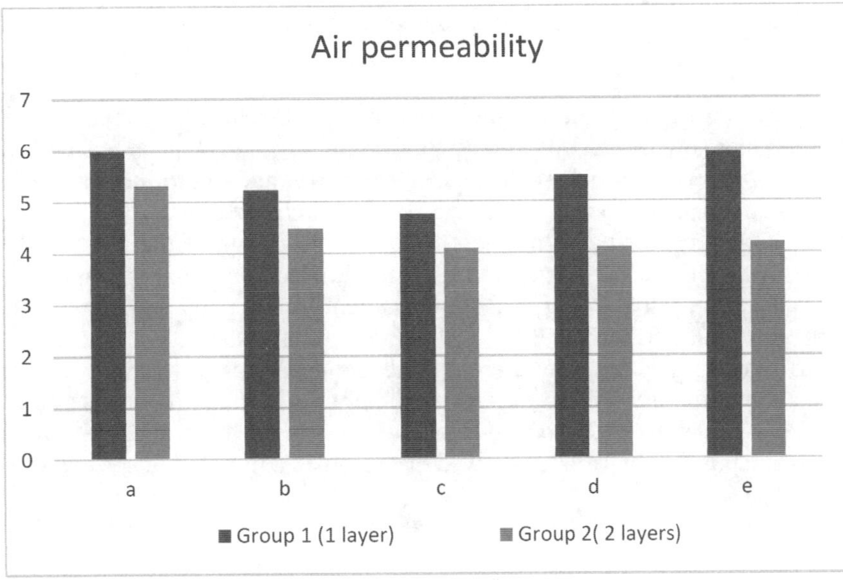


**Figure 5.** Air permeability of samples

From the above Tables 1 and 2, the thickness of each individual sample can be examined, with air permeability and the temperature of air passing through the sample. The thinnest sample was sample 1a (single layer) (Figure 6).

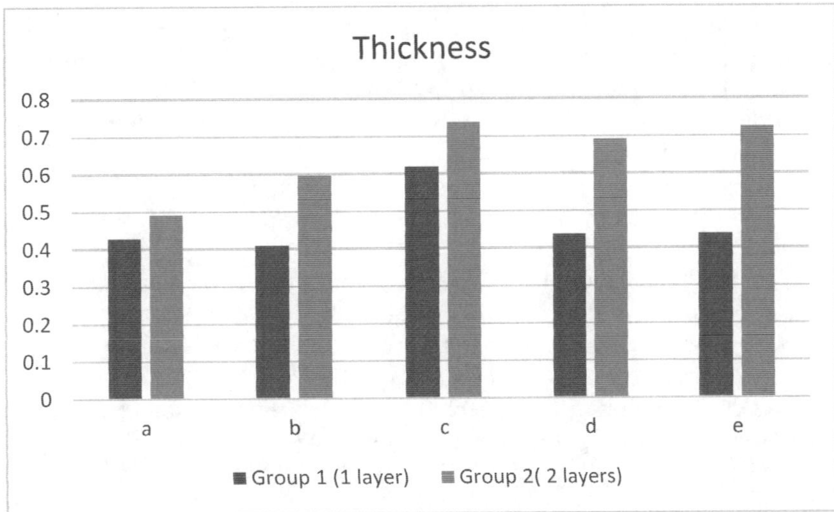


**Figure 6.** Sample thickness

A steady state, laminar, incompressible model has been adopted for the flow regime inside our virtual media. Implemented in the Fluent code [18] is used to solve continuity and conservation of linear momentum in the absence of inertial effects

$$\nabla \cdot \mathrm{v} = 0 \tag{2}$$
$$\nabla \mathrm{p} = \mu\, \Delta^2 \cdot \mathrm{v} \tag{3}$$

The grid size required to mesh the gap between two fibers around their crossover point is often too small.

## 4. Conclusion

The polymer melt electrospinning technique is becoming increasingly popular, with a number advantages over polymer solvent electrospinning techniques: more ecologically acceptable production conditions, ability to produce materials with specific geometry, and microfiber production, which has wide-ranging applications in biomedicine. In this study, microfiber materials were produced from polymer (PLA) melt electrospinning with a description of the design of the 2D model to make three groups of materials with differing geometry and filament density. The primary aim of this study was to determine the influence of geometry and density of filaments on porosity and permeability. By testing PLA material porosity and permeability, the following was established: 1) that increasing distance between filaments reduces the maximum force, corresponding strain at the same force and tensile strength, i.e. that a higher density of filaments gives higher tensile strength, but also deformation; 2) an analysis of the data in Tables 1 and 2 shows that sample 1a was the thinnest of the samples, in a single layer; 3) analysis of the porosity of samples showed that sample 2a in a double layer was the less porosity air permeability of the samples.

## References

[1] Nayak R, Padhye R and Arnold L 2017 Melt-electrospinning of nanofibers *Electrospun Nanofibers* 11–40
[2] Nazari T and Garmabi H 2017 The effects of processing parameters on the morphology of PLA/peg melt electrospun fibers *Polymer International* **67** 178–88
[3] Hochleitner G, Hümmer J F, Luxenhofer R and Groll J 2014 High definition fibrous poly(2-ethyl-2-oxazoline) scaffolds through melt electrospinning writing *Polymer* **55** 5017–23
[4] Dalton P D, Grafahrend D, Klinkhammer K, Klee D and Möller M 2007 Electrospinning of polymer melts: Phenomenological observations *Polymer* **48** 6823–33
[5] Dalton P D, Klinkhammer K, Salber J, Klee D and Möller M 2006 Direct in vitro electrospinning with polymer melts *Biomacromolecules* **7** 686–90
[6] Nayak R 2017 *Polypropylene nanofibers: Melt electrospinning versus meltblowing* (Cham: Springer International Publishing)
[7] Zhou H 2007 *Electrospun fibers from both solution and melt: Processing, structure and property*
[8] Anon 1969 Electrically driven jets *Proceedings of the Royal Society of London. A. Mathematical and Physical Sciences* **313** 453–75
[9] GAÑÁN-CALVO A L F O N S O M 1997 On the theory of electrohydrodynamically driven capillary jets *Journal of Fluid Mechanics* **335** 165–88
[10] Hohman M M, Shin M, Rutledge G and Brenner M P 2001 Electrospinning and electrically forced jets. II. applications *Physics of Fluids* **13** 2221–36
[11] Spivak A F, Dzenis Y A and Reneker D H 2000 A model of steady state jet in the electrospinning process *Mechanics Research Communications* **27** 37–42
[12] Yarin A L, Koombhongse S and Reneker D H 2001 Bending instability in electrospinning of nanofibers *Journal of Applied Physics* **89** 3018–26
[13] Fridrikh S V, Yu J H, Brenner M P and Rutledge G C 2003 Controlling the fiber diameter during electrospinning *Physical Review Letters* **90**
[14] Theron S A, Zussman E and Yarin A L 2004 Experimental investigation of the governing parameters in the electrospinning of Polymer Solutions *Polymer* **45** 2017–30
[15] Rutldge G C, Shin M Y, Warner S B, Bauer A, Grimler M, Ugbolue S C 2000 A fundamental investigation of the formation and properties of electrospun fibers *NTC Report*
[16] Luzhansky D M 2003 Quality Control in manufacturing of electrospun Nanofiber Composites *International Nonwovens Journal* **os-12**

[17] Berhan L, Yi Y B, Sastry A M, Munoz E, Selvidge M and Baughman R 2004 Mechanical properties of nanotube sheets: Alterations in joint morphology and achievable moduli in manufacturable materials *Journal of Applied Physics* **95** 4335–45
[18] FLUENT 6.2 User's Guide, www.fluent.com

IOP Conf. Series: Materials Science and Engineering 1208 (2021) 012005 doi:10.1088/1757-899X/1208/1/012005

# Wood Surface Roughness Prediction Modeling Depending on Influential Cutting Variables

**Atif Hodžić, Elmasa Aldžić and Damir Hodžić**

University of Bihac, Faculty of Technical Engineering Bihac, Bosnia and Herzegovina

E-mail: damir.hodzic@unbi.ba

**Abstract**. Paper presents the design of experiment and determining mathematical model to calculate roughness parameter of wood planned surface. For design of experiment three different types of solid wood were taken and processed on the planner with three different displacements and three different cutting speeds. After measuring the roughness parameter Rz, experimental results were obtained on the basis of which the central composite plan of the experiment was made. Based on that, a model of roughness parameter Rz was made, which is adequate and with high accuracy. The significance of the model coefficients was determined using the R software and the results were presented using the Design Expert software.

## 1. Introduction

Surface topography of wood has a dominant influence on the contact conditions of two or more surfaces and as such should be considered in three-dimensional form. However, in the analysis of the roughness of the treated surface, the problem is reduced to surface cross-sections, ie to the amplitudes of the spread of irregularities on the treated surface. The cause of such deformations occurs due to processing errors, incorrect position or vibration of tools or objects during processing. Wood as a specific basic material due to its porosity, ie anisotropy and orthotropy, in terms of its anatomical structure requires additional analysis when examining the roughness of the treated surface. The quality of treated wood surfaces depends on the workability of the wood type and the determined parameters of the processing parameters, above all the cutting speed, feed rate, cutting depth, and the thickness and type of sawdust. The basic numerical parameters of the roughness of the treated surface are determined in accordance with the ISO standard, system "M" which have 12 qualitative classes in relation to the arithmetical mean roughness value *Ra* or in relation to the mean roughness depth *Rz* [1,2,3].

In this paper, an experiment was performed in real conditions on samples of spruce, linden and beech, which were previously dimensioned in work pieces. During the experiment, in addition to the change in wood density $\rho$, the kinematic parameters of cutting speeds $v$ and feed rate $s$ at three different levels also changed. Measurement of the mean roughness depth *Rz* was performed using a contact profilometer Mitutoyo S201. ***Rz*** is the average value of the absolute values of the heights of five highest profile peaks and the depths of five deepest alleys within the evaluation length, Figure 1. We can also say that *Rz* is the difference between the tallest "peak" and the deepest "valley" in the surface [4,5].

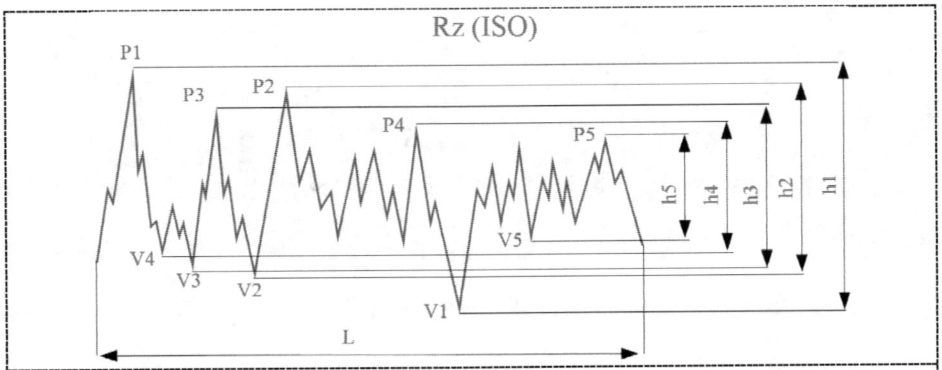


**Figure 1.** Rz value, distance between highest peak and deepest alley

Stochastic modeling of input kinematic parameters of cutting speed, feed rate, and wood density resulted in adequate models for output value of the mean roughness depth *Rz*, which were confirmed using ANOVA method.

## 2. Design of Experiment

Design of Experiment (DOE) is a powerful tool for achieving significant improvements in product quality and process efficiency. For experiment planning, one can also say that the methodology of applying statistics in the experiment execution process. As such discipline is not just a engineering problem but is used in different spheres of society.

A well-planned experiment allows you to get clear interpretations and avoid complicated analyzes. The poorly planned experiment gives us some wrong conclusions of a process. Experiment planning is of particular importance in all quantitative studies, especially when multiple factors are studied at the same time to examine and compare their effects.

For this research we choosed central composite experimental design (CCD) with 14 repetitions, of which 6 are in the central point [6,7].

### 2.1. Aim and object of research

The aim of the experimental research is to examine the significance of the influential kinematic parameters of the surface roughness, ie cutting speed ($v$), feed rate ($s$) and wood density ($\rho$) in milling solid wood elements of 24 mm thickness, and analysis of experimental data to achieve a lower mean roughness depth *Rz*, without compromising the quality and total production costs. The purpose of the experimental plan is to generate a mathematical model that describes the process. If the studied parameters in the experiment are really the ones that affect the process, and the data obtained by the experiment are of acceptable accuracy and precision, then it is possible to develop a model that credibly describes the process [7].

### 2.2. Experiment resources

For the 14 repetitions of experiment we prepared 4 spruce specimens, 4 beech specimens and 6 linden specimens for the central point of experiment. Dimensions of specimens are 620x80x24 [mm], Figure 2. To determine density of wood we used standard BAS EN 13061-2:2016 and the results are:

- Spruce; $\rho = 490$ [kg/m$^3$]
- Linden; $\rho = 590$ [kg/m$^3$]
- Beech; $\rho = 690$ [kg/m$^3$]

**Figure 2.** Labeling of specimens prepared for machining [8]

The wood samples were milled with cutting speed and feed rate which was available at the machine and according to the experiment. Cutting speeds were 12,5 [m/s], 25 [m/s] and 37,5 [m/s] and feed rates 2 [m/min], 3 [m/min] and 4 [m/min]. Table 1 shows input factors for experiment and their values and coded values. Samples were machined and mean roughness depth *Rz* was measured according to the experiment central composite design with 14 repetitions, Table 2 [8].

**Table 1.** Input factors for experiment [8]

| Effect | -1 | 0 | 1 |
|---|---|---|---|
| Factor A – Wood density ρ [kg/m$^3$] | 490 | 590 | 690 |
| Factor B – Feed rate, s [m/min] | 2 | 3 | 4 |
| Factor C – Cutting speed, v [m/s] | 12,5 | 25 | 37,5 |

**Table 2.** Experiment central composite design with 14 repetitions [8]

| Sample | Factor A – Density $\rho$ [kg/m$^3$] | Factor B Feed rate, s [m/min] | Factor C Cutting speed, v [m/s] | Rz, exp [μm] |
|---|---|---|---|---|
| **P 1** | 490 | 2 | 12.5 | 41.02 |
| **P 2** | 690 | 2 | 12.5 | 35.03 |
| **P 3** | 490 | 4 | 12.5 | 28.36 |
| **P 4** | 690 | 4 | 12.5 | 24.88 |
| **P 5** | 490 | 2 | 37.5 | 51.79 |
| **P 6** | 690 | 2 | 37.5 | 35.25 |
| **P 7** | 490 | 4 | 37.5 | 33.2 |
| **P 8** | 690 | 4 | 37.5 | 38.37 |
| **P 9** | 590 | 3 | 25.0 | 37.41 |
| **P 10** | 590 | 3 | 25.0 | 37.66 |
| **P 11** | 590 | 3 | 25.0 | 38.65 |
| **P 12** | 590 | 3 | 25.0 | 36.02 |
| **P 13** | 590 | 3 | 25.0 | 36.96 |
| **P 14** | 590 | 3 | 25.0 | 35.22 |

After measuring experimental values of mean roughness depth *Rz* for all 14 repetitions next step is to do analysis of variance (ANOVA) which will give us significance of input parameters and model for mean roughness depth *Rz*.

## 3. Analysis of Variance

By introducing the coded values according to the schedule of the experimental plan and the position of the coded quantity values and multiplying by the mean roughness depth *Rz* from Table 2, a linear equation of mean roughness depth *Rz* can be written:

$$Y = 36.416 - 2.605X_1 - 4.785X_2 + 3.665X_3 + 3.027X_1X_2 - 0.237X_1X_3 + 0.917X_2X_3 \quad (1)$$

To determine the significance of the regression coefficients for the linear model, an analysis of variance for a given experiment will be performed. Analysis of variance involves calculating the sum of squares (SS), degrees of freedom (dF) and mean squares (MS) for given input factors, interactions between them, lack of fit and error of the model, Table 3.

**Table 3.** Analysis of Variance (ANOVA) results

| Source of variation | Sum of squares | Degrees of freedom | Mean of squares | Fisher's coefficient | *p*-value | Significance |
|---|---|---|---|---|---|---|
| A | 54.288 | 1 | 54.288 | 6.672 | 0.036 | significant |
| B | 183.170 | 1 | 183.170 | 22.511 | 0.002 | significant |
| C | 107.458 | 1 | 107.458 | 13.206 | 0.008 | significant |
| AB | 73.326 | 1 | 73.326 | 9.012 | 0.019 | significant |
| AC | 0.451 | 1 | 0.451 | 0.055 | 0.821 | not significant |
| BC | 6.734 | 1 | 6.734 | 0.828 | 0.393 | not significant |
| Model | 425.428 | 6 | 70.905 | 8.714 | 0.005 | significant |
| Error | 56.958 | 7 | 8.137 | | | |
| Lack of fit | 49.503 | 2 | 24.751 | | | not significant |
| Pure error | 7.455 | 5 | 1.491 | | | |
| Total | 482.385 | 13 | 37.107 | | | |

Values from Table 3 are calculated according to statistical equations for those parameters. Fisher's coefficients were compared with coefficient from Fisher's distribution for level of significance $\alpha = 0,05$ (significance 5%).

In this example, $\alpha = 0.05$ or 5% was taken because this level is most often taken in engineering research.

Factor is significant if calculated F value is less than F value from Fisher table distribution.

To determine significance of the factors or interaction between them it can also be use statistical value p which can be calculated using software R.

After determining the significance of the input parameters only those regression coefficients whose values which are significant can be taken into account, so equation (1) will be transformed into:

$$Y = 36.416 - 2.605X_1 - 4.785X_2 + 3.665X_3 + 3.027X_1X_2 \quad (2)$$

After decoding the input values the final equation for the linear model of mean roughness depth *Rz* is:

$$R_z = 112.387 - 0.116\rho - 22.644s + 0.293v + 0.03\rho s \quad (3)$$

Now it is possible to calculate model results of mean roughness depth *Rz* and to compare them with experimental results, Table 4.

**Table 4.** Mean roughness depth *Rz* according to the experimental results and according to the model [8]

| Sample | Rz, exp [μm] | Rz, mod [μm] |
|---|---|---|
| **P 1** | 41.02 | 43.32 |
| **P 2** | 35.03 | 32.12 |
| **P 3** | 28.36 | 27.43 |
| **P 4** | 24.88 | 28.23 |
| **P 5** | 51.79 | 50.65 |
| **P 6** | 35.25 | 39.45 |
| **P 7** | 33.2 | 34.76 |
| **P 8** | 38.37 | 35.56 |
| **P 9** | 37.41 | 36.44 |
| **P 10** | 37.66 | 36.44 |
| **P 11** | 38.65 | 36.44 |
| **P 12** | 36.02 | 36.44 |
| **P 13** | 36.96 | 36.44 |
| **P 14** | 35.22 | 36.44 |

After obtaining the model equation, it is useful to check the homogeneity of the dispersions using the Cochran criterion, which is calculated for central point results, Table 5.

**Table 5.** Datas for calculating Cochran criterion [8]

| Run | Rz, exp [μm] | $\sigma_j^2 = (R_{zi} - \overline{R_z})^2$ |
|---|---|---|
| 9 | 37.41 | 0.179211 |
| 10 | 37.66 | 0.453378 |
| 11 | 38.65 | 2.766678 |
| 12 | 36.02 | 0.934444 |
| 13 | 36.96 | 0.000711 |
| 14 | 35.22 | 3.121111 |
| | $\overline{R_z} = 36.987$ | $\sum_{j=1}^{n} \sigma_j^2 = 7.4555$ |
| | | $max\sigma_j^2 = 3.1211$ |

According to the table 5 Cochran criterion can be calculated:

$$K_h = \frac{3.1211}{7.4555} = 0.418 \tag{4}$$

Table value for Cochran criterion is $K_t = 0.445$ which is to be compared with calculated one. In our example $K_h < K_t$ which means dispersion in the central point is homogeneous.

The adequacy of the model based on the multiple regression coefficient is determined by calculating the magnitude of R and based on the data from the table 6.

**Table 6**: Datas for calculating value of the regression coefficient R [8]

| Run | Rz, exp [μm] $= Y_j^E$ | Rz, mod [μm] $= Y_j^M$ | $\left(Y_j^E - Y_j^M\right)^2$ | $\left(Y_j^E - \overline{Y^E}\right)^2$ |
|---|---|---|---|---|
| 1 | 41.02 | 43.32 | 5.2969 | 21.19958 |
| 2 | 35.03 | 32.12 | 8.4593 | 1.920164 |
| 3 | 28.36 | 27.43 | 0.8584 | 64.8943 |
| 4 | 24.88 | 28.23 | 11.245 | 133.0724 |
| 5 | 51.79 | 50.65 | 1.3075 | 236.3691 |
| 6 | 35.25 | 39.45 | 17.6106 | 1.358856 |
| 7 | 33.2 | 34.76 | 2.4289 | 10.34073 |
| 8 | 38.37 | 35.56 | 7.9045 | 3.819288 |
| 9 | 37.41 | 36.44 | 0.9409 | 0.988632 |
| 10 | 37.66 | 36.44 | 1.4884 | 1.548282 |
| 11 | 38.65 | 36.44 | 4.8841 | 4.992096 |
| 12 | 36.02 | 36.44 | 0.1764 | 0.156578 |
| 13 | 36.96 | 36.44 | 0.2704 | 0.296262 |
| 14 | 35.22 | 36.44 | 1.4884 | 1.429698 |
| | $\overline{Y^E} = 36.41$ | | Σ = 64.3609 | Σ =482.386 |

Calculated value of the regression coefficient R is:

$$= \sqrt{1 - \frac{64.3609}{482.386}} = 0.9309 \qquad (5)$$

The calculated value of the regression coefficient is R = 0.9309 and that means that the linear model describes the accuracy of the experimental results with 93.09%, which is an excellent accuracy of the model.

## 4. Discussion of results

For the purposes of this paper and to confirm the obtained statistical results, the Design Expert software package was used.

Figure 3 show the dependence of the output values of the model, mean roughness depth *Rz* on the input factors, the density of the wood, feed rate and cutting speed.

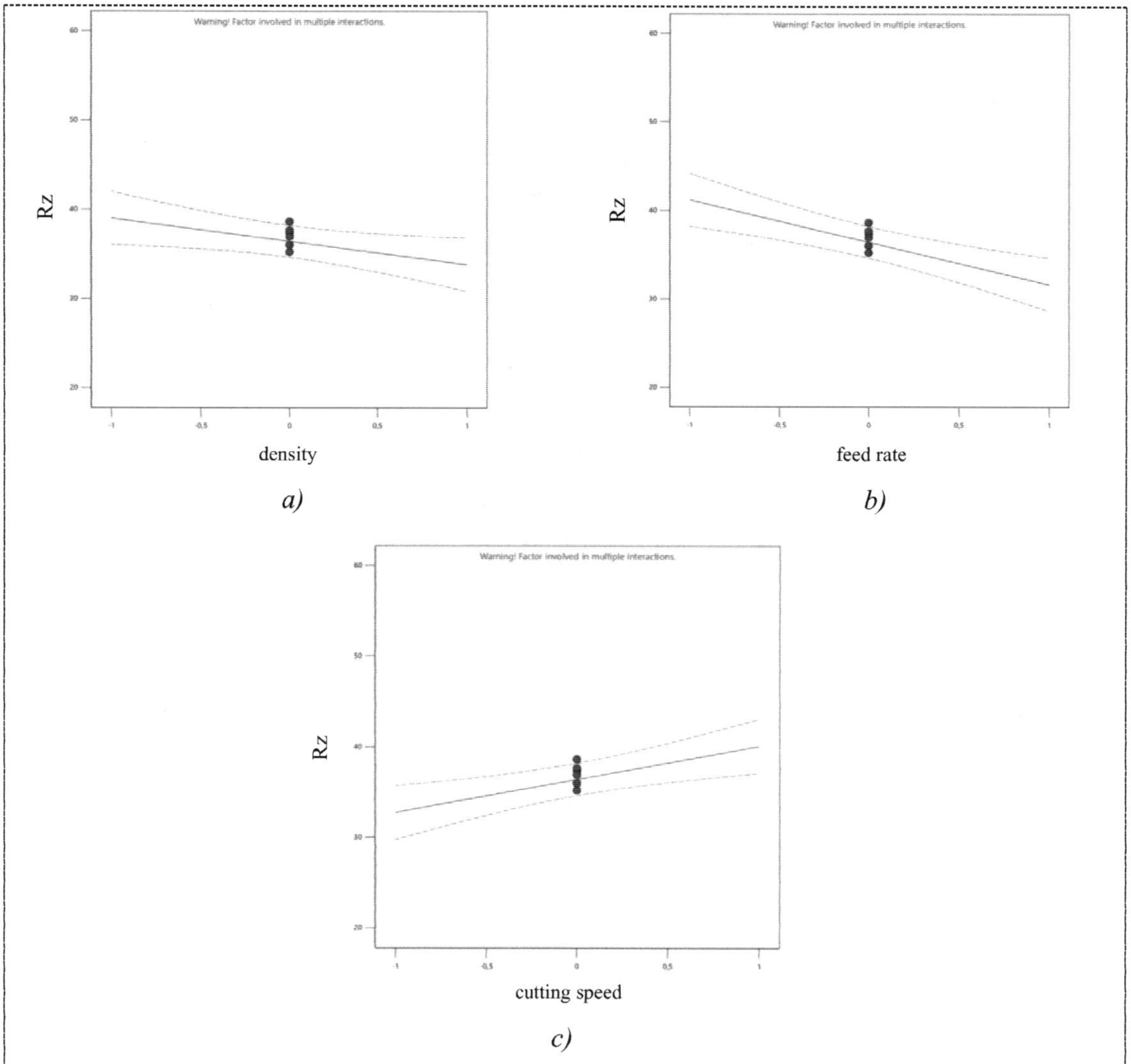


**Figure 3**: Dependence of the output value on the input factors a) density; b)feed rate; c)cutting speed

Figure 4 show the dependence of the output values of the model on the interaction of input values.

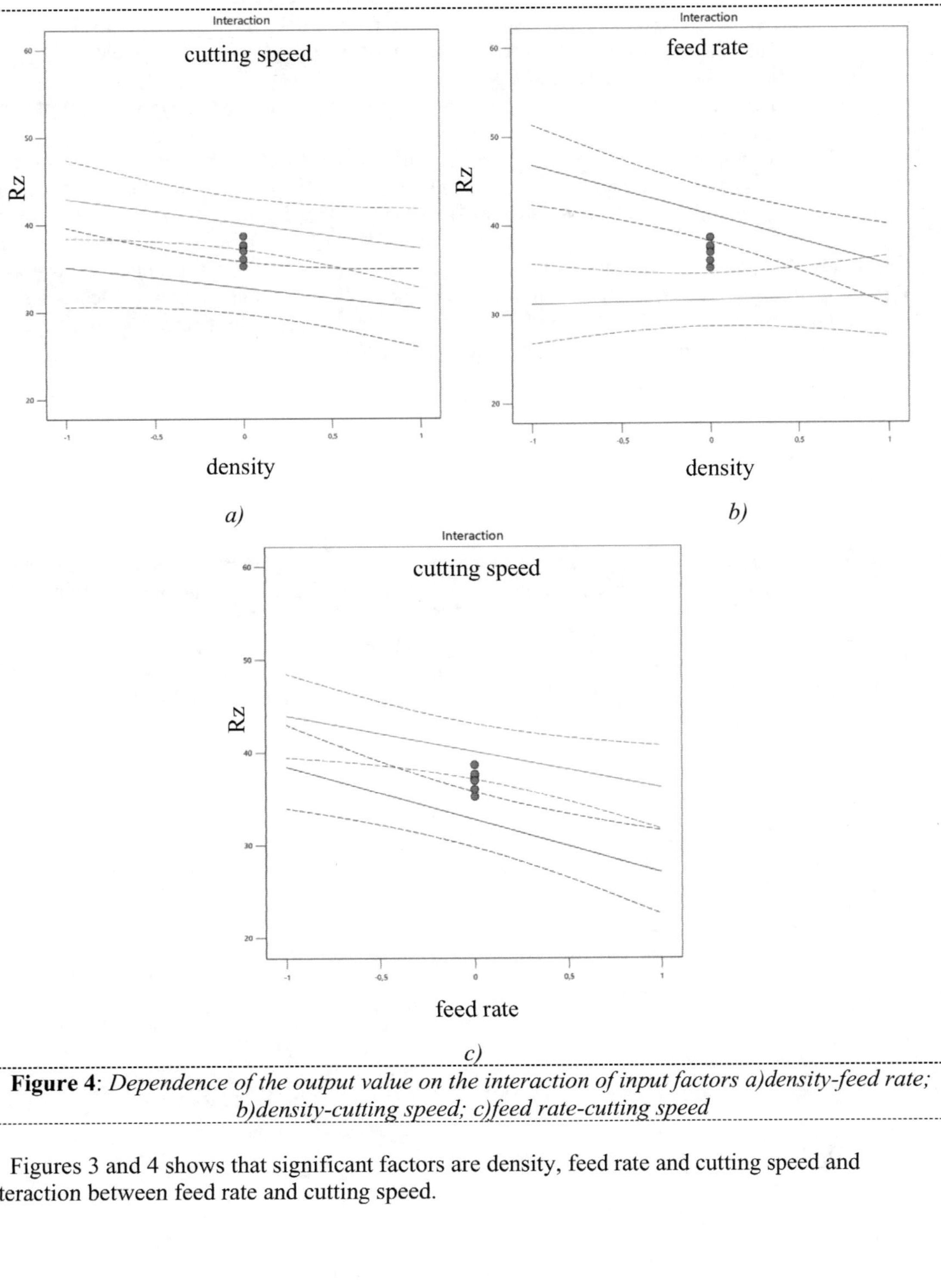


**Figure 4**: *Dependence of the output value on the interaction of input factors a)density-feed rate; b)density-cutting speed; c)feed rate-cutting speed*

Figures 3 and 4 shows that significant factors are density, feed rate and cutting speed and interaction between feed rate and cutting speed.

## 5. Conclusion

The quality of the treated wood surface is influenced by a large number of factors such as: physical and mechanical properties of wood, anatomical structure of wood, and technological parameters of the machining. Technological parameters that affect the quality of the processed surface are cutting speed, feed rate, cutting depth, thickness and type of sawdust.

One of the output characteristics of any production, in addition to economy, productivity, and profitability, is the quality of processing.

By statistical processing of the experimental results and analysis of the obtained models for mean roughness depth *Rz* using commercial software Design Expert, it can be concluded that the roughness of the treated surface by milling, both experimental and model results for the measured mean roughness depth *Rz* belongs to class N9 surface roughness according to ISO M.A1. 020. The kinematic parameter of the density, feed rate and cutting speed and interaction between feed rate and cutting speed proved to be significant parameters for the quality of the treated surface. Mean roughness depth *Rz* decreases with increasing density of wood and feed rate and increases with increasing cutting speed. The distribution of model values in relation to experimental values is satisfactory.

## References

[1] Skakic D 2002 *Finalna obrada drveta* (Univerzitet u Beogradu, Šumarski fakultet)
[2] Hasanagić, R., Hodžić, A., Bjelić, A. (2021). Mathematical modelling of the surface foughness in orthogonal cutting of the massive wood, Wood Materials Science & Engineering, 16, 2, 74-80, doi: 10.1080/17480272.2019.1608297
[3] Mikleċić, J., Jirouš-Rajković, V. (2021) The Relationship between Roughness of Finished Wood Floors and Slip Resistance, Drvna industrija: Znanstveni časopis za pitanja drvne tehnologije, VOL. 72 No. 1, 2021. doi:10-5552/drvind.2021Hodzic D and Dzanic A 2020 *Planiranje eksperimenta* (Univerzitet u Bihaću, Tehničkifakultet)
[4] Malkoçoğlu, A. (2007). Machining properties and suurface roughness of various wood species planed in different conditions, Building and Environment, 42, 2562-2567. doi: 10.1016/j.buildenv.2006.08.028
[5] https://upmold.com/surface-finish-ra-rz/
[6] Montgomery, D.C. (2015). Design and Analysis of Experiment, Arizone State University, John Wiley and Sons
[7] Hodžić, D., Džanić, A. (2020). Planiranje eksperimenta, University of Bihac, Faculty of Technical Engineering, ISBN 978-9958-533-19-8
[8] Aldžić, E. Aldžić, E. Experimental analysis and modeling of kinematic parameters influence on roughness of treated surface, Master thesis University of Bihac, Faculty of Technical Engineering

# Application of mathematical modelling in the process of design and production of pressure vessels

**Fadil Islamović, Mirzet Beganović, Esad Bajramović and Dženana Gačo**
University of Bihac, Faculty of Technical Engineering, Bosnia and Herzegovina

E-mail: fadil.islamovic@unbi.ba

**Abstract**. The paper presents the engineering practice, which the company ''Regeneracija'' Ltd. Velika Kladuša – Bosnia and Herzegovina uses to perform preliminary experimental testing and measurements, followed by mathematical modeling of critical pressure of these vessels, in order to obtain the projected quality of pressure vessels made of composite materials. The paper will confirm the hypothesis that it is possible to relate mathematical connection and dependence of the critical pressure of vessels of composite materials ($P_{kr}$) with mechanical characteristics of vessel material ($\sigma_M$), vessel diameter (D), and vessel wall thickness (s). In this way, by varying the mentioned parameters, it is possible to achieve the desired product quality in the production of composite material containers by achieving the projected critical and thus working pressure. Generally speaking, the mathematical model of critical pressure obtained in this way will be a good indicator for design engineers to know how much critical pressure a given vessel can withstand, and based on that to take quick control of working or projected pressure, but also for designing completely new vessels made of composite materials as a substitute for the expensive experimental testing.

## 1. Introduction

Today's modern structures require new materials with special properties and shapes that can meet specific operating conditions: increased strength, resistance to pressure, temperature, speed, impact, vibration, etc. The purpose of new construction materials is to respond to increasingly complex requirements in terms of durability and the performance of modern constructions, as well as increasing restrictions on their weight. In order for the set requirements to be met, a compromise is inevitable in achieving light construction and high mechanical properties of the material. The application of composite materials enables the fulfillment of such set requirements [1].

One of the important parameters for engineers who deal with the design of pressure vessels made of composite materials is to know at what pressure the failure or cracking of the vessels will occur in relation to the selected combination of materials. The values of this critical pressure also depend on the orientation of the reinforcement in the composite. As it is known that a composite material consists of a matrix (polyester resin) and reinforcement (two types of reinforcement, mat and mat-rowing), it is difficult to define the characteristics of the material without experimental methods.

Designers face major problems when designing composite vessels, as there are many manufacturers of resin as well as glass reinforcements on the market, thus different combinations give different material characteristics. Pressure vessels made of composite materials are mainly used for water treatment such as sand filters, water softeners, active carbon filters, etc., as well as pressure vessels in air purification plants that are made in horizontal and vertical versions with a diameter of 500 (mm) to 2400 (mm), in classes 2, 4, 5, 6 and 10 (bar) [1].

In order to obtain the necessary results, i.e., the input parameters required for mathematical modeling, experimental tests were performed. The first part of the experiment referred to the definition of the mechanical characteristics of specimens made of composite materials, while the second part of the experiment referred to the definition of the critical pressure at which the vessels will yield. After obtaining the experimental results, mathematical modeling of the critical pressure will be performed based on the input-output parameters of the process. Thus, experimental measurements and mathematical modeling will prove that the critical pressure of composite materials can be related to the mechanical characteristics of the material ($\sigma_M$), diameter (D) and vessel wall thickness (s), i.e., that $P_{kr} = f$ (s, D, $\sigma_M$, ...). In this way, by varying the mentioned parameters, it is possible to achieve the desired product quality in the production of vessels made of composite materials by achieving the projected critical and thus working pressure [1].

## 2. Analysis of the results of experimental testing

As already mentioned, different materials for making composites are offered on the market, so that when choosing different materials, it is possible to obtain different mechanical characteristics of the final composite. For these reasons, prior to making critical pressure test vessels, it is necessary to make composite tubes and test their mechanical properties. In the first experiment, tensile strength, modulus of elasticity and final elongation were obtained for two types of test tubes, and in the second experiment the values of critical pressures for eight tested vessels were obtained through previous research [2].

Table 1 shows the mean values of the obtained results, so that an increase in the mechanical properties of the material is seen when applying reinforcement in the form of rowing. Table 2 shows the results of experimental tests, i.e., the values of the measured achieved critical pressures for all eight tested vessels.

**Table 1**. Mean value of the obtained results [2].

| Tube mark | Rupture force $F_M$ (KN) | Tensile strength $\sigma_M$ (MPa) | Modulus of Elasticity E (MPa) | Final elongation $\varepsilon$ (%) |
|---|---|---|---|---|
| **A** | 9.74 | 118.09 | 6591.86 | 2.70 |
| **B** | 13.60 | 161.21 | 8177.09 | 2.56 |

**Table 2.** The results of measured critical pressure [2].

| No | D (mm) | s (mm) | $\sigma_M$ (N/mm$^2$) | Experimental results $P_{kr}$ (bar) |
|---|---|---|---|---|
| 1 | 400 | 3.2 | 118 | 22.0 |
| 2 | 800 | 3.2 | 118 | 9.0 |
| 3 | 400 | 6.4 | 118 | 30.0 |
| 4 | 800 | 6.4 | 118 | 17.0 |
| 5 | 400 | 3.2 | 161 | 26.5 |
| 6 | 800 | 3.2 | 161 | 12.5 |
| 7 | 400 | 6.4 | 161 | 35.0 |
| 8 | 800 | 6.4 | 161 | 20.5 |

The previous table shows the obtained results of critical pressure ($P_{kr}$) measurements for eight experimentally tested vessels depending on the diameter (D), wall thickness (s) and material characteristics ($\sigma_M$).

It should be emphasized that in the first four vessels, which were made with mat reinforcements, there was total damage during cracking, as shown in figure 1.a). Due to the action of circular stress, the cylindrical part of the vessel yielded, causing a crack along the entire vessel, i.e. along the cylindrical part of the vessel [2].

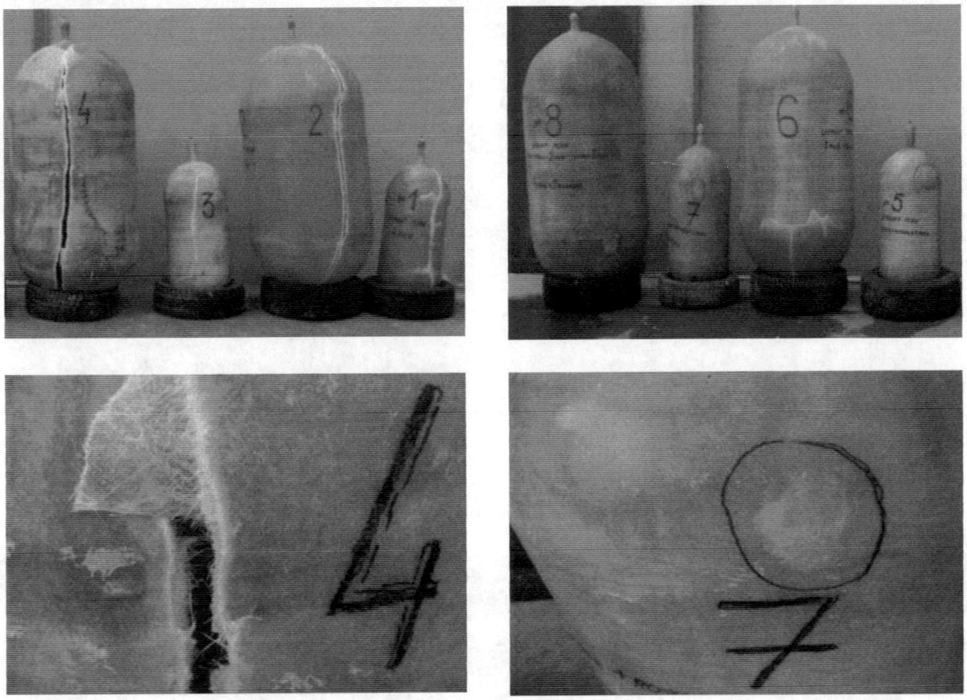

a) b)

**Figure 1.** a) Shape of yielding of vessels made of mat reinforcement (left)
b) Shape of yielding of vessels made of mat and rowing reinforcement (right).

The other four vessels, which have continuous rowing in their structure, yielded only in one place at the moment of critical pressure, as shown in figure 1.b). Due to the yielding of the vessel under a certain pressure, continuous fibers did not allow the crack to be transferred to the rest of the vessel but localized the crack, which gives vessels that have rowing in their structure a huge advantage for safety reasons [2].

According to table 2, it can be seen that increasing the diameter of the vessel decreases the value of the critical pressure, while the increasing the wall thickness increases its value. The table also shows the influence of the mechanical properties of the materials, i.e., the value of the critical pressure grows with their increase.

## 3. Mathematical modelling of critical pressure

Experimental tests are used to check, correct and verify numerical results, and mathematical modeling that most realistically describes machining processes and systems. Technological design and engineering of modern machining processes require the analysis of all technical and technological parameters of the process and the application of scientific methods in order to model and define the optimum conditions of processing systems and processes [3].

In this paper, it is necessary to obtain a mathematical model for critical pressure based on the input parameters and the results of experimental tests. The goal is to define the final critical pressure to which the vessel can withstand the load, i.e., pressure, by changing the input parameters.

**Table 3.** Variables and experiment results.

| No. | Physical variables $X_1$ D (mm) | $X_2$ s (mm) | $X_3$ $\sigma_M$ (N/mm$^2$) | Experimental results $P_{kr}$ (bar) |
|---|---|---|---|---|
| 1 | 400 | 3.2 | 118 | 22.0 |
| 2 | 800 | 3.2 | 118 | 9.0 |
| 3 | 400 | 6.4 | 118 | 30.0 |
| 4 | 800 | 6.4 | 118 | 17.0 |
| 5 | 400 | 3.2 | 161 | 2.5 |
| 6 | 800 | 3.2 | 161 | 12.5 |
| 7 | 400 | 6.4 | 161 | 35.0 |
| 8 | 800 | 6.4 | 161 | 20.5 |

### *3.1. Selection of influential process or system parameters*

In order to perform an experiment, it is necessary to limit it beforehand, i.e., it is necessary to isolate several influencing factors from a large number of them, as well as the output value to be measured. Since the task in this paper is to obtain a mathematical model for critical pressure, figure 2 shows the parameters that will be selected to influence this process.

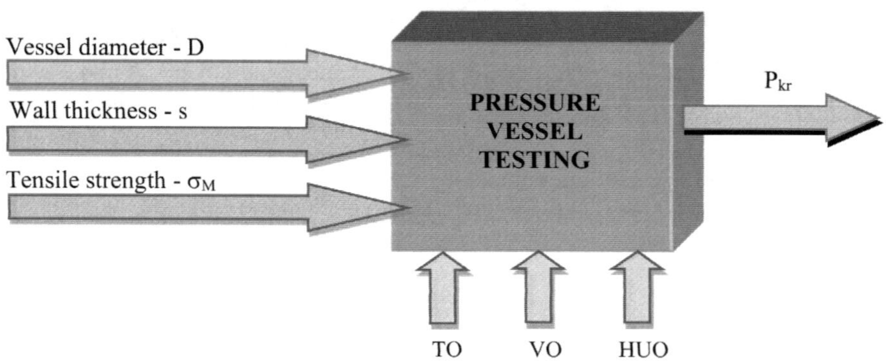


**Figure 2.** Scheme ofinput - output parameters [2].

Which are as follows:
TO ambient temperature;
VO ambient humidity;
HUO chemical impact of the environment.

These influential factors, as independent variables, are called the basic factors that vary during the performance of the experiment according to predetermined limits and the plan of the experiment. Other influential factors, whose influence is neglected or kept unchanged during the experiment, are called external factors [3].

The basic factors are selected so that their changes are not functions of external factors nor are they interrelated and it is relatively possible to perform their measurement in the process. The factors were varied within the limits shown in table 4.

**Table 4.** Influencing experiment factors.

| EXPERIMENT LEVEL | Influencing factors | | |
|---|---|---|---|
| Factor position | D (mm) | s (mm) | $\sigma_M$ (N/mm$^2$) |
| Minimum | 400 | 3.2 | 118 |
| Maximum | 800 | 6.4 | 161 |

In order to create a plan of the experiment matrix, it is necessary to encode the basic factors, and transport them into the space of coded coordinates. The factor variation interval represents half of the difference between the largest ($x_{i\ max}$) and smallest ($x_{i\ min}$) values of the i-th parameter (i = 1, 2, 3), and can be defined as follows:

$$\Delta x_i = \frac{x_{i\max} - x_{i\min}}{2} \quad (1)$$

The interval of variation of the i-th factor should be chosen so that it adequately represents the basic factor in the examined area. Coded values are always used in the paper because it makes it easier to calculate and form the plan of the experiment matrix. The coding is performed as follows:

$$X_i = \frac{x_i - x_{0i}}{\frac{x_{i\max} - x_{i\min}}{2}} = \frac{x_i - x_{0i}}{\Delta x_i} \quad (2)$$

Explained as:
$X_i$ coded value of independentvariables
$i$ number of independent variables ( $i$ =1, 2, 3, ....)
$x_i$ physical value of independentvariables at the upper or lower level
$x_{0i}$ physical value of independent variables in the center of the plan, i.e., zero mean value
$\Delta x_i$ interval of the limit of physical values of variables from the midpoint to the maximum, i.e., minimum values of the variable

The average level of physical value is determined by the expression:

$$x_{0i} = \frac{x_{i\max} + x_{i\min}}{2} \quad (3)$$

The coded values, as can be seen, are dimensionless quantities. Coding transfers the coded origin to the zero point of the experiment, as the central point of the experiment (which corresponds to the mean level of all three factors). Coding for the i-th factor is as follows:

$$X_i = +1 \quad \text{for factor} \quad x_i = x_{i\max}$$
$$X_i = -1 \quad \text{for factor} \quad x_i = x_{i\min}$$

### *3.2. Selection of influential process or system parameters*

When selecting the type of mathematical model, there is no generally valid rule. This means that for each researched process or system, a model should be chosen and its accuracy and adequacy in relation to the actual real process should be checked. In this experiment, three independent variables were selected, and thus there are three variable factors, i.e., k = 3, while the variation of the factor is at two levels r = 2 (min and max), which means that it has the character of a three-factor experiment plan where needed number of measurements is:

$$N = r^k = 2^3 = 8 \quad (4)$$

$r$ number of levels of basic factors (r = 2, min and max)
$k$ number of varied basic factors (k = 3)
$N = r\, k$ number of repetitions of the experiment with factor variation (N=8)

The initial mathematical model is a first-order three-factor polynomial:

$$Y = b_0 + b_1 X_1 + b_2 X_2 + b_3 X_3 \tag{5}$$

**Table 5.** Experiment plan matrix.

| No. | $X_0$ | $X_1$ | $X_2$ | $X_3$ | $y_j$ |
|---|---|---|---|---|---|
| 1 | 1 | -1 | -1 | -1 | 22.0 |
| 2 | 1 | 1 | -1 | -1 | 9.0 |
| 3 | 1 | -1 | 1 | -1 | 30.0 |
| 4 | 1 | 1 | 1 | -1 | 17.0 |
| 5 | 1 | -1 | -1 | 1 | 26.5 |
| 6 | 1 | 1 | -1 | 1 | 12.5 |
| 7 | 1 | -1 | 1 | 1 | 35.0 |
| 8 | 1 | 1 | 1 | 1 | 20.5 |

Coefficients of mathematical model:

$$b_0 = \frac{1}{N}\sum_{j=1}^{N} X_{0j} y_j ,....for...i = 1,2,3,...,k \tag{6}$$

$$b_i = \frac{1}{N}\sum_{j=1}^{N} X_{ij} y_j ,....for...1 \le i \prec m \le k , \tag{7}$$

Explained as:
$X_{ij}$ value of $X_i$ in j-th experiment
$y_j$ measured size in j-th experiment
$N$ number of experiments

$$\begin{aligned} b_0 &= 1/8( y_1 + y_2 + y_3 + y_4 + y_5 + y_6 + y_7 + y_8 ) \\ b_1 &= 1/8( -y_1 + y_2 - y_3 + y_4 - y_5 + y_6 - y_7 + y_8 ) \\ b_2 &= 1/8( -y_1 - y_2 + y_3 + y_4 - y_5 - y_6 + y_7 + y_8 ) \\ b_3 &= 1/8( -y_1 - y_2 - y_3 - y_4 + y_5 + y_6 + y_7 + y_8 ) \end{aligned} \tag{8}$$

**Table 6**. Model coefficients.

| | |
|---|---|
| $b_0$ | 21.5625 |
| $b_1$ | -6.8125 |
| $b_2$ | 4.0625 |
| $b_3$ | 2.0625 |

By including the coefficients from the previous table in the form (5), the following is obtained:

$$Y = 21{,}56 - 6{,}81 X_1 + 4{,}06 X_2 + 2{,}06 X_3 \tag{9}$$

### *3.3. Selection of influential process or system parameters*

To assess the adequacy of the model, i.e., to examine the relationship between the dependent variable quantity $y_j$ and the independent variable quantity $X_i$, the multiple regression coefficient is used as follows:

$$R = \sqrt{1 - \frac{\sum_{j=1}^{N}(y_j^E - y_j^R)^2}{\sum_{j=1}^{N}(y_j^E - \bar{y}^E)^2}} \tag{10}$$

Explained as:

$y_j^E$ values of experimental results

$y_j^R$ calculated values from he obtained model

$\bar{y}^E = \frac{\sum_{j=1}^{N} y_j^E}{N}$ arithmeticmeanof all experimental results

**Table 7.** Calculation of multiple regression components.

| No | $y_j^E$ | $y_j^R$ | $(y_j^E - y_j^R)^2$ | $(y_j^E - \bar{y}^E)^2$ |
|---|---|---|---|---|
| 1 | 22.0 | 22.3 | 0.090 | 0.19 |
| 2 | 9.0 | 8.7 | 0.090 | 157.75 |
| 3 | 30.0 | 30.4 | 0.160 | 71.23 |
| 4 | 17.0 | 16.8 | 0.040 | 20.79 |
| 5 | 26.5 | 26.4 | 0.010 | 24.40 |
| 6 | 12.5 | 12.8 | 0.090 | 82.08 |
| 7 | 35.0 | 34.5 | 0.250 | 180.63 |
| 8 | 20.5 | 20.9 | 0.160 | 1.12 |
| | $\bar{y}^E$= 21.56 | | Σ= 0.89 | Σ= 538.19 |

The values of the multiple regression coefficients are in the range $0 \le R \le 1$. When the value of the coefficient is R=1, the model fully describes the results of the experiment. The value R=0 shows that there is no interdependence between the variables $y_j$ and $X_i$ [3].

By including the obtained values in expression (10), the coefficient of multiple regression R=0,999 is obtained. The obtained value of the multiple regression coefficient indicates that the mathematical model (9) adequately describes the mentioned process, i.e., with 99.9 %.

Decoding:

$$X_1 = \frac{D - \frac{D_{max} + D_{min}}{2}}{\frac{D_{max} - D_{min}}{2}} = \frac{D - \frac{800+400}{2}}{\frac{800-400}{2}} = \frac{D-600}{200} \tag{11}$$

$$X_1 = (0.005D - 3)$$

$$X_2 = \frac{s - \frac{s_{max} + s_{min}}{2}}{\frac{s_{max} - s_{min}}{2}} = \frac{s - \frac{6.4+3.2}{2}}{\frac{6.4-3.2}{2}} = \frac{s-4.8}{1.6} \tag{12}$$

$$X_2 = (0.625s - 3)$$

$$X_3 = \frac{\sigma_M - \frac{\sigma_{M\,max} + \sigma_{M\,min}}{2}}{\frac{\sigma_{M\,max} - \sigma_{M\,min}}{2}} = \frac{\sigma_M - \frac{161+118}{2}}{\frac{161-118}{2}} = \frac{\sigma_M - 139.5}{21.5} \tag{13}$$

$$X_3 = (0.0465\sigma_M - 6.488)$$

By including the decoded values, the final mathematical form is obtained:

$$P_{kr} = 21.56 - 6.81(0.005D - 3) + 4.06(0.625s - 3) + 2.06(0.0465\sigma_M - 6.488) \quad (14)$$

That is:

$$P_{kr} = 16.445 - 0.034D + 2.538s + 0.096\sigma_M \quad (15)$$

The dependence of the critical pressure ($P_{kr}$) on the input parameters (vessel diameter - D, wall thickness - s, and material characteristics - $\sigma_M$) can be shown in the diagram, as shown in the following figure.

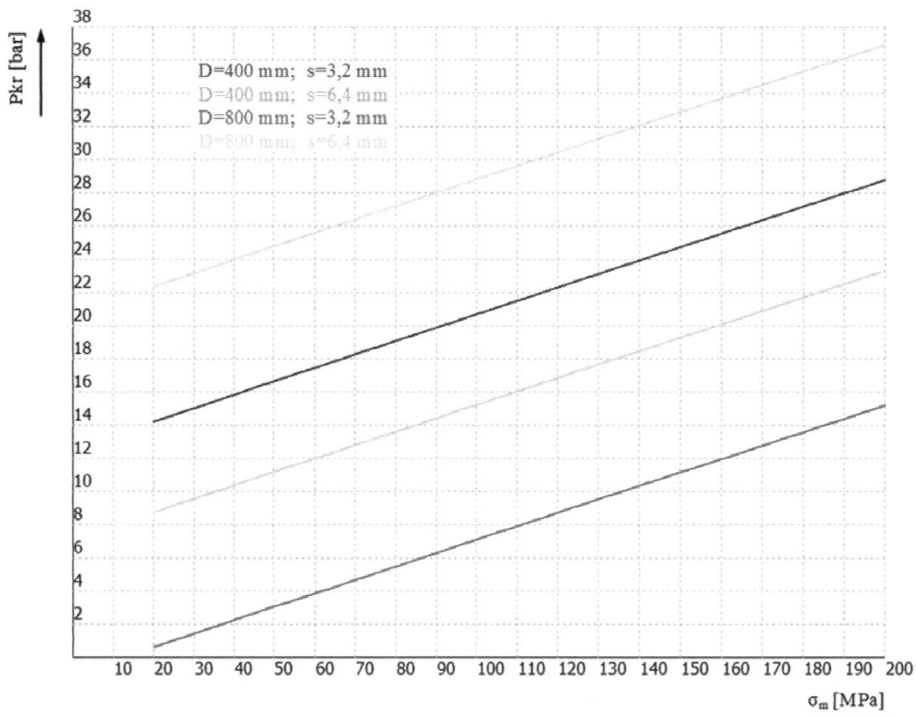


**Figure 3.** Dependence of critical pressure and tensile strength [4].

As an illustrative example, Figure 3 shows the dependence of the critical pressure and tensile strength of the material, with constant values of the diameter and wall thickness of the vessel.

*3.4. Selection of influential process or system parameters*

The obtained mathematical model of critical pressure (15) is a good indicator for designers to know which critical pressure the vessel can withstand, and base on that to make a quick control of the working or projected pressure.

The presented mathematical model for the tested pressure vessels gives results with a deviation of ±0.5 (bar) in relation to the experimental results, so that it can be said with accuracy at which pressure the total damage to the vessel will occur, i.e., failure or cracking of pressure vessels.

Table 8 shows the comparative results obtained according to mathematical modeling and according to experimental measurements.

**Table 8.** Results of critical pressure of experiment and mathematical model.

| | Physical variables | | | Experimental results | Resultsof mathematical model |
|---|---|---|---|---|---|
| N | $X_1$ D (mm) | $X_2$ s (mm) | $X_3$ $\sigma_M$ (N/mm²) | $P_{kr\,exp}$ (bar) | $P_{kr\,mat}$ (bar) |
| 1 | 400 | 3.2 | 118 | 22.0 | 22.1 |
| 2 | 800 | 3.2 | 118 | 9.0 | 8.5 |
| 3 | 400 | 6.4 | 118 | 30.0 | 30.3 |
| 4 | 800 | 6.4 | 118 | 17.0 | 16.7 |
| 5 | 400 | 3.2 | 161 | 26.5 | 26.3 |
| 6 | 800 | 3.2 | 161 | 12.5 | 12.7 |
| 7 | 400 | 6.4 | 161 | 35.0 | 34.4 |
| 8 | 800 | 6.4 | 161 | 20.5 | 20.8 |

The results of the mathematical model shown in table 8 for the tested pressure vessels show deviation of ±0.5 (bar) in relation to the experimental results, so that it can be known with great accuracy at which pressure the vessel will be completely damaged, i.e., the vessel will crack.

## 4. Recommendations for designers and production technologists

The European standard EN 13923 determines the material, design and testing of pressure vessels made of composite materials.

Design factor:

Permissible stress in each layer of load-bearing laminate is derived from characteristic elasticity and strength, material safety factor S, inhomogeneity ($A_1$), chemical influence ($A_2$), design temperature in relation to heat resistance ($A_3$), dynamic load effect ($A_4$) and long-term behavior ($A_5$).

The design factor is calculated according to the form:

$$K = S \cdot A_1 \cdot A_2 \cdot A_3 \cdot A_4 \cdot A_5 \tag{16}$$

The safety factor S must be a minimum of 2, while the partial factors $A_1$ to $A_5$ are given in table 9. It should be emphasized that the total value of the design factor (K) must not be less than 4.

**Table 9.** Partial design factors.

| Partial design factors | Min. value | Max. value |
|---|---|---|
| **$A_1$** | 1.00 | 2.00 |
| **$A_2$** | 1.10 | 1.80 |
| **$A_3$** | 1.00 | 1.40 |
| **$A_4$** | 1.00 | 1.10 |
| **$A_5$** | 1.25 | 2.00 |

Maximum allowed dilatation:

The maximum allowable dilatation is calculated on the basis of the limited dilatation $\varepsilon_{lim}$ and the partial factor of long-term behavior $A_5$.

$$\varepsilon_{max} = \frac{\varepsilon_{lim}}{A_5} \tag{17}$$

IOP Conf. Series: Materials Science and Engineering 1208 (2021) 012006 doi:10.1088/1757-899X/1208/1/012006

According to standard EN 13121-2, the dilatation must not exceed 0.25 % for all types of protective layer. Table 10 shows the limiting dilatation for three types of resins reinforced with spray rowing or with mat 900 (g/m$^2$).

**Table 10**. Limiting dilatation of the protective layer.

| Type of resin | $\varepsilon_{lim}$ |
|---|---|
| UP (unsaturated polyester resin) | min (0.1 x $\varepsilon_{resin}$ ili 0.20 %) |
| VE (vinyl ester resin) | min (0.1 x $\varepsilon_{resin}$ ili 0.25 %) |
| EP (epoxy resin) | min (0.1 x $\varepsilon_{resin}$ ili 0.30 %) |
| $\varepsilon_{resin}$- resin dilatation | |

Conditions of loads and dilatations:
Loads and dilatations must meet the following conditions:

$$\sigma_c = \frac{DP}{2s} \leq \frac{\sigma_M}{K} \tag{18}$$

$$\sigma_m = \frac{DP}{4s} \leq \frac{\sigma_M}{K} \tag{19}$$

$$\varepsilon_c = \frac{\sigma_c}{E_c} \leq \varepsilon_{max} \tag{20}$$

$$\varepsilon_m = \frac{\sigma_m}{E_m} \leq \varepsilon_{max} \tag{21}$$

Explained as:
$\sigma_c$ circular load
$\sigma_m$ meridian load
$P$ pressure in the vessel
$\sigma_M$ yield strength, in which case there are two types of materials:
- yield strength of mat composites, $\sigma_M = 118 (N/mm^2)$
- yield strength of composites with mat and rowing, $\sigma_M = 161 (N/mm^2)$

$E_c$ circular modulus of elasticity, in which case there are two types of materials:
- modulus of elasticity of mat composites, $E = 6592 (N/mm^2)$
- modulus of elasticity of composites with mat and rowing, $E = 8177 (N/mm^2)$

$E_m$ meridian modulus of elasticity (moduli of elasticity in circular and meridian directions are equal)
$\varepsilon_c$ circular dilatation
$\varepsilon_m$ meridian dilatation

Table 11 shows the obtained working pressures that were reached on the basis of the conditions shown in the previous expressions.

**Table 11.** Working pressures of vessels tested to critical pressure.

| N | Vessel diameter D (mm) | Wall thickness s (mm) | Tensile strength $\sigma_m$ (N/mm$^2$) | Operating pressure $P_{rad}$ (bar) |
|---|---|---|---|---|
| 1 | 400 | 3.2 | 118 | 4.72 |
| 2 | 800 | 3.2 | 118 | 2.36 |
| 3 | 400 | 6.4 | 118 | 9.44 |
| 4 | 800 | 6.4 | 118 | 4.72 |
| 5 | 400 | 3.2 | 161 | 6.44 |
| 6 | 800 | 3.2 | 161 | 3.22 |
| 7 | 400 | 6.4 | 161 | 12.88 |
| 8 | 800 | 6.4 | 161 | 6.44 |

Considering that there are various combinations of composite materials on the market, in order to be able to design the working pressure, the best option is to conduct experimental research, and based on that perform mathematical modeling of critical pressure, i.e., working pressure, as shown in this paper.

## 5. Conclusion

The research results in this paper have shown that experimental research and measurements can lead to very useful results that can serve as input data for the development of a mathematical model of the critical pressure of pressure vessels made of composite materials.

The development of a mathematical model confirmed the hypothesis that it is possible to bring into a mathematical relationship and dependence the critical pressure of vessels made of composite materials with mechanical characteristics of composite material of vessel ($\sigma_M$), vessel diameter (D) and vessel wall thickness (s). The obtained mathematical model for the tested pressure vessels gives results with a deviation of ±0.5 (bar) in relation to the experimental results, so that it can be determined with good accuracy at which pressure the vessel will fail or crack.

The mathematical model of critical pressure obtained in this way is a good indicator for designers to know how much critical pressure the specified vessel can withstand, based on which they can make a quick control of the working or projected pressure. In this way, by varying the mentioned parameters, it is possible to achieve the desired product quality by achieving the projected critical and thus working pressure in the production of pressure vessels made of composite materials.

## References

[1] Islamovic F, Beganovic M, Gaco Dz, Bajramovic E and Hodzic A 2013 Measurement and modeling to the quality of products from composite materials *XIV International Symposium on Quality QUALITY AGAINST RECESSION*

[2] Beganovic M 2012 *Theoretical and experimental analysis and mathematical modeling of critical pressure vessels made of composite materials* master's thesis (Bihać: University of Bihać)

[3] Jurkovic M 1999 *Mathematical modellingof engineering processes and systems* (Bihać: University of Bihać)

[4] Islamovic F, Gaco Dz, Bajramovic E, Beganovic M and Manjgo M 2014 Stochastic and genetic modeling of critical pressure of composite material vessels *3rd International Scientific Conference on Engineering Manufacturing and Advanced Technologies (MAT 2014)*

IOP Conf. Series: Materials Science and Engineering 1208 (2021) 012007 doi:10.1088/1757-899X/1208/1/012007

# Analysis of rope based mechanism for movement of telescopic conveyor segments

**Fuad Hadžikadunić**

Faculty of Mechanical Engineering, University of Zenica, Fakultetska 1, 72000 Zenica, Bosnia and Herzegovina

E-mail: fuad.hadzikadunic@unze.ba

**Abstract**. The subject of analysis in this paper are rope based mechanisms for the movement of the telescopic conveyors substructure. Experimental test of ropes in conjunction with pulleys on 2 telescopic conveyors were performed in order to assess the real cause of damage and cracking of ropes in a full sense, which occurred in the real application of the conveyor at the site. These are examples of two conveyors with the same characteristics on which combinations of ropes with characteristics of construction of cross section 7x19 and construction of cross section 19x7 are placed, as well as combinations of pulleys with two kinds of grooves. Some specific tests are performed that showed the best combination of construction driving system elements.

## 1. Introduction

Fig. 1 shows the basics of the telescopic belt conveyor construction analyzed. Dimensions and all the necessary details for the purpose of planning the experimental determination of forces in the ropes and other accompanying analyzes are not presented in the paper because of the professional regulations. Fig. 2 shows some sub-construction segments.

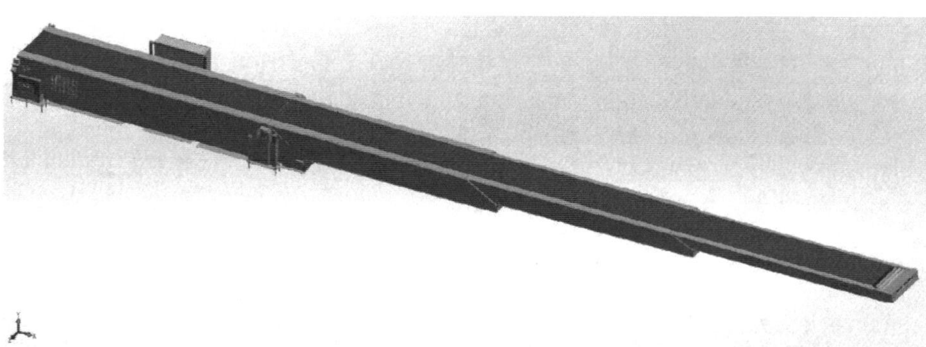

**Figure 1.** Telescopic belt conveyor.

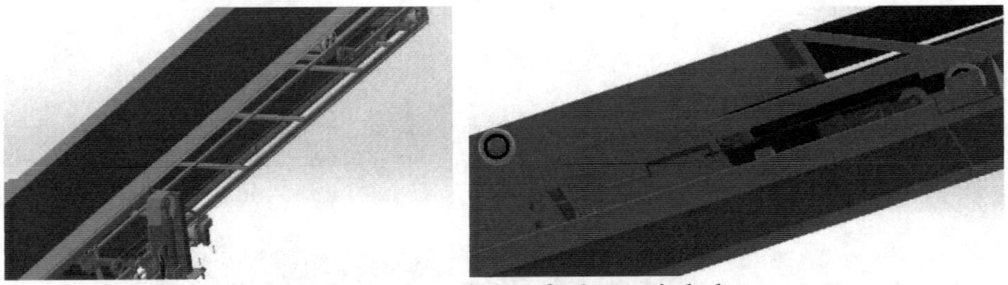

**Figure 2.** Some sub-constructions of telescopic belt conveyor system.

In the analysis of the mechanism, two rope constructions type 7x19 and rope construction type 19x7 are used. During the industry use there were some damages appeared related to the construction of ropes and its relation to the trolley pulleys, Fig. 3.

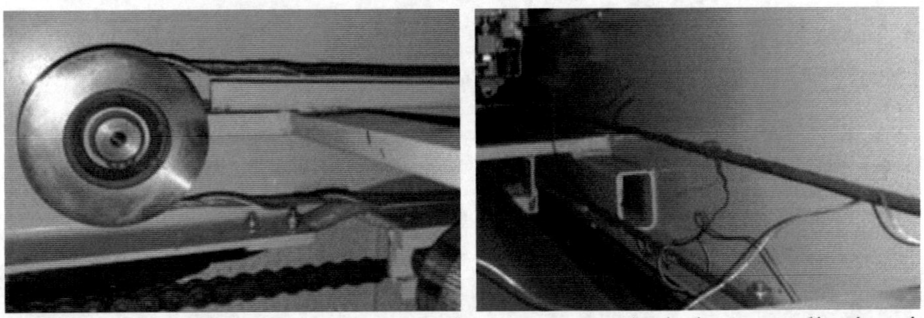

**Figure 3.** Examples of rope construction damage at the industry application site.

## 2. Experimental research program

There was a need to make a general plan to execute an appropriate experimental tests in the manner of the accepting all combinations of constructive elements for making the conclusion what was the optimum combination of the system elements, [1,2].

### *2.1. General description of the procedure for experimental tests*

In this part of the expertise, the aim was to investigate the reasons for the damage of ropes and their cracking in the application on the construction of a telescopic conveyor.

After a detailed consideration of the conveyor assembly and determining all the prerequisites for testing the forces in the ropes, it was found that no sensors can be used to directly measure the force, and since the ropes are mounted in a system with hydraulically pressed ends with screws, and for connection to the connection points on the structures of each of the 4 parts of the mobile segments.

Also, the very small spatial conditions available for the installation of the sensor were very complex, and modifications of the connecting links had to be performed, [3].
Therefore, 4 special sensors were made using strain gages, which were mounted on both sides in the selected section of the conveyor, at the points of connecting joints in the structures, on the second section of the conveyor, i.e. the first mobile.

For testing, 2 conveyors of the same characteristics were used, in which different combinations of pulleys and types of ropes were used for the purpose of testing.

The tests were performed without the presence of a conveyor belt and without loads on the conveyors, i.e. only load simulations were performed due to the change in the length and inclination of the conveyor.

Measuring equipment used: acquisition measuring system Spider 8, as well as created 4 force sensors with strain gages 6 LY-11 120 of manufacturer HBM Germany (Sensor 4; h=8,9 mm; P4 = 73,71 $mm^2$, Sensor 3; h=8,9 mm; P3 = 73,71 $mm^2$, Sensor 5; h=9,2 mm; P5 = 75,60 $mm^2$, Sensor 2; h=9,1 mm; P2 = 75,04 $mm^2$).

### *2.2. Experimental tests related to conveyor 1 and 2*

The initial installation of measuring sensors on both sides of the conveyor 1 was performed, and the test conditions are shown in Fig. 4 and Fig. 5.

The sensor placement methodology was symmetrical placement of the sensors on both sides of the second section, in order to monitor the behavior of the symmetry and the different combinations of pulley types and type of ropes.

**Figure 4.** Telescopic conveyor construction and measurement acquisition equipment used.

**Figure 5.** Experimental measuring sensors used – experiment 1.

Complete simulation of changes in the state of the conveyor 1 were performed in 6 phases, but in the Tab. 1 only 4 are presented because of professional relation conditions.

**Table 1.** Some of experimental test 1 results.

| Phase | Description | Time (s) | CH0 – F2 19x7 (kN) | CH1 – F5 19x7 (kN) | CH2 – F3 7x19 (kN) | CH3 – F4 19x7 (kN) |
|---|---|---|---|---|---|---|
| 0 | Conveyor 1 fully extended and maximally raised | - | - | - | - | - |
| 2 | Conveyor 1 retracted, hydraulics lowered | 185 - 375 | 6.06 | -0.72 | -1.41 | 0.22 |
| 4 | Raised hydraulics, conveyor 1 extraction | 520 - 620 | - [a] | -1.36 | -1.56 | -1.12 |
| 6 | Conveyor 1 extracted, hydraulic lift | 865 - 950 | - [a] | -1.65 | -1.6 | -1.4 |

[a] Note: the measuring connection cable 2 – CH0 broke.

Based on all the values (but not all presented in the paper), it is noted:

- in general, low values of working forces at all stages,
- in general, there was a redistribution of forces in the ropes, and unloading (except for rope 2),

- when ropes 2 and 4 are compared, a significant difference in forces is noticed, although the same constructions are 19x7,
- when compare ropes 3 and 5, a slightly higher force relief is noticed in the construction 7x19 in the first three phases of the simulation, and slightly less in the last three phases.

Some examples of test diagrams are showed in Fig. 6.

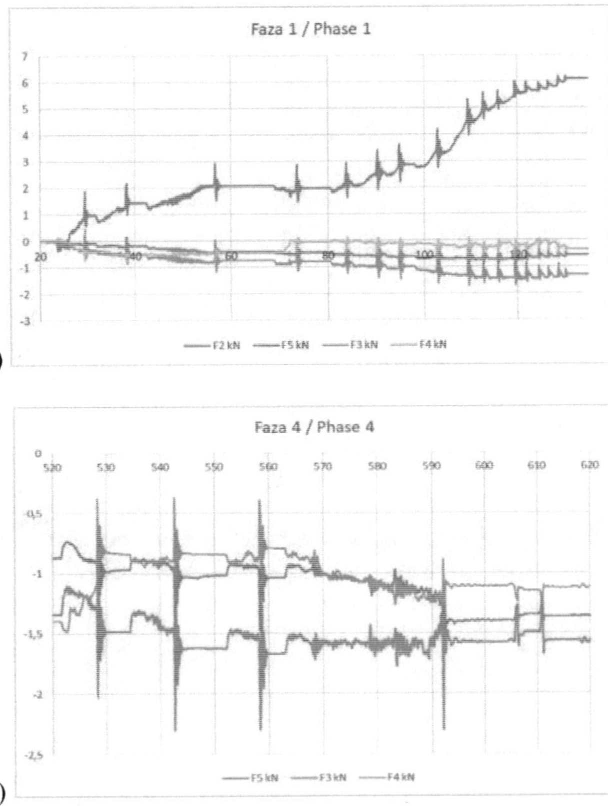


a) b)

**Figure 6.** Some experimental diagram examples of conveyor 1 testing: a) Phase 1, b) Phase 4.

Force investigations and installation of measuring sensors on both sides of the conveyor 2 were performed, completely identical to the procedure on the conveyor 1, Fig. 7.

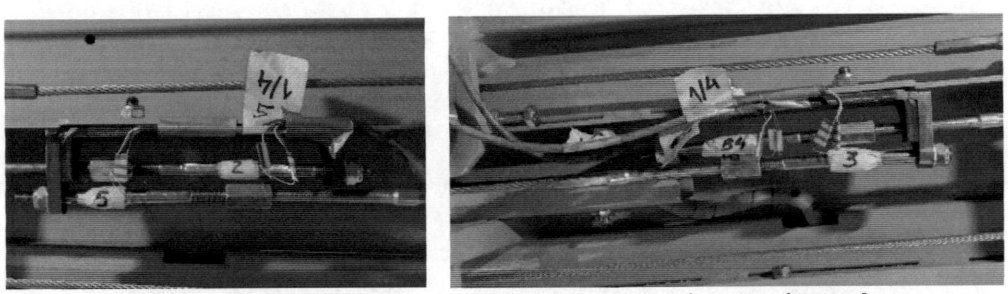

**Figure 7.** Experimental measuring sensors used – experiment 2.

Modifications of the connecting clamps were made in order to be able to insert sensors with strain gages. The implementation of the procedure on conveyor 2 was performed in 6 phase processes, in order to provide the same conditions for comparing the results and observing the mutual influences of the pulleys and ropes, but in the Tab. 2 only 4 are presented because of professional relation conditions.

**Table 2.** Some of experimental test 2 results.

| Phase | Description | Time (s) | CH0 – F2 7x19 (kN) | CH1 – F5 7x19 (kN) | CH2 – F3 7x19 (kN) | CH3 – F4 19x7 (kN) |
|---|---|---|---|---|---|---|
| 0 | Conveyor 2 fully extended and maximally raised | - | - | - | - | - |
| 2 | Conveyor 2 retracted, hydraulics lowered | 140 - 200 | -0.73 | 1.65 | -5.6 | 0.56 |
| 4 | Raised hydraulics, conveyor 2 extraction | 305 - 365 | -0.57 | -0.78 | -3.07 | 0.78 |
| 6 | Conveyor 2 extracted, hydraulic lift | 495 - 540 | -0.64 | -0.76 | -2.87 | 0.77 |

Based on all the values (but not all presented in the paper), it is noted:

- generally, low working forces values at all stages,
- there is a redistribution of forces in the ropes, and unloading in ropes 2 and 3, and 5 in the last three phases,
- in the rope 5 in the first three phases there is an increase in force, and in rope 4 there is an increase in force in the rope in all phases,
- when compare ropes 2 and 4, a difference in the character of forces is noticed, i.e. relief in the construction of 7x19, and an increase in force in the construction of 19x7 ropes,
- when compare ropes 3 and 5, a certain difference in behavior is noticed because in the first three phases with rope 5 there is an increase in force, and then it becomes unloading, while with rope 3 there is a significant relief, so there is asymmetry in behavior.

Some examples of test diagrams are showed in Fig. 8.

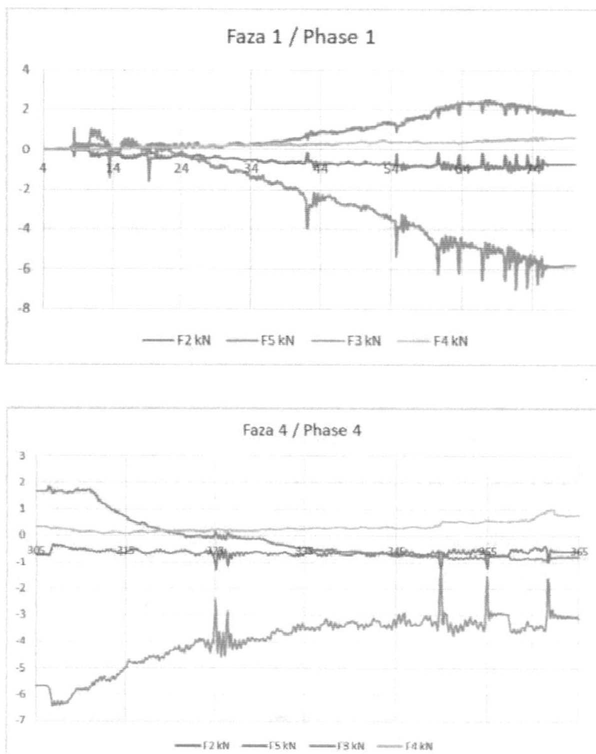


a)
b)

**Figure 8.** Experimental diagram examples of conveyor 2 testing: a) Phase 1, b) Phase 4.

### 2.3. *Results analysis*

Based on the total conducted and meaningful experimental tests, some of the following considerations and conclusions are given:

- The analysis of the values of forces in the ropes shows a better behavior of the rope construction 7x19, compared to the rope 19x7, so the values of forces show a more adequate behavior of the structure 7x19 in contact with the pulleys, with more flexibility,

- The tests were performed without the presence of a conveyor belt and without loads on the conveyors, i.e. only load simulations were performed due to the change in the length and inclination of the conveyor, which means that higher loads are expected in the ropes at the site.

- Rope failure is not caused by tensile loads, because lower values of working forces are generally present, than due to bending dynamic loads. The values of the obtained absolute forces are below the breaking limit (for these test conditions).

- The 19x7 rope is not adequate for this conditions because its structure is different, mainly for more significant tensile loads and lower bending loads, they have less stiffness (which was proven with another experimental testings), and less pronounced elasticity because residual deformations are also present. The fact that Young's modulus of elasticity is smaller is also present.

- The durability of the rope is closely related to the bending radius around the pulleys. It increases with increasing ratio of pulley diameter to rope diameter. The magnitude of the tensile load of the rope is essential for the increase in durability. Industrial practice shows that even with a small D / d ratio, high rope durability can be achieved, but at lower loads. To achieve sufficient durability of the rope, the norms set minimum D / d ratios. This ratio also keeps under control the magnitude of the normal stress in the wires due to the bending of the rope around the pulleys. It would be advantageous to increase the diameter of the pulley radius, if the design allows, in order to achieve more favorable permanent bending strength, [4,5,6].

- It would be good for the project solution, among other things, to prescribe the approximate values of rope prestressing forces in conveyor sections, given the logic that, for example, the prestressing force of the conveyor belt must be calculated and realized as such during installation and before workloads. The recommendation of the prestressing force of the ropes is also important in order to avoid asymmetry in the load of the ropes on both sides of the conveyor.

- After changing the structure of the rope, it is necessary to check the behavior of the system in practice, i.e. to go to meet the recommendations related to the diameters of the reels, if a further need to improve the conveyor structure is determined.

## 3. Conclusion

The paper presented a specific procedure for investigating the causes of damage to the rope based mechanism for extracting and retracting components of the telescopic conveyor structure.

Experimental research was conducted in very specific circumstances because measuring sensors were specially created only for this research.

Measurements have determined the optimal combinations of rope and pulley construction that ensure further reliability of the system without further damage and interruption in use.

During these studies, somewhat broader experimental trials were applied. The paper presents only a part of the overall procedure and the obtained data.

The significance of the research lies in the fact that modern measuring technology was applied, as well as in the fact that the obtained results indicated an optimal combination of mechanism elements without subjective approaches of researchers in decision making, which solved significant dilemmas of equipment manufacturers.

## References

[1] Vukojević D 1998 *Teorija elastičnosti sa eksperimentalnim metodama* (Zenica: Mašinski fakultet)

[2] Ostrić D and Tošić S 2005 *Dizalice* (Beograd: Mašinski fakultet)

[3] Hoffmann K 1987 *Eine Einfürung in die Technik des Messens mit Dehnungsmeßstreifen* (Darmstadt: Hottinger Baldwin Messtechnik GmbH)
[4] FEM 9.661 *Rules for the Design of Serial Lifting Equipment, Dimensions and Design of Rope Reeving Components*
[5] DIN 5881 *Drahtseile und Seiltriebe*
[6] DIN 15061 *Rillen profile fur Seilrollen*.

IOP Conf. Series: Materials Science and Engineering 1208 (2021) 012008 doi:10.1088/1757-899X/1208/1/012008

# Aircraft hydraulic axial piston pump fundamental pulsation and simulation

**Želimir Husnić[1*], Remzo Dedić[2], Faris Ustamujić[3**] and Zlata Jelačić[4]**

[1]The Boeing Company, Philadelphia, USA
[2]Faculty of Mechanical Engineering, Computer Science and Electrical Engineering, University of Mostar, B&H
[3]MSc, Airbus DS GmbH, Germany
[4]Faculty of Mechanical Engineering, University of Sarajevo, B&H

E-mail: jelacic@mef.unsa.ba

**Abstract.** The axial piston pump for aircraft hydraulics systems and other high pressure hydraulic system applications is presented. This paper discusses the pump's pressure pulsation and the fundamental frequency. Pressure pulsation associated with single piston failure is explained in relation to its fundamental frequency. A predictive approach in maintenance and pump sub system health monitoring is proposed, using numerical modelling and applicable software.

## 1. Introduction
The aircraft hydraulics and mechanical systems are complex due to requirements, which are specific for aerospace applications. Aircrafts hydraulic systems must comply with very demanding requirements. These requirements often include a brief and sudden flow and pressure change as well as steady state run with constant pressure or constant flow.

An axial piston pump is the most common solution for delivering a flow at specified pressure, through the aircraft hydraulics system. Various types of pumps could be used (Piston, Gear, Vane, Fixed Displacement), but it is common to use variable-flow axial piston pumps for the aircrafts hydraulics system.

In any case, the pump of choice for high demanding hydraulics system is the axial piston pump. Such a pump is a critical component in the system overall operation. Its failure due to damage introduced by debris is the subject of this paper.

## 2. Description of the axial piston pump

### *2.1. Overview*
Axial piston pump is probably the most used variable displacement type of pump. This type is capable of continuously changing fluid discharge, based on load requirements and maximum pressure cut-off settings. This is achieved by using different compensation methods [1], [3].

Axial piston in-line pumps, which make use of the swashplate principle (fixed and adjustable displacement) have a quality that is approximately the same as the bent axis model. They have the advantage of being more compact in design and have a longer service life. In addition, the pumps can be manufactured in an easier and more economical manner. By using different compensation methods, the variable displacement type of these pumps can continuously change fluid discharge based on load requirements and maximum pressure cut-off settings.

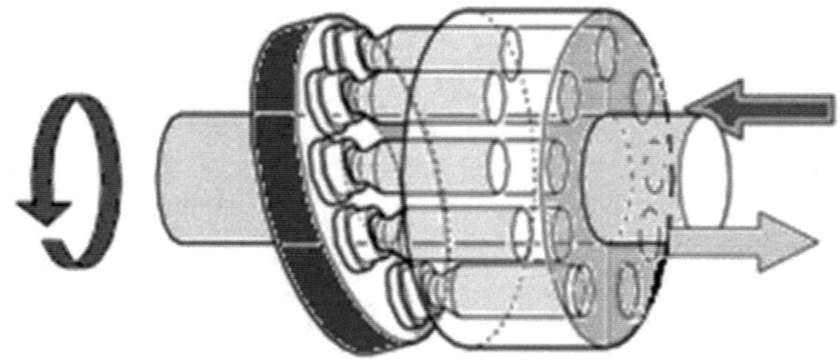

**Figure 1.** Axial piston in-line pump

*2.2. Pump piston design*

The axial piston hydraulic pump´s major characteristics are: pressure and speed, displacement, flow, pulsation, temperature, case drain pressure, case drain flow, power requirements, inlet pressure, inlet flow, envelope and material.

Pressure pulsation is one of the axial piston pumps' features that needs to be considered during design of the hydraulics system, the design of the pump, system health monitoring and troubleshooting. Fluctuation of pressure in the hydraulic fluid is generated during process of transforming input power (rotation) to flow and pressure. These pulsations are critical because of the high-energy potential they generate when the system loading requirements change. Pulsations are a result of each piston within the pump transferring a discrete amount of fluid to the system, leading to an overall pulsed input in the hydraulic system.

Lower pressure pulsation levels are important characteristic related to the working life of the pump and are needed for better hydraulics system performance. The pump with an even number of pistons has a higher flow variation than pump with an odd number of pistons. Common choice for aircraft hydraulics systems are axial piston pumps with 9 or 11 pistons [2], [3].

For the analysis of piston/cylinder pair it is assumed that the piston possesses only two degrees of freedom. One is the translational motion along the axis of cylinder bore, while the other is the rotation about the axis of main shaft.

The true motion of piston is much more complicated. It has radial micro-motion due to the existence of clearance and lateral forces. Therefore, the piston and cylinder bore are linked through a contact force instead of a cylindrical joint. In addition to the pressure field of lubricating oil film, the reaction forces from cylinder bore are acting onto piston only when the piston and cylinder get in contact at the positions where the oil film is too small as depicted in Figure 2.

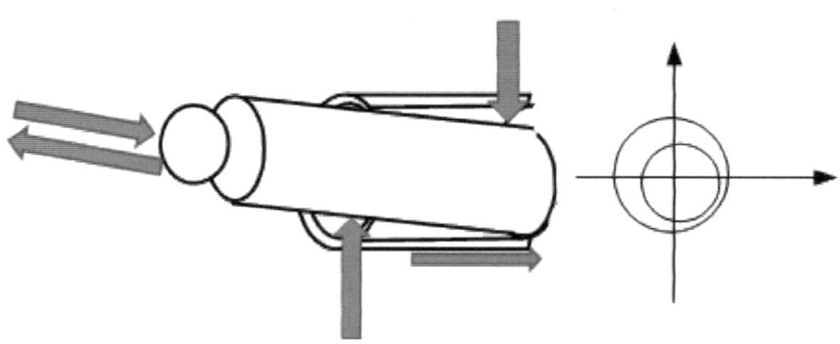

**Figure 2.** Pump piston in bore

### 2.3. Pump piston frequency

As explained in the previous paragraph, pulsations are the result of each piston within the pump transferring a discrete amount of fluid to the pump outlet, producing a pulsed input in the hydraulic system [4]. The amplitude of the torque or flow rate variation for odd numbered pumps is much less than that for the even numbered devices.

Variations in pressure will influence the pulsation amplitude (it will increase approx. linear in proportion to the pressure). The product of shaft rotational frequency and the number of the pump pistons defines pump´s frequency. This is expressed mathematically by the following formula:

$$\mathrm{f} = \frac{n}{60} \cdot N_{pis} \tag{1}$$

Where is:
$N_{pist}$ is the number of pistons [-], $f$ is the pump frequency [Hz] and $n$ is the pump rotational speed [rpm].

Angular frequency is defined as:

$$\mathrm{f} = \frac{\omega}{2\pi} \tag{2}$$

where ω is angular frequency [rad/s].

The Piston Pump Frequency for the Pump used for Test in Hydraulics Laboratory are respectively $f$ = 930 Hz and $\omega$ = 5843 rad/s when $N_{pist}$ = 11 and $n$ = 5073 rpm.

### 2.4. Fundamental frequency

Fundamental frequency is the lowest frequency of pressure (or flow) ripple considered in a theoretical analysis or measured by the frequency analysis instrument.

Flow ripple is the fluctuating component of flow rate in a hydraulic fluid, caused by interaction with a flow ripple source within the system. Pressure ripple is fluctuating component of pressure in a hydraulic fluid, caused by interaction with a flow ripple source within the system. Pressure pulsations are oscillations of the discharge pressure, occurring during nominal steady operating conditions, at a frequency equal to or higher than the pump drive shaft speed.

The pump test is used to determine the amplitude of pressure pulsations. When the frequency (speed) and harmonics of the pump is equal to the natural frequency of the hydraulic fluid column/tubing, a resonance condition occurs. A tubing line length whose natural frequency is resonant with pulsation frequency shall be avoided.

In this work, the single piston frequency as the lowest/fundamental frequency of pressure will be studied and its effect on the overall pump performance. The impact of the single piston frequency is often undervalued in relation to the pump's working life.

## 3. Axial piston pump test

### 3.1. Test results

The pump catastrophic failure was demonstrated by pump's housing crack during testing in laboratory. The fundamental frequency amplitude change is main suspect for the pump failure. The pump internal overload and fatigue was cause for failure due to the pump's outlet pressure fundamental frequency amplitude change.

An axial piston pump test was performed at the hydraulics lab when a pump failure occurred. Failure investigation considered the pump pulsation as one of the causes of failure.
During the test, in addition to all other data, the pump pressure at outlet and case drain ports is measured and recorded. A Pump Case Drain pressure values were used in the experiment. The Pump Case normal pressure for this type of pump is 140 psi.

The pump case pressure measurements shown in Figure 3 demonstrate pressure values after 40 hours of operation for healthy pump and pressure values after 3200 hours of operation for failed pump.

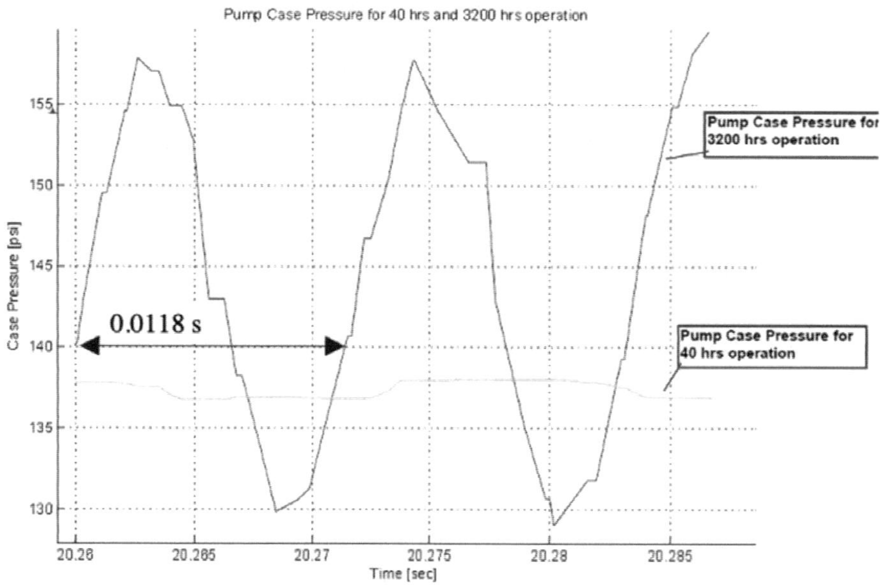


**Figure 3.** Pump case pressure

In this paper, a pump with speed of 5073 rpm was subject of study. As already listed in paragraph 2.3, this pump has the following characteristics:

$$f_1 = 84.55 \text{ Hz when } N_{pist} = 1 \text{ and } n = 5073 \text{ rpm.}$$

The single piston time for one cycle in seconds is

$$t_1 = \frac{1}{f_1} = \frac{1}{84.55} = 0.0118\ s \tag{3}$$

The healthy units demonstrate pulsation like the pump case pressure after 40 hours of operation shown in Figure 3. The single piston frequency is considered as noise and without significant impact to operation or the unit work life. In available literature, overall pressure pulsation with amplitude shown like in Figure 3 is not considered as harmful pulsation to the pump or system. This is true until the single piston frequency changes as shown in Figure 3 for the pump case pressure after 3200 hours of operation.

The pump output pressure corresponding to the case pressure changes in pressure frequency and amplitude. Frequency after approx. 3200 hours in Figure 3 corresponds to the single piston frequency with significant amplitude change. In most cases, the available literature does not recognize the single piston frequency and amplitude impact. Whereas it is largely apparent that 3200 hours of operation have large pressure ripple (oscillation), as shown in Figure 3.

This can be directly traced back to the single piston operation. The pump could have had single piston frequency change as described here, if during operation one of piston´s conditions changed significantly compared to other pump's pistons.

If these conditions are considerably different for one of piston compared to other pistons, then a significant change of the single piston frequency can be expected.

For the pump used in test, contamination with fine particles was identified as the root cause for failure and piston condition change.

The piston clearance and friction between piston and cylinder would have had significant effect on the piston motion and pressure pulsation, as illustrated in Figure 4.

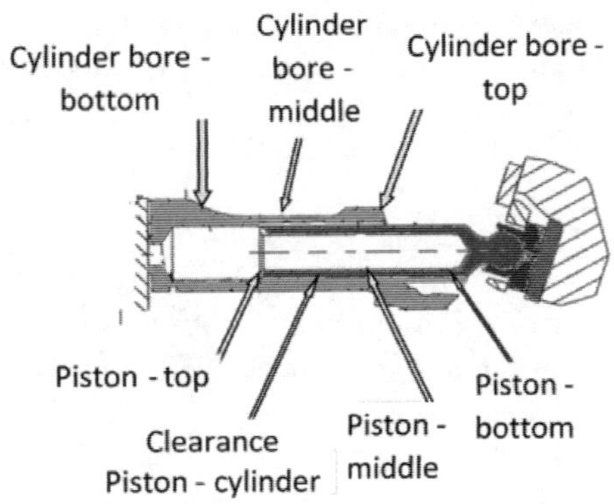


**Figure 4.** Piston and cylinder bore clearance

The pump output pressure specified by national standards [1], sets limits for the pump without outlet pressure ripple attenuator.

The pressure ripple Helmholtz resonator is often used for attenuation of pressure pulsation. The monitoring of the outlet pressure changes and comparison with a healthy pump, shall lead to conclusion regarding this failure mode, rather than simple comparison to the national standards requirements.

In case of the pressure pulsation increase the Helmholtz resonator will protect he downstream hydraulics installation and components. The Helmholtz resonator cannot protect the pump. This paper, therefore, also makes a proposal on how to protect the pump as well, using observations discussed previously.

## 4. Maintenance

In this chapter a predictive maintenance approach with respect to the described failure mode is proposed. Predictive maintenance method should help to determine the pump condition and predict when maintenance should be performed. As such, pump´s catastrophic failure could be prevented. This type of maintenance is based on forecast if problems would increase to an operationally damaging level.

Measurements of pressure related to the single piston failure could be done at the case pressure or the outlet pressure. Most axial piston pump applications have a system that includes pressure measurement at either both or one of these two locations. These data should be analysed to enable decision making regarding the single piston failure and predictive maintenance.

The pump sub-system predictive maintenance involves a lot of data and more than one failure mode. The best option to analyse a lot of complex data is a model-based system and use of applicable software. Matlab [5], Amesim [6] or other similar software could be used for predictive maintenance and data analysis. To simulate and analyse failures using such software, engineers with knowledge about the system, components and failure modes must be involved.

This work offers better understanding of the single piston failure mode which could be used in modelling and predictive maintenance analysis.

## 5. Conclusions

The axial piston pump´s failure mode related to the single piston failure is explained. The understanding of this failure mode and how it can be detected by the outlet pressure change and relative comparison with healthy pump pressure characteristics has been shown to be crucial.

Most axial piston pump applications already have pressure measurement sensors that could be used for monitoring of the outlet pressure change. By implementing automated comparison with healthy pump state outlet pressure a feasible and practical solution is provided, which could prevent pump failure. Furthermore, such sensing capability also enables predicative maintenance introduction, which then reduces operational costs and down-time due to pumps failures.

The intensive use of computers and software in design, modelling, digital simulation, provide significant help regarding predictive maintenance, the system and components availability and reduce maintenance cost.

## References

[1] SAE AS19692 *Aerospace – Military Type Variable Delivery, Pressure Compensated Hydraulic Pump*, Aerospace Standards, SAE, Warrendale, PA

[2] SAE AIR 1922 *Aerospace - System Integration Factors That Affect Hydraulic Pump Life*, Aerospace Standards, SAE, Warrendale, PA

[3] ISO 8278 *Aerospace - Hydraulic, pressure compensated, variable delivery pumps - General requirements*, International Organization for Standardization

[4] Husnić Z, Dedić R 2016 *Uljna Hidraulika* Univerzitet „Džemal Bijedić" Mašinski fakultet Mostar

[5] Simulink - MATLAB, MathWorks (mathworks.com)

[6] Simcenter Amesim, Siemens PLM Software (plm.automation.siemens.com)

Note:
*The Boeing Company, Ridley Park, PA, USA
Note: The Boeing Company is not associated with this paper
**Airbus DS GmbH, Taufkirchen, Germany
Note: Airbus DS GmbH is not associated with this paper

# The Methodology of Distinguish Between Random and Chaotic Machine Tool Oscillations

**Sanel Gredelj**

Faculty of Technical Engineering, University of Bihać, Bosnia and Herzegovina

E-mail: sanel.gredelj@unbi.ba

**Abstract**. Machine tool oscillations are irregular or aperiodic. Most often, these oscillations are chaotic but, in some cases, they can be quasi-periodic or random. The methodology for characterizing oscillations in the first of two steps uses the nonparametric hypothesis tests which the observed oscillations confirmed as irregular. The methodology for the final characterization of oscillations is based on chaos quantifiers. A time series defined as the measured values of oscillations in the time domain is the basis for calculating the quantifiers of chaos. There are four quantifiers of chaos: the Lyapunov exponent, Kolmogorov entropy, fractal dimension and correlation dimension. The correlation dimension and Kolmogorov entropy are important for distinguishing between random and chaotic oscillations. Other quantifiers of chaos are not used for this purpose. The methodology requires a multidisciplinary approach based on combining Nonlinear Dynamics and Probability Theory and Statistics. The methodology can be applied to many oscillating phenomena. Therefore, the paper mainly used the term oscillations, not vibrations, chatter, etc.

## 1. Introduction

The difficulty of predicting machine tool oscillations was already recognized and described by Frederick Winslow Taylor [1856-1915], published in a scientific paper "On the Art of Cutting Metals" in 1907. Since then, several attempts have been made to explain this phenomenon. The introduction of the regenerative effect and the corresponding mathematical models -Delay Differential Equations (DDEs) resulted a breakthrough in the modelling of machine tool oscillations. After the extensive work of Tlusty, Tobias, Kudinov, Thompson, Werner, Stépán, Insperger, Moon, Kalmár-Nagy, Warmiński, Litak and others, regenerative effect became the most commonly accepted explanation for machine tool oscillations. [1-5]

The time series analysis method is very popular to analyse machine tool oscillations. This method is based on the use of a time-series of experimentally generated data (acceleration, velocity or displacement of oscillation). The time-series analysis method is turning out to be one of the major contributions of nonlinear dynamics. This method is based on the notion that even for a multidimensional dynamical system the time series record of a single variable is often sufficient to determine many of the properties of the full dynamics of the system. The paper used the time series data of a single variable to create a multidimensional embedding space. The use of a single time series to generate a reconstruction space to characterize nonlinear dynamical systems was put on a firm mathematical basis by Floris Takens. The mathematical basis is Takens' theorem embedding (reconstruction) theorem which is a bridge between the nonlinear dynamics and the time-series analysis method. The time-series is the basis for calculating the quantifiers of chaos. The quantifiers of chaos can be specified quantitatively whether or not a system's apparently erratic behavior is indeed chaotic. This method was criticized by Moon and Kalmár-Nagy because it contained no physics and several other important reasons. However, in combination with an experimental design stochastic modeling, the time-series analysis method gives good results in the analysis machine tool oscillations. [6-11]

## 2. Random and Chaotic Machine Tool Oscillations

Any oscillatory motion of a mechanical system about its equilibrium position is called vibration or mechanical oscillation. The oscillations may be periodic, random or chaotic. Periodic oscillations repeat motion in equal intervals of time. The sufficient and necessary conditions for the existence of chaotic oscillations are [7]: nonlinearity, energy dissipation, external excitation and sensitivity on the initial conditions. Machine tools as dynamic systems meet the above-mentioned conditions so that machine oscillations are usually chaotic. Any chaotic behavior is unpredictable with the possibility of predicting declining over time. An example is the weather forecast.

Machine tool oscillations are self-excited vibrations. The mathematical model of machine tool oscillations is a Retarded Functional Differential Equations - RFDE. The physical basis of this model is the regenerative effect. This effect is related to the cutting force variation due to the wavy workpiece surface cut one revolution ago (in turning) or one tooth ago (in milling). In grinding, based on doubly regenerative theory, the grinding force is proportional to the grinding depth which involves both the wheel and the workpiece regenerations. Retarded Functional Differential Equations - RFDEs are mathematical terminologies. In engineering RFDEs called Delay Differential Equations and abbreviation as DDEs. DDEs are also called time-delay systems. [2, 5, 12]

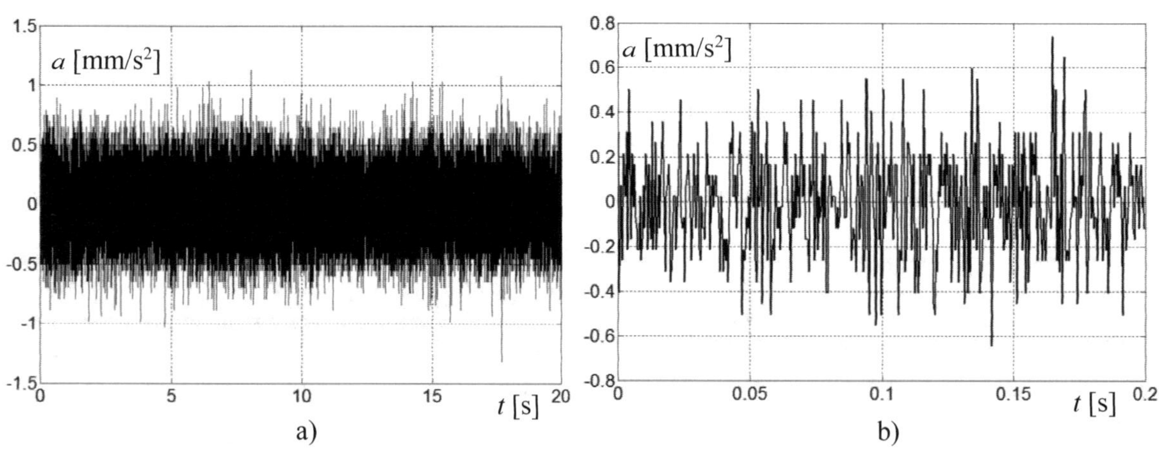


**Figure 1**. The acceleration diagram, complete and on short time

Both the random and the chaotic machine tool oscillations are irregular. Figure 1 shown the acceleration diagram of experimentally measured grinding machine oscillations during cylindrical (transversal) grinding.

Machine tool oscillations measurement system was shown in figure 2. The piezoelectric accelerometers were installed on the machine and they were connected to a "Spider" device that registered the voltage change in the accelerometers. "Spider" device was connected to a computer. The computer with the "Catman" software converted the voltage changes to the acceleration values (time series), one monitored the measurement, the type of processed data and saved data in MS Excel [6].

IOP Conf. Series: Materials Science and Engineering 1208 (2021) 012009 doi:10.1088/1757-899X/1208/1/012009

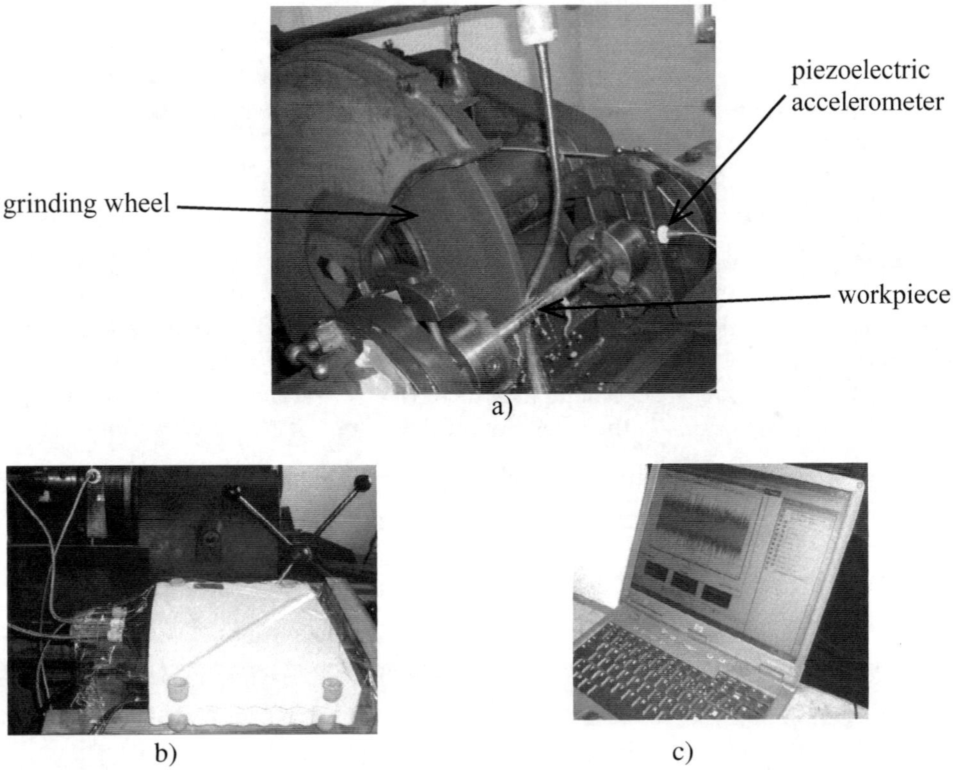


**Figure 2**. Machine tool oscillations measurement system

It was obvious that the oscillations were aperiodic or irregular. The qualitative analysis of the diagram cannot determine whether the observed oscillations were random or chaotic. Therefore, it was necessary to introduce a quantitative method.

## 3. Correlation Dimension

The quantifiers of chaos are often the only method for reliably recognition the types of dynamics. The quantifiers of chaos - the correlation dimension and the Kolmogorov entropy are important for distinguishing the random and chaotic oscillations [6].

Grassberger and Procaccia 1983. introduced a dimension based on the behavior of a correlation sum. This dimension is called the correlation dimension $D_c$. For the scalar time series $y(i)$, $i = 1, 2, \ldots, N$, with $N$ points and time delay $\tau$, the correlation sum is [7, 16]:

$$C(\rho)=\frac{1}{N(N-1)}\sum_{i=1}^{N}\sum_{j=1,j\neq i}^{N}\Theta\left(\rho-\left|y_i-y_j\right|\right). \qquad (1)$$

The scalar time series $y(i)$, $i = 1, 2, \ldots, N$, is experimentally generated data some of the oscillation's characteristic or real data set. Figure 1 shown diagram of experimentally generated data the acceleration of the grinding machine oscillations during cylindrical grinding. $N$ points is actually the number of experimentally generated data and in the measurement shown above is about $N \approx 20.000$.

In equation (1) $\Theta(x)$ is the Heaviside step function:

$$\Theta(x) = \begin{cases} 1 \ za \ x \geq 0 \\ 0 \ za \ x < 0 \end{cases}. \tag{2}$$

In equation (1) the distance $\rho$ is assigned value, which the $\Theta(x)$ compare with the absolute values of the differences $|y_i - y_j|$, except $j = i$, which means that it does not compare $y_i$ with itself. If it is $\rho \geq |y_i - y_j|$, then it is $\Theta(x) = 1$, and if there is $\rho < |y_i - y_j|$, then there is $\Theta(x) = 0$. Any real data set consists of a finite number of points. There is some minimum distance between trajectory points. When $\rho$ is less than that minimum distance, the correlation sum, equation (1), is equal to 0. When there is $\rho$ larger than the size of the attractor and all points are within $\rho$ of each other and $C(\rho)$ is equal to 1, not depending to $\rho$ [7].

If there is $y(i)$, $i = 1, 2, \ldots, N$, are periodic, then there is $C(\rho)$ = const. provided $0 < C(\rho) < 1$, not depending to $\rho$. The reason for this is the repetition of the values of $y_j$ and $y_i$, which causes that the same number of differences $|y_i - y_j|$ are inside and outside the distance $\rho$.

The correlation dimension $D_c$ is then defined to be the number that satisfies [16]:

$$C(\rho) = \lim_{\rho \to 0} g\rho^{D_c}. \tag{3}$$

or, more precisely, after taking logarithms the correlation dimension $D_c$ is:

$$D_c = \lim_{\rho \to 0} \frac{\log C(\rho)}{\log \rho}. \tag{4}$$

The correlation dimension $D_c$ is practically the slope of the regression line logarithmic value of $C(\rho)$ and $\rho$. The range of $D_c$ values is $0 \leq D_c \leq 1$, as the maximum slope angle of the regression line is $\alpha = 45°$.

The correlation dimension distinguishes between random and chaotic data provided that the number of data $N$ is large ($N > 1000$.). The second condition is that the value of the correlation dimension is not equal to the number of degrees of freedom of the state space (attractor) $d$. The second condition is met if $d >> 1$ [7]. When the above conditions are met, then the following criterion can be used:

- if the regression line is constant, $\alpha \approx 0° \Rightarrow D_c \approx 0$, the oscillations are periodic;
- if the slope of the regression line is $0 < \alpha < 45° \Rightarrow 0 \leq D_c \leq 1$, the oscillations are chaotic;
- if the slope of the regression line is $\alpha \approx 45° \Rightarrow D_c \approx 1$, the oscillations are random.

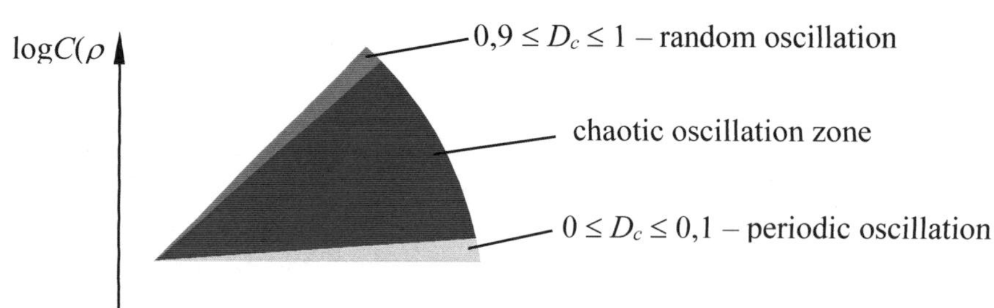


**Figure 3**. The oscillation slope

IOP Conf. Series: Materials Science and Engineering 1208 (2021) 012009 doi:10.1088/1757-899X/1208/1/012009

For practical application, the above-mentioned criterion is too restrictive and there is necessity to introduce the concept of oscillation zones, figure 3.

For experimentally measured acceleration of the grinding machine oscillations (figure 1), using "Matlab", a regression line was achieved whose equation was $\log C(\rho) = 0{,}73455 \cdot \log \rho - 0{,}18$. Therefore, the correlation dimension value is $D_c$ = 0,73455, which means that the oscillations were chaotic.

## 4. Kolmogorov Entropy

The Kolmogorov entropy (K-entropy) is important quantifier of chaos which describes the degree of chaoticity of the systems. It gives the rate of information loss about a position of the phase point on the attractor. The values of the K-entropy belong to the set of real positive numbers, $K \in \mathbf{R}+$ or $0 \leq K < \infty$. Depending on the value of the K-entropy for the characterization of oscillations, where a well-known criterion can be used also [17]:

- if there is $K = 0$,then the oscillations are periodic and there is no loss of information;
- if there is $0 \leq K < \infty$, then the oscillations are chaotic and information loss is exponential;
- if there is $K \rightarrow \infty$ then the oscillations are random and information loss is instantaneous.

Obviously, there is a correlation between the criteria based on the correlation dimension and the K-entropy. One of the criteria is sufficient to characterize the oscillations. Criteria based on the correlation dimension is more practical to apply. The reason is the possible large values of the K-entropy which can even take on infinite values. However, it is practical to use both of these criteria, due to mutual complementarity and control.

Time-series are also the basis for calculating the K-entropy. To calculate the K-entropy, it is necessary to calculate the generalized correlation sum of the vector time series firstly [7]:

$$C_q^{(d)}(\rho) = \left\{ \frac{1}{N_d} \sum_i \left[ \frac{1}{N_d - 1} \sum_{j, j \neq i} \Theta(\rho - \rho_{ij}) \right]^{q-1} \right\}^{\frac{1}{q-1}}, \tag{5}$$

where there are as follows:

$\rho_{ij}$ - a difference between two of the embedding space $d$-dimensional "vectors" $\vec{y}_i$ ,which were determined from a scalar time serie as the "Euclidean length" or the "maximum coordinate difference";

$q$ - the order of the generalized correlation sum;

$d$ - the dimension of the vectors $\vec{y}_i$ .

The generalized correlation sum is related to the $K$ - entropy by the equation ($q = 2$):

$$C_2^{(d)}(\rho) = g\rho^{D_c} e^{-d \cdot K}. \tag{6}$$

Taking the natural logarithm of both sides of the previous equation gives the following:

$$\ln C_2^{(d)}(\rho) \approx D_2 \ln \rho - d \cdot K, \tag{7}$$

where the unimportant ln($g$) term ($g$ is a constant) has been dropped. From equation (7), it is visible how to find $K$: with $\rho$ fixed $-\ln C_2^{(d)}(\rho)$ was plotted as a function of the embedding dimension $d$. The slope of that curve should be $-K$ [18].

For experimentally measured the acceleration of the grinding machine oscillations (figure 1), using "Matlab", a regression line whose equation is $-\ln C_2^{(d)}(\rho) = 0{,}3918 \cdot d + 0{,}39$ was achieved, figure 4. Therefore, the K-entropy value is $K$= 0,3918 which also means that the oscillations are chaotic and the informationloss is exponential.

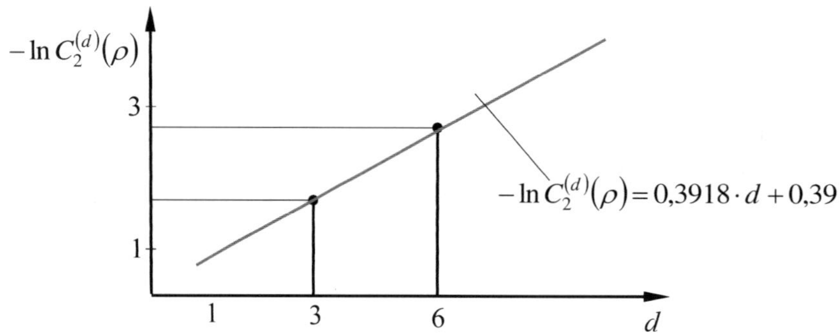


**Figure 4**. The K-entropy calculation for $\rho = 0{,}3$

The values of $K$ that distinguish chaotic and random oscillations are a bit messy, especially for $K \to \infty$. It is therefore necessary to redefine and explain this criterion. Thus, if the attractor of the dynamical system (machine tool) is $d$-dimensional ($d$ is also the number of degrees of freedom of the state space) and if $K > 0$, then it claims that the oscillations are chaotic. Otherwise, if the value of $K$ is a relatively large number and if $K >> d$, then the oscillations are random. It was based upon the claim from Pesin's equation according to which the K-entropy was equal to the sum of the positive Lyapunov exponents [19].

## 5. Non-Parametric Hypothesis Test

A statistical test is a formal technique that relies on the theoretical probability distribution for reaching the conclusion concerning the accepting of the hypothesis. These hypothetical testing related to differences were classified as parametric and nonparametric tests. The parametric test is one which has information about the population parameter. Non-parametric test, as its name tells, is statistical test without parameters [20].

### *5.1. Finding empirical and theoretical distribution*

The first step is to find an empirical and theoretical distribution. The Weibulll theoretical distribution is to be used, because the Weibull distribution is related to a number of other theoretical distributions: the exponential, hyperexponential, normal and Rayleigh one. The procedure is as follows (figure 5) [21]:

a) some value is to be added to all data so that they are all positive (the diagram is to be translated into the first quadrant of the coordinate system, this is possible because the data values at this stage are not important but only their randomness), figure 5- columns B and C;
b) sort the data by values from smaller to bigger (this is possible because the time of data creation is not important, but also their randomness), figure 5- column C;
c) find the empirical distribution by the equation (figure 5 - column D):

$$F_e(i) = \frac{i - 0{,}3}{n + 0{,}4}, \qquad (8)$$

where there are: $i$- ordinal number of the single data and $n$-total number of data (1000 in the example).

d) the shape parameter $\beta$ and the scale parameter $\eta$ in the Weibull distribution:

$$F(z)=1-e^{-\left(\frac{z}{\eta}\right)^{\beta}}, \tag{9}$$

was found by the method of least squares (figure 6):

$$a=\frac{n\cdot\sum_{i=1}^{n}(x_i y_i)-\sum_{i=1}^{n}x_i\cdot\sum_{i=1}^{n}y_i}{n\cdot\sum_{i=1}^{n}x_i^2-\left(\sum_{i=1}^{n}x_i\right)^2}, \tag{10}$$

$$b=\frac{\sum_{i=1}^{n}x_i^2\cdot\sum_{i=1}^{n}y_i-\sum_{i=1}^{n}x_i\cdot\sum_{i=1}^{n}(x_i y_i)}{n\cdot\sum_{i=1}^{n}x_i^2-\left(\sum_{i=1}^{n}x_i\right)^2}, \tag{11}$$

and the parameters in the Weibull distribution were estimated by the equations (Figure 6):

$$\beta=a, \tag{12}$$

$$\eta=e^{-\frac{b}{a}}. \tag{13}$$

At the end of this section, the values of the cumulative Weibull distribution were inserted for the calculated parameter values and the independent variable $z$, figure 5 - column D. A theoretical distribution with that was found.

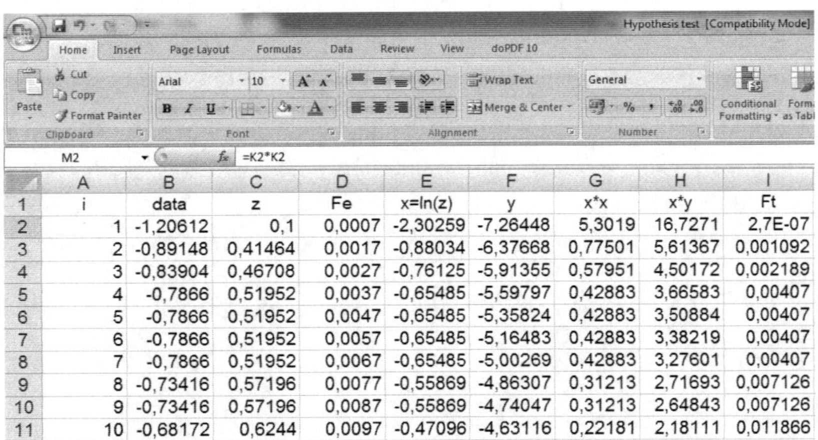


| | A | B | C | D | E | F | G | H | I |
|---|---|---|---|---|---|---|---|---|---|
| 1 | i | data | z | Fe | x=ln(z) | y | x*x | x*y | Ft |
| 2 | 1 | -1,20612 | 0,1 | 0,0007 | -2,30259 | -7,26448 | 5,3019 | 16,7271 | 2,7E-07 |
| 3 | 2 | -0,89148 | 0,41464 | 0,0017 | -0,88034 | -6,37668 | 0,77501 | 5,61367 | 0,001092 |
| 4 | 3 | -0,83904 | 0,46708 | 0,0027 | -0,76125 | -5,91355 | 0,57951 | 4,50172 | 0,002189 |
| 5 | 4 | -0,7866 | 0,51952 | 0,0037 | -0,65485 | -5,59797 | 0,42883 | 3,66583 | 0,00407 |
| 6 | 5 | -0,7866 | 0,51952 | 0,0047 | -0,65485 | -5,35824 | 0,42883 | 3,50884 | 0,00407 |
| 7 | 6 | -0,7866 | 0,51952 | 0,0057 | -0,65485 | -5,16483 | 0,42883 | 3,38219 | 0,00407 |
| 8 | 7 | -0,7866 | 0,51952 | 0,0067 | -0,65485 | -5,00269 | 0,42883 | 3,27601 | 0,00407 |
| 9 | 8 | -0,73416 | 0,57196 | 0,0077 | -0,55869 | -4,86307 | 0,31213 | 2,71693 | 0,007126 |
| 10 | 9 | -0,73416 | 0,57196 | 0,0087 | -0,55869 | -4,74047 | 0,31213 | 2,64843 | 0,007126 |
| 11 | 10 | -0,68172 | 0,6244 | 0,0097 | -0,47096 | -4,63116 | 0,22181 | 2,18111 | 0,011866 |
| 12 | 11 | -0,68172 | 0,6244 | 0,0107 | -0,47096 | -4,53254 | 0,22181 | 2,13466 | 0,011866 |
| 13 | 12 | -0,68172 | 0,6244 | 0,0117 | -0,47096 | -4,44269 | 0,22181 | 2,09235 | 0,011866 |
| 14 | 13 | -0,62928 | 0,67684 | 0,01269 | -0,39032 | -4,36017 | 0,15235 | 1,70186 | 0,018936 |
| 15 | 14 | -0,62928 | 0,67684 | 0,01369 | -0,39032 | -4,28387 | 0,15235 | 1,67208 | 0,018936 |
| 16 | 15 | -0,62928 | 0,67684 | 0,01469 | -0,39032 | -4,21292 | 0,15235 | 1,64439 | 0,018936 |
| 17 | 16 | -0,62928 | 0,67684 | 0,01569 | -0,39032 | -4,1466 | 0,15235 | 1,6185 | 0,018936 |
| 18 | 17 | -0,57684 | 0,72928 | 0,01669 | -0,3157 | -4,08434 | 0,09966 | 1,28942 | 0,029127 |
| 19 | 18 | -0,57684 | 0,72928 | 0,01769 | -0,3157 | -4,02568 | 0,09966 | 1,2709 | 0,029127 |

**Figure 5**. Finding empirical and theoretical distribution in MS Excel

I1001 =WEIBULL(C1001;$F$1009;$F$1011;TRUE)

| | A | B | C | D | E | F | G | H | I | J | K | L | M |
|---|---|---|---|---|---|---|---|---|---|---|---|---|---|
| 989 | 988 | 0,41952 | 1,72564 | 0,98731 | 0,5456 | 1,47397 | 0,29768 | 0,8042 | 0,989115 | 0,00181 | 0,48731 | 3,2754E-06 | 0,237466239 |
| 990 | 989 | 0,41952 | 1,72564 | 0,9883 | 0,5456 | 1,49258 | 0,29768 | 0,81435 | 0,989115 | 0,00081 | 0,4883 | 6,5642E-07 | 0,238441459 |
| 991 | 990 | 0,47196 | 1,77808 | 0,9893 | 0,57553 | 1,51247 | 0,33124 | 0,87048 | 0,99541 | 0,00611 | 0,4893 | 3,7282E-05 | 0,239418677 |
| 992 | 991 | 0,47196 | 1,77808 | 0,9903 | 0,57553 | 1,53386 | 0,33124 | 0,88279 | 0,99541 | 0,00511 | 0,4903 | 2,6075E-05 | 0,240397893 |
| 993 | 992 | 0,47196 | 1,77808 | 0,9913 | 0,57553 | 1,55706 | 0,33124 | 0,89614 | 0,99541 | 0,00411 | 0,4913 | 1,6865E-05 | 0,241379108 |
| 994 | 993 | 0,47196 | 1,77808 | 0,9923 | 0,57553 | 1,58246 | 0,33124 | 0,91076 | 0,99541 | 0,00311 | 0,4923 | 9,6543E-06 | 0,242362321 |
| 995 | 994 | 0,5244 | 1,83052 | 0,9933 | 0,6046 | 1,61065 | 0,36554 | 0,9738 | 0,998305 | 0,005 | 0,4933 | 2,5021E-05 | 0,243347533 |
| 996 | 995 | 0,57684 | 1,88296 | 0,9943 | 0,63285 | 1,64243 | 0,40049 | 1,0394 | 0,99946 | 0,00516 | 0,4943 | 2,6603E-05 | 0,244334743 |
| 997 | 996 | 0,57684 | 1,88296 | 0,9953 | 0,63285 | 1,67907 | 0,40049 | 1,06259 | 0,99946 | 0,00416 | 0,4953 | 1,7291E-05 | 0,245323952 |
| 998 | 997 | 0,57684 | 1,88296 | 0,9963 | 0,63285 | 1,72273 | 0,40049 | 1,09022 | 0,99946 | 0,00316 | 0,4963 | 9,9768E-06 | 0,246315158 |
| 999 | 998 | 0,62928 | 1,9354 | 0,9973 | 0,66031 | 1,77748 | 0,43601 | 1,17369 | 0,999854 | 0,00255 | 0,4973 | 6,5184E-06 | 0,247308364 |
| 1000 | 999 | 0,62928 | 1,9354 | 0,9983 | 0,66031 | 1,85278 | 0,43601 | 1,22342 | 0,999854 | 0,00155 | 0,4983 | 2,4134E-06 | 0,248303567 |
| 1001 | 1000 | 0,7866 | 2,09272 | 0,9993 | 0,73846 | 1,98304 | 0,54533 | 1,46441 | 0,999999 | 0,0007 | 0,4993 | 4,8838E-07 | 0,249300769 |
| 1002 | | -1,20612 | | 0,5 | 188,666 | -575,873 | 80,4159 | 153,104 | | | | 1,18642742 | 83,26662338 |
| 1003 | | 1,20612 | z | Fe | x | y | x*x | x*y | Ft | | | | |
| 1004 | | | | | | | | | | | | | |
| 1005 | | | | | | 261752 | | | | | | 0,01424854 | |
| 1006 | | | | | | 44820,9 | | | -75194,9 | | R*R | 0,98575146 | |
| 1007 | | | | | a | 5,83996 | | | | | R | 0,99285017 | |
| 1008 | | | | | b | -1,67768 | | | | | | | |
| 1009 | | | | | α | 5,83996 | | | | | | | |
| 1010 | | | | | | 0,28728 | | | | | | | |
| 1011 | | | | | β | 1,33279 | | | | | | | |
| 1012 | | | | | | | | | | | | | |

**Figure 6**. Calculation of the parameters and cumulative Weibull distribution in MS Excel

### *5.2. Fitting empirical to theoretical probability distribution*

In this chapter it was determined how well theoretical Weibull distribution (figure 5 - column I) fitted to the empirical distribution, (figure 5 - column D). Regarding the example (figure 5), it was actually tested how well the theoretical oscillations fitted to the empirical (experimentally measured) oscillations. In this purpose, non-parametric statistical hypothesis tests were used. First, the hypotheses for performing the test was checked. The two types of hypotheses are null and alternate. The null hypothesis ($H_0$) is a statement that indicates no difference exists between an empirical and theoretical distribution. The alternate hypothesis ($H_A$), also called a research hypothesis, is the statement that predicts a difference between an empirical and theoretical distribution. There are several popular types of non-parametric hypothesis tests used in research nowadays: chi-square $\chi^2$ test, Friedman one, Romanovsky, Kolmogorov-Smirnov one and others. Also, for a goodness of fit test the coefficient of determination between an empirical and theoretical distribution can be used. [15, 20, 24]

In this paper the null hypothesis ($H_0$) is - machine tool oscillations were random. The alternate hypothesis ($H_A$) is - machine tool oscillations were not random, i.e. the oscillations were chaotic or regular. It was assumed for the moment that the null hypothesis was true. To test the null hypothesis, chi-square $\chi^2$ test and the correlation coefficient are to be used.

Second, the values of empirical (figure 5 - column D) and theoretical distribution were specified (Figure 5 - column I).

Third, the statistic test was computed. In 1900, Karl Pearson [1857-1936] defined the chi-squared $\chi^2$ statistic [20]:

$$\chi^2 = \frac{(f_1 - f_{t1})^2}{f_{t1}} + \frac{(f_2 - f_{t2})^2}{f_{t2}} + \ldots + \frac{(f_r - f_{tr})^2}{f_{tr}} = \sum_{i=1}^{r} \frac{(f_i - f_{ti})^2}{f_{ti}}, \quad (14)$$

where there are the following:

$f_i$ - frequency of the $i^{th}$ class of the empirical distribution;

$f_{ti}$ - frequency of the $i^{th}$ class of the theoretical distribution;

$r$ - number of classes.

It was recommended that the theoretical frequencies be greater than 5. If $f_{ti} < 5$, adding those classes to the adjacenting ones. It was also recommended that the frequency values be $f_{ti} > 10$.

Fourth, the value of the chi-square distribution was determined $\chi_{\alpha}^{2(k)}$, where $k$ was the number of degrees of freedom. The $k$ was calculated by the equation [22]:

$$k = r - l - 1, \tag{15}$$

where there are as follows:
$r$ - number of classes (from equation (14));
$l$ - number of the parameters in the theoretical distribution. $l = 2$ in the Weibull distribution, from equation (9).

The significance level $\alpha$ is the probability of rejecting the null hypothesis [23]:

$$P(\chi^2 > \chi_{\alpha}^{2(k)}) = \alpha . \tag{16}$$

The common alpha values $\alpha = 0{,}05$ and $\alpha = 0{,}01$ are based on tradition.

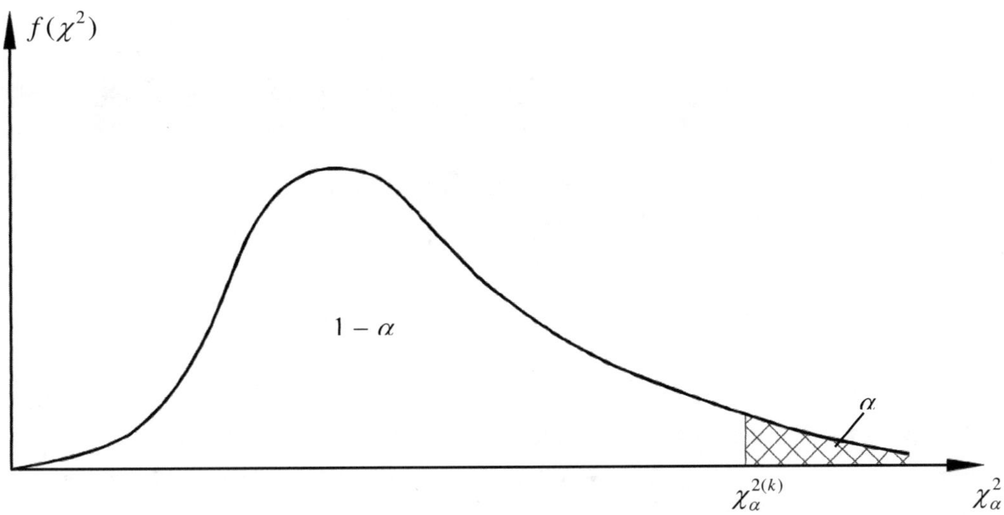


**Figure 7**. The chi-square $\chi^2$ distribution and significance level $\alpha$ [23]

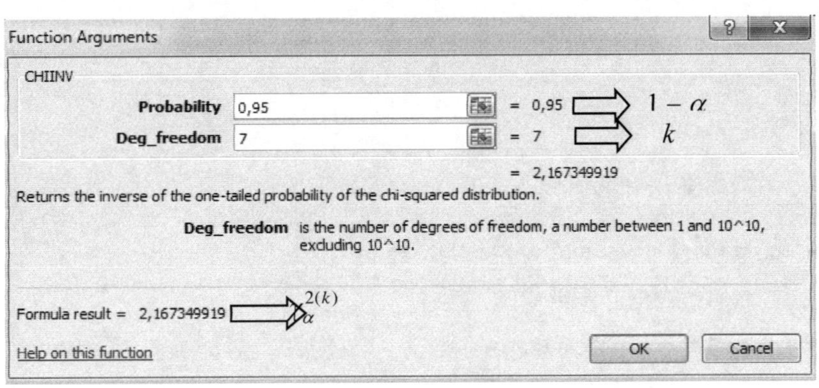


**Figure 8**. Finding the value of $\chi_{\alpha}^{2(k)}$ in MS Excel

For the significance level $\alpha$ ($\alpha = 0{,}05$ or $\alpha = 0{,}01$) and the number of degrees of freedom $k$, the valueof $\chi_{\alpha}^{2(k)}$ was found in MS Excel, figure 8.

If there is $\chi^2 > \chi_{\alpha}^{2(k)}$, then the probability, equation (16), is very low and the null hypothesis was rejected. There is a significant difference between the empirical and theoretical distribution.

If there is $\chi^2 \le \chi_{\alpha}^{2(k)}$, then the probability, equation (16), is high and the null hypothesis was accepted. There is not a significant difference between the empirical and theoretical distribution.

In the example, it was determined how well the theoretical Weibull distribution (figure 5 - column I) fitted to the empirical distribution, (figure 5 - column D), for randomly selected 1000 experimentally generated data (total number of data $N \approx$ 20.000), figure 1. These 1000 data were classified in 10 classes and determined the frequencies classes. Next, the chi-squared $\chi^2$ statistic was computed by the equation (14), figure 9.

| | C | D | E | F | G | H | I | J | K | L | M |
|---|---|---|---|---|---|---|---|---|---|---|---|
| 1 | Fte | Ft | | | | | | | | | |
| 2 | 0,0007 | 2,7E-07 | | | | | | | | | |
| 3 | 0,0017 | 0,00109 | | | | | | | | | |
| 4 | 0,0027 | 0,00219 | | | | | | | | | |
| 5 | 0,0037 | 0,00407 | | | | | | | | | |
| 6 | 0,0047 | 0,00407 | | | | f | ft | fi | fit | | |
| 7 | 0,0057 | 0,00407 | | | f1 | 0,0097 | 0,011866351 | 0,08996401 | 0,109494539 | 0,00038144 | 0,00348366 |
| 8 | 0,0067 | 0,00407 | | | f1+f2 | 0,09966 | 0,12136089 | 0,09996002 | 0,092097672 | 6,1816E-05 | 0,00067121 |
| 9 | 0,0077 | 0,00713 | | | f1+f2+f3 | 0,19962 | 0,213458562 | 0,09996002 | 0,060161935 | 0,00158389 | 0,02632707 |
| 10 | 0,0087 | 0,00713 | | | f1+f2+f3+f4 | 0,29958 | 0,273620497 | 0,09996002 | 0,06931347 | 0,00093921 | 0,01355019 |
| 11 | 0,0097 | 0,01187 | | | f1+f2+f3+f4+f5 | 0,39954 | 0,342933967 | 0,09996002 | 0,160242374 | 0,00363396 | 0,02267791 |
| 12 | 0,0107 | 0,01187 | | | f1+f2+f3+f4+f5+f6 | 0,4995 | 0,503176341 | 0,09996002 | 0,085612763 | 0,00020584 | 0,00240436 |
| 13 | 0,0117 | 0,01187 | | | f1+f2+f3+f4+f5+f6+f7 | 0,59946 | 0,588789104 | 0,09996002 | 0,084384659 | 0,00024259 | 0,00287483 |
| 14 | 0,01269 | 0,01894 | | | f1+f2+f3+f4+f5+f6+f7+f8 | 0,69942 | 0,673173762 | 0,09996002 | 0,079033923 | 0,0004379 | 0,00554068 |
| 15 | 0,01369 | 0,01894 | | | f1+f2+f3+f4+f5+f6+f7+f8+f9 | 0,79938 | 0,752207685 | 0,09996002 | 0,127838551 | 0,00077721 | 0,00607964 |
| 16 | 0,01469 | 0,01894 | | | f1+f2+f3+f4+f5+f6+f7+f8+f9+f10 | 0,89934 | 0,880046236 | 0,09996002 | 0,119952883 | 0,00039971 | 0,00333226 |
| 17 | 0,01569 | 0,01894 | | | | 0,9993 | 0,999999119 | | | | |
| 18 | 0,01669 | 0,02913 | | | | | | | | | |
| 19 | 0,01769 | 0,02913 | | | χ2 | | 0,086941807 | | | | |
| 20 | 0,01869 | 0,04337 | | | | | | | | | |
| 21 | 0,01969 | 0,04337 | | | χ2(k,α) | | 2,167349919 | | | | |
| 22 | 0,02069 | 0,04337 | | | k | | 7 | | | | |
| 23 | 0,02169 | 0,04337 | | | α | | 0,05 | | | | |
| 24 | 0,02269 | 0,04337 | | | | | | | | | |
| 25 | 0,02369 | 0,04337 | | | Conclusion | | The null hypothesis (H0) is accepted | | | | |

**Figure 9**. Hypothesis testing in MS Excel

The value of $\chi_{\alpha}^{2(k)} = 1{,}635$ has already been found in MS Excel, figure 8.

The null hypothesis ($H_0$) was accepted because $\chi^2 \le \chi_{\alpha}^{2(k)}$. It can be concluded that machine tool oscillations were random or the theoretical Weibull distribution fitted very well to the empirical distribution.

The coefficient of determination calculated $R^2$ by the equation [24]:

$$R^2 = 1 - \frac{\sum_{j=1}^{N} (F_{tj} - F_{ej})^2}{\sum_{j=1}^{N} (F_{ej} - \overline{F}_e)^2}, \tag{17}$$

where there were:

$F_{ej}$ - value of the empirical distribution for $i^{th}$ data;

$F_{tj}$ - value of the theoretical distribution for $i^{th}$ data;

$\overline{F}_e$ - the average value of the empirical distribution.

IOP Conf. Series: Materials Science and Engineering 1208 (2021) 012009 doi:10.1088/1757-899X/1208/1/012009

$R^2$ is a statistic that gives the information about the goodness of fit of an empirical and theoretical distribution. $R^2$ normally ranges from 0 to 1. $R^2 = 1$ indicates that the theoretical distribution perfectly fits to the empirical distribution, while $R^2 = 0$ indicates that the calculation fails to accurately model the data at all. The correlation coefficient $R^2$ = 0,98575 was calculated in MS Excel, figure 6. The obtained value of $R^2$ confirmed that the theoretical Weibull distribution fitted very well to the empirical distribution. Recifically, $R^2$ = 0,98575 suggests that 98, 57% of the empirical distribution is predicted by the theoretical distribution.

In chapters 3. and 4. $D_c$= 0,73455 and $K$= 0,3918 were calculated, using identical data as in chapter 5. According to $D_c$ and $K$, the oscillations are chaotic, which is not consistent with the accepted null hypothesis and the correlation coefficient. This contradiction was checked for several times and have similar results were obtained.

It can be concluded that a non-parametric hypothesis test does not distinguish well between chaotic and random vibrations. Anyway, the nonparametric hypothesis test only confirms that the oscillations are irregular. The final confirmation can be given based upon the quantifiers of chaos.

## 6. Conclusion

The above statement can be represented by the algorithm for distinguishing between random and chaotic oscillations, figure 10.

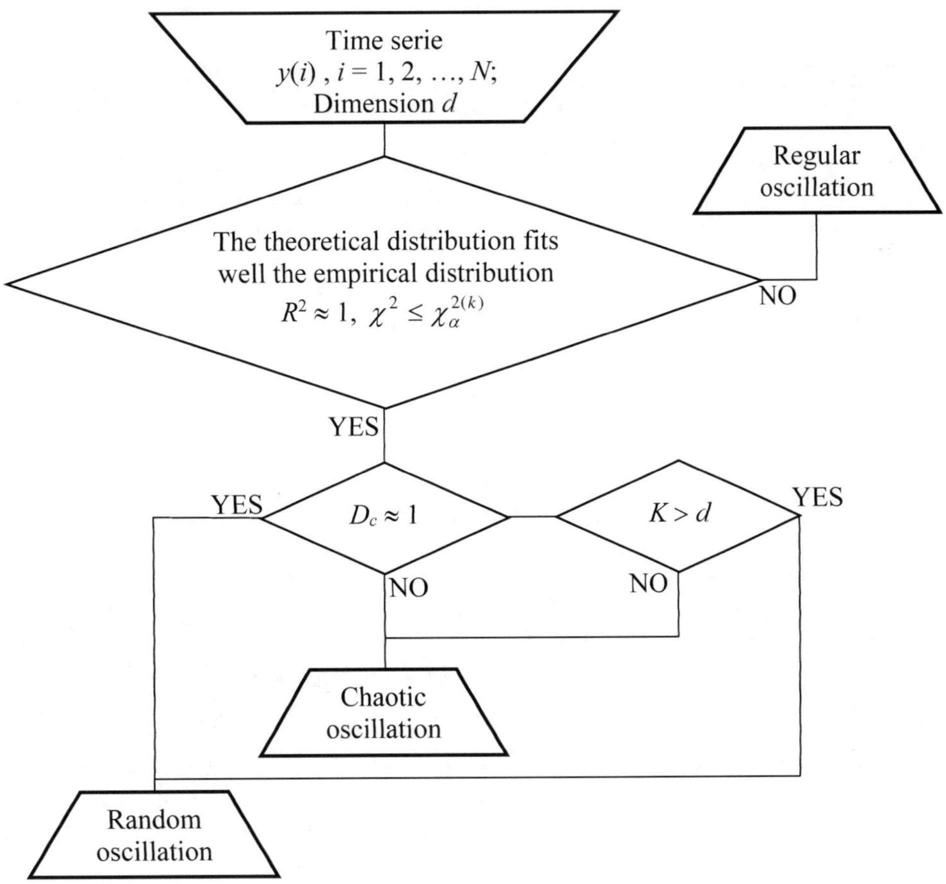


**Figure 10**. The algorithm for distinguishing between random and chaotic oscillations

The time series is experimentally generated oscillations data and the input data for distinguishing between random and chaotic oscillations. The methodology for distinguishing between random and chaotic oscillations has two steps. In the first step the nonparametric hypothesis tests are to be used, which the observed oscillations confirm as irregular. The nonparametric hypothesis test determines how well theoretical distribution fits to the empirical distribution. The chi-square test is a very popular type of nonparametric test. The correlation coefficient between the empirical and theoretical distributions can also be used for a goodness of fit.

In the second step, the quantifiers of chaos are to be used. The quantifiers of chaos are often the only method for reliable recognition of the types of dynamics. The methodology for the final characterization of oscillations was based on chaos quantifiers. The quantifiers of chaos - the correlation dimension and the Kolmogorov entropy are crucial for distinguishing between random and chaotic oscillations. The calculated value of the correlation dimension or the Kolmogorov entropy determined exactly whether the oscillations were chaotic or random.

The methodology can be applied to many oscillating phenomena, not only on mechanical oscillations, e.g. temperature oscillations. Therefore, in this paper the term oscillations was mainly used, not vibrations, chatter, etc. In combination with an experimental design stochastic modeling, this methodology gives good results in distinguishing between random and chaotic oscillations.

The methodology requires a multidisciplinary approach based on combining Nonlinear Dynamics and Probability Theory to the Statistics and the successful co-operation among mathematicians, physicists, engineers and experts in various different fields in which irregular oscillations are studied.

**References**

[1] Gegg C B, Suh C S and Luo C J A 2011 Machine Tool Vibrations and Cutting Dynamics *Springer New York USA*

[2] Insperger T and Stépán G 2004 Vibration Frequencies in High-Speed Milling Processes *Journal of Manufakturing Scieince and Engineering* Vol. 126 pp 481-487

[3] Stépán G 1989 Retarded dynamical systems: stability and characteristic functions *Longman Scientific & Technical GB*

[4] Warmiński J, Litak G et. al 2002 Approximate Analytical Solutions for Primary Chatter in The Nonlinear Metal Cutting *Journal of Sound and Vibration* 259(4) pp 917-933

[5] Yan Y, Xu J and Wiercigroch M 2014 Chatter in a transverse grinding process *Journal of Sound and Vibration* 333 pp 937-935

[6] Gredelj S 2017 Istraživanje utjecaja vibracija obradnih sistema na stohastično i haotično ponašanje procesa obrade *doctoral dissertation Faculty of Technical Engineering University of Bihać BiH*

[7] Hilborn C R 2000 Chaos and Nonlinear Dynamics 2nd edition *Oxford University Press GB*

[8] Huke P J 2006 Embedding Nonlinear Dynamical Systems: A Guide to Takens' Theorem *The University of Manchester GB*

[9] Kantz H Schreiber T 2004 Nonlinear Time Series Analysis *Oxford University Press GB*

[10] Moon F C and Kalmár-Nagy T 2001 Nonlinear models for complex dynamic in cutting materials *Philosophical Transactions of the Royal Society A London GB* pp 695-711

[11] Small M 2005 Applied Nonlinear Time Series Analysis *World Scientific Publishing Hong Kong China*

[12] Insperger T 2002 Stability Analysis of Periodic Delay Differential Equations Modeling Machine Tool Chatter *PhD dissertation Technical University of Budapest Hungary*

[13] Doleček V, Voloder A and Isić S 2009 Vibracije *Faculty of Mechanical Engineering University of Sarajevo BiH*

[14] Kelly G S 2012 Mechanical Vibrations: Theory and Applications SI *The University of Akron Ohio USA*

IOP Conf. Series: Materials Science and Engineering 1208 (2021) 012009 doi:10.1088/1757-899X/1208/1/012009


[15] Sheskin J D 2000 Handbook of Parametric and Nonparametric Statistical Procedures *Chapman & Hall/CRC Washington USA*

[16] Grassberger P and Procaccia I 1983 Characterization of Strange Attractors *Physical Reviev A The American Physical Society* pp 364-349

[17] Zmeskal O, Dzik P and Vesely M 2013 Entropy of fractal systems *Computers and Mathematics with Applications Elseiver Ltd.* 135-146

[18] Grassberger P and Procaccia I 1983 Estimation of the Komogorov entropy from a chaotic signal *Physical Reviev A The American Physical Society* pp 2591-2593

[19] Mañé R 1980 A proof of Pesin's formula, Ergodic Theory and Dynamical Systems *Cambridge University Press GB* pp 95-102

[20] Corder W G and Foreman I D 2014 Nonparametric statistics: a step-by-step approach – Second edition *John Wiley & Sons Inc. Hoboken New Jersey USA*

[21] Milčić D 2005 Pouzdanost mašinskih sistema *Faculty of Mechanical Engineering University of Niš Serbia*

[22] Adnroić V et. al 2006 Inženjerstvo pouzdanosti *IA Projektiranje Zagreb Croatia*

[23] Gupta C S and Kapoor K V 2000 Fundametals of Mathematical Statistic Tenth Revised Edition *University of Delhi India*

[24] Jurković M 1999 Matematičko modeliranje inženjerskih procesa i sistema *Faculty of Mechanical Engineering University of Bihać BiH*

IOP Conf. Series: Materials Science and Engineering 1208 (2021) 012010 doi:10.1088/1757-899X/1208/1/012010

# Kinematics of steam film wetting while quenching cylindrical samples in thermal oils

**Šefik Behrem and Bahrudin Hrnjica**

University of Bihac, Faculty of Technical Engineering, 77000 Bihac, Bosnia and Herzegovina

E-mail: sefik.behrem@unbi.ba

**Abstract**. The paper presents the determination of the velocity of the vapor front along the outer wall of a cylindrical specimen in the process of two-dimensional axisymmetric quenching of the samples in thermal oils. One thermal oil is Isorapid 277 HM heated to 40°C and the other thermal oil is Marquench 722 heated to 90°C. The experimental setup of the work consists of heating to a temperature of 850°C, then quenching three dimensionally different cylindrical probes. The dimensions of the probe were: *ϕ25x100*, ϕ50x150 and ϕ75x225 mm. All quenchings were done in strictly controlled conditions of the flow rate of the quenchant around the cylinder as well as maintaining the temperature of the quenchant within the maximum *40±2.5°C* or *90±3°C*, during the quenching of the samples. The velocity of movement of the steam front on the outer surface of the cylinder was determined from the time-measured values of temperatures at the marked points of each sample. The analysis of the steam front movement velocity along the cylinder wall starts from the moment the lower base of the probe touches the quenchant. During the process of immersing the test probe in the quenchant, in addition to measuring the temperatures in time, the time of lowering the probe to contact with the quenchant sample was also measured. The approximate average velocity of the vapor front was determined based on the indications of the lower and middle thermocouples located 1.5 mm below the outer surface of the cylinder wall. Based on the distance of one half the height of each probe and time, the velocity of the steam film movement or the kinematics of the steam film wetting was obtained. The obtained results were compared with the results of quenching in water and aqueous solutions of the same probes under the same strictly controlled conditions.

## 1. Introduction

Quenching process represents a fast cooling of a workpiece in quenchant in order to gain an certain material properties. However, the quenching process also prevents undesired crystal grain size and other similar reaction. During quenching three different phases can be identified: vapor blanket (full -film boiling), nucleate boiling and convective heat transfer. Most vaporized liquids which can be found in the quenching are water, oil and aqueous solutions [1]. The cooling conditions can be categorized by three groups factors (i) workpiece characteristics (composition, mass, geometry, surface roughness and condition) (ii) quenchant characteristics (density, viscosity, specific heat, thermal conductivity, boiling temperature) and (iii) quenching facility (bath temperature, agitation rate, flow direction). The cooling behavior of thermal oil during quenching can be identified similar like water which involves three stages of cooling namely mentioned previously. However, the cooling performance of thermal oil is much lower than water [2].

The estimation of the velocity of the steam front on the surface of the test sample was made on the basis of cooling curves formed by measuring the temperatures at the indicated points of the test probe according to [4] and [5]. The theoretical description of the problem is shown in Figure 1.

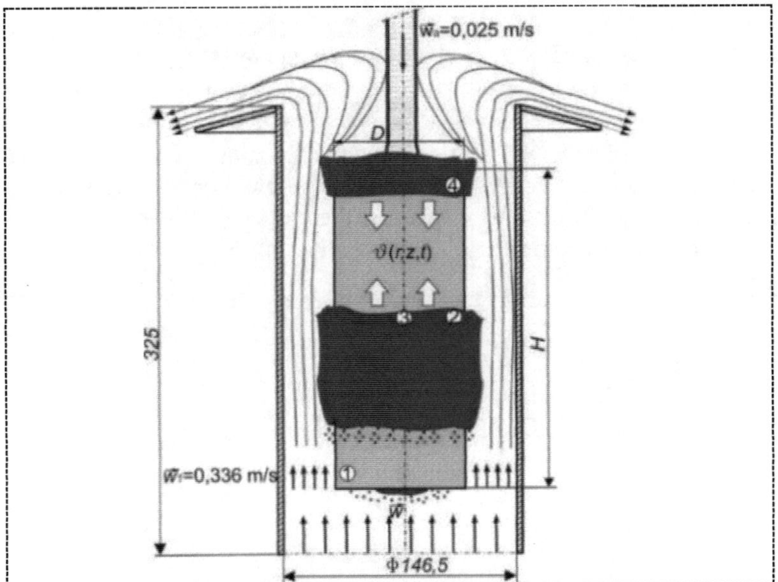


**Figure 1.** Theoretical description of the problem

Figure 1 shows a cylinder of diameter D, height H, initial temperature 850 °C, which was vertically immersed at a velocity of $w_a$=0,025 m/s in the bottom-up cooling fluid, which flows to the cylinder at a velocity of $w_l$*=0,336 m/s*. During immersion of the probe in the quenchant, four probe temperatures were measured continuously over time using built-in thermocouples. Based on the obtained values of temperatures in the probe, the velocity of movement of the vapor front on the surface of the probe was estimated. The experimental setup of the work consisted of quenching three dimensionally different cylindrical probes. The dimensions of the probes were: *ϕ25x100*, *ϕ50x150* and *ϕ75x225 mm*, and the material of the probes was AISI 304 steel. Quenching agents were two thermal oils. One thermal oil was Isorapid 277 HM and the other thermal oil was Marquench 722. A detailed description of the experimental setting of the problem as well as the experimental results were given in [3] and [6]. The task of the work was solved at the level of heat conduction through the cylinder wall. As the solving algorithm did not know the events in the fluid that surrounds the probe, it was not able to give a current picture of the movement of steam on the probe surface. Accordingly, the rate of wetting, i.e. the rate of disappearance of the vapor film was estimated on the basis of experimental results of temperature measurements at points 1,2, in the probe ϕ25x100, then 5,6 in the probe *ϕ50x150* and 9,10 in the probe *ϕ75x225 mm*, which were 1,5 mm below the outer surface of the probe.

## 2. Experimental determination of steam front velocity

The analysis of the velocity of the vapour film along the cylinder wall began from the moment the lower base of the probe touched the quenshant. At places 1,2,3,4, then 5,6,7,8 and 9,10,11,12, in the mentioned tests, temperatures in time were measured in parallel, on the basis of which the average velocity of the steam front was estimated. In [3] and [4] in addition to the cooling curves, the results based on video recordings were used because pure water is transparent for video recordings. In this paper, the analysis of the velocity of the steam front was given only on the basis of the measured cooling curves because thermal oils were not transparent to video rays. The physical properties of both thermal oils were given in detail in [3]. The recommended quenching temperature of steel in Isorapid 277 HM oil was from 50 to 80 $^0$C and max up to 130 $^0$C in a protective atmosphere, while the recommended quenching temperature in Marquench 722 oil was from 60 to 130 $^0$C and in a closed furnace up to 150 °C. In this paper Isorapid 277 HM was heated and maintained at 40 $^0$C, while Marquench 722 was heated and maintained at 90 $^0$C.

### *2.1. Determination of steam front velocity based on cooling curves*

The velocity of the steam front was determined on the basis of showing the temperatures in the time of thermocouples at half the height of each probe, i.e.: 2 in the probe $\phi$25x50 mm, 6 in the probe $\phi$50x150 and 10 in the probe $\phi$75x225 mm. Each thermocouple was located 1,5 mm below the outer surface of the corresponding probe. A vapour film was formed at the moment of contact of the probe base with the quenchant. This moment was determined by measuring the movement time of each probe from the center of the furnace to the water contact indicated on each cooling curve.

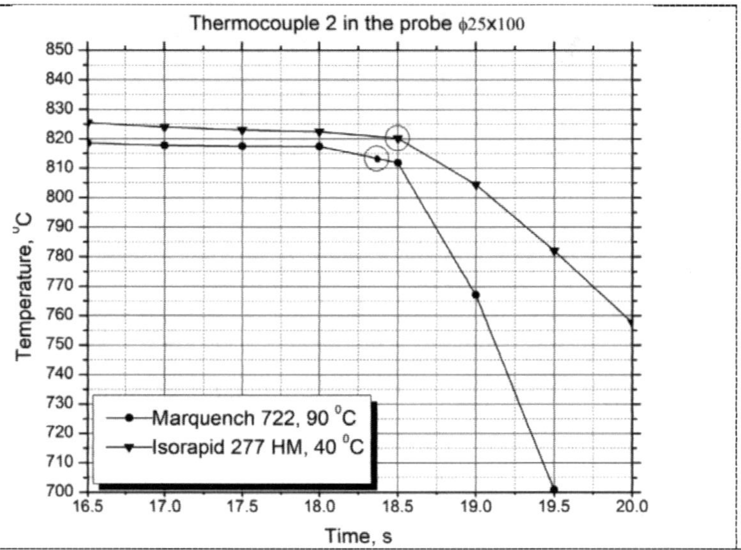


**Figure 2.** Movement time of the steam front along the way up to half of the *$\phi$25x100* probe height

The second time or position of the vapor front was its breaking in the middle of each probe indicated by a sharp rise in the temperature gradient of cooling, read from the cooling curves. The approximate mean velocity of the vapour front over the surface of the probe was determined with two necessary assumptions: the speed of the steam front or the speed of the Leidenfrost front (disappearance of the steam film) was constant and the delay of the central thermocouple was also compensated by the delay of the lower thermocouple (1,5 and 9).

Figure 2 shows an enlarged view of the cooling curves at the location of thermocouple 2 in the probe $\phi 25x100$mm for both thermal oils. The contact of the lower base of the probe with the oil was in 16,5 seconds, corresponding to the moment of formation of the steam front or a sudden rise in temperature gradient at position 1, and the breaking of the steam front (Leidenfrost effect), in the middle of the probe height was indicated in Figure 2.

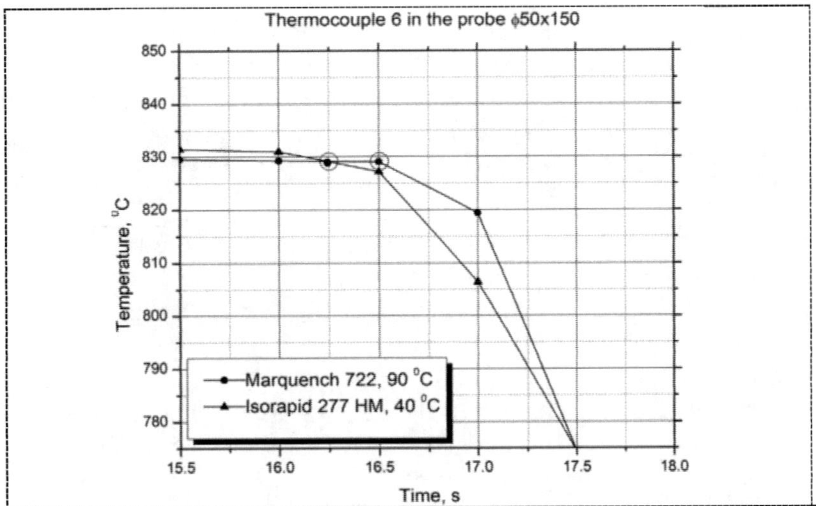


**Figure 3.** The steam front movement time on the way to half of the *ϕ50x150* probe height

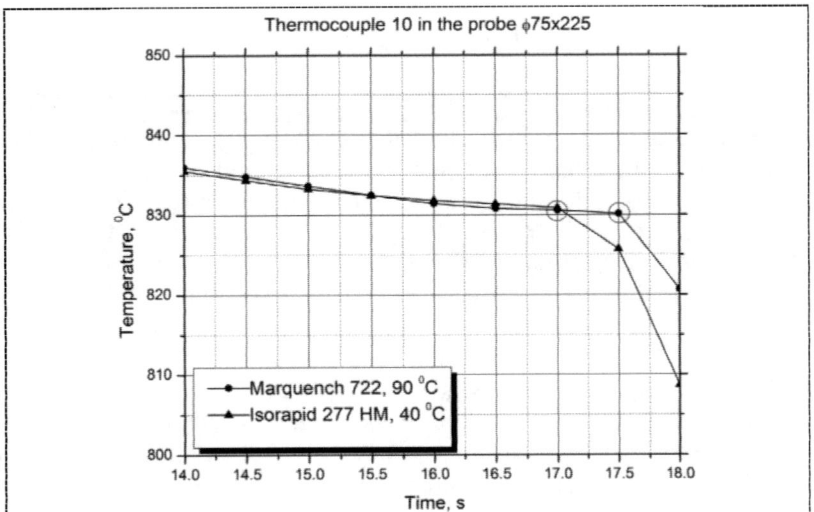


**Figure 4.** The steam front movement time on the way to half of the *ϕ75x225* probe height

The difference between these two times defines the movement time of the steam front along the way of half the probe height. Cooling curves at the location of thermocouple 6 in the probe $\phi 50x150$ mm for both thermal oils were given in Figure 3. Contact of the lower base of the probe with thermal oils was in 15,5 seconds, corresponding to the time of steam front formation or sudden rise in temperature gradient at location 5 and the breaking of the steam front was in the middle of the probe height, marked in the same Figure. The time temperature response of thermocouple 10 in the probe $\phi 75x225$ mm for both thermal oils was given in Figure 4. The contact time of the lower base of the probe with the oils was 14 seconds corresponding to the time of vapour front formation or a sudden rise in temperature gradient at position 9, and the Leidenfrost effect was in the middle of the probe height, indicated in Figure 4.

**Table 1.** Mean velocity of the vapour front on the outer surfaces of individual probes quenched in Marquench 722 thermal oil preheated at 90 $^0$C.

| Probe | Time to probe contact with Marquench oil $t_1$, s | Steam film breaking time, $t_2$, s | Temperature 2, 6 and 10, at the $t_2$, $^0$C | The way of steam film, *h*, mm | Steam velocity mm/s, $\overline{w}$ |
|---|---|---|---|---|---|
| *ϕ25x100* | 16,5 | 18,3 | 815 | 50 | 27,8 |
| *ϕ50x150* | 15,5 | 16,5 | 827 | 75 | 75 |
| *ϕ75x225* | 14 | 17,5 | 830 | 112,5 | 32,14 |

**Table 2.** Mean velocity of the vapour front on the outer surfaces of individual probes quenched in Isorapid 277 HM thermal oil at a temperature of 40 $^0$C

| Probe | Time to probe contact with Marquench oil $t_1$, s | Steam film breaking time, $t_2$, s | Temperature 2, 6 and 10, at the $t_2$, $^0$C | The way of steam film, *h*, mm | Steam velocity mm/s, $\overline{w}$ |
|---|---|---|---|---|---|
| *ϕ25x100* | 16,5 | 18,3 | 815 | 50 | 27,8 |
| *ϕ50x150* | 15,5 | 16,5 | 827 | 75 | 75 |
| *ϕ75x225* | 14 | 17,5 | 830 | 112,5 | 32,14 |

According to the assumptions introduced in Table 1, the mean velocity of the steam front was calculated. The quenching results from Table 2 shown the same tendency as the quenching results from Table 1. The mean temperature read at the thermocouples 2, 6 and 10 was 826,7 $^0$C.

## 3. Results comparison with previously published results

Previously published papers describing the velocity of the vapor front with the same test probes and under the same experimental conditions, but with different quenchants which were water at a temperature of 40 °C [4] and aqueous solutions at a temperature of 40 °C, [5]. Cooling rates in thermal oils Isorapid 277 HM and Marquench 722 are about 3.5 to 10 times higher than quenching rates in water, and 15 to 40 times higher than aqueous solutions. Range values mostly depend on the probe diameter, as shown in Figure 5. The conclusion for the given results would be that thermal oils belongs to different category, while water and aqueous solutions in another. The results shown that the obtained cooling rates in water and aqueous solutions are far less than the cooling rates in thermal oils, which is nonobvious since the heat capacities of water and aqueous solutions are twice as high as the heat capacities of thermal oils. The reason for this is the stability of the vapor film on the wall of the probe which acts as an insulator. Figure 5 shows the cooling curves at point 2 in the □25x100 probe.

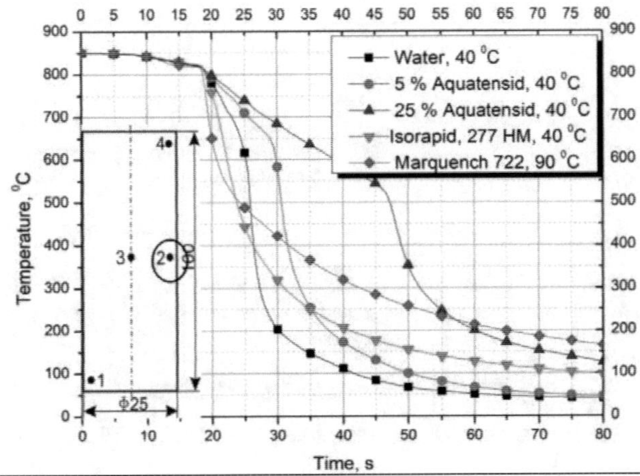


**Figure 5.** Cooling curves at point 2 of the 25x100 probe for five different quenchants

A comparison of the corresponding cooling rates for five different extinguishing agents is shown in figure 6.

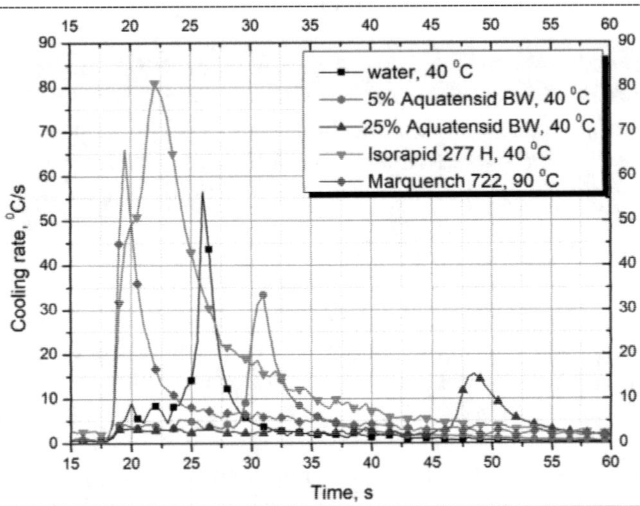


**Figure 6.** Cooling rates at point 2 of the 25x100 probe for five different extinguishing agents

The maximum values of the temperature gradients are related to the moment of vapor front decay or the period of the droplet evaporation phase. Figure 7 shown the comparative values of the mean velocities of the steam front depending on the diameter of the test probe. Quenching results in: pure water was reported in [4], 5% aquatenside BW in water and 25% aquatenside BW in water were reported in [5].

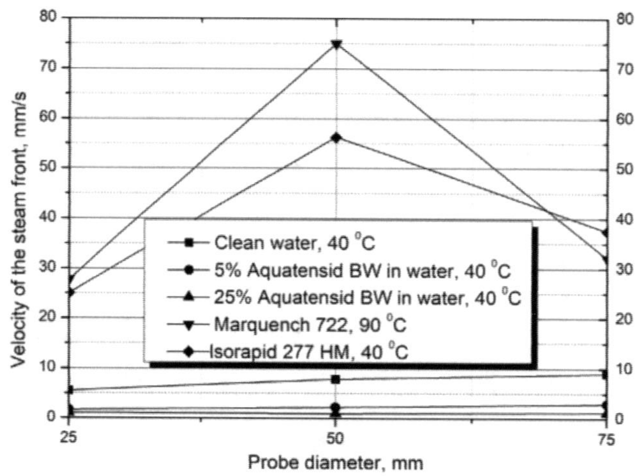


**Figure 7.** Velocity of the steam front on the surface of the probes

It can be seen from the figure that the velocities of the vapour front when quenching the probes in thermal oils were the highest and not comparable with water and aqueous solutions. The average temperature read at the site of thermocouples 2, 6 and 10 was 824 $^{0}$C.

## 4. Conclusion

The mean temperatures at the thermocouple sites 2, 6 and 10 in the middle of the height of each probe at the time of the steam front breaking were approximately 824 $^{0}$C, for quenching in Marquench 722 the temperature was 90 $^{0}$C, while the mean temperatures at the same positions in Isorapid were 277 HM 826,7 $^{0}$C. The results show a short duration of the vapour film during quenching in thermal oils, and thus the values of surface temperatures at the time of breaking of the vapour front remain high. The reason for this was the high resistance of thermal oils to evaporation. Previously published results of quenching in water and aqueous solutions [4] and [5] shown lower velocities of the vapour front, and the residence of the vapour film on the sample wall was longer with a reduced surface temperature at the time of vapour breaking. The final contribution of this paper is reflected in the comparison of the results obtained during quenching in five different steel quenchants.

## References

[1] Totten G E and Tensi H M 2002 Quenching Fundamentals: Using Conductance Data to Characterize Quenchants, Heat Treating Progress, Volume 2, Issue, August 2002 (ASM International), https://www.asminternational.org/

[2] Ramesh G and Prabhu K N 2014 Wetting and cooling performance of mineral oils for quench heat treatment of Steels ISIJ International 54 1426–35.

[3] Behrem Š and Hrnjica B 2018 Estimate of heat transfer coefficient during quenching steel in water Transactions of FAMENA 42 61–73, doi:10.21278/TOF.42Si106

[4] Behrem Š and Hrnjica B 2017 Kinematics of steam film wetting while quenching cylindrical samples in water 11th International Scientific Conference on Production Engineering 1 vol 1 (Bihać, Bosnia anf Herzegovina: University of Bihać) pp 30–5

[5] Behrem Š and Hrnjica B 2019 Kinematics of steam film wetting while quenching cylindrical samples in aqueous solutions 12th International Scientific Conference on Production Engineering 1 vol 1 (Bihać, Bosnia and Herzegovina: University of Bihać) pp 50–5

[6] Behrem Š and Hrnjica B 2021 Results of heat transfer coefficient estimation during the quenching of cylindrical samples in aqueous solutions Transactions of FAMENA 45 49–58, doi:10.1520/JAI101981

# Assessment of integrity and remaining working life of welded steel structures

**Ermin Bajramović and Fadil Islamović**

University of Bihac, Faculty of Technical Engineering, Bosnia and Herzegovina

E-mail: ermin.bajramovic@unbi.ba

**Abstract**. Analyzing the period of exploitation of welded steel structures it can be concluded that they are predominantly exposed to the action of variable load. The welded joint as the largest stress concentrator due to the heterogeneity of structural, mechanical and operational properties is a key problem that is further complicated by the possible and realistically probable presence of crack-type faults. The assessment of integrity largely depends on a comprehensive analysis of the welded joint as the most critical place of welded steel structures. Integrity assessment is a necessary obligation for extending the working life, as well as revitalization, as a way to keep the structures in operation, despite the long period of exploitation. This paper presented an analysis of the process of fatigue crack initiation and growth, i.e. an assessment of the of the welded steel structures' integrity and remaining service life under the influence of variable load.

## 1. Introduction

Welded steel structures in most cases represent significant and responsible systems whose integrity and service life must be viewed from several aspects. The conditions in which the structure was exposed, i.e. in which it works, significantly affect its integrity and service life, therefore the determination of the service life is imposed as a necessity. Operating conditions imply a set of parameters that affect the structure by directly or indirectly affecting the three basic performances: reliability, safety and durability. [1]

The causes of structural failures can generally be divided into four groups:

- causes or design errors,
- manufacturing and/or assembly errors,
- errors caused by deviation of operating conditions, and
- errors as a result of accidental circumstances.

Prevention of failure of welded structures due to weakened material by changing its physical and chemical properties in long-term operation or increasing errors above the allowable level can be achieved by systematic and continuous monitoring of material properties and the occurrence and development of defects in base material and welded joints.

It is known that certain components, when it comes to process equipment operating in conditions of low and high temperatures and pressures such as pressure vessels, are critical places due to specific operating parameters. This is especially significant, knowing that many process equipment plants around the world have been in use for so long time that individual components have been in use for longer than the projected service life. Eventual failure of these components would pose a danger not only to the operation of the plant but also to the environment surrounding them. [2]

Welded components are also a problem, where the possibility of crack-type defects must not be ruled out. In case of such errors, it is necessary to accurately assess the integrity of the component and make a decision on its further exploitation. The integrity of welded joints is a very important factor in the exploitation of welded structures due to their heterogeneity and tendency to the appearance and growth of cracks as a consequence of the welding process. Therefore, it can be said that welded joints represent the most critical part of welded structures. [2]

In practice, in most cases, the failure of welded structures occurs due to fatigue. Research and analysis of the impact of operating conditions on the behaviour of the base material and components of the structure welded joint intended for operation under conditions of variable load, all in order to assess the integrity and remaining life of the structure, i.e. extending the service life have been increasingly present today. [3]

## 2. Analysis of welded structures

### *2.1. Materials and technology of construction*

The service life of all welded structures has been conditioned by the choice of materials of appropriate properties and quality, well-chosen and implemented production and welding technology, provided that the loads were in line with expectations.

As a starting point on the applicability of the materials used in structures are the characteristics obtained by tensile testing such as yield stress, tensile strength, elongation at break, and modulus of elasticity. It is these characteristics that describe the mechanical behavior of materials and are the basis for the construction of parts. In addition to the basic characteristics of the material, one should certainly know additional data on the behavior of materials obtained by impact testing such as refractive energy which has been most often used for comparative analysis in the selection of materials. [3]

In order to ensure that the integrity of the structure remains undamaged during the expected service life, in addition to the choice of material, the activities that precede its commissioning, defining the geometry and the welding process itself should also be taken into account. When defining these activities, the most important thing is to avoid potential stress concentrations which are the most favorable places for cracks to start (and later grow). Furthermore, the welded joint due to the heterogeneity of structural, mechanical and operational properties as well as due to the presence of defects (blowhole, slag, cracks) caused that for the operational safety of welded steel structures, the most important micromechanical characteristics describing the occurrence and growth of cracks under variable load. [3,4,5]

### *2.2. Influence of cracks on welded structures*

Problems occurring in the structure during the exploitation period most often happen due to insufficient resistance of the material to the formation and growth of cracks, especially in welded joints. In solving this problem, the classical fracture mechanics has been applied by comparative analysis of the crack growth force on the one hand and the resistance of the material on the other hand. Fracture mechanics aims to significantly reduce the risks of fractures of structures, which can often have catastrophic consequences (especially in welded pressure vessels).

This procedure includes detecting the location of defects in the structure and determining their impact on its integrity, i.e. determining the moment when these defects (primarily cracks) would exceed the standards prescribed by the standards and lead to failure. [6]

The application of fracture mechanics has been based on the determination of appropriate parameters (stress intensity factor, crack tip opening and J-integral) on the basis of which the remaining service life of the structure was determined. These parameters are related to the presence of cracks which in the case of welded joints have been inevitable, to a greater or lesser extent. [7]

Depending on their position and the loads acting in their environment, these cracks can start to grow and when they reach a certain length they can lead to failure and fracture of the structure. Therefore, it is very important to monitor their behaviour during exploitation as well as to prevent their occurrence (although it is impossible to do it fully), and/or to keep them within the permitted limits which depend primarily on the conditions of exploitation.

### *2.3. Cracks under the influence of variable loads*

Parts of structures and machines can be exposed to variable loads during operation, both in size and in the direction of action. Due to the long-term effect of the variable load there is a gradual destruction of the material, i.e. fatigue, which can result in a fracture called fatigue fracture. It was estimated that 50-90 % of all structural damage in operation occurs due to material fatigue.

Fatigue represents the gradual destruction of a material until the moment when its cross section can no longer bear the load and due to changes in the material caused by a load that is cyclical in nature, i.e. whose value changes over time. In such conditions and after a sufficiently large number of cycles, the fracture of the structure can occur at voltages significantly lower than the yield strength, in contrast to static fracture, which is a special problem. It is generally considered that the total lifetime consists of the crack initiation period and the crack propagation period until the final fracture. With this in mind, researches have been mainly focused on establishing independent procedures for the formation and growth of cyclic cracks. [8]

The collected knowledge on fatigue crack growth made it possible to determine with sufficient certainty the remaining life of the crack component and thus assess whether the component can operate until the next inspection.

## 3. Integrity of welded structures

The integrity of structures, i.e. their ability to function in an inefficient and safe way in the intended conditions of exploitation is a very important aspect from the point of view of safety.

The term structural integrity assessment means the following:

- analysis of the existing condition of the structure including diagnostics of behaviour and failure with an assessment of the remaining strength and service life;
- the structure revitalization with the application of measures in order to preserve functionality.

Based on the input data, most often obtained experimentally, it is possible to determine the actual behaviour of the structure during operation, to estimate the service life, but also the causes of bad behaviour or even the failure of the structure. Based on this knowledge, the parameters of choices and decisions were obtained, the realisation of which improves the construction, and thus prolongs the service life. [10]

The application of failure analysis methods and the determination of the estimated service life of a welded structure imply knowledge of fracture mechanics, methods of stress and strain testing and knowledge of mathematical and statistical methods in data analysis and processing. Taking into account the operating conditions, the following parameters characterising the welded structure and its joints are to be usually checked: fatigue strength, resistance to brittle fracture and permanent strength (often in combination with fatigue strength if the operating temperature causes creep). [11]

Estimation of the service life of structures loaded with variable loads is also needed to define the control interval for detecting and eliminating damage caused in operation. The procedure for determining the fatigue characteristics and estimating the service life of a structure exposed to a variable load was shown in figure 1.

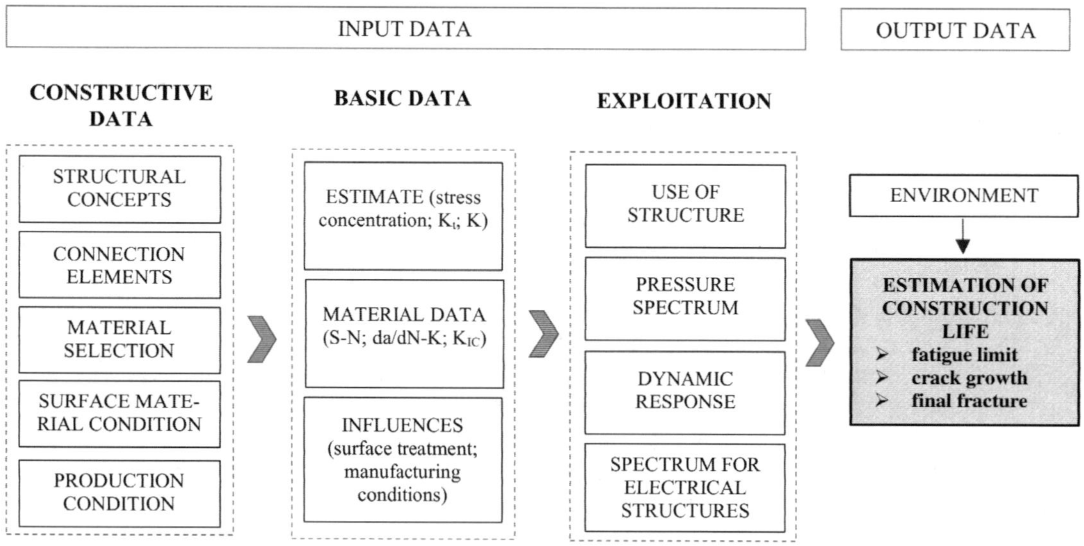


**Figure 1.** Methodology for determining fatigue characteristics and estimating the service life of structures exposed to variable loads

The process of fatigue and fracture can be divided into three time phases:

- The phase of micro-crack formation ai and its further growth to a size called an engineering crack, $a_0$. This may be the minimum crack length that can be detected, $a_{pr}$.
- Phase of stable macro-crack growth until instability occurs at length $a_{cr}$.
- Phase of unstable growth and final fracture.

Experience gained by testing aircraft structures with simulated loads corresponding to flight conditions has shown that the crack formation time for most structural details was relatively short and that the life of the structure was determined by the time required for crack growth to fracture, which was about 95 % of life. Therefore, it is common to view the entire process of damage accumulation only as a crack growth process.

In order to ensure the safety of welded components, it is necessary to clarify their behaviour under the action of low-cycle fatigue and in the presence of crack type error (fatigue crack growth rate da/dN, and the fatigue threshold $\Delta K_{th}$), as well as the behavior under the action of shock (impulse) load on parts that contain a notch-type stress concentration.

Analysis of the stress state and deformation at the top of the rising fatigue crack using linear-elastic fracture mechanics (LEML) led to the formulation of the Paris equation, which relates the fatigue crack growth rate to the range of stress intensity factors at the crack tip. [11]

Although the Paris equation of crack growth was not valid in the whole range, between low velocities near the fatigue threshold, $\Delta K_{th}$ in figure 2, and high velocities ($K_{Ic}$) the large linear mean part of the curve covered by the Paris relation proved to be by far the most important from a practical point of view. At the same time it allows a distinction to be made between the initiation and growth of a fatigue crack. From sharp stress concentrators, under conditions of variable load, after a certain number of cycles, a crack will be initiated and its growth will occur if the fatigue threshold $\Delta K_{th}$ is exceeded.

As the structure will not be endangered under certain conditions until the crack reaches a critical size, it is possible, with previous analyses, to allow the exploitation of the structure with a crack during the crack growth period. An important piece of information for the decision on further exploitation of the structure is the knowledge of the crack growth rate depending on the load that acts on it.

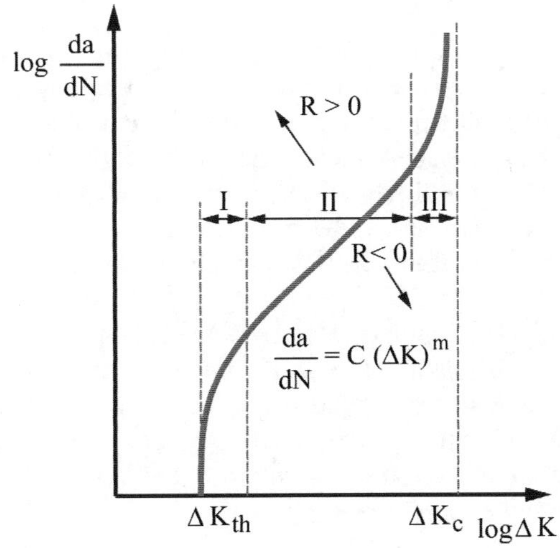


**Figure 2.** Functional dependence of fatigue growth rate cracks and stress intensity ranges; *da/dN=f(ΔK,R)* [2]

If the Paris equation for determining the crack growth rate:

$$\frac{da}{dN} = C \cdot (\Delta K)^m \tag{1}$$

is presented in the following form:

$$\frac{da}{dN} = \frac{\Delta a}{\Delta N} = f(K_{max}\ or\ \Delta K, R...) \tag{2}$$

The structure service life can be determined by integrating the previous equation from the initial ($a_0$) to the final crack length ($a_{cr}$):

$$T_k = \int_{a_o}^{a_{cr}} \frac{da}{f(K_{max}, ...)} \tag{3}$$

Estimating the fatigue life under conditions of constant load amplitude was given by the precise integration of the Paris equation written in the form:

$$\left(\frac{da}{dN}\right)_i = C_i (\Delta K)^{m_i} \tag{4}$$

where: $C_i$ and $m_i$ are constants, i.e. exponents for certain values of the stress intensity range ΔK, for which the equation also holds.

Therefore the lifespan of the construction is:

$$T_k = \sum \Delta T_i = \int_{a_0}^{a_1} \frac{da}{C_1(\Delta K)^{m_1}} + \int_{a_1}^{a_2} \frac{da}{C_2(\Delta K)^{m_2}} + \ ... \ + \int_{a_j}^{a_{cr}} \frac{da}{C_{cr}(\Delta K)^{m_{cr}}} \tag{5}$$

Models for predicting fatigue life under variable amplitude (load spectrum) originate from various researchers. These models have been based on the crack growth rate delay due to overload and provide an estimate for certain materials and load conditions.

## 4. Conclusion

This paper presented a general analysis and presentation of the influence of crack-type defects that occur in welded structures and the process of assessing the integrity and service life of welded steel structures based on the application of fracture mechanics. Fracture mechanics is extremely suitable for determining the development of damage which occurs as a result of crack growth. In cases where the life depends on the crack growth the fracture mechanic is able to determine the residual strength and life at any point in exploitation. However, the assessment of the condition of the structure was not based solely on the examination of cracks, as this may be too late from the point of view of preventing sudden fractures.

For successful and economical maintenance of components in order to prevent sudden failures and fractures, the reduction of strength parameters and service life of structures should be determined and taken into account. This paper can be a good basis for more detailed research of the influence of variable loads on the process of initiation and crack growth in welded steel structures, the results of which can be the basis for reliable assessment of integrity and remaining life of welded steel structures exposed to variable loads during exploitation.

## References

[1] Lukić B, Đurić S and Đuričić M 2009 Zavarene konstrukcije - pouzdanost, sigurnost i otkazi *Zavarivanje i zavarene konstrukcije* **2** 69-5

[2] Islamović F 2006 *Poboljšanje integriteta stacionarnih višekomornih tankostijenih posuda za tečna goriva* dissertation (Tuzla: Univerzitet u Tuzli)

[3] Burzić Z 2000 Ispitivanja promjenljivim opterećenjem glatkih i zareznih epruveta *Eksperimentalne i numeričke metode mehanike loma u oceni integriteta konstrukcija* Joint edition: TMF Beograd - Institute GOŠA Smederevska Palanka - JSZ Beograd

[4] Islamovic F, Gaco Dz, Hodzic D and Bajramovic E 2020 Determination of stress-strain state on elements of cylindrical tank strukture *9th International scientific conference on defansive tehnologies (OTEH 2020)*

[5] Arsić M 2001 Procena veka zavarene konstrukcije sa prslinom u uslovima promenljivog opterećenja *Eksperimentalne i numeričke metode mehanike loma u oceni integriteta konstrukcija* Joint edition: TMF Beograd - Institute GOŠA Smederevska Palanka - JSZ Beograd

[6] Sedmak S A, Burzić Z, Perković S, Jovičić R, Aranđelović M, Radović L and Ilić N 2019 Influence of welded joint microstructures on fatigue behaviour of specimens with a notch in the Heat Affected Zone *Engineering Failure Analysis* **106** 104162 DOI:10.1016/j.engfailanal.2019.104162

[7] Manai A 2020 A framework to assess and repair pre-fatigued welded steel structures by Tig Dressing *Engineering Failure Analysis* **118** 104923 DOI:10.1016/j.engfailanal.2020.104923

[8] Gačo, Dž, Burzić Z, Islamović F and Bišćević N 2014 Analiza zamora zavarenog spoja čelika NN-70 u prisustvu greške tipa prsline *X Naučno/stručni simpozij sa međunarodnim učešćem METALNI I NEMETALNI MATERIJALI*

[9] Burzić M, Manjgo M, Bernetič J, Burzić Z and Arsić M 2015 Effect of variable load on crack initiation microalloyed steel S 690-QL *METALURGIJA* **54** 1 55-8

[10] Sedmak S, Jovičić R, Sedmak A, Aranđelović M and Đorđević B 2018 Influence of multiple defects in welded joints subjected to fatigue loading according to sist EN ISO 5817:2014 *Structural integrity and life* **18** 1 77–1

[11] Paris P and Erdogan F 1963 A critical analysis of crack propagation laws *Journal of Basic Engineering* **85** 528–33 DOI:10.1115/1.3656900

# Investigation of the influence of electrode manufacturers 'recipes on the production and chemical composition of welding fume particles

**Razija Begić**
Faculty of Technical Engineering, University of Bihać, Bosnia and Herzegovina

E-mail: razijabegic@yahoo.co.uk

**Abstract.** The quality of the welded joint, production and chemical composition of welding fume particles in the MMAW process, in addition to the welding parameters, mostly depend on the quality of the coated electrode. As part of the preliminary experimental research of the optimal technological composition of the electrode coating, from the aspect of minimizing welding fume, an experimental research of one class of electrodes, E 42 4 B 32 H5, EN 499/94, standard production program of three manufacturers of welding consumables. The test was performed to examine the influence of the manufacturer's recipe on the production and chemical composition of welding fume particles.

## 1. Introduction

To test the production and chemical composition of welding fume particles, a basic coated electrode marked, E 42 4 B 32 H5, with controlled hydrogen content, was selected for welding in all positions, except vertically downwards, carbon and medium-stress steels and mild steels under the condition of high loads. The initial and re-establishment of the arch is very good, and the slag is easily separated. It can be used for all general purposes where control of the hydrogen content is necessary. The basic materials are, general construction steels, pipe steels, steels for pressure equipment, steels for shipbuilding and steel castings. The main goal of this research is to experimentally check some literature data and to narrow the research range to a certain type of electrode, for which, through the implementation of the main experiment, we would try to determine the optimal composition of the electrode coating, which will, in addition to satisfactory welding chemical composition of welding fume particles and have good welding characteristics in every respect. [1,2]

## 2. Performance of experiments

The amount and chemical composition of welding fumes is directly related to the composition of the coating and the electrode core. Each electrode manufacturer has its own recipe for a particular type of coating, which can result in the production of different amounts and different chemical composition of welding fume particles for the same type of electrode. On the way from the establishment of the electric arc to the bath, the coated electrode fulfills several functions, namely physical, electrical, metallurgical and economic function. The greatest role during welding with a coated electrode is played by the electrode coating, so the functions to be fulfilled by the coated electrode will also be listed as functions for electrode coating. [3,4,5]

Low-carbon steel was used as the base material in the experiment, and electrodes for welding low-carbon steels were used accordingly. No significant amounts of certain chemical elements are expected in the particles of welding fume, such as e.g. Cr, Mn, Mo, Ni, etc., which are otherwise found in high-alloy CrNi electrodes or alloy steels as alloying elements. Other chemical elements detected in the tests are considered regular impurities and steel companions. They appear in insignificant quantities and do not have a great impact on health and the environment. [1]

### *2.1. Equipment and procedure for conducting tests for the collection of welding fume particles*

Standard EN ISO 15011, in the first part, defines a method for taking evaporation samples and describes a laboratory procedure for determining the emission levels of corrections generated during arc welding and suggests possible analytical techniques or chemical analysis to determine the chemical composition of consumer-generated evaporation. The emission level and evaporation composition depend on the welding process, welding parameters, coating, work surface size, etc. In this case, the welding was performed by the MMAW process. Welding equipment consisting of a welding chamber, glass fiber filters, traction fan, welding apparatus, precision scales for measuring the weight of welding fume particles was used to collect samples of welding fume from these electrodes. Among the 2 models of the welding chamber containing the standard EN ISO 15011, model 1 was selected and made, shown in Figure 1. [6]

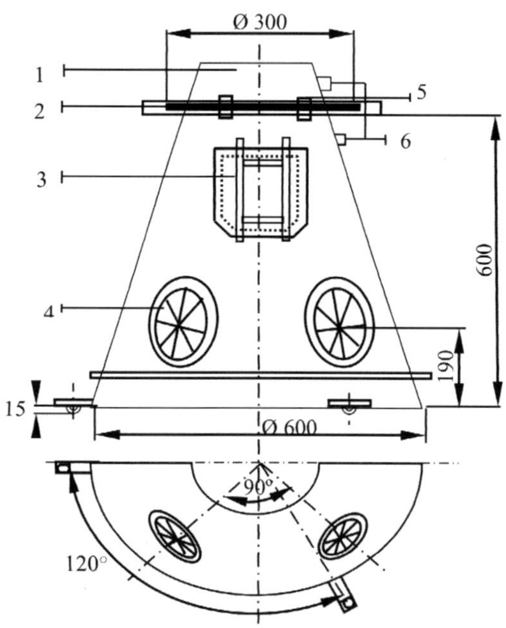


1. the upper part of the chamber
2. filter
3. protective glass and welding monitoring hole
4. welder arm openings
5, 6. chamber closing system

**Figure 1.** Model 1 of the welding chamber standard EN ISO1511

The test welding chamber for collecting welding fume should have a process chamber and a ventilation duct in which the filter is located. The process chamber should be suitable for the selected welding process, which means that it must enclose the entire MMAW process and be large enough to allow the collection of the complete emission of welding fumes. The flow of ventilation through the process chamber must enhance the thermal movement of air which means that the ventilation outlet should be above the welding process to enhance the natural flow of heated air, while the air inlet should be below the welding process. The air flow capacity of the pump or fan should be sufficient to capture the fully emitted gases from the chamber but should not affect the integrity of the process [6].

Sampling of welding fumes will be performed at the outlet of the evaporation chamber where the appropriate filter should be located. The filter should be placed at a distance from the welding process or air inlet, 5 × larger than the diameter of the air outlet from the chamber. All equipment that comes into contact with welding fumes must be made of inert metals to prevent absorption, accumulation or chemical reaction with fume components. PVC pipes are also suitable for use. It is not necessary to use heat-resistant materials or to use cooling devices. It is necessary to regularly clean the inside of the chamber. Also the ambient air in the room where the chamber is located should meet the quality requirements prescribed by ISO 8756. It is a procedure for adjusting the air quality with changes in temperature and humidity used in the report of results. The requirements of ISO 8756 should be taken into account in the measurement methods specified in this standard [6].

Two experiments were performed, measurement of welding fume production and chemical analysis of welding fume particles. For the needs of both experiments, it is necessary to collect welding fume particles. The collection of welding fume particles was performed in a welding chamber according to the procedure described in standard EN ISO 15011 [6,1].

### *2.2. Experiment and experimental results for determining the production of welding fumes*

Standard ISO 15011, describes the problems of the laboratory method for collecting samples of evaporation and gases generated during electric arc welding processes, for estimating the rate of exhaust gases (production) stipulates that three tests are made and the mean value is calculated. [6]

For the purposes of this experiment, three basic electrodes, of the same class E 42 4 B 32 H5, from three different manufacturers, whose chemical composition of pure weld metal is shown in Table 1, were selected.

**Table 1.** Chemical composition of pure weld metal

| % | C | Mn | Si | S | P |
|---|---|---|---|---|---|
| **B1** | 0,07 | 0,80 | 0,50 | 0,02 | 0,02 |
| **B2** | 0,07 | 1,00 | 0,60 | - | - |
| **B3** | 0,08 | 1,00 | 0,50 | - | - |

The data were taken from the catalog of three manufacturers. Electrodes from three manufacturers are marked B1, B2 and B3. Differences in the chemical composition of pure weld metal are minimal. The recommended welding current for the electrode diameter Ø 3.25 mm and this class of electrodes is I = (110 ÷ 140) A. [1]

During the experiment for measuring fume production, a total of nine tests were planned and performed, three tests for each type of electrode marked B1 (B11; B12; B13), B2 (B21; B22; B23) and B3 (B31; B32; B33). Standard electrodes with a diameter of 3.25 mm and a length of 350 mm were used. The welding current for all three electrodes was 120 A during the experiment. Fume production tests were performed by single electrode welding, recording the active duration of the arc burning. Filters were weighed before and after the test. [1]

The data were taken from the catalog of three manufacturers. Electrodes from three manufacturers are marked B1, B2 and B3. Differences in the chemical composition of pure weld metal are minimal. The recommended welding current for the electrode diameter Ø 3.25 mm and this class of electrodes is I = (110 ÷ 140) A. [1]

During the experiment for measuring fume production, a total of nine tests were planned and performed, three tests for each type of electrode marked B1 (B11; B12; B13), B2 (B21; B22; B23) and B3 (B31; B32; B33). Standard electrodes with a diameter of 3.25 mm and a length of 350 mm were used. The welding current for all three electrodes was 120 A during the experiment. Fume production tests were performed by single electrode welding, recording the active duration of the arc burning. Filters were weighed before and after the test. [1]

**Table 2.** Welding fume production

| Marking | Welding time t (s) | Particle weight M (mg) | Fume production P (mg/s) | Mean values productions P (mg/s) |
|---|---|---|---|---|
| **B11** | 93 | 610 | 6,56 | |
| **B12** | 98 | 560 | 5,71 | 6,37 |
| **B13** | 95 | 650 | 6,84 | |
| **B21** | 91 | 580 | 6,37 | |
| **B22** | 88 | 590 | 6,71 | 6,69 |
| **B23** | 90 | 630 | 7,00 | |
| **B31** | 94 | 550 | 5,85 | |
| **B32** | 89 | 510 | 5,73 | 6,03 |
| **B33** | 86 | 560 | 6,51 | |

A graphical representation of the results of welding fume production for three electrodes and three samples for each electrode, according to Table 2 is given in Figure 2 [1].

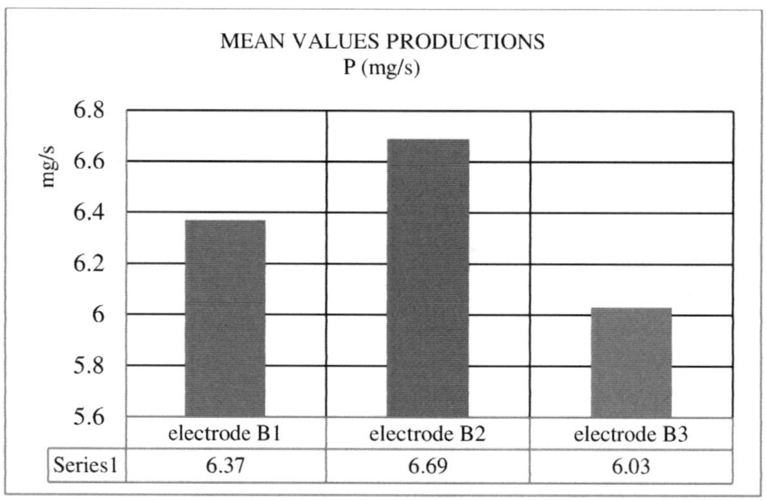


**Figure 2.** Production of welding fume for electrode E 42 4 B 32 H5 three manufacturers

The highest production of welding fume particles was generated by welding with electrodes of the manufacturer of electrodes B2 and amounts to 6.69 mg/s, then the manufacturer of electrodes B1 and amounts to 6.37 mg/s and the lowest production is generated by the manufacturer of electrodes B3 and amounts to 6.03 mg/s . Based on the values of the mean productions of welding fume particles, it can be concluded that there are differences in productions for different electrode manufacturers, but that in this case they are not so significant. However, when it comes to high consumption of coated electrodes in production, higher production of welding fume particles means both higher total emission and amount of particles in the workspace. [6,1]

*2.3. Experiment and results of chemical analysis of welding fume particles*

During the experiment, welding fumee particles were collected to examine the effect of the manufacturer's recipe on the chemical composition of the particles. In each sample, a mass of fume particles generated from three electrodes (a total of 27 electrodes) was collected on one filter, which gives a sufficient amount of particles for chemical analysis. A total of nine tests were performed, three tests for each type of electrode marked B1, B2 and B3. The results of chemical analysis of welding fujme particles are shown in Table 3. [1]

**Table 3.** Results of chemical analysis of welding fume particles

| Ordinal number | Chemical elements | Chemical composition of welding fume particles for the electrode E 42 4 B 32 H5 % Electrode designation and sample designation | | | | | | | | | | | |
|---|---|---|---|---|---|---|---|---|---|---|---|---|---|
| | | B1 | | | | B2 | | | | B3 | | | |
| | | B11 | B12 | B13 | $\overline{B1}$ | B21 | B22 | B23 | $\overline{B2}$ | B31 | B32 | B33 | $\overline{B3}$ |
| 1. | Ca | 1,99 | 1,94 | 1,96 | 1,963 | 1,66 | 2,18 | 1,88 | 1,91 | 1,68 | 1,75 | 1,71 | 1,713 |
| 2. | Mg | 0,30 | 0,30 | 0,33 | 0,31 | 0,33 | 0,36 | 0,30 | 0,33 | 0,16 | 0,15 | 0,16 | 0,156 |
| 3. | K | 1,35 | 0,14 | 0,16 | 0,55 | 0,81 | 0,13 | 0,10 | 0,35 | 0,10 | 0,10 | 0,12 | 0,11 |
| 4. | Na | 3,50 | 3,76 | 3,51 | 3,59 | 0,82 | 0,72 | 0,75 | 0,76 | 1,97 | 1,67 | 1,68 | 1,77 |
| 5. | Fe | 16,76 | 15,60 | 18,96 | 17,11 | 20,06 | 25,92 | 22,2 | 22,71 | 20,68 | 18,07 | 19,90 | 19,55 |
| 6. | Cu | 0,01 | 0,01 | 0,01 | 0,01 | 0,02 | 0,02 | 0,01 | 0,016 | 0,05 | 0,05 | 0,02 | 0,04 |
| 7. | Mn | 3,16 | 3,06 | 3,30 | 3,173 | 4,20 | 4,00 | 4,05 | 4,08 | 5,27 | 5,02 | 5,10 | 5,13 |
| 8. | Zn | 0,01 | 0,01 | 0,01 | 0,01 | 0,01 | 0,01 | 0,008 | 0,01 | 0,01 | 0,009 | 0,01 | 0,01 |
| 9. | Cr | 0,001 | 0,001 | 0,001 | 0,001 | 0,001 | 0,001 | 0,001 | 0,001 | 0,001 | 0,001 | 0,001 | 0,001 |
| 10. | Cd | 0,001 | 0,001 | 0,001 | 0,001 | 0,005 | 0,005 | 0,001 | 0,004 | 0,001 | 0,001 | 0,001 | 0,001 |
| 11. | Pb | 0,032 | 0,029 | 0,030 | 0,030 | 0,026 | 0,031 | 0,027 | 0,028 | 0,017 | 0,018 | 0,018 | 0,02 |
| 12. | Ni | 0,001 | 0,001 | 0,001 | 0,001 | 0,001 | 0,001 | 0,001 | 0,001 | 0,001 | 0,001 | 0,001 | 0,001 |
| 13. | Co | 0,001 | 0,001 | 0,001 | 0,001 | 0,002 | 0,001 | 0,001 | 0,0013 | 0,002 | 0,002 | 0,001 | 0,002 |

According to the procedure prescribed by the ISO 15011 standard, chemical analysis of fume particles was performed according to the AAS method (atomic absorption spectrometry) in the Laboratory for Chemical Analysis of the University "Džemal Bijedić" in Mostar. Chemical analysis of fume particles was performed for 13 chemical elements. The table also gives the mean values of the content of chemical elements for three samples of one electrode and are marked with: $\overline{B1}$, $\overline{B2}$ i $\overline{B3}$. All conditions of the experiments of this experiment were the same, so that the results were mutually comparable. [1]

Higher Mn content in fume particles can be found when basic electrodes are used because they contain ferrous manganese in the coating which serves as a deoxidizer. Manganese is prone to oxidation states so that its content in pure weld metal is 0.8%, table 1, and in fume particles its content ranges from 3.17% to 5.13%, for electrode B1 it is 3.17% , for electrode B2 is 4.08% and for electrode B3 the Mn content is 5.13%. [1]

From the aspect of Fe content in fume particles, electrode B2 contains the most Fe and is 22.71%, electrode B3 contains 19.55% Fe and electrode B1 has the lowest Fe content and is 17.11% [1].

From the aspect of Pb content in fume particles, electrode B3 contains the least Pb and is 0.02%, electrode B2 has 0.028% Cd and electrode B1 has the highest Cd content and is 0.03% [1].

From the aspect of Cu content in fume particles, electrode B1 has the lowest Cu content and is 0.01%, electrode B2 has 0.016% and electrode B3 has the highest Cu content and is 0.04% [1].

Other extremely harmful components, Cr, Ni, Co, are present in much smaller quantities and their contents are approximately the same for all three electrodes. As these are electrodes for welding general structural steels that do not contain alloying elements, such chemical elements are not expected in large quantities in fume particles and welding fumes generated from electrodes for welding general structural steels are much less dangerous, so the health of welders and environment.

The difference in the content of individual chemical elements for the three tested electrodes of different manufacturers exists, but analyzing the mean values of the percentages of 13 analyzed chemical elements, it can be concluded that these differences are not significan. [7,1]

## 3. Conclusion

The coated base electrode E 42 4 B 32 H5 for MMAW process from three different manufacturers of additional material of the same class, from the aspect of production and chemical composition of welding fume particles, was tested. The production of welding fume particles is different for each coated electrode from different manufacturers. The lowest production of welding fume particles was obtained by welding with electrode B3 and is 6.03 mg/s, and electrode B3 is the best in terms of production of welding fume. It was also found that the chemical composition of welding fume particles is different for each manufacturer of welding consumables. As it is welding of general structural steel with the mentioned coated basic electrode, in the particles of welding fume, as expected, there are no large amounts of the most harmful chemical elements, such as Cr, Mn, Ni, Mo, etc., which cause severe diseases. From the aspect of Mn content, which is very harmful to health, electrode B1 is the most favorable because it contains the least Mn in particles and amounts to 3.173%. From the aspect of Fe content in welding fume particles, because Fe causes long-term harmful effects on health but much less in relation to Cr, Mn, Ni, Mo, etc., the most favorable electrode is B1 because it has the lowest Fe content in particles and is 17.11% . Which electrode the consumer will choose depends on the set requirements that must be met. In any case, it is necessary to apply personal protective equipment and space protection with adequate ventilation equipment.

## References

[1] Razija B 2011 Bihać B&H Technical Faculty Doctoral thesis *Investigation of the optimal technological composition of the electrode coating from the aspect of minimizing welding fume*

[2] https://www.masinskiprirucnik.com/uporedni-*pregled-oznaka-elektroda-poznatih-proizvodjaca/* 19.10.2021

[3] Robert W M Jr, 2004 NY *Principles of welding*

[4] Omer P 1998 IP „SVJETLOST” D.D. Sarajevo *Zavarivanje*

[5] Klaus-J, Erhard R 2008 Leipzig Germany *Schweißtechnik*

[6] EN ISO 15011-1 april 2002 *Health and safety in welding and allied processes-Laboratory method for sampling fume and gases generated by arc welding*-Part 1: *Determinaton of emission rate and sampling for analysis of particulate fume*

[7] M. Tayyab Khan, B. Sc., E. A. Karpinski, M. Sc., P. Eng September 28 2018 *WELDING AND ALLIED PROCESSES, a guide to health hazards and hazard control measures*, ESDC Labour Program – Technical Services Unit

# Pressure stress modeling on expanded polystyrene materials using genetic programming

**Ermin Bajramović, Bahrudin Hrnjica, Redžo Hasanagić and Emir Bajramović**
University of Bihac, Faculty of Technical Engineering, Bosnia and Herzegovina

E-mail: ermin.bajramovic@unbi.ba

**Abstract**. In the production of expanded polystyrene, the standards are very high in terms of thermal, fire, dimensional, and mechanical characteristics, because each of the characteristics is a condition for achieving quality that allows competitiveness in the market. To ensure high-quality products, it is necessary to achieve optimal performance and product quality through carefully adjusted input parameters of production. Since the production of expanded polystyrene is specific in several ways, an experimental study was conducted in which the basic parameters affecting product quality were detected and through which a series of experiments were performed to prove product quality. Experimental research for this work was conducted on three types of expanded polystyrene samples whose purpose is to insulate floors exposed to pressure. The samples were made of the same material of different densities and aging times for which the pressure stress at a deformation of 10% was tested. After the experimental phase, the modeling of the output parameters was performed. Modeling involved the development of a model that describes a given problem and the obtained modeled values were analyzed and compared with the experimental one. The modeling method used genetic programming using the GPdotNET software package. The goal of modeling with the GpdotNET tool is to obtain a realistic model that would give the value of the compression stress at a deformation of 10% as an output variable in materials made of expanded polystyrene.

## 1. Introduction

Depending on the purpose, products made of expanded polystyrene (EPS) need to be subjected to various tests in order to prove their quality. The most common tests being performed are those on thermal, fire, dimensional and mechanical characteristics such as compression strength, bending, tensile and shear strength. These tests are usually being performed and controlled by experimental methods directly in laboratories using specialised equipment with appropriate software.

As experimental methods require high costs there is a need to determine the properties and indirect methods. Thus, the standard EN 13163 specifies indirect methods of mathematical calculation through input parameters as well as other forms of mathematical proofs.

Mathematical modeling is not excluded as one of the methods also if the obtained model gives a high degree of reliability for the problem under research.

With the development of computer science artificial intelligence also developed, which found its application in solving engineering problems. Artificial intelligence is a broad field in which methods based on biological evolution have their place. All methods based on the biological principle of survival and reproduction are called evolutionary algorithms (EA). By using them in different areas of research they have evolved into special methods for solving engineering problems, so they are also called evolutionary methods. One of the evolutionary methods is a genetic programming, which will be used in this paper.

Automatic problem solving with the help of computers is a central part of artificial intelligence; machine learning is wide area covered machine intelligence, as Turing called it (Turing 1950). Genetic programming and evolutionary programs were mentioned at the later stage. In the book Genetic Programming - On the Programming of Computers by Means of Natural Selection, 1992 John R. Koza talks about computer-aided programming through natural selection.[1]

The development of new methods was accompanied by the development of tools for modeling engineering problems, so the authors M. Jurković and B. Hrnjica wrote about modeling and optimisation of tool stresses in the drilling process using evolutionary algorithms [2]. B. Hrnjica talks in his paper about mathematical modeling of engineering problems using the method of genetic programming with GPdotNET tools. In his doctoral dissertation he used a numerical-evolutionary approach to determining the parameters of vessels' fracture mechanics under pressure [3, 4].

Moreover the authors E. Nezirević et al. in their paper Modeling of Pressing Forces using Genetic Programming demonstrated the application of genetic programming using the GPdotNET tool [5]. Authors B. Hrnjica and A. Danandeh Mehr explored new possibilities of application of optimised application for genetic programming [6]. R.Hasanagić presented modeling of fracture force on tensile strength of solid wood elements by genetic programming [7].

M. Kovacic and U. Župerl talked about the application of genetic programming in the steel industry to predict the diversity of serial processes and equipment that affected the properties, quality and final price of the product [8]. M. Semenkina et al. in their paper they wrote about the evolution of a model based on genetic programming [9]. All these papers indicate that genetic programming is being increasingly used in all areas of engineering and respective problems.

Many authors have used various methods in their research to analyse, model and simulate the mechanical properties of expanded polystyrene. Authors N. Tang et al. in Mechanical performance of polystyrene foam (EPS): Experimental and numerical analysis performed a series of tests to characterise the mechanical properties of polystyrene foam (EPS), including uniaxial compression tests, uniaxial stress tests and three-point bending tests [10].

L.G. Sun et al. In the paper Mechanical analysis and numerical simulation of EPS exterior decorative element, did perimental studies covering tensile, compression and bending tests of expanded polystyrene foam material used as a component of EPS decorative line thus obtaining the basic parameters of mechanics [11].

A. Krundaeva et al. in the paper Dynamic compressive strength and crushing properties of expanded polystyrene foam for different strain rates and different temperatures, performed static and dynamic tests of compression and crushing on expanded polystyrene foam to characterise the material at high stress rates in order to obtain a stress-strain curve for different temperatures and densities. They presented a methodology for processing experimental data for use in finite elements' modeling (FE).[12]

This paper aims to present the possibility of application and reliability of the genetic programming method using GPdotNET tools for creating GP models based on the obtained experimental values.

## 2. Materials and methods

### *2.1. Materials and equipment*

The experimental research was carried out on three types of products made of expanded polystyrene marked EPS 100, EPS 150, and EPS 200 of different densities. The aim of the test was to check the pressure stress S (k Pa) at a deformation of 10% as a function of input parameters. Density ρ (kg/m$^3$), aging time ω (d), relative humidity φ (%) and temperature T ($^\circ$C) were taken as input parameters.

EPS 100, EPS 150 and EPS 200 of different densities were depending on their purpose. The aim of the research was to test the pressure stress S (k Pa) at a deformation of 10% as a function of input parameters. The density ρ (kg/m$^3$), aging time ω (d), relative humidity φ (%) and temperature T ($^\circ$C) were taken as input parameters.

The test was performed in laboratory controlled conditions on highly sophisticated equipment, electromechanical press labelled ES10 connected to the monitoring program, digital scale, thermometer labelled PCE-T311 with probe for measuring temperature in the product and hygrometer labelled FMC with probe for measuring humidity in the product. The following methods were used in the preparation and performance of the experiment: normative, empirical, measurement method and experimental method. The test was performed on 35 samples of different densities within 15 days.

### *2.2. Genetic programming method*

Genetic programming (GP) is a method based on artificial intelligence with the ability to solve complex engineering problems generally. The real goal of genetic programming is to create computer programs that solve given problems.

The genetic algorithm creates a population of computer programs and through a number of generations improves individuals of the population using conventional genetic operators. The quality of an individual program is reflected in the efficiency in performing the desired task and the best rated program is presented as a result. The quality of the obtained solution is expressed by a predefined fitness function. Based on various parameters it determines the fitness of an individual on which the efficiency of the individual later depends. In evolutionary algorithms and thus genetic programming the most commonly used genetic operators are: reproductions, selections, crossovers and mutations.

In some cases, operators of permutation, swap, inversion, etc. also appear.

The GP algorithm can be controlled by nineteen parameters, as follows:

- two basic parameters that make up the population size and the maximum number of the algorithm generations performance,
- eleven auxiliary parameters that control the performance of the GP,
- six qualitative variables are selected differently from the performance of the genetic programming algorithm.

The optimal selection of GP control parameters is reflected both in the quality of the obtained results and in the performance of the algorithm. Quality results will not always be obtained if all parameters were included in the algorithm. Moreover, a fast and quality result will not be obtained if the number of individuals in the population was set large while solving a simple problem.

Optimal use of GP control parameters is a complex task depending primarily on the experience of the researcher.

## 3. Experimental research

Measurement and testing of expanded polystyrene products marked EPS 100, EPS 150, and EPS 200 yielded the results used as a substrate in this study, shown in table 1.

**Table 1.** Experimental test results of stress pressure.

| | **Input values** | | | | | **Output values** |
|---|---|---|---|---|---|---|
| No of exp. | Type | Density $\rho$ (kg/m$^3$) | Aging time $\omega$ (d) | Humidity $\varphi$ (%) | Temperature T (° C) | Pressure stress S (k Pa) |
| 1 | EPS 100 | 20 | 5 | 10.5 | 24.1 | 134.03 |
| 2 | | 20 | 6 | 10.5 | 24.5 | 138.44 |
| 3 | | 20 | 7 | 10.4 | 24.8 | 138.69 |
| 4 | | 20 | 8 | 10.4 | 24.8 | 143.09 |
| 5 | | 20 | 9 | 10.4 | 25.4 | 144.70 |
| 6 | | 20 | 10 | 10.4 | 25.5 | 146.83 |
| 7 | | 20 | 11 | 10.3 | 26.8 | 147.21 |
| 8 | | 20 | 12 | 10.3 | 27.3 | 149.48 |
| 9 | | 20 | 13 | 10.3 | 27.6 | 149.59 |
| 10 | | 20 | 14 | 10.3 | 27.8 | 149.02 |
| 11 | | 20 | 15 | 10.3 | 28.0 | 152.02 |

| | | | | | | |
|---|---|---|---|---|---|---|
| 12 | | 25 | 6 | 10.6 | 23.8 | 166.84 |
| 13 | | 25 | 7 | 10.6 | 24.5 | 166.07 |
| 14 | | 25 | 8 | 10.6 | 24.6 | 168.02 |
| 15 | | 25 | 9 | 10.5 | 24.8 | 167.78 |
| 16 | | 25 | 10 | 10.4 | 25.1 | 169.92 |
| 17 | EPS 150 | 25 | 11 | 10.4 | 25.3 | 171.27 |
| 18 | | 25 | 12 | 10.3 | 25.4 | 173.64 |
| 19 | | 25 | 13 | 10.3 | 25.7 | 176.91 |
| 20 | | 25 | 14 | 10.3 | 25.8 | 189.41 |
| 21 | | 25 | 15 | 10.3 | 25.8 | 188.04 |
| 22 | | 25 | 16 | 10.3 | 26.0 | 190.92 |
| 23 | | 25 | 17 | 10.3 | 26.1 | 194.25 |
| 24 | | 30 | 8 | 10.6 | 24.1 | 200.17 |
| 25 | | 30 | 9 | 10.6 | 24.5 | 199.41 |
| 26 | | 30 | 10 | 10.6 | 24.1 | 202.98 |
| 27 | | 30 | 11 | 10.5 | 24.8 | 202.42 |
| 28 | | 30 | 12 | 10.5 | 25.3 | 204.30 |
| 29 | EPS 200 | 30 | 13 | 10.4 | 25.8 | 210.24 |
| 30 | | 30 | 14 | 10.4 | 26.1 | 209.22 |
| 31 | | 30 | 15 | 10.3 | 25.9 | 211.49 |
| 32 | | 30 | 16 | 10.3 | 26.0 | 214.82 |
| 33 | | 30 | 17 | 10.3 | 26.3 | 214.46 |
| 34 | | 30 | 18 | 10.3 | 26.3 | 218.08 |
| 35 | | 30 | 19 | 10.3 | 26.6 | 223.51 |

The table shows the values of input parameters, sample density (kg/m$^3$), sample aging time (d), humidity in the sample before the test (%),sample temperature before the test ($^o$ C) as well as output pressure stress values (k Pa) for all 35 examined samples over a period of 15 days.

The values of input and output parameters varied in the following intervals:

- density ρ, in the interval from 20 ÷ 30 (kg/m$^3$)
- aging time ω, in the interval from 5 ÷ 19 (d),
- relative humidity φ, in the interval from 10.3 ÷ 10.6 (%),
- temperature T, in the interval from 23.8 ÷ 28.0 ($^o$ C),
- pressure stress S, in the interval from 134.03 ÷ 223.51 (k Pa).

The organisation of the data was shown by scatter diagrams that showed the relation between the input and output parameters as well as the input parameters with each other.

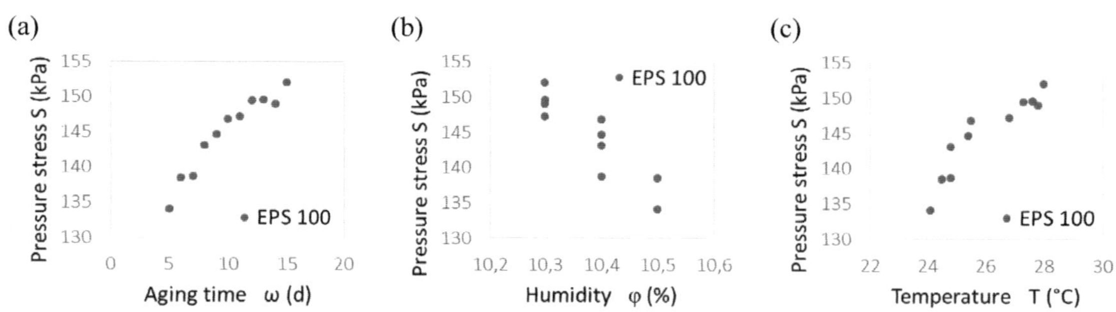


**Figure 1.** Scatter diagram EPS 100: (a) pressure stress-aging time; (b) pressure stress-humidity; (c) pressure stress-temperature.

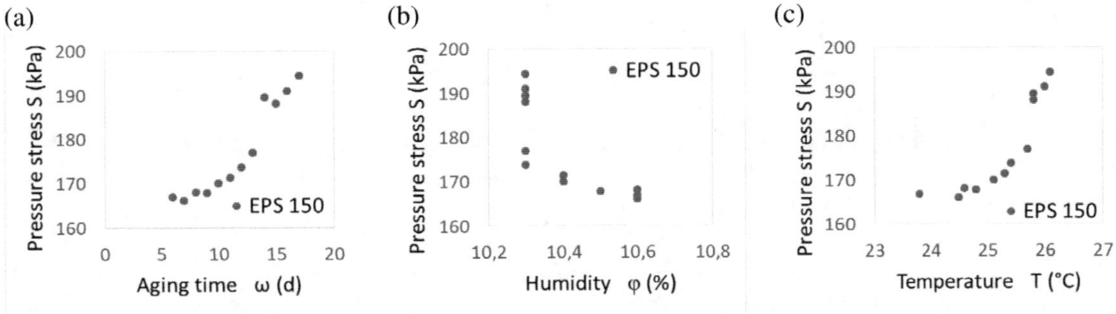


**Figure 2.** Scatter diagram EPS 150: (a) pressure stress-aging time; (b) pressure stress-humidity; (c) pressure stress-temperature.

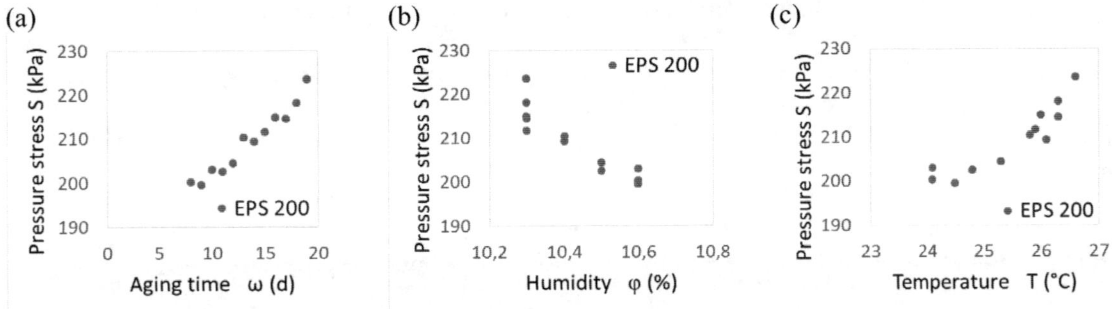


**Figure 3.** Scatter diagram EPS 200: (a) pressure stress-aging time; (b) pressure stress-humidity; (c) pressure stress-temperature.

The diagrams shown in figures 1, 2 and 3 showed the pressure stress as a function of aging time, relative humidity and temperature from which it can be seen that there was an interdependence.

The diagram in figure 4 showed the interdependence of the input parameters temperature as a function of aging time, temperature as a function of relative humidity and relative humidity as a function of aging time for all tested samples.

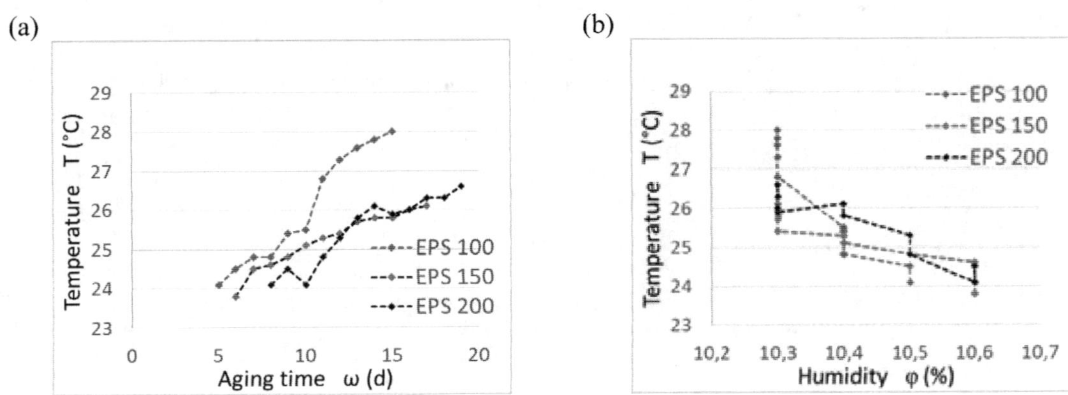

(c)

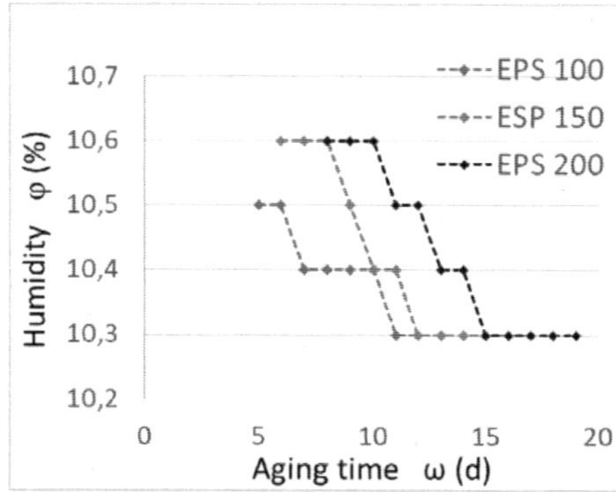


**Figure 4.** Diagram of the parameters' interdependence: (a) temperature-aging time; (b) temperature-humidity; (c) humidity-aging time.

For a complete picture of the interdependence, a correlation analysis was performed for all input parameters with output parameters as well as input parameters with each other. Correlation analysis for all examined types of EPS showed that the input parameters were in a very significant correlation with each other as well as with the output parameters.

Thus, the regression coefficient was for EPS 100 r above 0.90563, for EPS 150 r in the range from 0.78340 to 0.97080 while for EPS 200 r it was above 0.90942, which showed that over 90% of the variability was achieved by influence and by the action of input parameters (variables).

## 4. Modeling by GPdotNET tool

In GPdotNET-V5 is a modeling tool currently in the fifth version indicating that it is constantly being improved to make it as easy as possible for use. It contains a set of very complex algorithms for normalising input parameters, algorithms for manipulating binary and multiclass types of input parameters and algorithms to manage missing data.

The data management module is one of the most important modules in GPdotNET. The quality of the obtained solution will depend on which parameters and from which set of functions the genetic programming algorithm will look for solutions. The obtained solution represents a mathematical model that is usually nonlinear.

GPdotNET consists of several user interfaces into which modeling data is to be entered. For GP modeling it is necessary to perform data selection and randomly select data for model training and model testing. The first step is to load test and training data followed by determining the mathematical operations with which the program will work. The most important thing is to set up a data management module in which the parameters of genetic programming are entered based on which the program will offer the best solution, shown in figure 5.

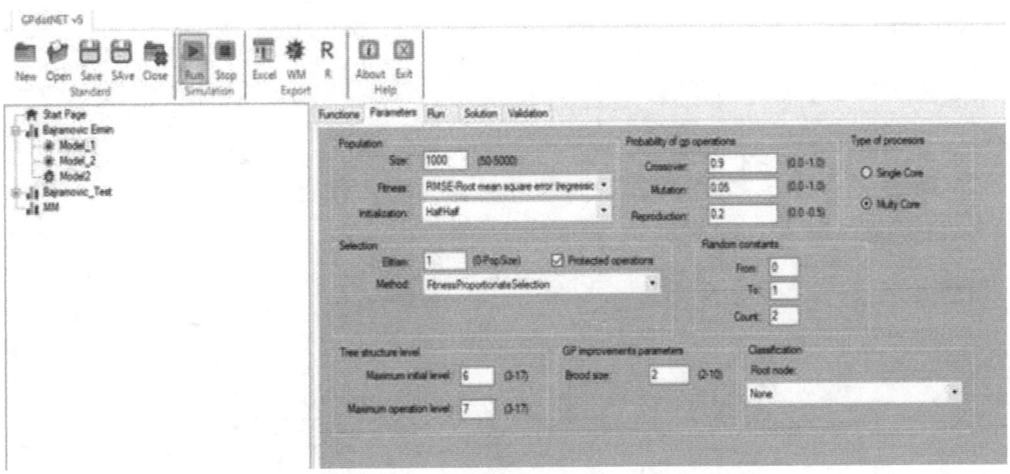

**Figure 5.** GPdotNET interface (input of parameters).

After selecting the functions, entering the input parameters and defining the variable to be modeled, the algorithmic programming program is to be started following the process of obtaining the optimal model that will describe the given problem best, figure 6.

A change in fitness function and prediction as well as a change in the correlation coefficient gives a picture when the model is good. Sometimes it takes more time and attempts with different functions and parameters to get the appropriate model for a given problem.

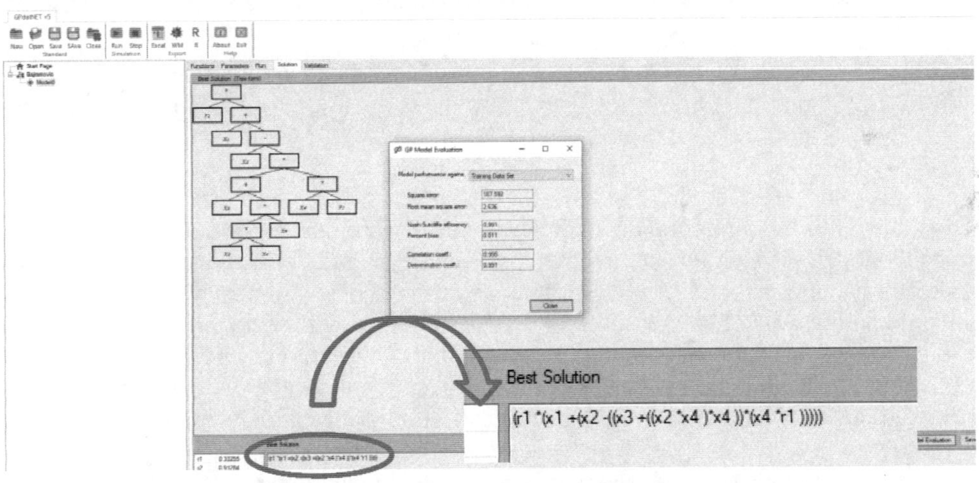


**Figure 6.** Interface display of the best model solution in GPdotNET.

IOP Conf. Series: Materials Science and Engineering 1208 (2021) 012013 doi:10.1088/1757-899X/1208/1/012013

The resulting GP model shown as an expression tree looks as follows:

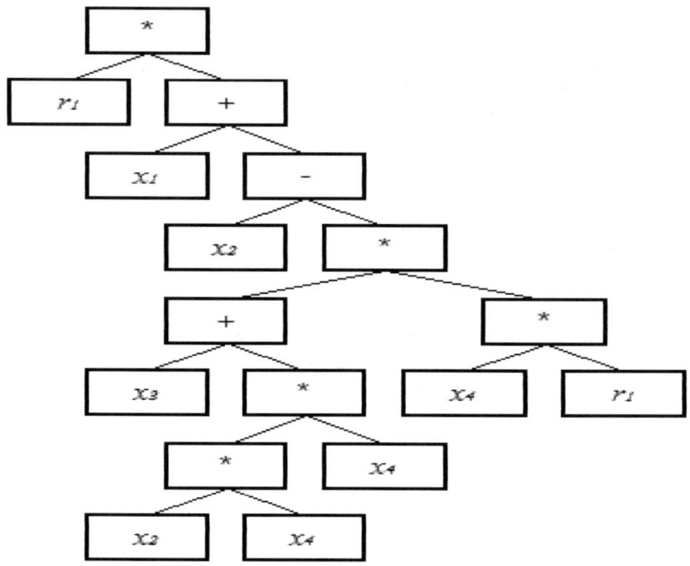


**Figure 7.** GP model shown as expression tree.

Taking into account that the constants were $r_1 = 0.33255$ and $r_2 = 0.91284$, the GP model in figure 7 after the transformations of the initial expression was reduced to the form:

$$S(\rho,\omega,\varphi,T) = 138.44 + 28.29\left[\rho + 1/13(-6+\omega) - 0.08\left(\frac{3.33(-10.3+\varphi)+0.004}{(-6+\omega)(-23.8+T)^2}\right)(-23.8+T)\right] \quad (1)$$

The GP pressure stress model obtained for a particular GP configuration depends on the input parameters, the set of functions and the parameters under which the algorithm would be processed. For the above GP model, the addition, subtraction, multiplication and division functions were used.

The default parameters for which the program found the best solution were: population size 1000, mutation probabilities 0.05, initial level of tree structure 5 and working level 6, and number of generations 100. Change of fitness function showed that 54 iterations were needed and last change on fitness function occurred in the 48th iteration with a stated value of 969.95 out of a maximum of 1000. It also showed in which iteration the GP model that best describes the given problem was obtained.

The parameters can have a wide range of values, which can significantly affect the operation and quality of the solution. On the other hand, if the selection pressure is too low, time is spent on useless iterations because the convergence is too slow in that case. The probability of mutation is a key parameter allowing GP new genetic material to be introduced into the population, and thus the possibility of obtaining better solutions. Therefore, the number of generations should be positively correlated with the quality of the solution and negatively with the speed of the program performation.

## 5. Results analysis and discussion

Based on the obtained mathematical model, the pressure stress ($S_{GP}$) was calculated for all input parameters in tables 2 and 3.

**Table 2.** Experimental and GP modeling results for Training data.

| Training data | | | |
|---|---|---|---|
| No | $Y_E$ ($S_E$) | $Y_{GP}$ ($S_{GP}$) | $Y_E$ - $Y_{GP}$ ($S_E$ - $S_{GP}$) |
| 1 | 149.02 | 150.85 | -1.83 |
| 2 | 218.08 | 219.30 | -1.22 |
| 3 | 169.92 | 174.38 | -4.46 |
| ... | ... | ... | ... |
| ... | ... | ... | ... |
| 25 | 147.21 | 148.00 | -0.79 |
| 26 | 200.17 | 198.70 | 1.47 |
| 27 | 149.48 | 148.98 | 0.50 |

**Table 3.** Experimental and GP modeling results for Testing data.

| Testing data | | | |
|---|---|---|---|
| No | $Y_E$ ($S_E$) | $Y_{GP}$ ($S_{GP}$) | $Y_E$ - $Y_{GP}$ ($S_E$ - $S_{GP}$) |
| 1 | 211.49 | 213.79 | -2.30 |
| 2 | 143.09 | 142.03 | 1.06 |
| 3 | 166.07 | 167.33 | -1.26 |
| 4 | 202.98 | 203.05 | -0.07 |
| 5 | 134.03 | 135.82 | -1.79 |
| 6 | 176.91 | 181.49 | -4.58 |
| 7 | 190.92 | 187.45 | 3.47 |
| 8 | 167.78 | 171.74 | -3.96 |

It can be seen from the above that the experimental results and the results obtained by the method of genetic programming were close, and the maximum result was -5.91, which is an error of less than 2.75%. The results obtained experimentally and by genetic programming can be presented in the form of graphs showing the pressure stress distribution in figure 8. The graphs also show close experimental values and GP model values.

(a)

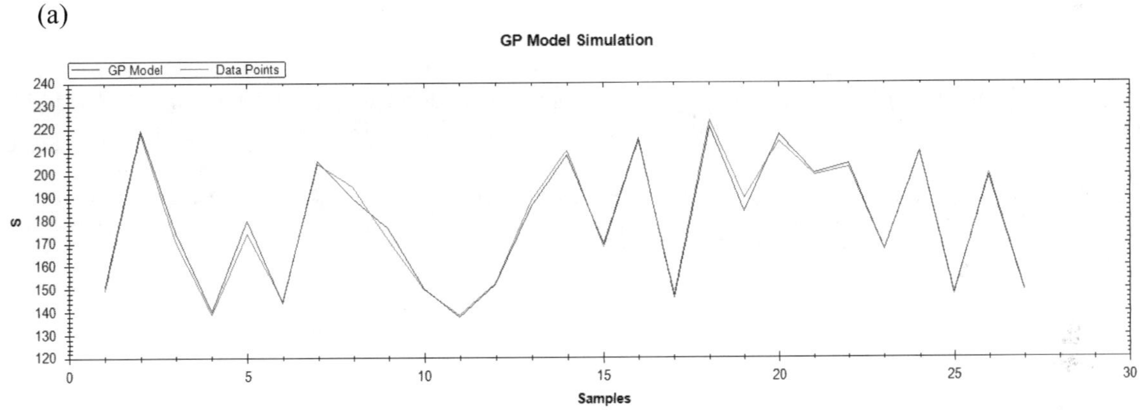


(b)

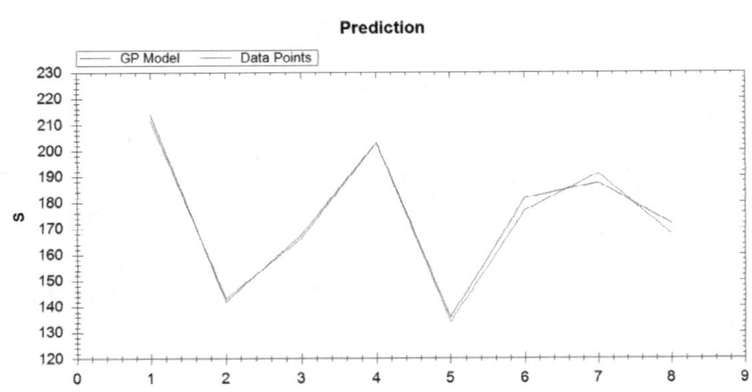


**Figure 8.** Comparison of experimental and GP modeling results of (a) training data ($Y_E$-$Y_{GP}$); (b) testing data ($Y_E$-$Y_{GP}$).

From the diagram shown in figure 8 it can be seen that the measured pressure stress by experimental methods (blue line) is close to the values of pressure stress calculated according to GP model (1) (red line).

(a) (b)

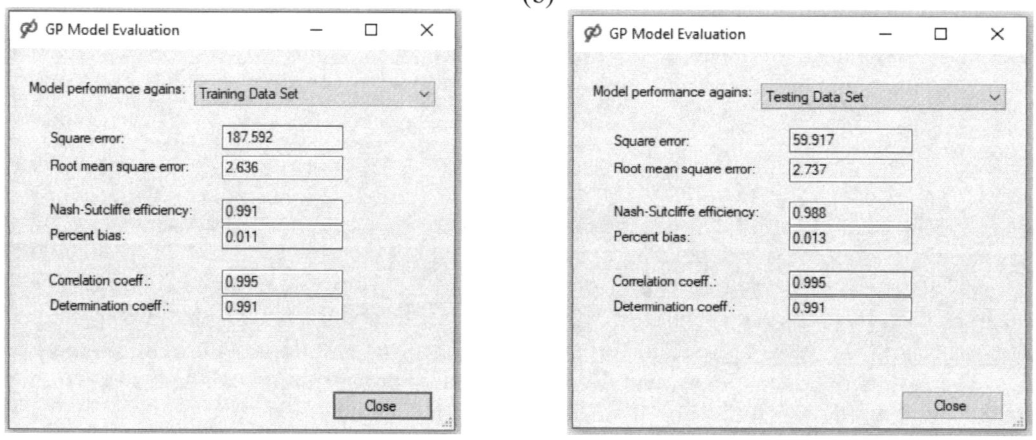


**Figure 9.** Performance models for (a) training and (b) testing data.

After the evaluation of the model, figure 9 shows the basic statistical indicators of the model, where it can be seen that the root mean square error for the trained set was 2.636 while for the validation set was 2.737. The correlation coefficient for the trained and validation set was the same and it was 0.995 and the coefficient of determination 0.991 which showed that the model was well trained.

All the above indicates that the model describes very well the given problem in the observed area. The deviations of the absolute values of the experimental results pressure stress in relation to the results obtained by genetic programming range in intervals in the validation set $S_{GP}$=0.07÷4.58 [k Pa] or in the trained set $S_{GP}$=0.11÷5.91 [k Pa].

From this it can be seen that the experimental results and the results obtained by the method of genetic programming were close and the maximum result was -4.58 for the trained set and -5.91 for the validation set, which was an error of less than 2.75%.

## 6. Conclusion

In this paper, experimental research of products from expanded polystyrene was performed, and several input/output parameters were detected. Due to the specificity of the product, the experiments were performed in the laboratory and several parameters were monitored. The experimental part included the measurement of pressure stress for several types of products with different input parameters (density, aging time, relative humidity and temperature). The obtained experimental results were taken as input data for modeling. The modeling phase involved the modeling of pressure stresses during deformation of 10%. Pressure stress modeling was done using the GPdotNET software package. The results of the obtained GP model show close values to the experimental ones. The correlation coefficient between the model and the training set was 0.995 and the correlation coefficient between the model and the validation set was also 0.995 which gave a high degree of reliability to this model. Given this correlation coefficient, the model obtained by the method of genetic programming showed that it was reliable for the researched area.

It can be seen from the presented that genetic programming is a very efficient method of model making, and unlike classical regression analysis there are no restrictions in terms of the degree of polynomials. This method of genetic programming is based on the concept of computer programs and represents an intuitive and simple way of building a model over experimental data that can then be used in future research as a starting point.

## References


[1] Koza J R 1992 *Genetic Programming: On the programming of computers by means of natural selection* (London: Cambridge Massachusetts)

[2] Jurković M, Hrnjica B 2007 Modeling and optimization of the tool stress in driling process by evolution algorithms *3th International conference on manufacturnig science and education (MSE)*

[3] Hrnjica B 2016 Matematičko modeliranje inženjerskih problema korištenjem metode genetskog programiranja *VI Međunarodni naučni skup o ekonomskom razvoju i životnom standardu (EDASOL 2016)*

[4] Hrnjica B 2014 *Numeričko-evolucijski pristup određivanja parametara mehanike loma posuda pod pritiskom* Doktorska disertacija (Bihać: Univerzitet u Bihaću)

[5] Nezirević E, Hrnjica B and Hodžić A 2016 Modeling of Pressing Forces using Genetic Programming *20th International Research/Experts Conference (TMT 2016)*

[6] Hrnjica B and Danandeh Mehr A 2019 Optimized genetic programming applications *Advances in Medical Technologies and Clinical Practice* DOI:10.4018/978-1-5225-6005-0

[7] Hasanagić R 2018 Modeling and prediction of fracture force to tighten solid wood elements by genetic programming *Tehnika* **73** 653–7 DOI:10.5937/tehnika1805653H

[8] Kovačič M and Župerl U 2020 Genetic programming in the steelmaking industry *Genetic Programming and Evolvable Machines* **21** 99–128 DOI: 10.1007/s10710-020-09382-5

[9] Semenkina M, Burlacu B and Affenzeller M 2020 Genetic programming based evolvement of models of models *Computer Aided Systems Theory – EUROCAST 2019* 387–95 DOI:10.1007/978-3-030-45093-9_47

[10] Tang N, Lei D, Huang D and Xiao R 2019 Mechanical performance of polystyrene foam (EPS): Experimental and numerical analysis *Polymer Testing* **73** 359-65 DOI:10.1016/J.POLYMERTESTING.2018.12.001

[11] Sun L G, Du C B and Wang D P 2013 Mechanical analysis and numerical simulation of EPS exterior decorative element *Journal of Building Materials* **16**-4 637-41+56 DOI:10.3969/j.issn.1007-9629.2013.04.015

[12] Krundaeva A, De Bruyne G, Gagliardi F and Van Paepegem W 2016 Dynamic compressive strength and crushing properties of expanded polystyrene foam for different strain rates and different temperatures *Polymer Testing* **55** 61-8 DOI:10.1016/j.polymertesting.2016.08.005

[13] Bajramović E and Đuzelić R 2018 Influence of production and products of Expanded Polystyrene (EPS) on leaf and living organisms *6th International Scientific and Professional Conference "5 June - World Environmental Day" Proc. 6-2018* ISSN 2566-4530

[14] Awol T 2012 A Parametric Study of Creep on EPS Geofoam Embankments, *Norwegian*

[15] European standrard EN 13163 2012 Thermal insulation products for buildings – Factory made expanded polystyrene (EPS) products – Specification

IOP Conf. Series: Materials Science and Engineering 1208 (2021) 012014 doi:10.1088/1757-899X/1208/1/012014

# Measurement of welding fume production in welding with high-alloy electrodes

**Razija Begić**
Faculty of Technical Engineering, University of Bihać, Bosnia and Herzegovina

E-mail: razijabegic@yahoo.co.uk

**Abstract**. Due to their good structural properties, stainless Cr-Ni steels have a very wide application in various branches of technology. During the welding of stainless Cr-Ni steels with high-alloy coated electrodes, welding fumes of complex chemical composition are generated, which is very harmful for welders and the environment. For the purposes of this experiment, two variants of one rutile Cr-Ni commercial electrode, designated E 23 12 2 LR 12, were designed and fabricated. Higher production of welding fume particles also means greater danger to humans and the environment. In order to show the influence of the base material on the production of welding fume particles, an experiment for measuring the production of welding fume particles was performed in which two different steels were used as the base material, general structural steel S235JRG2 and stainless steel X6CrNiTi18.10.

## 1. Introduction

Welding fumes, which are a complex mixture of fumes and gases, are an inevitable harmful emission in most electric arc welding processes. The largest amount of welding fumes is generated from the electrode coating and it also represents the place where the reduction of the amount of welding fumes can be most influenced. For these reasons, the assessment of electrode quality, from the aspect of health protection, can be performed by measuring the production of welding fume particles. Measurement of the production of welding fume particles, in addition to the stated reasons, is performed in order to determine the influence of the coating character (rutile, basic, cellulose, oxide), the influence of current change, the influence of base material and other parameters on welding fumeproduction. The production of welding fumes, their chemical composition, as well as the quantities of harmful ingredients in the particles, are subject to adherence to the permitted exposure limits and norms prescribed by legal norms. Public health institutes periodically perform measurements for all economic entities, whose production results in harmful emissions into the environment. [1]

Additional materials, in this case coated electrodes for MMAW process, can be competitive on the market only if their harmful emissions in terms of quantity and chemical composition meet the regulations. Only when the set environmental requirements are met, it is possible to move on to the welding characteristics of the additional material. [1]

The experiment for measuring fume production can be reported in several ways, and the results can be expressed using the weight of welding fume particles per unit time, the weight of welding particles per unit mass of deposit and the weight of welding fume particles per unit mass of electrode. measurements of welding fume particle production per unit time (mg/s) [1,2].

## 2. Description of an experiment to measure the production of welding fume particles

The prerequisite for performing this experiment was to create a functional welding chamber for collecting welding fume particles in which welding will be performed and welding fume particles will be collected, which was done according to EN ISO 15011 and described in the preliminary experimental part of the doctoral thesis [2,3 ].

For the purposes of this experiment, two electrodes of the Cr-Ni type were used, made in the laboratory of the factory for the production of additional welding material "ELEKTRODA ZAGREB", according to a predetermined composition of the lining and the core. The electrodes marked A and B represent two variants of one commercial electrode, marked E 23 12 2 LR 12 in which, within the allowed limits from the aspect of weld quality, the contents of the most harmful chemical elements Mn, Mo, Cr and Ni are varied. [2]

Electrode E 23 12 2 LR 12 is a rutile coated electrode intended for welding identical and similar stainless steels, as well as for mutual welding of dissimilar steels in the REL welding process. It is suitable for plating non-alloy and low-alloy steels. The low-carbon Cr – Ni – Mo steel weld is stable at temperatures up to 350 °C. The weld metal structure is austenite with delta ferrite. The chemical composition of pure weld metal is given in Table 1. [4]

**Table 1.** Chemical composition of pure weldmetal of a commercial electrode E 23 12 2 LR 12

| Chemical element | C | Mn | Si | Cr | Ni | Mo |
|---|---|---|---|---|---|---|
| % | ≤ 0,03 | 0,9 | 0,8 | 22,5 | 12,5 | 2,7 |

The recipe of the chemical composition of the electrode coating (internal name of the electrode manufacturer, the so-called formula), raw materials and their origin used for electrode coating differ from manufacturer to manufacturer and are known only to the electrode manufacturer, Table 2. [2]

**Table 2.** Orientation composition of the electrode coating

| Material type | Weight share % |
|---|---|
| Rutile | 25 |
| Silicates (feldspar, quartz…) | 25 |
| Fe alloys (FeCr, FeMn, FeSi…) | 8 |
| Metal powder (Ni, Cr…) | 25 |
| Carbonates (limestone) | 10 |
| Other (binder, liquid silicates) | 7 |
| Σ | 100 % |

Two types of wire, marked 316 L and 308 L, were used to make the core of the laboratory electrodes. Standard EN ISO 14343 prescribes the quality of the chemical composition of the drawn wire from which the electrode core is made. The chemical composition of the wire for the electrode core by the wire manufacturer "Rodacciai" Italy, is shown in Table 3 [5].

IOP Conf. Series: Materials Science and Engineering 1208 (2021) 012014 doi:10.1088/1757-899X/1208/1/012014

**Table 3.** Chemical composition of core wire according to the manufacturer

| Chemical composition of the wire according to the manufacturer % | | | |
|---|---|---|---|
| r.br. | Chemical element | 316L | 308L |
| 1. | C | 0,011 | 0,012 |
| 2. | Mn | 1,680 | 1,630 |
| 3. | Si | 0,080 | 0,080 |
| 4. | S | 0,011 | 0,012 |
| 5. | P | 0,012 | 0,011 |
| 6. | Cr | 18,240 | 20,160 |
| 7. | Ni | 11,280 | 10,100 |
| 8. | Mo | 2,590 | 0,040 |
| 9. | Al | 0,003 | 0,004 |
| 10. | Cu | 0,080 | 0,040 |
| 11. | Co | 0,027 | 0,024 |
| 12. | N | 0,048 | 0,044 |
| 13. | Ti | 0,005 | 0,005 |
| 14. | Nb | 0,010 | 0,010 |
| 15. | B | 0,001 | 0,001 |
| 16. | W | - | 0,010 |
| 17. | V | 0,039 | 0,037 |

The most influential alloying chemical elements in electrodes A and B are: Mn, Mo, Cr and Ni. The percentages of chemical elements Mn, Mo, Cr and Ni in wire, coatings and electrodes A and B, as well as the percentages of chemical elements Mo, Cr and Ni in pure weld metal for electrodes A and B are given in Table 4.

**Table 4.** Contents of main alloying chemical elements in electrodes A and B

| electrode A<br>*wire 316 L* | | elecktrode B<br>*wire 308 L* | |
|---|---|---|---|
| *mass fraction (w) of the chemical element in the coating %* | | | |
| Mn | 1,49 | Mn | 1,40 |
| Mo | 1,04 | Mo | 4,31 |
| Cr | 20,82 | Cr | 18,14 |
| Ni | 4,97 | Ni | 6,44 |
| *mass fraction (w) of chemical element in the wire - core%* | | | |
| Mn | 1,68 | Mn | 1,63 |
| Mo | 2,59 | Mo | 0,04 |
| Cr | 18,24 | Cr | 20,160 |
| Ni | 11,28 | Ni | 10,100 |
| *mass fraction (w) of chemical element in electrode%* | | | |
| Mn | 1,62 | Mn | 1,69 |
| Mo | 1,96 | Mo | 1,62 |
| Cr | 19,29 | Cr | 18,99 |
| Ni | 8,71 | Ni | 8,49 |
| *mass fraction (w) of chemical element in the weld%* | | | |
| Mo $\approx$ 2,6 | | | |
| Cr $\approx$ 22,5 | | | |
| Ni $\approx$ 11,5 | | | |

As part of the research of welding fumes in the welding of high-alloy steels, experiments were performed to measure the production of welding fume particles depending on the change in current and base material, as well as experiments to determine the chemical composition of welding fume particles [2].

*2.1. Results of measurements of welding fume particle production of high-alloy electrodes*

According to the literature, the basic material participates in the formation of welding fume particles with about 10 to 20% in relation to the electrode, which participates in the formation of welding fume particles with about 80 to 90% [6]. In order to show the influence of the base material on the production and chemical composition of welding fume particles, an experiment for measuring the production of welding fume particles was planned and performed. Carbon steel, S235JRG2, of non-guaranteed chemical composition with limited impurity content and partially prescribed content of C, Si and Mn, is used for welded and riveted structures in cases where there is no danger of brittle fracture as well as at temperatures up to 100 °C. Stainless steels are characterized by high resistance to corrosion, which is achieved by the formation of a thin, hard and compact surface layer, which protects the metal from further corrosion. In addition to high corrosion resistance, these steels are also fireproof, so they are used to work at temperatures above 550 ° C, instead of low-alloy steels. Stainless steels are high-alloy, and in addition to at least 10.5% chromium, they also contain nickel, molybdenum, copper, titanium, silicon, manganese and niobium. The chemical composition of steel X6CrNiTi18.10 is given in Table 5.

**Table 5.** Chemical composition of steel X6CrNiTi18.

| Chemical element | C | Si | P+S | Cr | Mo | Ni | Mn | The others |
|---|---|---|---|---|---|---|---|---|
| content % | 0,06 | 2,0 max | 0,045 % P<br>0,030 % S | 17÷19 | - | 9÷11,5 | 2,0 max | Ti<br>5×C min |

The production of welding fume particles was measured depending on the change in current and the change in the base material. The recommended welding current for electrode E 23 12 2 LR 12, diameter 3.20 mm is (80 ÷ 110) A. The test was performed for three different current values, 80, 95 and 110 A. [4] The plan of the experiment according to which twelve tests were performed, the measured values during the experiment and the obtained values of the production of welding fume particles are shown in Table 6. [2]

**Table 6.** Experimental plan and values of welding fume particle

| *s.n test* | *Label electrode wire* | *Basic material* | *Test mark* | *Current A* | *The duration of the arc s* | *Particle weight M (mg)* | *Production particle P (mg/s)* |
|---|---|---|---|---|---|---|---|
| 1. | | | A1 | 80 | 85 | 290 | 3,41 |
| 2. | | S235JRG2 | A2 | 95 | 69 | 252 | 3,65 |
| 3. | A | | A3 | 110 | 59 | 228 | 3,86 |
| 4. | 316 L | | A4 | 80 | 84 | 228 | 2,71 |
| 5. | | X6CrNiTi18.10 | A5 | 95 | 71 | 253 | 3,56 |
| 6. | | | A6 | 110 | 61 | 291 | 4,77 |
| 7. | | | B1 | 80 | 85 | 256 | 3,01 |
| 8. | | S235JRG2 | B2 | 95 | 70 | 244 | 3,48 |
| 9. | B | | B3 | 110 | 60 | 288 | 4,85 |
| 10. | 308 L | | B4 | 80 | 81 | 237 | 2,95 |
| 11. | | X6CrNiTi18.10 | B5 | 95 | 69 | 284 | 4,12 |
| 12. | | | B6 | 110 | 59 | 292 | 4,95 |

The calculated values of welding fume particle production, for three values of welding current, 80, 95 and 110 A, two basic materials S235JRG2 and X6CrNiTi18.10 and for electrodes A and B, are shown in the diagram, Figure no. 1. On the diagram, certain laws of change in the value of production of welding fume particles can be observed.

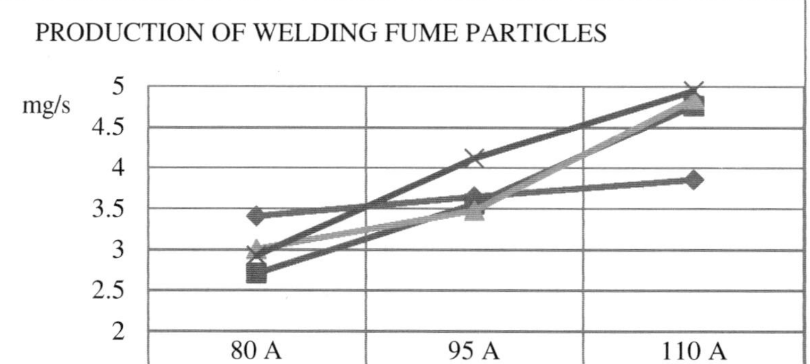


| | 80 A | 95 A | 110 A |
|---|---|---|---|
| S235JRG2-A | 3.41 | 3.65 | 3.86 |
| X6CrNiTi 18.10-A | 2.71 | 3.56 | 4.77 |
| S235JRG2 -B | 3.01 | 3.48 | 4.85 |
| X6CrNiTi 18.10-B | 2.93 | 4.12 | 4.95 |

**Figure 1.** Welding fume particle production

All values of welding fume production show a growth trend with increasing current strength. The highest production of welding fume particles is generated when welding X6CrNiTi18.10 alloy steel with electrode B at a current strength of 95A in the value of 4.09 mg/s and 110A in the value of 4.95 mg/s. The values of fume production for current 80A are higher for both electrodes when welding carbon steel, while the values of production of welding fume for current of 110A are higher when welding stainless steel. This result can be explained by the presence of alloying elements in high-alloy steel, which evaporate at higher temperatures and generate welding fume particles. Alloying chemical elements from the base material evaporate harder and less because the temperature of the welding bath is significantly lower than the temperature of the electric arc. The alloying elements in the electrode (coating and wire core of the electrode) at the moment of transmission through the electric arc, where the highest temperature is, evaporate intensively and the influence of the electrode on the production and chemical composition of welding fume is much greater.

## 3. Conclusion

The obtained results defined the influence of variables on the obtained results. It was found that the production of welding fume does not depend significantly on the type of base material, in welding with lower currents. All values of welding fume production show a growth trend with increasing current strength. The fume production values for current 80 A are higher for both electrodes, A and B, when welding carbon steel, while the values of welding fume production for current 110 A are higher when welding stainless steel. Higher temperatures that develop when welded with higher currents cause evaporation of alloying chemical elements that are known to have higher evaporation points. From this aspect, it is recommended to work with lower welding currents within the limits recommended by the manufacturer of the welding consumable. The production of welding fume particles in the value of 4.95 mg / s, when welding steel X6CrNiTi18.10 with electrode B, represents the third degree of harmful emission of a total of 4 degrees, which requires the use of a certain type of ventilation. The production of welding fume particles acquires the right meaning when expressed in mg/m3 of closed working space. Higher production of welding fume particles does not mean a greater danger to the health of welders. The real danger is posed by harmful ingredients in welding fume particles that can be determined by various methods of chemical analysis. The electrode has the greatest influence on the production and chemical composition of welding fume particles, namely the electrode coating, which participates in the generation of welding fume with (80 to 90)%. From this aspect, the activities on the development of the electrode coating are always present, significant and interesting for reducing the production and harmful components in the welding fume particles.

**References**

[1] Razija B, Azra I 2015 Dubrovnik *Methods for measuring the production of welding fume particles*, 10th International Scientific Conference on Production Engineering, Development and modernization production
[2] Razija B 2011 Bihać Doctoral thesis Faculty of Mechanical Engineering, University of Bihać, *Investigation of the optimal technological composition of the electrode coating from the aspect of minimizing welding fume*
[3] EN ISO 15011-1, april 2002 *Health and safety in welding and allied processes-Laboratory method for sampling fume and gases generated by arc welding*-Part 1: *Determinaton of emission rate and sampling for analysis of particulate fume*
[4] *Catalog of additional factory material* 2018 „ELEKTRODA ZAGREB“
[5] Rodacciai 05.2008 Italia *Certificato di collaudo*
[6] *Manual Metal Arc welding* 2005 Aachen ISF

# Analysis of the influence of condensation temperature and compressor efficiency on heat pump system efficiency

**Nemanja Koruga[1], Mirko Dobrnjac[1], Dušan Golubović[2] and Nemanja Dobrnjac[3]**

[1]Faculty of Mechanical Engineering, University of Banja Luka, Stepe Stepanovica 73, Banja Luka, Bosnia and Herzegovina,
[2]Faculty of Mechanical Engineering, University of East Sarajevo, Vuka Karadzica 30, East Sarajevo, Bosnia and Herzegovina,
[3]Faculty of Mechanical Engineering, University of Belgrade, Kraljice Marije 16, Belgrade, Republic of Serbia

E-mail: nemanjakoruga@yahoo.com

**Abstract.** In heat pump cycles, heat is supplied to the working fluid from a certain group of low-temperature bodies and transferred to a group of high-temperature bodies, i.e. the heat source is at a lower temperature and the heat sink at a higher temperature. Using the method of circular processes, in synergy with the possibility of mutual conversion of thermal and mechanical interactions, the process of heat transfer from a lower temperature level to a higher temperature level is enabled. Mechanical work, which, as compensation, should be given by the environment to the system (working substance), is a difference between heat removed and heat supplied. The efficiency of the heat pump mostly depends on the temperature interval at which the process takes place, however, the efficiency of the heat pump is also affected by the thermodynamic parameters of its parts: compressor, condenser, throttle valve, and evaporator. In this paper, the influence of condensing temperature and compressor efficiency on the efficiency of the system as a whole is examined. The calculation was performed for two working substances, R123 and R134a, using the EES software package (Engineering Equation Solver) which is used for numerical modeling of thermodynamic systems, process optimization, and making process diagrams.

## 1. Introduction

Recently, there is a growing awareness of the dangers of increased global warming and environmental pollution as a result of the use of energy, which is mainly obtained from fossil fuels. Precisely for these reasons, as well as the accelerated growth of fossil fuel prices predicted by their disappearance in the future, the world is increasingly turning to the use of renewable energy sources.

Heat pumps are known in systems for saving and using renewable energy sources. Their recent development has yielded significant results. In the past, the ratios of invested electricity and obtained heat energy using a heat pump were 1:2, while today these ratios have been brought up to 1:6, all thanks to significant investments in the development of this energy system. Heat pumps are environmentally friendly devices, which is why there is a tendency to research and improve them.

Thermal transformers work on the principle of the heat pump cycle, in which the heat source is at a lower and the heat sink at a higher temperature level. The beneficial effect (useful heat) of these devices can be the heat taken from the cycle to the heat sink or the heat brought into the cycle from the heat source depending on whether they are used for heating or cooling. A heat pump is a device whose basic function is heating with heat that is removed from the cycle. According to the energy balance, the amount of heat dissipated to the heat sink, at a higher temperature level, is the sum of the added mechanical work from the environment and the heat brought from the heat source at a lower temperature level. [1] [2]

The efficiency of a heat pump is expressed by the Coefficient of Performance (*COP*):

$$COP = \frac{Q_{cond}}{W_{comp}} \tag{1}$$

$Q_{cond}$ – the part of the heat that is transferred from the working substance to the heat sink (beneficial effect),

$W_{comp}$ – the work required to drive the compressor.

Changes in the state of the working substance in the heat pump cycle according to Figure 1:

1) evaporation: 5 – 1,
2) compression: 1 – 2,
3) cooling of steam to condensation temperature: 2 – 3,
3) condensation: 3 – 4,
4) expansion: 4 – 5

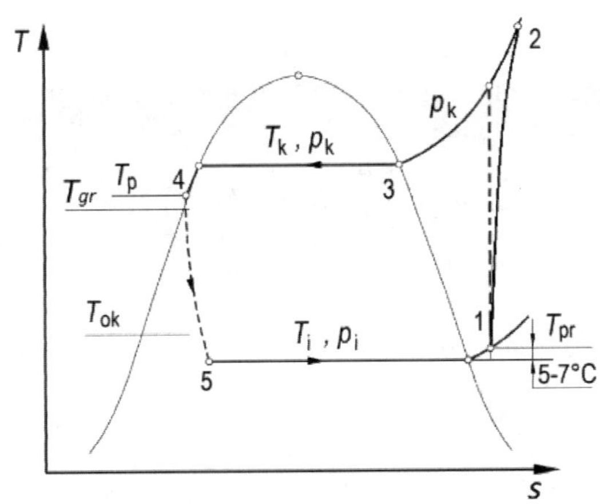


**Figure 1.** Representation of a real heat pump cycle in a T- s diagram [3].

$T_{gr}$ – heat sink temperature, $T_{ok}$ – heat source temperature, $T_p$ – subcooling temperature, $T_k$, $T_i$ – condensation temperature, evaporation temperature, $T_{pr}$ – superheated steam temperature.

The working substance in the heat exchanger (evaporator) takes energy from the heat source, during which it passes from the liquid to the vapor phase. To be able to transfer heat to the consumer (in heating mode), it is necessary to increase the pressure and temperature of the working substance, which is achieved by compression using a compressor. At a higher temperature than the sink temperature, the working substance can transfer heat to the sink in the exchanger (condenser) where it returns to the liquid phase (condenses). To return its pressure to the pressure of evaporation, the refrigerant is led from the condenser to the expansion valve, where its pressure is reduced, thus closing the cycle and repeating the procedure.

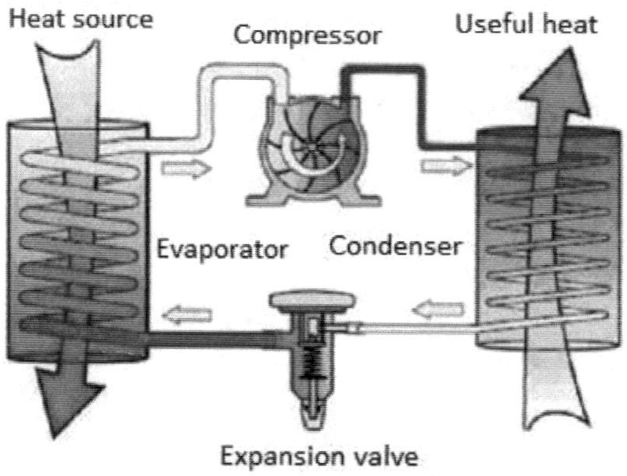


**Figure 2.** The principle of operation of the heat pump.

It is obvious that heat pumps transfer heat energy from a body of lower to a body of higher temperature, so there is a question does this contradicts the second law of thermodynamics which says that heat spontaneously passes from a warmer to a cooler body. The answer is no, heat pumps do not contradict the second law of thermodynamics due to the nature of heat pump processes and the energy consumption in the compressor. By compression, in addition to pressure, the working substance also increases the temperature, which must be higher than the temperature of the heated medium, thus enabling the spontaneous transfer of heat from a warmer to a cooler body (the second law of thermodynamics is satisfied when exchanging heat in a condenser). By expansion, the working substance temperature is decreased below the temperature of the heat source, so that the evaporator also enables the spontaneous transfer of heat from a warmer body (source) to a cooler body (working substance). We can say that heat pumps transfer heat from a cooler to a warmer body at a macro-level, but at a micro-level, they still transfer heat from a warmer to a cooler body, all under the second law of thermodynamics.

Inside the heat pump, as a heat transformer, we have a flow of working fluid which is a mediator in the process of heat exchange between two heat tanks (heat source and heat sink) which are at different temperatures. Thermophysical properties of the working substance significantly affect the thermodynamic efficiency of the heat pump itself, due to different behavior during heat transfer, flow resistance, differences in viscosity, different boiling points, and condensation, etc.

Depending on the chemical composition, working substances can be divided into:

- pure hydrocarbons (R290 – $CH_3CH_2CH_3$; R600a – $C_4H_{10}$),
- halogenated hydrocarbons (R134a – $C_2H_2F_4$, R123 – $C_2HF_3Cl_2$; R142b – $CH_3CClF_2$),
- azeotropic mixtures,
- zeotropic mixtures (R404a – $C_2H_3F_3$, R407a, R407c) and
- inorganic substances (R744, R764).

Despite many theoretical assumptions that indicated a significant improvement in thermodynamic efficiency using zeotropic mixtures, experiments did not confirm this. The reason for this is the poorer heat transfer in exchangers when using these working materials. Changes in the composition of the two-phase flow of the working substance in the evaporator and condenser lead to deviations to the one-component working fluids. Weaker heat transfer is caused by an increase in the content of a weaker volatile component with heated surfaces, less suitable physical properties (conductivity and viscosity), and difficult bubble evaporation.

By choosing the working fluid and changing the condensing temperature, we directly influence some of the parameters (fluid density, heat capacity, dynamic viscosity, etc.), on which the heat transfer coefficient and friction resistances in the system depend, which ultimately determines the heat efficiency of the pump.

In the following, a comparison of two working fluids, R134a (HFC) and R123 (HCFC) will be made, in terms of their impact on the efficiency of the heat pump system.

## 2. Review of previous research

Energy, exergy and characteristics of heat pumps with different refrigerants have been analyzed in numerous scientific publications, of which we will mention only a few.

A comparison of the energy and exergy efficiency of a cascade cooling system using R170-R161 and R41-R404A as refrigerant pairs was analyzed in the paper by Roy and Mandal. The calculation code for this analysis was developed in the power system software that uses the basic equations of energy and exergy of the cascade cooling system. The influence of evaporator temperature on basic parameters, such as COP, compressor power and total exergy loss, was analyzed. Analysis of the results shows that for each evaporation temperature, an optimal point exists where the system shows maximum performance. Efficient reduction of compressor operation and total exergy loss is observed in systems using R170-R161 than in systems with R41-R404A. This ultimately results in a higher optimal COP as well as exergetic efficiency for the system with R170 - R161 compared to the system R41 - R404A. [5]

Saeidi, V. and Mafi, M. use Engineering Equation Solver (EES) in their work, examining the behavior of the thermodynamic system when using working tools R152a, R600 and R717 and comparing with the behavior results of conventional refrigerants R22 and R134a for geothermal heat pumps with horizontal and vertical heat exchangers. Fluids, R152a, R600 and R717, were introduced as replacement refrigerants to replace synthetic agents in the above-mentioned geothermal heat pump systems, increasing by 2.97 to 5.8% of the cooling coefficient, which means a reduction in the operating costs of the system relative to conventional working fluids. [6]

The influence of different cooling fluids on the efficiency of a geothermal heat pump was analyzed in the work of Dashtebayaz and Maddah. The analysis was performed for five HFC refrigerants R125, R134A, R404A, R407C and R507A used in the geothermal heat pump. Critical parameters, such as performance coefficient (COP), exergetic efficiency and exergy destruction for different components were calculated. The results show that in the geothermal heat pump cycle, R134A and R125 refrigerants have the highest and lowest COP as well as exergetic efficiency. For the mentioned refrigerants, the destruction of exergy by the compressor, as the primary equipment for energy consumption, is achieved between 26.7 and 27.3% concerning the entire system. [7]

## 3. Description of the research

In this paper, it was interesting to perform a thermodynamic analysis to see the effects of the abovementioned substances on the overall operation of the heat pump system. For this analysis, the EES (Engineering Equation Solver) program was used, specifically to examine the influence of condensation temperature on the efficiency of the heat pump system, i.e. to determine the amount of energy losses in individual components concerning changes in the temperature of condensation. A simulation calculation was performed for two working fluids R123 and R134a to conclude which working fluid is better, i.e. which requires lower mass flow with the highest heat and energy transfer. Also, heat pumps use compressors with different degrees of efficiency, so the impact of the same on the overall efficiency of the cycle, and its impact on energy loss is analyzed. The evaluation of the quality of operation of the heat pump is shown by the heating factor.

The calculation was performed for heating a building with a volume of 11 x 10 x 3 m, assuming the supply of 75 W/m$^3$ of heat flow to maintain a room temperature of 20°C, at an outdoor temperature of –5°C. The compressor sucks in dry saturated steam at a temperature of –10°C and compresses it isentropically to the condensing pressure, assuming the efficiency of the compressor from 0,7 to 1. In the condenser, the steam is completely condensed and the condensate is cooled by 5°C. Condensation temperatures vary in values: 30°C, 40°C, and 50°C.

IOP Conf. Series: Materials Science and Engineering 1208 (2021) 012015 doi:10.1088/1757-899X/1208/1/012015

*3.1 Results analysis*

Figures 3, 4, and 5 show diagrams for working substance R123. The analysis was also performed for working substances R123 and R134a and their mutual comparison at the condensing temperature $T_{kond}$ = 40°C is shown in Figures 7, 8, and 9. [8]

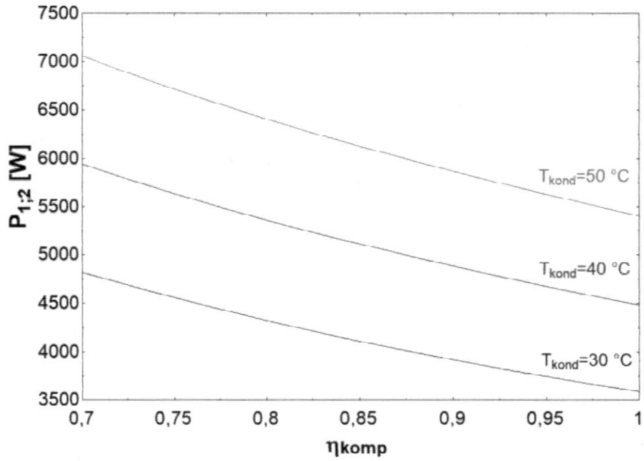


**Figure 3.** Dependence of compressor power on the compressor efficiency and condensation temperature for the working substance R123.

Where is:
$P_{1,2}$ – required compressor power,
$\eta_{komp}$ – the degree of usefulness of the compressor,
$T_{kond}$ – condensation temperature.

The previous figure shows a diagram for the working material R123, which shows the dependence of the compressor power on the condensing temperature and the efficiency of the compressor. As the condensing temperature increases, the required compressor power increases, and as the compressor efficiency increases, the required compressor power decreases.

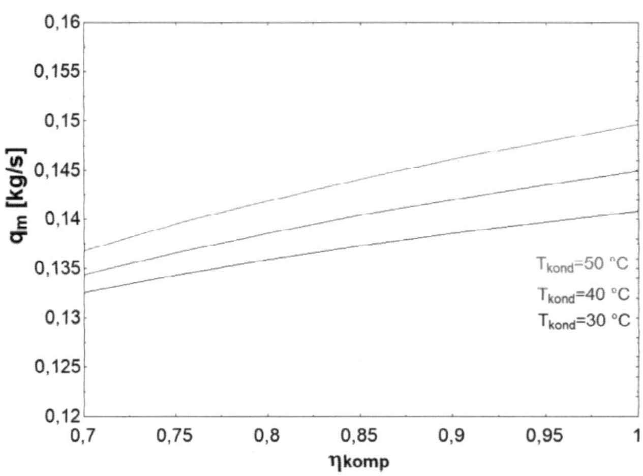


**Figure 4.** Dependence of mass flow on the compressor efficiency and condensation temperature for the working substance R123.

Where is:
$q_m$ – required mass flow.

The figure shows a diagram for the working material R123 which shows the dependence of the mass flow ($q_m$) on the condensation temperature ($T_{kond}$) and the efficiency of the compressor ($\eta_{komp}$). As the condensing temperature increases, the required mass flow increases, and as the efficiency of the compressor decreases, the required mass flow decreases.

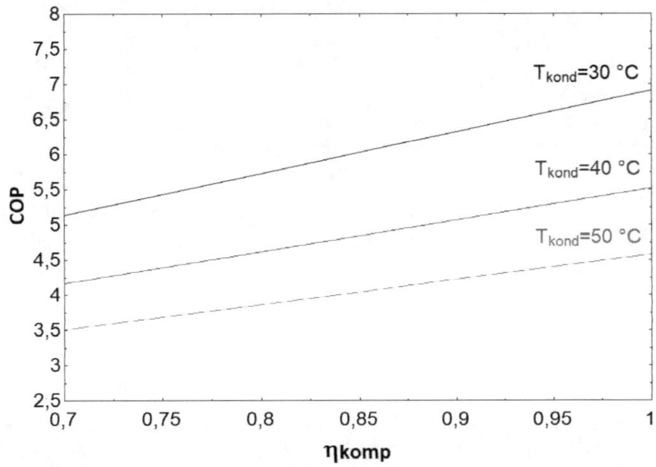


**Figure 5.** Dependence of COP on the compressor efficiency and condensation temperature for the working substance R123.

Where is:
*COP* – Coefficient of Performance.

The analysis was repeated for the working substance R134a.

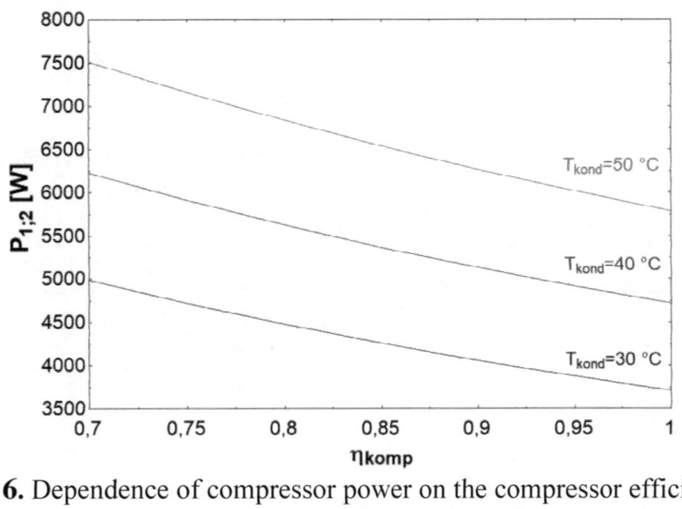


**Figure 6.** Dependence of compressor power on the compressor efficiency and condensation temperature for the working substance R134a.

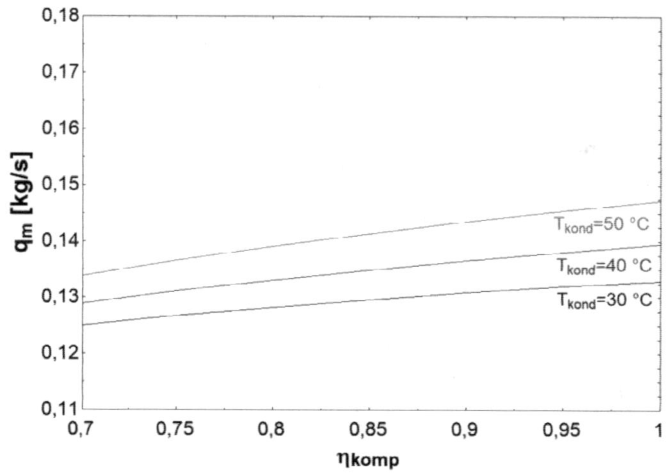


**Figure 7.** Dependence of mass flow on the compressor efficiency and condensation temperature for the working substance R134a.

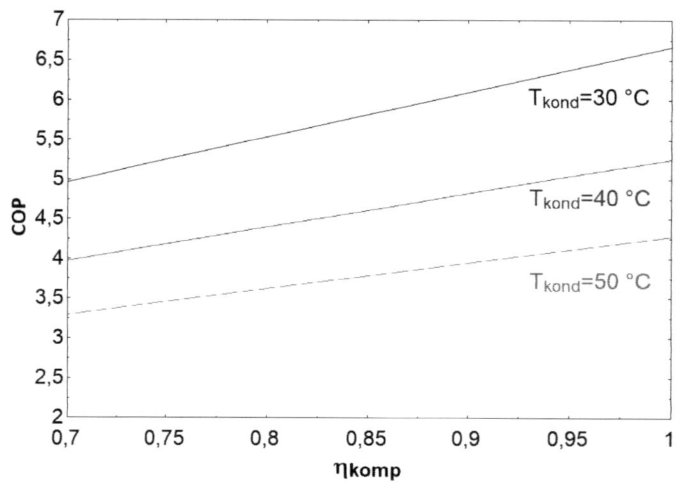


**Figure 8.** Dependence of COP on the compressor efficiency and condensation temperature for the working substance R134a.

The previous figure shows the diagram for the working substance R134a, which shows the dependence of the *COP* on the condensation temperature and the efficiency of the compressor. As the condensing temperature increases, the *COP* decreases, and as the compressor efficiency increases, the *COP* increases.

From the obtained results for the working substance R134a, it can be concluded that the increase of condensation temperature adversely affects the operation of the system because it increases the required mass flow of working material and the required compressor power and reduces the *COP*.

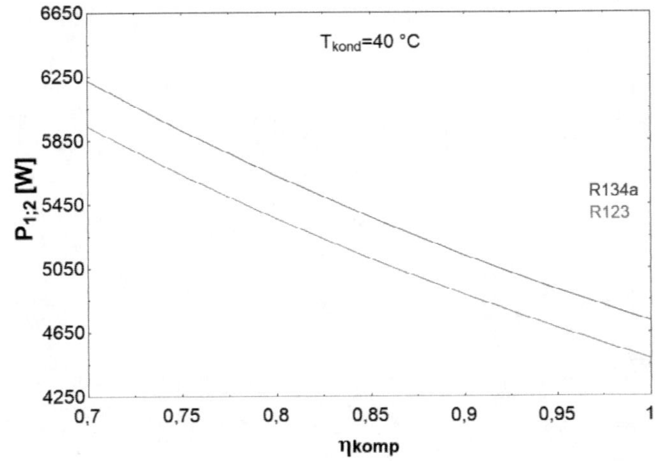


**Figure 9.** Comparison of the power curves for refrigerants R123 and R134a [5].

From the obtained, it can be concluded that for the working substance R134a for the given parameters the maximum power of the compressor is required, while for the working material R123 the minimum power of the compressor is required under the given process conditions.

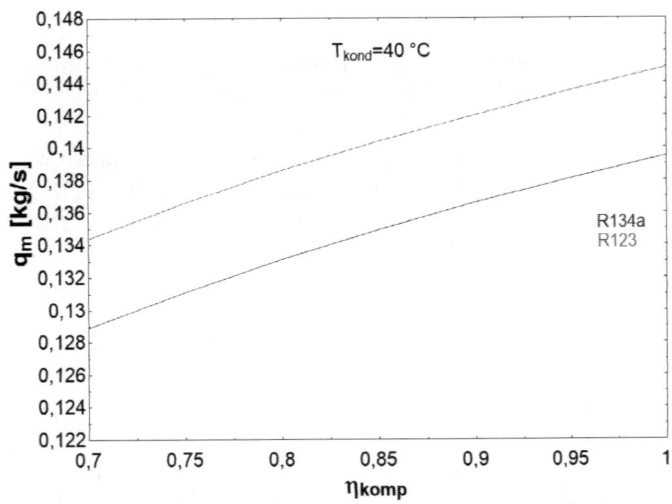


**Figure 10.** Comparison of mass flow curves for refrigerants R123 and R134a [5].

From the obtained diagram it can be seen that for the working substance R123 the highest parameters of the required mass flow are required for the given parameters, while for the working material R134a.

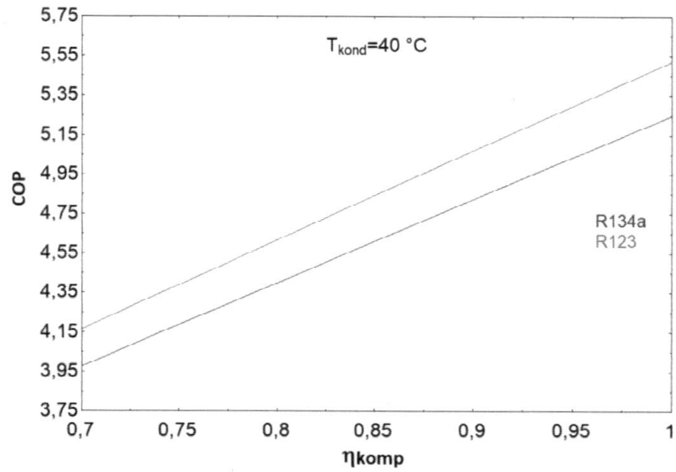


**Figure 11.** Comparison of the COP curves for refrigerants R123 and R134a [5].

The previous figure shows that the heating coefficient is higher in the case of the use of the working substance R123.

## 4. Conclusion

The operation of the heat pump depends on the condensing temperature and the efficiency of the compressor, which was shown by the calculation and obtained graphs. With selected working fluids, it is seen that a large influence on the results of mass flow, compressor power, and *COP* has a change in condensation temperature and the efficiency of the compressor. The diagram of the dependence of the compressor power on the efficiency of the compressor and the temperature of condensation in both working materials shows how the power of the compressor increases with increasing condensing temperature and decreases with increasing the efficiency of the compressor. Also, based on the diagram of the dependence of mass flow on the condensing temperature and the compressor efficiency, it can be concluded that the required mass flow increases with increasing condensing temperature and decreases with decreasing compressor efficiency. The dependence of the *COP* on the efficiency of the compressor and the condensing temperature, shown in the diagram for given working materials shows that the value of the heating factor decreases with increasing condensing temperature and the value of the heating factor increases with increasing degree of compressor efficiency.

In the compressor power comparison diagram between the refrigerants, the working fluid R134a proved to be worse because it has higher irreversibility. In the mass flow comparison diagram for the given working fluids, the working fluid R123 proved to be worse, that is, the highest mass flow is required for the given conditions. In the *COP* comparison diagram, the working material R123 proved to be better since the *COP* is higher for the given conditions. As the condensation temperature increases, the losses in the system increase due to the larger temperature difference and the distance from the ideal Carnot process. Heat pumps should have a condenser that contains the refrigerant with the lowest possible condensing temperature for the device to be more efficient and consume less energy. In practice, it is difficult to achieve such low condensation temperatures with environmentally friendly working fluids, so today many types of research of working fluids are performed to find them with more optimal properties.

## References

[1] Guzović Z *Geothermal energy and heat pumps* (Zagreb: University of Zagreb)
[2] Đuričković V 2000 *Heat transformers* Thermodynamic and Engineering bases (Belgrade: University of Belgrade)

[3] Group of authors 2010 *Manual for Energy certification of buildings* (Zagreb: University of Zagreb) chapter 9
[4] https://zpenergy.co.za/what-are-heat-pumps-and-how-they-can-assist-you-to-save-energy-for-water-heating/
[5] Roy R and Mandal B 2018 Energetic and exergetic performance comparison of cascade refrigeration system using R170-R161 and R41-R404A as refrigerant pair *Springer-Verlag GmbH Germany*, part of Springer Nature
[6] Saeidi V Mafi M and Mohammadi A 2018 Development of geothermal heat pumps by using environment friendly refrigerants *Journal of Renewable and Sustainable Energy*
[7] Dashtebayaz D and Maddah S 2020 Investigation of the effect of using various HFC refrigerants in geothermal heat pump with residential heating applications *Journal of Thermal Analysis and Calorimetry*
[8] Klein S and Nellis G 2012 Mastering EES *F-Chart Software* (Madison)

# Optimization of injection parameters from the aspect of dimension accuracy

**Amir Ćoralić**

Faculty of Mechanical Engineering, Computing and Electrical Engineering, University of Mostar

E-mail: amir@ebi.ba

**Abstract**. During the injection molding process, the polymers are subject to shrinkage and bending, which causes an unwanted change in geometry. These deficiencies can be caused by many factors: injection pressure, overpressure, overpressure time, cooling, etc. Within this work, the influence of overpressure and overpressure time on the accuracy of product dimensions is presented. Dimension control was performed by measuring on a 3D coordinate machine. The products are injected in 2 different combinations of parameters. For each combination of parameters, measurements were performed in five injection cycles.

## 1. Introduction

Injection molding is a manufacturing process that creates plastic products. The injection molding process itself is when the plastic mass in the form of granules melts to the extent that it is injected by pressure into the injection molding tool, where it is cooled to a given shape. When the product is sufficiently cooled, the tool is opening, and the product is ejected from the tool. The whole sequence of events when injecting one piece is called the cycle [1].

The main goal of the process is to get a part in one place and in one operation with exact dimensions and properties for use.

The injection parameters of such systems are based on iterative procedure, ie on the "Try and fail" method. The most important parameters for injection molding are: injection pressure, overpressure, overpressure time, cooling time that affect the accuracy of the dimensions of the finished product [2].

In the control of the accuracy of construction, three-coordinate measuring devices that serve for the spatial measurement of the measuring piece have a great role. The measurement is based on the identification of points from the surface of the piece, and they are identified with respect to the position in X, Y and Z coordinates.

Within this work, the influence of overpressure and overpressure time on the accuracy of product dimensions is presented. Dimension control was performed by measuring on a 3D laser machine. The products were injected in 2 different combinations of parameters. Measurements were performed in five injection cycles for each parameter combination.

## 2. Injection molding

Injection molding is the most important process in the processing of plastic products, because today more than a third of thermoplastics have been reshaped by this process. The process is very suitable for making a large number of pieces of complex shapes and achieving narrow dimensional tolerances.

IOP Conf. Series: Materials Science and Engineering 1208 (2021) 012016 doi:10.1088/1757-899X/1208/1/012016

Plastic injection molding machines are used for the manufacture of objects in the automotive industry, for plastic parts of household appliances and the manufacture of simple plastic elements, etc. [1].

**Figure 1.** Injector display.

The sequence of events when injecting a single product is called the injection cycle. The cycle begins when the tool closes with a certain closing force. This is followed by injecting the polymer melt into the tool engraving with the injection pressure caused by the screw force (F). When the engraving is filled, the application of overpressures is followed, which achieves the geometric and weight accuracy of the piece. After cooling, the tool is opening, and the piece is ejected [3]. After the last cycle, the spraying unit moves away from the tool, Fig. 2.

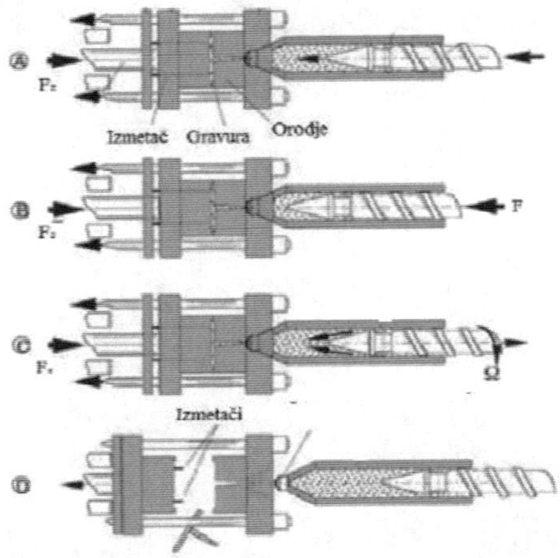


**Figure 2.** Injection molding cycle.

The filling of the mold is done under the influence of pressure, while the resistances that arise due to the high injection speed are reduced. As the melt cools, the mass begins to schrink, and the resulting space is filled with additional melt. For subsequent dosing of the melt, we need a much lower pressure in the tool, which is also called the subsequent pressure. Therefore, the subsequent pressure strives to keep the size of the product the same as the engraving. The subsequent pressure is set at 40 to 70% of the value of the achieved injection pressure.

Subsequent pressure is only effective if the following basic conditions are met:

- the filling opening must be large enough not to freeze too quickly,

- the „cushion" of the melt in front of the screw must be large enough to provide sufficient mass for the subsequent injecting of the melt into the engraving,
- the size of the subsequent pressure, the operating time of the subsequent pressure, the temperature of the tool and the melt must be correctly selected [4].

Causes of excessive post-pressure: higher internal stress, larger dimensions, higher weight, tool damage. Causes of too low post-pressure: possession, non-flooding, bubbles, high shrinkage, lower mass [4].

### *2.1. Injection molding materials*

Plastics are polymers to which various additives have been added. With regard to the behavior during processing and application, they are divided into two groups: plastomers and duromers [5].

Plastomers are polymeric materials with linear and branch structures. They dissolve in solvents and are meltable. By heating to the softening or melting temperature, they do not change their structure and therefore their processing involves only reversible changes. At these temperatures they can be formed into the desired shapes. The structure of the plastomer can be amorphous or crystalline.

Durometers are soft at first, but during heating at elevated temperature, irreversible 3D crosslinking into a solid material occurs. Further heating leads to their thermodegradation.

Today, polymer materials are used in all areas, they are indispensable for industry, because due to good physical properties such as good design, easy processing, light weight and satisfactory mechanical properties, they are increasingly displacing classic materials such as metal, wood and fabrics. Polymer products are found mainly in the automotive industry, packaging production, electrical industry, construction, furniture, and sports industry, but also in space technology and aviation due to their light weight.

The main methods of plastic processing are injection molding, gas injection, thermoforming, extrusion, blowing and 3D printing. In particular, injection molding is very widespread, because the polymers thus produced are suitable for mass processing as well as for individual test products [6].

## 3. Experimental work

In order to perform the experiment, the material PA6 GF 25 was injected. It is characterized by light processing and relatively good mechanical properties, which is maintained even at higher temperatures. It can also be recycled [7]. It is most often used in the automotive industry, and for various plastic products of higher rigidity and strength.

The tool is made with two nests, and an inflow point system. Filling points shall be set reasonably in accordance with the given functional requirements and the selected product geometry, Figure 3.

**Figure 3.** Injection tool, ejection side, nozzle side

During the injection molding process, the polymers are subject to deformation, which causes an unwanted change in geometry. The accuracy of the geometry can be caused by many factors such as: injection pressure, overpressure, overpressure time, cooling, etc. For the experimental, we selected two parameters: overpressure and overpressure time. Since it was necessary to determine higher and lower parameter values, we took the existing settings on the machine, according to the recommendations from the literature. For the first test, series 1, injection into the mold, a subsequent pressure between thirty and fifty percent of the injection pressure was used, and the time was four seconds, according to the literature. For Series 2, we reduced the overpressure and the overpressure time. Selected parameters and values are shown in Table 1.

**Table 1.** Injection parameters.

***Injection parameters***

| | Series 1 | | | | Series 2 | | | |
|---|---|---|---|---|---|---|---|---|
| Profile | Profile | 1 | 2 | 3 | Profile | 1 | 2 | 3 |
| Overpressure | | 700 | 600 | 100 | | 500 | 430 | 100 |
| Overpressure time | | 4 | 4 | 0,5 | | 2 | 3 | 0,5 |
| Cycle time | | 20.4 sec | | | | 20,4 sec | | |
| Total cycle weight | | 37,09 | | | | 36,92 | | |
| Gate weight | | 0,88 | | | | 0,82 | | |
| The weight of the piece | | 16,13 /20,08 | | | | 16,07/20,01 | | |

Based on the injection parameters, we can see that due to the variation of overpressure and overpressure time, there was a change in weight in both the workpiece and the gate.

In order to perform an analysis of the influence of parameters on the dimension, shape and errors of the product, it is necessary to measure the dimensions of the product. The measurement was performed on a three-coordinate machine manufactured by TESA MICRO-HITE 3D, Figure 4.

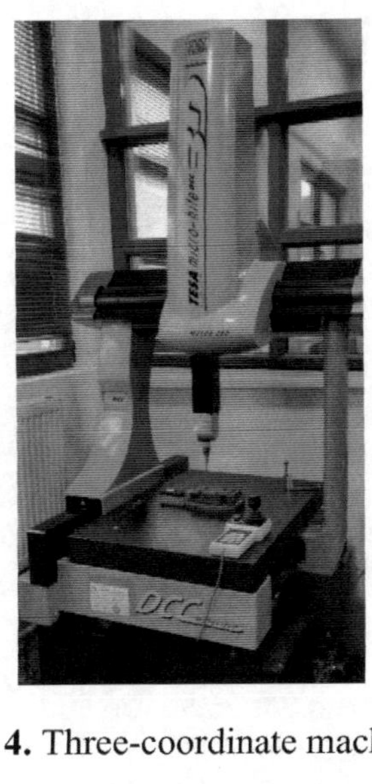

**Figure 4.** Three-coordinate machine [8].

The measurement is performed by placing the measuring product on the desk. The machine is positioned at the zero-point depending on the shape of the piece, and the given dimensions are measured. The measurement was performed on both workpieces from the nest, Figure 5, and Figure 6.

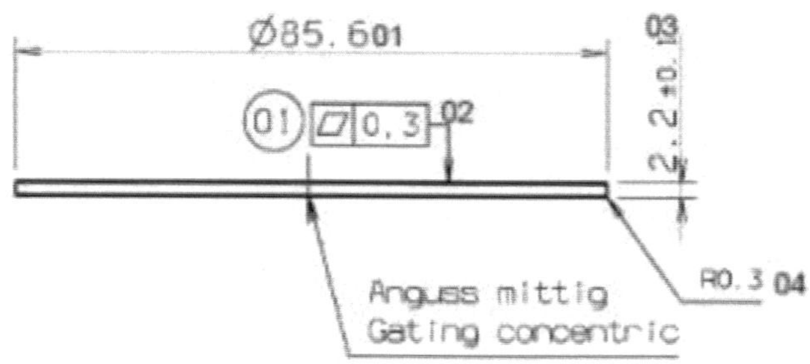


**Figure 5.** Workpiece 1.

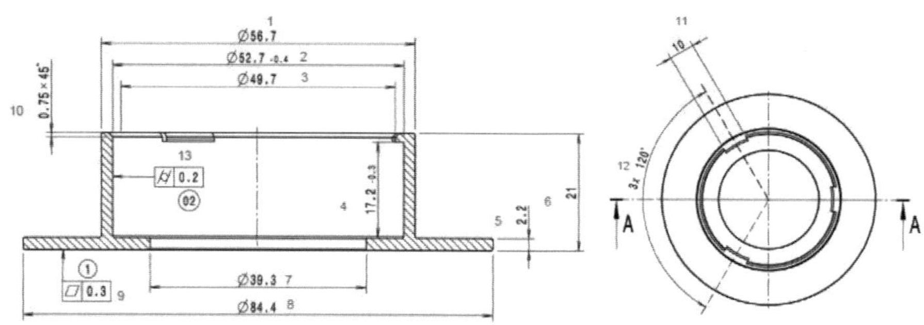


**Figure 6.** Workpiece 2.

The measurement results are given in a Table 2.

**Table 2.** Measurement results Series 1- Workpiece 1.

| | | *Workpiece 1* | | |
|---|---|---|---|---|
| Ref.br. | Specifications | Measured | Yes | No |
| 01 Fach1 | 85,6±0,340 | 85,97; 85,99; 85,97; 85,99; 85,98 | | |
| 02 | 2,20±0,100 | 2,35; 2,36: 2,36: 2,35; 2,36 | | |
| 03 | 0,3 | 0,3; 0,3; 0,3; 0,3; 0,3 | | |

**Table 3.** Measurement results Series 2- Workpiece 1.

| | | *Workpiece 1* | | |
|---|---|---|---|---|
| Ref.br. | Specifications | Measured | Yes | No |
| 01 Fach1 | 85,6±0,340 | 85,63; 85,62; 85,64 ; 85,62 ; 85,63 | | |
| 02 | 2,20±0,100 | 2,19; 2,20; 2,20; 2,19; 2,20 | | |
| 03 | 0,3 | 0,3; 0,3; 0,3; 0,3; 0,3 | | |

**Table 4.** Measurement results Series 1- Workpiece 2.

| | | *Workpiece 2* | | |
|---|---|---|---|---|
| Ref.br. | Specifications | Measured | Yes | No |
| Fach1 | Ø56,70±0,28 | 56,96; 56,95; 56,97;56,97; 56,95 | • | |
| 02 | Ø52,70-0,4 | 52,86; 52,85; 52,87; 52,86; 52,87 | | • |
| 03 | Ø49,7±0,25 | 49,93; 49,93; 49,94; 49,92; 49,93 | • | |
| 04 | 17,20-0,30 | 17,36; 17,37; 17,36; 17,36; 17,37 | | • |
| 05 | 2,2±0,09 | 2,29; 2,29; 2,28; 2,29; 2,28 | • | |

| | | | | |
|---|---|---|---|---|
| *06* | 21,00±0,15 | 21,22; 21,24; 21,23; 21,22; 21,24 | | • |
| *07* | Ø39,30±0,20 | 39,52; 39,53; 39,54; 39,52; 39,53 | | • |
| *08* | Ø84,40±0,34 | 84,39; 84,38; 84,37; 84,38; 84,39 | • | |
| *09* | ▱ 0,3 | O,09; 0,10; 0,06; 0,10; 0,09 | • | |
| *10* | 0,75 | 0,8; 0,8; 0,8; 0,8; 0,8 | | |
| *10* | 45° | 45; 45; 45; 45; 45 | | |
| *11* | 10,00±0,11 | 9,94; 9,95; 9,94; 9,93; 9,94 | • | |
| *12* | 120° | 120,0,8; 120,09; 120, 07; 120,06; 120,07 | | |
| *13* | /o/ 0,2 | 0,18; 0,19; 0,17; 0,18; 0,19 | • | |

**Table 5.** Measurement results Series 2- Workpiece 2.

| | | *Workpiece 2* | | |
|---|---|---|---|---|
| Ref.br. | Specifications | Measured | Yes | No |
| Fach1 | Ø56,70±0,28 | 56,71; 56,72; 56,70; 56,71; 56,70 | | |
| 02 | Ø52,70-0,4 | 52,67, 52,65; 52,69; 52,67; 52,69 | | |
| 03 | Ø49,7±0,25 | 49,55; 49,54; 49,54; 49,55; 49,54 | | |
| 04 | 17,20-0,30 | 17,11; 17,14; 17,13; 17,13; 17,14 | | |
| 05 | 2,2±0,09 | 2,18; 2,19; 2,19; 2,18; 2,19 | | |
| 06 | 21,00±0,15 | 20,94; 20,91; 20,90; 20,91 | | |
| 07 | Ø39,30±0,20 | 39,28; 39,26; 39,25; 39,27; 39,25 | | |
| 08 | Ø84,40±0,34 | 84,19; 84,20; 84,18; 84,19; 84,18 | | |
| 09 | ▱ 0,3 | 0,05; 0,03; 0,04; 0,05; 0,04 | | |
| 10 | 0,75 | 0,8; 0,8; 0,8; 0,8; 0,8 | | |
| 10 | 45o | 45; 45; 45; 45; 45 | | |
| 11 | 10,00±0,11 | 9,94; 9,95; 9,94; 9,93; 9,94 | | |
| 12 | 120o | 120,05; 120,04; 120,05; 120,05; 120,05 | | |
| 13 | /o/ 0,2 | 0,15; 0,14; 0,15; 0,13; 0,15 | | |

## 4. Conclusion

Based on the obtained results it can be concluded:

At higher overpressure and overpressure time, series 1, and at the same cycle time of 20.4 sec, there was a deviation of dimensions. All deviations were in positive tolerances. In workpiece 1 there was a deviation in all dimensions, while in workpiece 2 some dimensions were in the given tolerances that can be attributed to the place or point of pouring.

Also, by changing these parameters, there was a change in the weight of the workpiece and the weight of the gate by about 0.4%.

Based on this, we can conclude that the values of overpressure as well as the overpressure time itself affect the optimization of the product. In addition to these parameters, it is necessary to consider other parameters that have an impact on the optimization of the injection molding process, so it would be necessary to investigate these parameters.

## References

[1] Vračević M 1997 Prerada plastičnih masa brizganjem *Novi Sad*
[2] Siminiati D, Pomenić L and Oršić F 2008 The base parameter adjustment for injection moulding of plastic Eng. Rev. 28-1 pp 119-127
[3] Dežman J 2017 Analiza brizganja polipropilena s prototipnim konformno hlajenim polimernim orodjem *Ljubljana*
[4] Privšek 2015 Umetnost brizganja *PROFIDTP* Škofljica
[5] Europe's Association for plastics and rubber machinery manufacturers (www.euromap.org), Date: September 2021

[6] Polymer Technology & Services (www.ptsllc.com), Date: September 2021
[7] Navodnik J and Kopčič M 1998 Plastik – orodjar *Navodnik* Velenje
[8] Three Dimensional Coordinate Measuring Machine Machine catalog TESA MICRO-HITE 3D Bugnon **38** – CH-1020 *Renens*

IOP Conf. Series: Materials Science and Engineering 1208 (2021) 012017 doi:10.1088/1757-899X/1208/1/012017

# Concept design and design optimization of the transfemoral prosthetic leg prototype

**Remzo Dedić[1], Faris Ustamujić[2*], Zlata Jelačić[3], Želimir Husnić[4**] and Vlado Kvesić[1]**

[1]Faculty of Mechanical Engineering, Computer Science and Electrical Engineering, University of Mostar, B&H
[2]Airbus DS GmbH, Germany
[3]Faculty of Mechanical Engineering, University of Sarajevo, B&H
[4]The Boeing Company, Philadelphia, USA

E-mail: jelacic@mef.unsa.ba

**Abstract.** The biggest challenge with today's lower leg prosthetic devices is to climb the stairs in a biologically inspired way. When climbing the stairs, due to the large moments in the knee joint, great forces in the knee occur. Because of the magnitude of these forces, which can be up to six times the weight of a person, it is necessary to use a hydraulic drive with an external power source in the knee and ankle joint. However, most commercially available above-knee prosthetic legs only have actuation in the form of damping in the knee joint. Hence, an amputee is forced to use a healthy leg to advance to each step first while amputated leg follows.

## 1. Introduction

The project of optimization of the transfemoral (above-knee) modified prosthesis RZ-0 includes the development and production of a new type of above-knee prosthesis RZ-W with double drive integrated in the lower leg, i.e., independent management of the knee and ankle joints together with separate drive units. By enabling a more natural gait and the ability to master activities such as getting up and climbing the stairs, this prosthesis aims to improve the quality of life of people with above-knee amputations. Section 1 contains a brief description of the history of the development of the RZ-0 (initial model) upper leg prosthesis project. There is an overview of the first four development phases and the tasks are listed, i.e., the goals of the current, fifth phase of work. Section 2 presents a series of activities undertaken to enable the first part of the fifth phase of the project. The following is section 3 with details of performing the activity, which is, assembling the components into a working prototype RZ-W. The conclusions are presented in Section 4.

## 2. Background

The activities related to this project have been going on for several years and consist of several phases. The project has been multidisciplinary from the very beginning. Social workers, orthopaedic doctors, and a group of engineers consisting of specialists in biomechanics, statics, hydraulics, mechanics, sensors, and robotics have been involved [1].

The first phase was to select the appropriate group of amputees and try to define all the parameters of the terms of reference that would be represent to the whole group.

The second phase was to perform the necessary calculations related to the design of the structure.

The third phase was the optimization of the construction of the above-knee prosthesis, where based on the obtained data the most favourable position for the installation of a hydraulic linear actuator in the knee joint was obtained by synthesis, and then its stroke and other structural dimensions were defined.

In addition, it was necessary to solve the method of supplying the starter with oil, and select all the necessary hydraulic components to control and regulate the oil flow.

The fourth phase was the optimization of an artificial foot with an ankle joint. Experimental tests have shown that the time of movement of linear actuators, as well as the stroke of linear actuators in the knee and ankle joint are not equal. Therefore, it was concluded that the management of these linear actuators should be separated [2,3,4,5].

This article describes the initial activities of the fifth phase of the project and has as a starting point the observations and conclusions collected during the implementation of the fourth phase. One of the important limitations for this project, in addition to reducing the volume of prosthesis aggregates as a whole, is the financial accessibility of prostheses to a wider population of users, which is reflected in the use of exclusively commercially available components [6].

The following points have been identified as primary objectives:

• Detailed analysis of the prototype of the above-knee prosthesis RZ-0 through an analytical model in CAD software. Special attention is paid to the verification of prosthesis kinematics using available, commercial actuators.

• Rapid construction of a load-bearing structure based on wood in order to build and verify the feasibility of a physical prototype. This approach allows the verification of the intended actuators and the design of the housings for the drive actuators in the lower leg.

• Translation of acquired observations into the design and production of 3D printed components of the entire structure.

## 3. Design and development of the new prosthesis leg concept

The commercial aspect of the project leads to limitations for prosthesis design, which is reflected in the use of exclusively commercially available components. The following figure shows the working prosthesis made at the end of the fourth phase of the project in which the actuators of the upper and lower part of the prosthesis are visible. These cylinders are made to order, for prototype testing purposes, and do not represent available components for the commercial development of a robust prosthesis model (Figure 1).

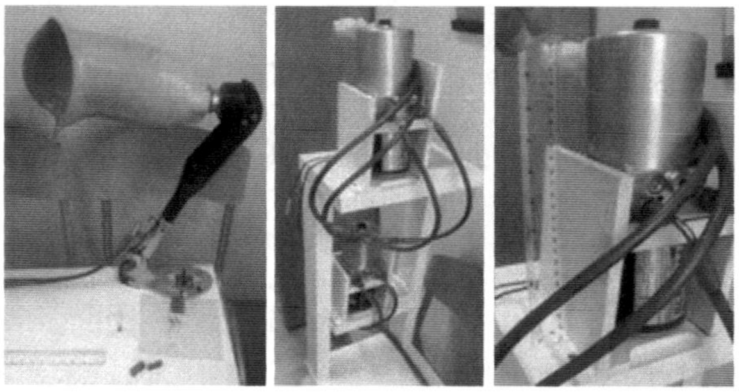

**Figure 1.** Layout of the RZ-0 prototype prosthesis at the end of the fourth phase of the project of development of the upper leg prosthesis (left) and used aggregates (right)

From the conclusions of the previous phases of the project, it was established that the hydraulic actuators of the company SNC, type CHNC20-25 and CHNC20-35, best meet the technical requirements of the imagined prosthetic actuators and are commercially available. Electric motors and hydraulic pumps from JUNG were selected for the unit, which with an individual total weight of 1.5 kg are currently the smallest and lightest versions available on the market. At the same time, these aggregates are half as light as those used in the previous prosthesis and with their smaller volume for the first time offer the possibility of integration into the lower leg part of the prosthesis.

As mentioned in the introductory chapter, the idea was for the very beginning to make a prosthesis construction out of wood to get a tangible working prototype as soon as possible for further design testing. The reason for this is that wood is easy to work with, and it is not expensive to fix something if it goes wrong or ultimately make a new prosthesis. Drawings according to which parts for a wooden prosthesis were made were drawn in the Solidworks program. The original idea was to make a lower leg and upper leg that would connect with each other. According to the literature, the following dimensions were taken [1]:

• Length of the lower leg (from the axis of the ankle joint to the axis of the knee joint) - 425 mm

• Length of the upper leg (from the axis of the knee joint to the end of the pelvis) - 575 mm.

The dimensions of the wooden bar are 60X40 mm. The lower leg would be connected to a metal foot that was already ready for use.

The following were used as input data for the kinematics of the prosthesis components:

• The foot should be maximally straightened at an angle of 90 degrees to the ground,

• In the lowered position, when the lower leg is at a 90 degree angle to the ground, the upper leg should be slightly less than 45 degrees to the ground.

These parameters were achieved with the help of Solidworks (Figure 2) to later make the physical wooden version of the prosthetic leg.

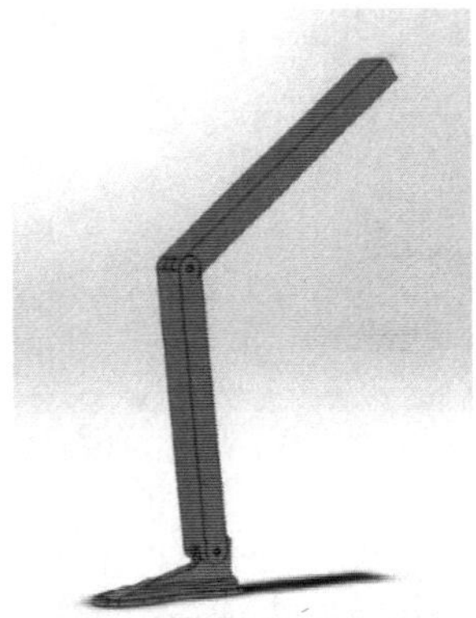

**Figure 2.** 3D model of the RZ-W prosthesis concept in Solidworks

The next step was the installation of cylinders that were selected based on the conclusions from the previous phases of the project. Two cylinders were procured from the US and SMC. The cylinders are from the CHN series (Figure 3). Some features of these cylinders are:

• relatively light weight
• Nominal pressure: 7 MPa
• Drill size (mm): 20.25, 32, 40
• Standard strokes (mm): 25 to 800
• 5 types of mounting

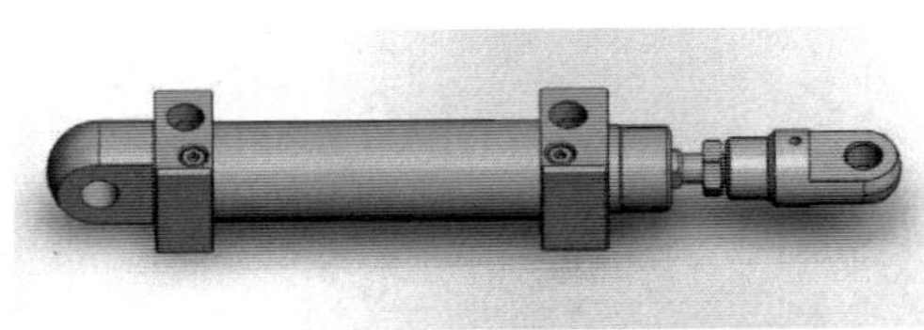

**Figure 3.** Cylinder of the CHN series

The design of the two cylinders available to the project is type C. The size of the drill is 20 mm. The shorter cylinder has a stroke (working path), i.e. 25 mm can be pulled out, and a longer cylinder 35 mm. On the SMC website, there is a possibility to download ready-made CAD models of these cylinders from the database, so they were downloaded for the Solidworks program. For both cylinders comes the equipment, i.e. two small holders named I0-2 (I-shaped single knuckle joint) that are wound on one side of the cylinder. They are like a CAD model, also taken from the SMC database (Figure 4).

When checking the dimensions of the CAD model on the basis of the physical models of the cylinders, a small mismatch of the drawings from the prospectus was noticed, i.e. actual dimensions and those in the CAD model, so they are aligned with the physical model. When this was done, the next step was to install a shorter cylinder (CHNC20-25) in the CAD model of the prosthesis. The idea was to install it in a prosthesis so that the cylinder is connected on one side to the hole in the foot provided for it. The upper side of the shorter cylinder would then be connected to the upper leg taking into account a maximum working distance of 25 mm. It was decided to connect it so that the centre is in the axis of the lower leg, i.e. in the middle. Therefore, it was necessary to make an opening in the wood where the

IOP Conf. Series: Materials Science and Engineering 1208 (2021) 012017 doi:10.1088/1757-899X/1208/1/012017

cylinder should fit and holes on the sides where the cylinder will be connected with a screw, which was done in the Solidworks program (Figure 4).

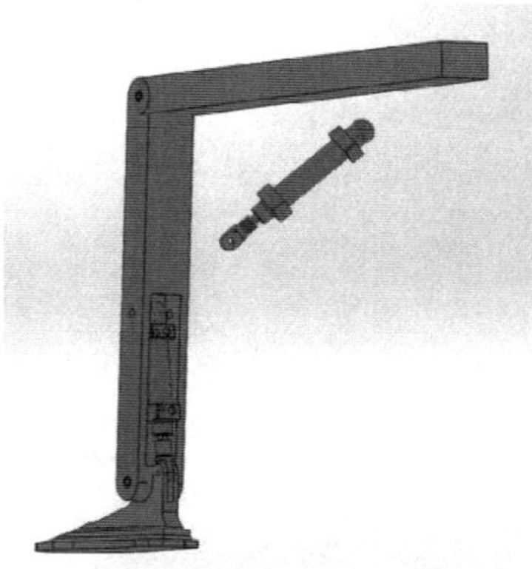

**Figure 4.** Drawing of the opening for the shorter cylinder in the knee joint

The longer cylinder (CHNC20-35) should be tied to the lower leg on the underside and to the upper leg on the upper side. Since the longer cylinder should also be tied to the central axis of the lower leg, an opening in the should be made for it as well. On the basis of the kinematics and working paths of these cylinders they come into a position very close to each other, it was necessary to make an opening along the entire lower leg as the most efficient solution. For the upper connection point of the longer cylinder, it was decided to make a metal part that would be fastened with bolts and to which the cylinder would be attached. The metal part is drawn in Solidworks and it is made of aluminium in the local company.

When the wooden structure was fully modelled and the cylinders were virtually connected to the foot and the metal part for the upper leg, the resulting virtual RZ-W prototype looked like in Figure 5.

**Figure 5.** Wooden prototype after installation of the cylinders and metal fixation

The next, and at the same time the biggest challenge, was the realization of the necessary kinematics of the prosthesis within the given limits of the cylinder. The movement of the lower leg was not a major challenge and the stroke of 25 mm of the lower cylinder was sufficient to achieve the desired movement of the lower leg. The challenge was to achieve sufficient thigh rotation. The lower leg has a given movement in the range of 95 degrees. In order to achieve this, it was necessary to adjust the position of the metal part to which the upper cylinder is connected. Mathematical calculation and displacement in the CAD model yielded a satisfactory solution which is shown in Figure 6. The effective operating lever of the cylinder is 24 mm in the obtained model.

With the successful completion of the design in CAD and the virtual verification of the kinematics, it is decided to move on to the development of a physical prototype of a wooden prosthesis with all the listed, available components.

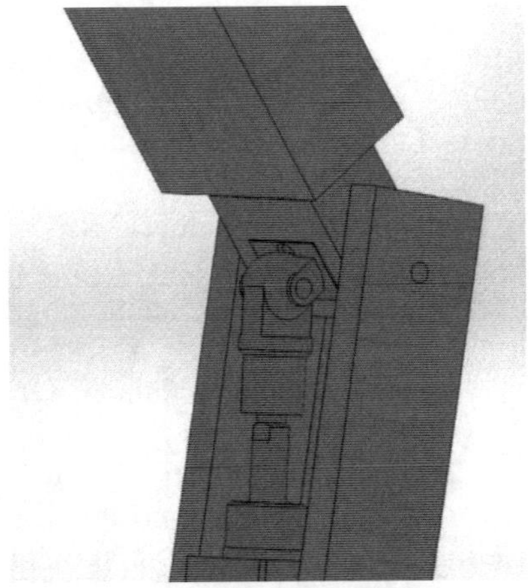

**Figure 6.** Wooden prototype after installation of the cylinders and metal fixation

## 4. Development and assembly of the physical prototype

By making parts of the structure from wood, cylinders were installed as shown in Figure 7.

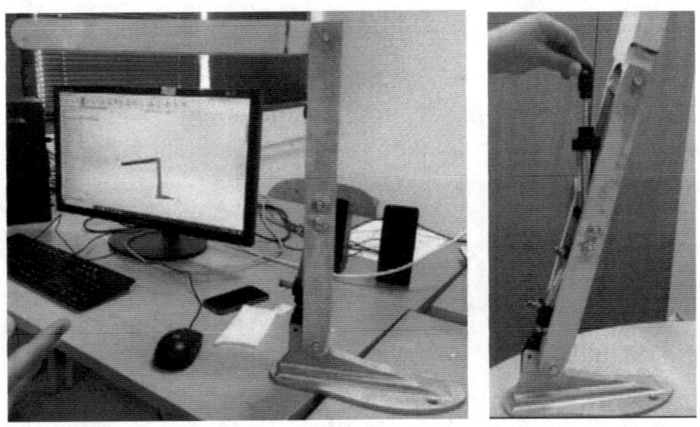

**Figure 7.** Wooden prototype RZ-W with both cylinders and metal foot

With the successful completion of the physical prototype movement check, the next step was to install the units (pumps and electric motors) purchased from the German company JUNG-FLUIDTECHNIK GMBH. The prosthesis with all its parts was taken to the company HANSA-FLEX where all hydraulic components and lines are integrated in the prosthetic housing. The units (electric motors with pumps) are connected to small plates on the sides of the lower leg, and the oil tank is installed a little higher on the plate on the front side of the lower leg. The final first version of the prototype prosthesis with a wooden structure is shown in Figure 8. The electrical, drive control system is shown in the figure to the right.

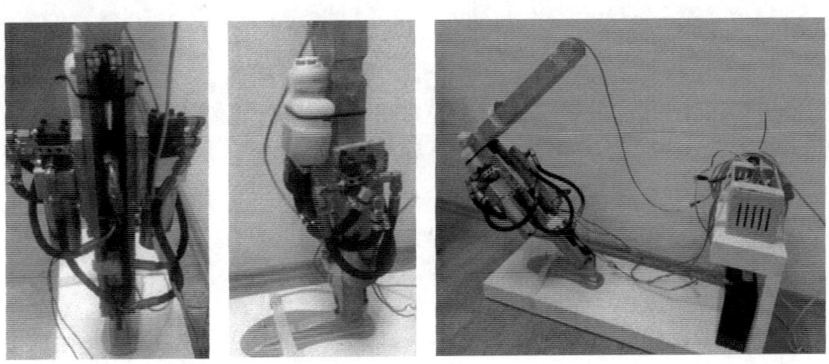

**Figure 8.** Appearance of the prototype prosthesis RZ-W with cylinders and wooden construction and metal foot. The electrical, drive control system is shown in the figure to the right.

## 5. Discussion and conclusion

The previous sections show a fast process of designing the components of the RZ-W prototype of the prosthesis based on the load-bearing parts of the wooden structure.

The following points are described in detail:

• Detailed analysis of the prototype of the upper leg prosthesis RZ-W through an analytical model in CAD software. Special attention was paid to the verification of prosthesis kinematics using available, commercial actuators. The observed deviations in the available CAD models of the cylinders were eliminated through the inspection of the physical models. The assembly of components into a virtual prosthesis is performed by complying with all given constraints.

• Rapid prototyping of a load-bearing structure based on wood in order to build and verify the feasibility of a physical prototype. The production itself was done relatively quickly, which confirmed the validity of the choice of wood as the primary material for the rapid and efficient production of a working prototype of the prosthesis. The flexibility of the wood and the possibility of subsequent constructions are thus recognized.

• The installation of available and commercially available (off-the-shelf) components (aggregates, rubber hoses, manifolds and oil tank) was also done relatively quickly and a verification, a functional, working prototype ready for further testing, took place.

The entire process was done in less than 30 days. Observations gained in this process are taken for the next step in the fifth phase of prosthesis development, which is the design and manufacture of 3D printed components of the entire structure for easier physical construction and more favourable integration of hydraulic and control system within the prosthesis.

Development of the model in which all the components are integrated brings us a step closer to the goal of creating a low-cost solution for transfemoral amputees.

## References

[1] Zlata Jelačić, Remzo Dedić, Haris Džindo 2020 *Active above-knee prosthesis*. Elsevier Academic Press, ISBN 978-0-12-818683-1

[2] Jelačić Z & Dedić R 2019 *Multimodal optimal trajectory planning of a prosthetic leg*, Proceedings of the 30th DAAAM International Symposium, pp.0878-0885, B. Katalinic. DOI: 10.2507/30th.daaam.proceedings.

[3] Jelačić Z, Dedić R 2020 Real Time Control of Above-Knee Prosthesis with Powered Knee and Ankle Joints. In: Karabegović I. (eds) New Technologies, Development and Application II. NT 2019. Lecture Notes in Networks and Systems, vol 76. Springer, Cham. https://doi.org/10.1007/978-3-030-18072-0_33

[4] Jelačić Z, Dedić R 2020 Modelling and reference tracking of the robotic above-knee prosthetic leg with actuated knee and ankle joints during stair climbing. Health Technol. 10, 119–134 https://doi.org/10.1007/s12553-019-00383-8

[5] Jelačić Z, Dedić R 2020 Reference Tracking of the Robotic Above-Knee Prosthetic Leg with Actuated Knee and Ankle Joints Using Robust Control. In: Badnjevic A., Škrbić R., Gurbeta Pokvić L. (eds) CMBEBIH 2019. CMBEBIH 2019. IFMBE Proceedings, vol 73. Springer, Cham. https://doi.org/10.1007/978-3-030-17971-7_44

[6] Dedić R, Husnić Ž, Ustamujić F, Jelačić Z 2021 Development of the Concept of the Integrated Hydraulic System of the Knee Prosthesis. In: Karabegović I. (eds) New Technologies, Development and Application IV. NT 2021. Lecture Notes in Networks and Systems, vol 233. Springer, Cham. https://doi.org/10.1007/978-3-030-75275-0_11

Note:
*Airbus DS GmbH, Taufkirchen, Germany
Note: Airbus DS GmbH is not associated with this paper
**The Boeing Company, Ridley Park, PA, USA
Note: The Boeing Company is not associated with this paper

# Measurement analysis of braking forces and braking coefficient at different loads of a passenger vehicle

**Ramo Halilagić[1] and Nedim Hodžić[2]**

[1]University of Bihać, Faculty of Technical Engineering, Bosnia and Herzegovina
[2]University of Zenica, Faculty of Technical Sciences, Bosnia and Herzegovina

E-mail: halilagic.r@gmail.com

**Abstract**. This paper presented an analysis of the braking force efficiency, i.e. its intensity in the case. It presented an analysis of the efficiency of braking forces, i.e. their intensity in the case when the load of a passenger vehicle was gradually changing. The change in the load of the vehicle was done by varying the number of passengers in it from one to five which is a feature of a higher number of M1 category vehicle. By increasing in the number of passengers the weight of the vehicle varied by approximately 75kg per passenger. The experimental test was performed in the technical inspection station, i.e. on standard rollers for the brake system testing. The test results shown that the relative difference of braking forces occurred on the front axle during intensive braking and that the braking coefficient was the lowest of about 69.94% in the case when only the driver was in the vehicle. It increased with increasing load, i.e. the number of passengers in the vehicle up to 85.91% and then dropped by increasing the rear passenger behind the driver and front passenger to 80.43% which can affect the stability of the vehicle when braking in the event of a dangerous traffic situation.

## 1. Introduction

The process of motor vehicle braking has been a preoccupation of scientific circles in the automotive industry and beyond for a long time and accordingly it emphasizes that the greatest interest has been expressed in motor vehicle braking systems. Taking into account only the information and analysis of statistical indicators a wrong conclusion can be made which means that the technical malfunction of the vehicle contributes a small percentage (4-6%) to the immediate occurrence of traffic accidents [1].

However, it should be emphasized that these statistical indicators have been having low values due to the fact that they were very difficult to identify during the analysis of the cause of the accident compared to the obvious ones (such as damaged tire, brake system malfunction, steering system malfunction, etc.). Following research into the causes of accidents Volvo company conducted certain tests immediately after the accident asking whether there was a loss of stability and handling of the vehicle involved in the accident immediately before the accident or there was an another factor [2]. After the analysis of traffic accidents it was found that more than 50% of vehicles performed braking immediately before the accident, 14% lost stability due to vehicle skidding, 16% of vehicles that did not brake lost stability in another way. Such an analysis requires a more rigorous approach to the control of the technical correctness of motor vehicles in both regular and extraordinary technical inspections. When performing a technical inspection special attention should be paid to the inspection of brake systems.

Unfortunately, nowadays given that there were a lot of competition at technical inspection stations, the quality of the detailed approach to performing technical inspection tests has been at a very low level so that sometimes completely defective vehicles receive a certificate of technical correctness and readiness for its operation and that endangers the safety of traffic on the roads. International regulations that are directly related to the issue of braking to the greatest extent derive from the ECE R13 Regulation [3]. In a large number of traffic accidents with fatal consequences for both drivers and passengers that occur in Bosnia and Herzegovina but also in the wider region, the cause of their occurrence was the loss of stability of the vehicle during braking, the research of which has been actual on a daily basis [6]. In addition to the braking force, the braking coefficient has been taken as a key factor. In previous research, the determination of the braking coefficient was performed in the phases of pressure on the brake pedal for three different braking modes, i.e.: sudden braking, gradual one and braking, where it was shown that the best efficiency of braking forces was achieved with sudden braking [4]. Also, according to similar studies, a study was conducted in order to examine the braking coefficient of a set of vehicles (tractor trailer+semi-trailer) when changing the load distribution on a semi-trailer and finding the best load distribution to achieve the highest braking coefficient [5].

The research of improving the vehicle braking coefficient was performed on the basis of the change in the diameter of the hydraulic brake hoses [7]. In this paper, the efficiency of braking force in a classic passenger motor vehicle was tested with a variation in the number of passengers from 1 to 5 with a standard weight of 75 kg.

## 2. Braking coefficient testing methodology

### *2.1. Force and moments on the braked wheel*

As the total moment of inertia was reduced to the dynamic radius of the braked wheel $r_d$ it was necessary to perform an analysis of the forces and moments on the braked wheel. The analysis was performed on the example of brakes with friction discs and a hydraulic brake mechanism possessed by the ABS system. During braking, friction forces appeared between the tire and the ground. Braking was achieved when a moment was brought to the point of the vehicle in the opposite direction of its movement, and this was achieved precisely by pressing the brake pad on the brake disc. This creates a braking torque $M_k$ that opposed the movement of the wheels as shown in Figure1.

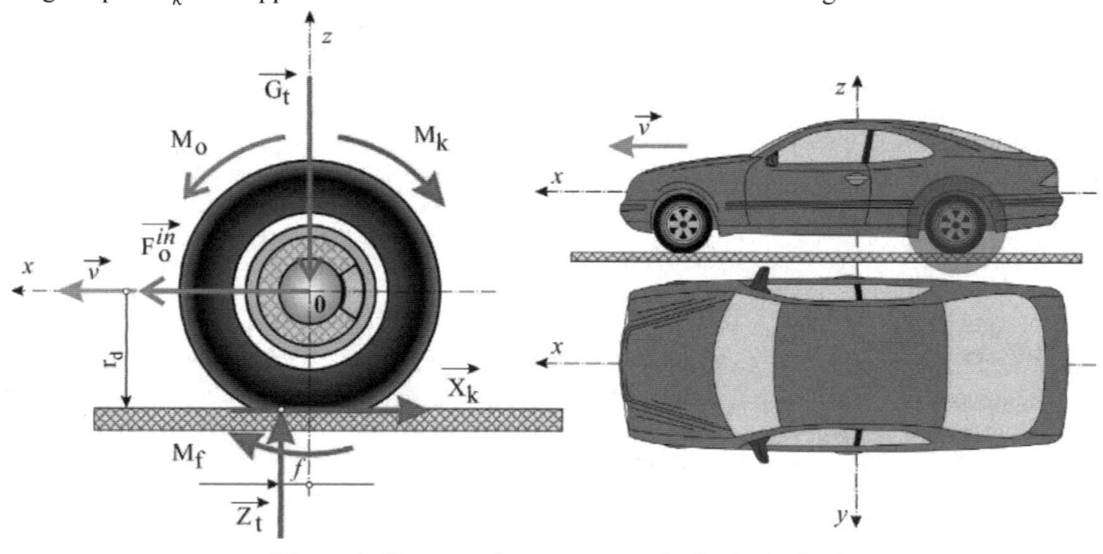


**Figure 1.** Forces and moments on the braked wheel

During intensive braking, i.e. the pressure of the pedal by the driver of the vehicle caused the brake linings or brake pads to stick to the brake disc, as a result of which a braking moment $M_k$ occurred which created a tangential force $\vec{X}_k$ of the ground opposite to the direction of movement of the vehicle or opposite to the speed vector $\vec{v}$. In addition to this braking torque, the wheel is also affected by the rolling moment $M_f$ which also helps to stop the vehicle. In the case of a drive wheel the rolling moment $M_o$ obtained from the engine via the half-axle was calculated according to the form

$$M_o = \sum J \frac{d\omega_o}{dt} - M_{rt} \,. \tag{1}$$

Where there was:

- $\sum J \frac{d\omega_o}{dt}$ moment of inertia of all rotating parts connected to the wheel axle,
- $M_{rt}$ reduced moment due to friction in the transmission mechanism.

*2.2. Braking force efficiency test method*
The experimental part of determining the intensity of braking forces as well as their efficiency was performed in an authorised vehicle technical inspection station, on a line owned by MAHA EUROSYSTEM for testing the technical correctness of vehicles with standard brake rollers. The test vehicle was a passenger motor vehicle of the SEAT brand, the technical characteristics of which were given in table 1.

**Table 1**. Technical characteristics of the test vehicle

| | |
|---|---|
| Manufacturer | SEAT, S.A. |
| Category | M1- Passenger car |
| Brand | SEAT |
| Typ | NH |
| Model | TOLEDO |
| Body type | Hatchback |
| The year of production | 2017 |
| Power (*kW*) | 66 |
| Engine displacement (*cm*³) | 1197 |
| Top speed (*km/h*) | 186 |
| Acceleration from 0-80 (*km/h*) | 7,3 |
| Acceleration from (*km/h*) | 11,3 |
| Fuel Type | Petrol (Gasoline) |
| Emission standard | EURO 6 |
| Doors | 5 |
| Kerb Weight (*kg*) | 1081 |
| Seats | 5 |
| Length (*mm*) | 4482 |
| Width (*mm*) | 1715 |
| Height (*mm*) | 1466 |
| Wheelbase (*mm*) | 2602 |
| Front overhang (*mm*) | 876 |
| Rear overhang (*mm*) | 1004 |
| Front track (*mm*) | 1463 |
| Rear track | 1500 |
| Number of Gears | Manual / 5+1 |
| Drive wheel | Front wheel drive |
| Tires size | 185/60R15 84 H |
| Turning radius (*m*) | 10,2 |

The brake force test was performed in five cases taking into account the masses of vehicles and passengers:

| | |
|---|---|
| I | - vehicle + driver, |
| II | - vehicle + driver + co-driver, |
| III | - vehicle + driver + co-driver+ right rear passenger, |
| IV | - vehicle + driver+ co-driver + right and middle rear passenger, |
| V | - vehicle + driver+ co-driver + right, middle and left rear passenger. |

### 2.3. Results of brake force intensity testing and analysis of results

*2.3.1. Results of tested vehicle with a driver*

Figure 2. shows the results of the brake force test when there was only a driver with an approximate mass of about 75 kg in the vehicle.

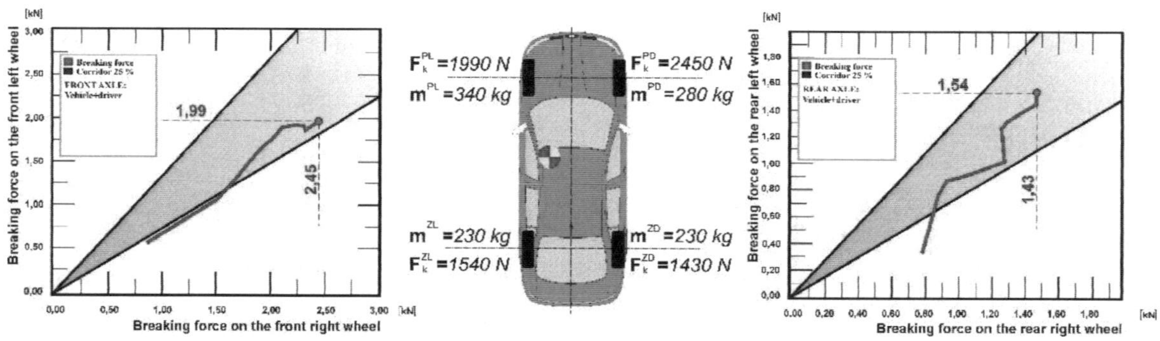


**Figure 2**. Test results (vehicle + driver)

In the first test, when only the driver was in the vehicle, an increased difference in braking force intensity was seen on the front axle wheels, which can be seen in the diagram when the maximum braking force approached the limit line, while the difference in braking forces on the rear axle wheels was significantly smaller.

*2.3.2. Test results on the vehicle with driver and co-driver*

Figure 3 shows the test results of the braking forces when the driver and co-driver were in the vehicle, i.e. when there was a slow increase in the total mass of the vehicle, which was distributed on the axles and wheels.

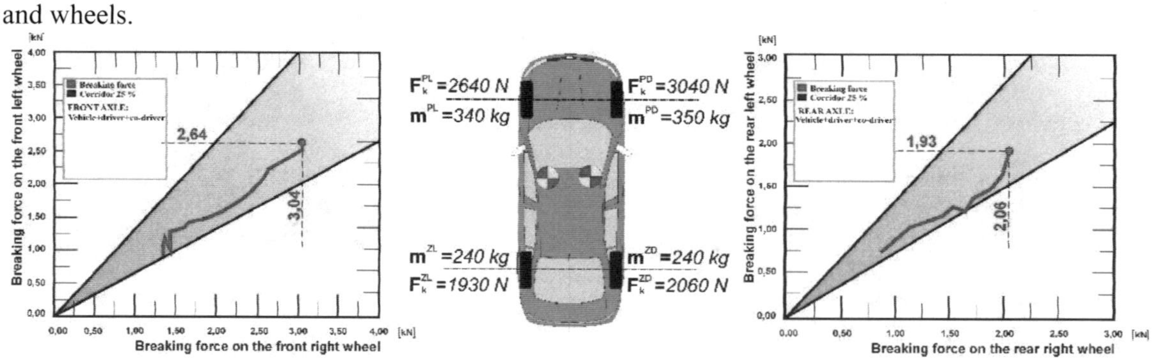


**Figure 3**. Test results (vehicle + driver + co-driver)

In the second test, when the driver and co-driver were in the vehicle, an increase in the braking forces was seen on both axles, i.e. on the front and rear, but their difference also increased.

*2.3.3. Test results on a vehicle with driver, co-driver and right back passenger*

Figure 4. shows the test results of the braking force when the driver, co-driver and right rear passenger were in the vehicle, i.e. when there was a slow increase in the total mass of the vehicle which was distributed on the axles and wheels.

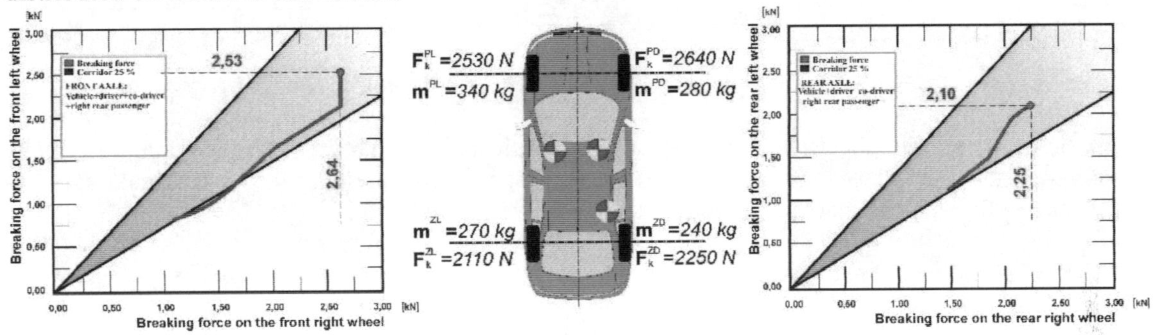


**Figure 4**. Test results (vehicle + driver + co-driver + right rear passenger)

In the third test when the driver, co-driver and right rear passenger were in the vehicle an increase in braking forces on the rear axle was visible, while their differences on the wheels of the same axle were significantly reduced as a result of changing the position of the vehicle centre.

*2.3.4. Test results on a vehicle with driver, co-driver, right and middle rear passenger*

Figure 5. shows the results of testing the braking forces when in the vehicle there were: driver, co-driver, right and middle rear passenger, or when there was a slow increase in the total mass of the vehicle which was distributed on the axles and wheels.

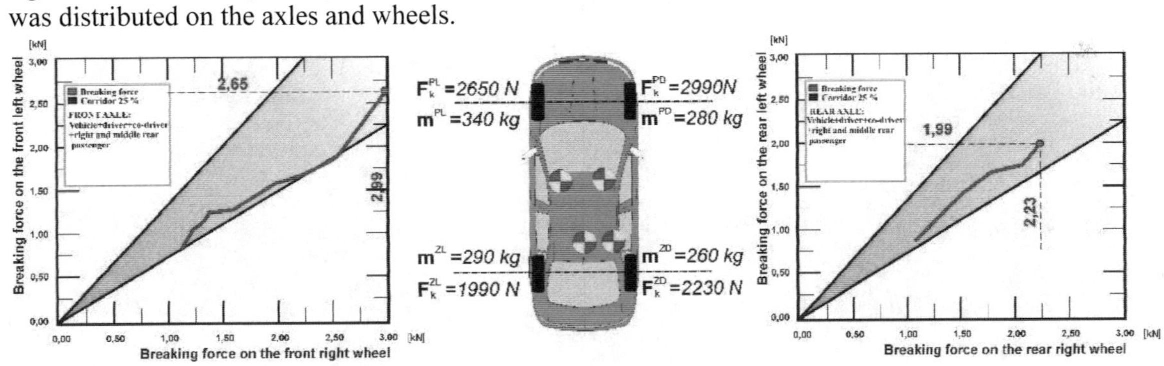


**Figure 5**. Test results (vehicle + driver + co-driver + right and middle rear passenger)

In the fourth test when the driver, co-driver, right and middle rear passenger were in the vehicle, an increase in the braking forces on the rear axle and their smaller differences on the wheels of the same axle was visible.

*2.3.5. Test results on a vehicle with driver, co-driver, right, middle and left rear passenger*

Figure 6. shows the results of testing the braking forces when in the vehicle there were: driver, front passenger, right, middle and left rear passenger, or when there was a slow increase in the total mass of the vehicle which was distributed on the axles and wheels.

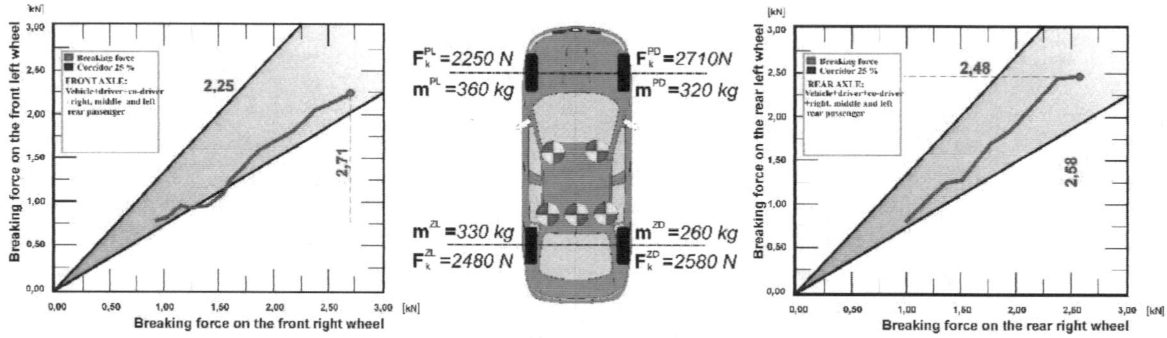


**Figure 6**. Test results (vehicle + driver co-driver + right, middle and rear passenger)

In the fifth test when the driver, co-driver, right, middle and left rear passenger were in the vehicle, a decrease in the intensity of the braking forces of the front axle wheels and an increase in braking forces with a more intense difference on the front axle wheels were visible.

*2.3.6. Analysis of the obtained results*

It has been known that the quality of the braking system was measured by the braking coefficient prescribed by law for certain categories of vehicles, which in the ideal case is to be calculated according to the expression

$$z = \frac{\sum_{i=1}^{4} F_{ki}}{\left(\sum_{i=1}^{4} m_i\right) \cdot g} \cdot 100\ \% \quad , \tag{2}$$

where there was:

- $\sum_{i=1}^{4} F_{ki}$ total braking force on all wheels,
- $\left(\sum_{i=1}^{4} m_i\right) \cdot g$ the total weight of the vehicle which was equal to the sum of all normal reactions on the braked wheels [8].

For the given test results in the developed form of expression (2)

$$z_n = \frac{F_{kn}^{pl} + F_{kn}^{pd} + F_{kn}^{zl} + F_{kn}^{zd}}{\left(m_n^{pl} + m_n^{pd} + m_n^{zl} + m_n^{zd}\right) \cdot g} \cdot 100\ \% \quad . \tag{3}$$

Based on the expression (3) the braking coefficients were obtained:

$$z_I = 69,94\ \%$$
$$z_{II} = 84,25\ \%$$
$$z_{III} = 85,97\ \%$$
$$z_{IV} = 85,91\ \%$$
$$z_V = 80,43\ \%$$

The obtained results of braking coefficients were shown by the diagram in Figure 7.

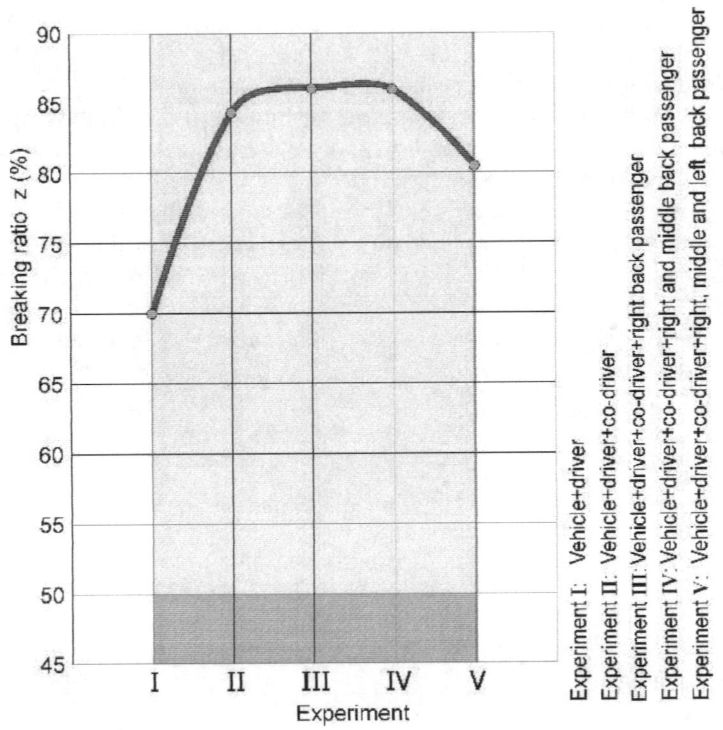


**Figure 7**. Diagram of the braking coefficient of the experimental values

Figure 7. shown that the braking coefficient of the test vehicle was above the minimum legally allowed 50% because it was a relatively newer vehicle that had less time in technical operation [8]. What was characteristic was the fact that the braking coefficient was changed with the change of the total mass of the vehicle and the distribution of these masses in the vehicle, as in this case the passengers.

## 3. Conclusion

The braking force test was performed in five cases, i.e. by gradually increasing the total mass of a passenger motor vehicle of category M1 by the mass of one passenger with an approximate mass of 75 kg. The experiment showed the deviations of the partial braking forces of individual axles and thus the deviation of the braking coefficient which was the basic technical characteristic of the quality of the vehicle brake mechanism. The test vehicle was relatively newer so that the differences in the intensity of the braking forces of the front and rear axles remained below the legal maximum 25% and the braking coefficient above the legal 50% for a given vehicle category [8]. However, as the experiment showed deviations of both parameters with increasing vehicle weight and mass position on the vehicle itself which caused a change in the centre of gravity of the vehicle, i.e. its inertial center of mass, it was very important to consider the actual schedule of passengers. other masses that were at a given moment just before its formation in order to adequately arrive at its cause.

## References

[1] Todorović J 1972 *Istraživanje osnovnih karakteristika puzdanosti frikcionih obloga kočnica u uslovima eksploatacije teških motornih vozila u našoj zemlji [disertacija]* (Beograd: Univerzitet u Beogradu, Mašinski fakultet)
[2] AB Svensk Bilprovning, 1969. Weak points of cars; Stockholm.
[3] Regulation No 13 of the Economic Commission for Europe of the United Nations (UN/ECE) — Uniform provisions concerning the approval of vehicles of categories M, N and O with regard to braking [2016/194] doi: 10.18356/ddf30969-en
[4] Firdaus R M, Supriyo B and Suharjono A 2019 Analysis of braking force efficiency measurements for various braking strategy applied for vehicle tested on roller brake tester *Journal of Physics: Conference Series* **1517** 012079 doi: 10.1088/1742-6596/1517/1/012079
[5] Šarkan B, Jaśkiewicz M and Kiktová M. 2020 The impact of the truck loads on the braking efficiency assessment *Journal* **10** doi. 10.1515/eng-2020-0014
[6] Rivers R W 2006 *Evidence in traffic crash investigation and reconstruction: identification, interpretation and analysis of evidence, and the traffic crash investigation and reconstruction process* (Springfield: Charles C Thomas)
[7] Saeed Abu Alyazeed Albatlan 2012 Automotive brake pipes characteristics and their effects on brake performance *Ain Shams Engineering Journal* 3(2012)3 doi: 10.1016/j.asej.2012.02.002
[8] Pravilnik o dimenzijama, ukupnoj masi i osovinskom opterećenju vozila, o uređajima i opremi koju moraju imati vozila. Službeni glasnik BiH. 2007. Broj 23. str. 2776-2812. [ 02.04.2007.]

# Mechanical properties comparison of PLA, tough PLA and PC 3D printed materials with infill structure – Influence of infill pattern on tensile mechanical properties

**Adi Pandzic and Damir Hodzic**

Faculty of Mechanical Engineering Sarajevo, University of Sarajevo, Vilsonovo Setaliste 9, Sarajevo, Bosnia and Herzegovina

E-mail: pandzic@mef.unsa.ba

**Abstract**. One of the advantages provided by fused deposition modelling (FDM) 3D printing technology is the manufacturing of product materials with infill structure, which provides advantages such as reduced production time, product weight and even the final price. In this paper, the tensile mechanical properties, tensile strength and elastic modulus, of PLA, Tough PLA and PC FDM 3D printed materials with the infill structure were analysed and compared. Also, the influence of infill pattern on tensile properties was analysed. Material testing were performed according to ISO 527-2 standard. All results are statistically analysed and results showed that infill pattern have influence on tensile mechanical properties for all three materials.

## 1. Introduction

3D printing technology, also known as additive manufacturing, has developed rapidly in the manufacturing, engineering, automotive, civil, aerospace, medical and other fields. AM is a fast-growing technology that enables the production of products of both simple and complex geometries [1], [2]. Additive manufacturing is relatively new technology which uses a deposition of material layer by layer to produce parts using a CAD model. The additive manufacturing process is based on the generating of the virtual CAD model of product, then slicing this model into 2D cross-sections using "slicer" software and translating these 2D data to the 3D printers in order to manufacture the physical product layer by layer (Figure 1) [3].

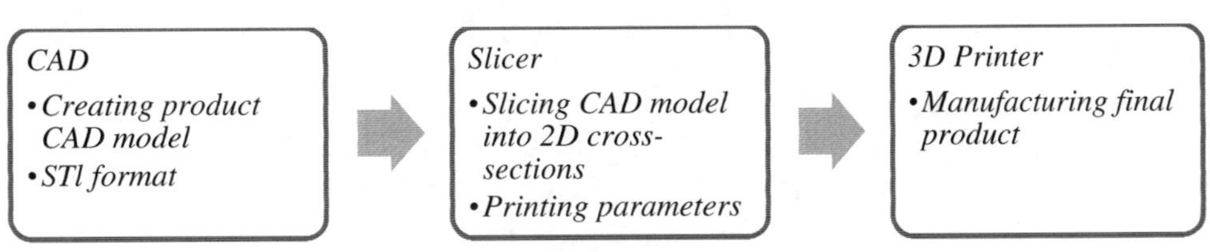


**Figure 1.** Schematic representation of the additive manufacturing process

During the last decade, 3D printing has made significant developments in industrial products. One of the main advantages of 3D printing is converting industrial products from a proposal to an actual application in minimum time. On todays market, there are many different AM technologies, but one of the widely used technology is Fused Deposition Modelling (FDM), and mostly due to relatively low cost and its user-friendliness [4], [5].

FDM as one of the most popular but also complex additive manufacturing (AM) technology that provides a number of advantages over conventional manufacturing processes. FDM process, also known as "Fused Filament Fabrication, FFF" and "Material Extrusion", is based on the extrusion of molten material layer by layer on working bed and producing part on working print bed, showed on picture bellow [6], [7].

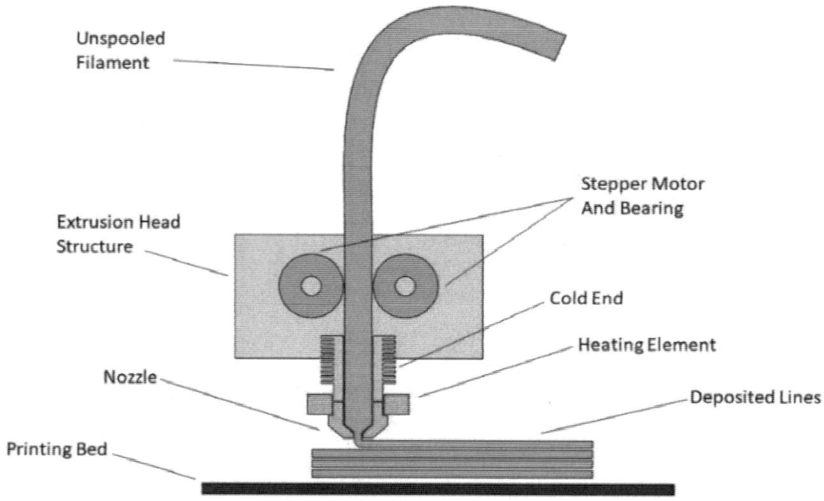


**Figure 2.** Schematic representation of FDM process [8]

The 3D printing parts fabrication with FDM technology begins with the material also known as a filament, most often the material is thermoplastic, that is conveyed into the extrusion head via a feeder. In the extrusion head, the filament is melted in a heating element and through the nozzle it is applied in the form of deposited lines on the printing bed, where the final geometry of the product is formed. The material is in the form of wire, known as filament, and usually materials for FDM are thermoplastics, but also materials can be different types of composites, ceramics and even metals. Some of the commonly used materials for producing parts with FDM technology are PLA (Polylactic Acid), PC (Polycarbonate), PET (Polyethylene terephthalate), PP (Polypropylene), ABS (Acrylonitrile Butadiene Styrene), TPU (thermoplastic polyurethane) and Nylon presented in the below (Figure 3) [7], [9], [10].

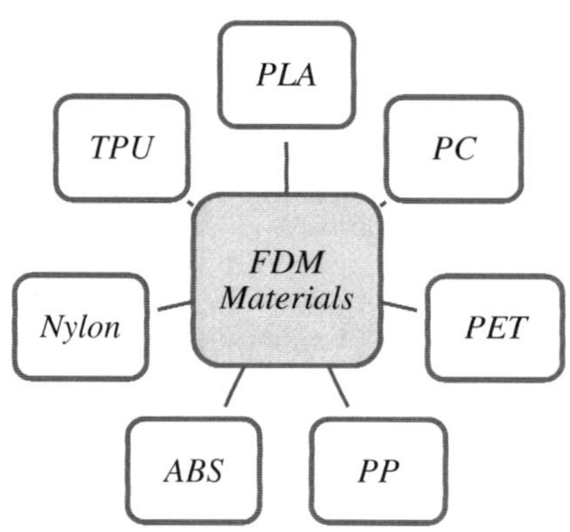


**Figure 3.** Most commonly used materials in FDM technology

Product quality and mechanical properties of printed materials depends on a large number of parameters (Figure 4) that need to be analyzed in order to make the most of the advantages of this technology. Based on the literature analysis of previous research, the parameters that are most often examined and that have a significant impact on the mechanical properties of FDM printed materials are: layer thickness, infill density, infill pattern, infill angle, build orientation, raster angle, printing speed, printing temperature, airgap, and more [8], [11], [12].

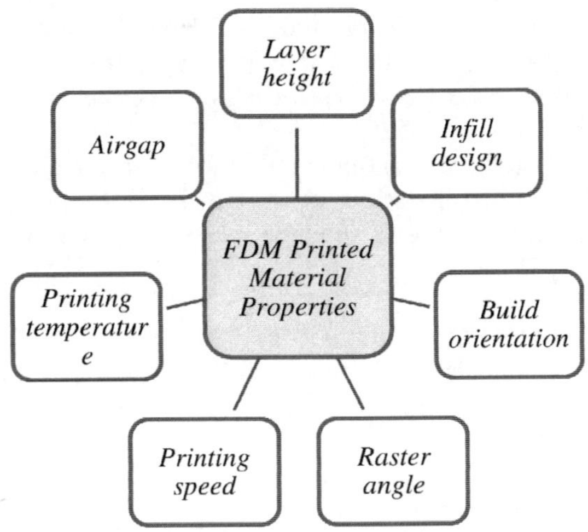


**Figure 4.** Parameters with significant impact on FDM printed materials mechanical properties

As meintoined before, FDM printed material with an infill structure provides several advantages such as reduction of printing time and the amount of consumed material, which directly affects the reduction of the price, also lighter weight of final product. To make the most of this, we need to understand how infill design affects the mechanical properties of the printed material in order to provide required strength as well as other properties of the finished product.

The main influencing factors of infill design (Figure 5), also the most frequently examined are "Infill density", "Infill pattern" and "Infill angle", and through literature review it can be seen that all three have an impact on different mechanical properties of printed materials.

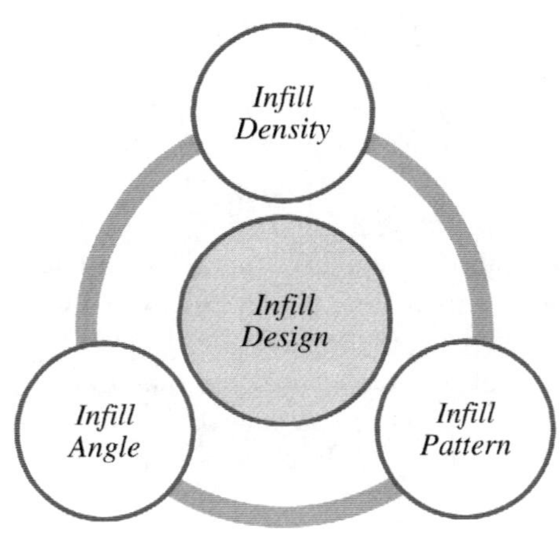


**Figure 5.** The main influencing factors of infill design

The importance of studying this topic is evidenced by previous research of infill design impact on the mechanical and other properties of FDM printed parts, such as that of Mustafa et al. [13], where effect of infill density on tensile and flexural properties of PLA material was analysed; that of Miguel and Rismalia et al. [14], [15], the influence of infill pattern and density on ABS and PLA FDM printed material tensile mechanical properties; that of Jaya [16], the impact of infill pattern, density and angle on the printing time and the filament length were analysed, that of Nuha et al. [17], the influence of infill density on the flexural properties of PLA printed material; and that of Pushpendra et al. [18], strength and surface characteristics analyse of FDM printed PLA parts for multiple infill patterns. In addition to this, master thesis on this topic were written, where impact of infill design on mechanical strength and production cost in FDM was researched [19] and observing the effects of infill shapes on tensile properties of 3D printed polymeric parts [20].

In this paper, the focus is on the analysis of the influence of infill pattern on the mechanical properties of three different FDM 3D printed materials: PLA, Tough PLA and PC. For every material, seven different infill patterns were examined, all results are statistically analysed and presented in further text.

## 2. Materials

For this study, three FDM printing material types, PLA, Tough PLA and PC were used (Figure 6). All materials were from same manufacturer "Ultimaker" and same colour (black). This is important because in research [5] and [10] has concluded that mechanical properties differ from one material manufacturer to another, but also colour affects mechanical properties even though the material is from the same manufacturer.

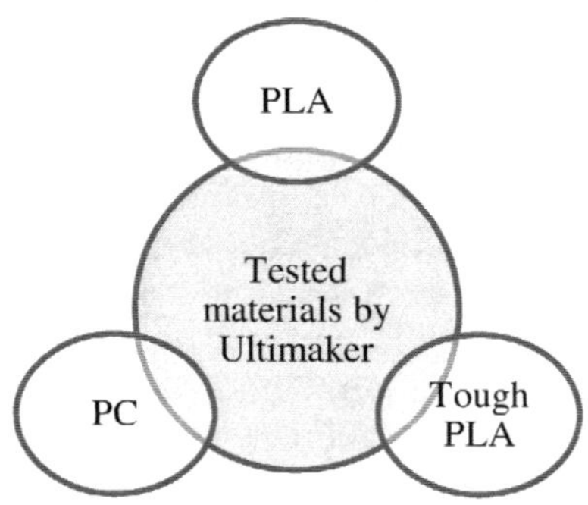


**Figure 6.** Tested materials by manufacturer Ultimaker: PLA, Tough PLA and PC

Ultimaker PLA material is a polymer named polylactic acid, made from renewable and organic resources. It is biocompatible thermoplastic that provides easy printing and good surface quality. It is safe and nontoxic material and is used in a wide range of applications for novice and advanced users. It has relatively good tensile and flexural strength, higher than ABS, easy to work with at high printing speeds, user-friendly for home and office environments with a wide range of colour options available. It has a low coefficient of thermal expansion, which limits its applications in which the printed part is exposed to temperatures higher than 50 °C [7], [21].

Ultimaker Tough PLA material is a technical PLA material with better toughness comparable to ABS. It is material good for printing a "large" sizes technical models, and also offers the same safe and easy use as regular Ultimaker PLA material. It has stronger impact resistance than PLA and similar to ABS, also with higher stiffness compared to ABS. Tough PLA bends before it breaks, which makes it well suitable for many engineering applications, especially where high-wear and impact resistance are required. It is less brittle than regular PLA and gives a more matte surface finish quality. With heat resistance similar to standard PLA materials, so printed products should not be exposed to temperatures above 60 °C [7], [22], [23].

PC material is polycarbonate thermoplastic material for manufacturing strong and tough products. In general, PC material has high thermal stability, up to 100 °C. Ultimaker PC material is engineered to be printed at moderate temperatures compared to other PC filaments and shows minimized warping and good printability. Mechanical properties differs from the colour of the material, so that the transparent colour gives the higher tensile and flexural properties compared to black and white colour. In general, material is with high toughness, flame retardant characteristics, temperature resistance, strong interlayer bonding, dimensionally stable (up to 100 °C) and with good bed adhesion. It is used for molds, engineering parts, tools, functional prototyping and short-run manufacturing [24], [25].

The mechanical properties of these materials defined by the manufacturer are presented in Table below. What should be noted is that the values of these mechanical properties from the Table below may differ because they depend on the 3D printing parameters and other factors, as shown by many studies [1], [2], [3], [11]. Also, the method of material testing can affect, because one of the influencing factors on tensile mechanical properties is strain rate during material testing [26].

**Table 1.** Material specification of PLA, Tough PLA and PC materials defined by manufacturer

| Material properties | *Standard* | PLA | *Tough PLA* | PC |
|---|---|---|---|---|
| Colour | - | Black | Black | Black |
| Diameter [mm] | - | 2,85 | 2,85 | 2,85 |
| Tensile modulus [GPa] | ISO 527 (1 mm/min) | 2,35 | 1,8 | 1,9 |
| Yield stress [MPa] | ISO 527 (50 mm/min) | 49,5 | 37 | - |
| Tensile at break [MPa] | ISO 527 (50 mm/min) | 45,6 | 37 | 53,7 |
| Elongation at brake [%] | ISO 527 (50 mm/min) | 5,2 | 3,1 | 5,9 |
| Flexural strength [MPa] | ISO 178 | 103 | 78 | 95.5 |
| Hardness [Shore D] | Durometer | 83 | 79 | 80 |

## 3. Methodology

In this study, three different FDM printed materials with infill structure were tested. The aim was to analyse influence of infill pattern on tensile mechanical properties. Experiment methodology is presented on diagram below (Figure 7), and details are presented in further text.

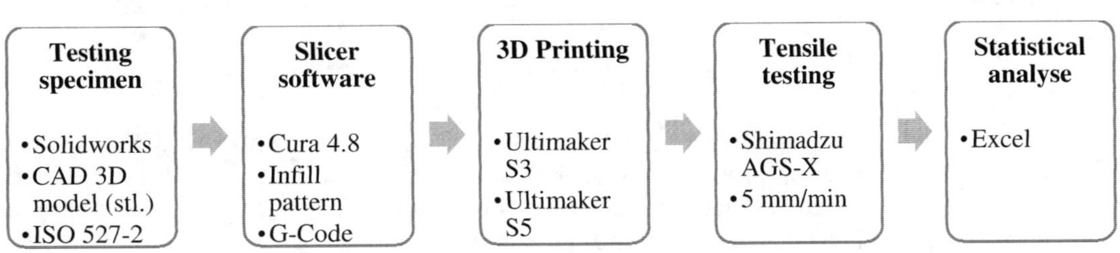


**Figure 7.** Experiment methofology

CAD 3D model of testing specimens are designed in Solidworks, according to ISO 527-2 in dog-bone shape, and prepared in "stl." format. Specimen geometrical characteristics and dimensions are presented on Figure 8.

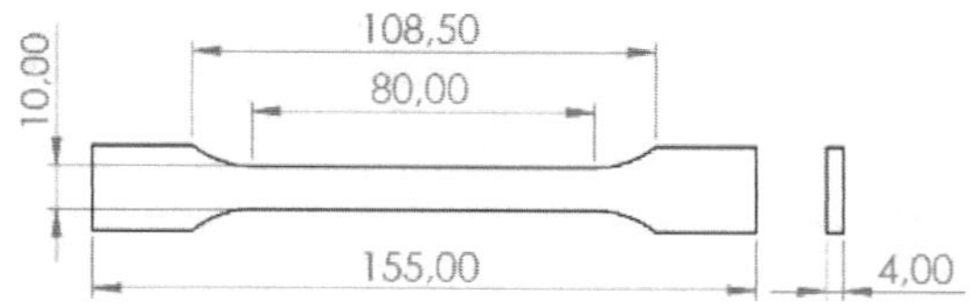


**Figure 8.** Tensile testing specimen dimensions according to ISO 527-2

Cura slicer software was used to prepare specimens for 3D printing and defining printing parameters. All specimens are printed with predefined printing parameters for Ultimaker materials in black colour and with "Normal 0,15 mm" printing profile. Main printing parameters for every material are presented in Table 2.

**Table 2.** Main 3D printing parameters for PLA, Tough PLA and PC materials

| 3D printing parameter | PLA | Tough PLA | PC |
|---|---|---|---|
| Cura printing profile | Normal, 0,15mm | Normal, 0,15mm | Normal, 0,15mm |
| Nozzle diameter, [mm] | 0,4 | 0,4 | 0,4 |
| Layer height, [mm] | 0,15 | 0,15 | 0,15 |
| Printing temperature, [°C] | 200 | 215 | 280 |
| Build plate temperature [°C] | 60 | 60 | 110 |
| Printing speed, [mm/s] | 70 | 45 | 50 |
| Infill print speed, [mm/s] | 50 | 50 | 50 |
| Fan speed, [%] | 100 | 100 | 0 |
| Infill density, [%] | 60 | 60 | 60 |
| Infill angle, [°] | 45 | 45 | 45 |

For every material 7 different infill patterns were analysed. In 4.8 version of Cura slicer, there are 13 types of infill available for use, but for this research, only patterns that were named as "strong" by the manufacturer Ultimaker were selected, and these are: Cubic Subdivision (CS), Cubic (CU), Grid (G), Octet (OC), Quarter Cubic (QC), Triangles (T) and Tri-Hexagon (TH), presented on Figure 9.

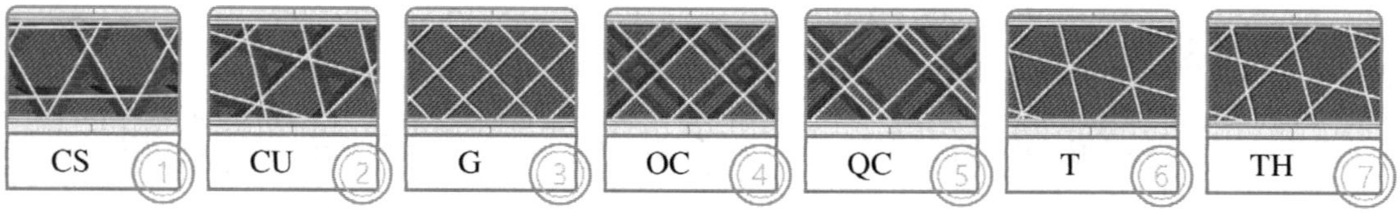


**Figure 9.** Examined infill patterns from slicer Cura 4.8

Tensile specimens were manufactured on Ultimaker S5 and S3 desktop FDM 3D printers (Figure 10). Three replicas of tensile specimens were printed for each infill pattern. In order to reduce the influence of external "negative" factors on the results, the same 3D printer was always used and the specimens were printed in the same position on the build plate. Also, as already mentioned, material from the same manufacturer and the same colour was used.

**Figure 10.** 3D printing of tensile specimens on Ultimaker S5 and S3 3D printers

A total of 63 specimens were printed and tensile tested using Shimadzu AGS-X tensile testing machine (max. force 10kN). The tensile test was performed according to the ISO 527-2 standard, where the strain rate was 5 mm/min, at room temperature. Figure 11 shows tensile testing process.

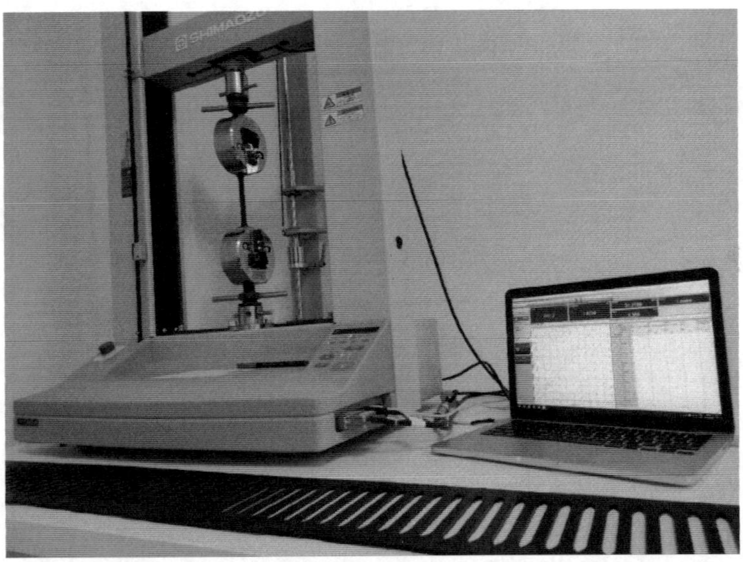

**Figure 11.** Process of tensile testing according to ISO 527-2

Monitoring and recording of mechanical properties of the tested materials during test were done using the program Trapezium X (Shimadzu). All results were statistically analysed in Excel and presented in the next chapter.

## 4. Results and discussion

Tensile testing of PLA, Tough PLA and PC printed materials with infill structure was carried out for all specimens at the same loading conditions. In general, from Stress-Strain diagrams it can be seen that infill pattern have influence on mechanical properties of FDM printed materials with infill structure. Also, Stress-Strain diagrams with all and average results for all three tested materials are presented on Figure 12.

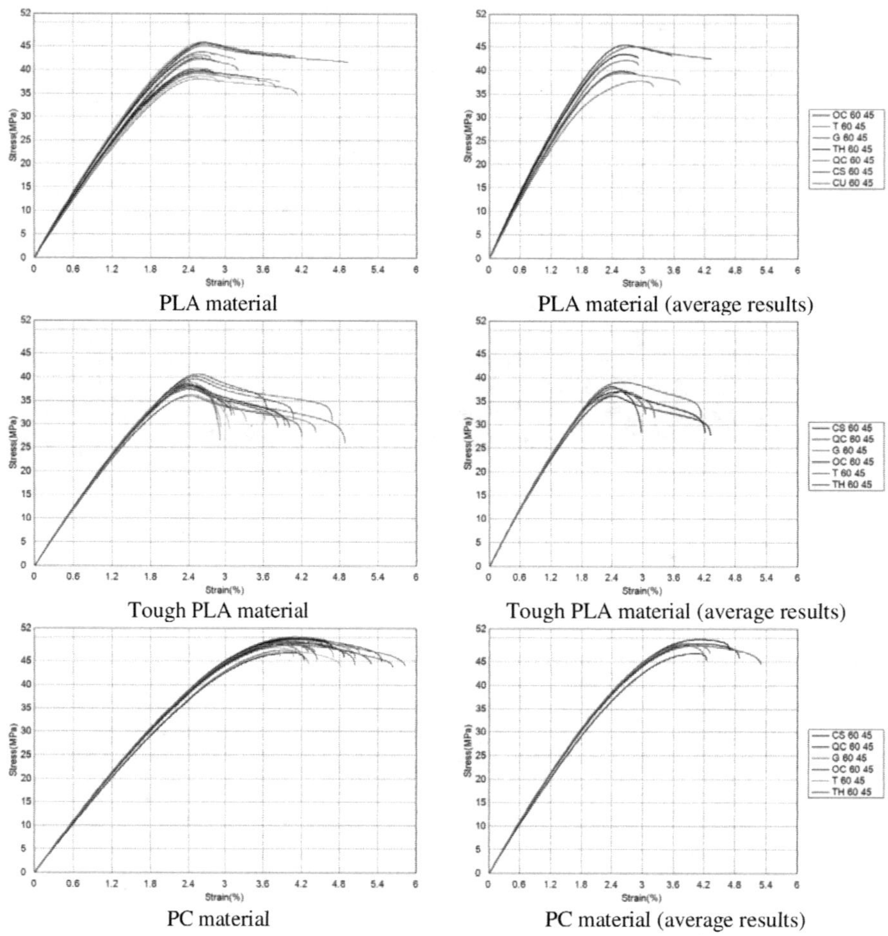


**Figure 12.** Stress-Strain diagrams for all three tested FDM printed materials with infill structure

In the following text, the results of tensile strength (Rm) and elastic modulus (E) for tested materials are presented and commented, which are also presented trough diagrams. Also, all results with average values of Rm and E are presented in Table 3.

**Table 2.** Tensile strength and elastic modulus results, with average values, for PLA, Tough PLA and PC materials with infill structure

| Infill design | | | PLA | | Tough PLA | | PC | |
|---|---|---|---|---|---|---|---|---|
| Pattern | Angle | Density | Rm [MPa] | E [GPa] | Rm [MPa] | E [GPa] | Rm [MPa] | E [GPa] |
| CS | 45 ° | 60 % | 42.50 | 2.20 | 38.13 | 2.1 | 47.20 | 1.8 |
| CU | 45 ° | 60 % | 39.47 | 2.20 | 38.33 | 2.1 | 49.67 | 1.9 |
| G | 45 ° | 60 % | 45.53 | 2.30 | 40.10 | 2.1 | 49.17 | 1.9 |
| OC | 45 ° | 60 % | 45.57 | 2.40 | 37.77 | 2.1 | 50.03 | 1.9 |
| QC | 45 ° | 60 % | 38.40 | 2.10 | 36.40 | 2 | 49.50 | 1.9 |
| T | 45 ° | 60 % | 39.90 | 2.20 | 38.40 | 2.1 | 48.60 | 1.9 |
| TH | 45 ° | 60 % | 43.57 | 2.30 | 38.23 | 2.1 | 48.97 | 1.9 |

Full name of infill pattern according to Cura slicer: Cubic Subdivision (CS), Cubic (CU), Grid (G), Octet (OC), Quarter Cubic (QC), Triangles (T) and Tri-Hexagon (TH).

Analysing the tensile strength of printed PLA material with infill structure, it can be seen that the value of tensile strength is from 45,57 MPa to 38,40 MPa, depending on the infill pattern (Figure 13). The highest tensile strength is shown by "Octet" and "Grid" patterns with 45,5 MPa, and the lowest tensile strength is shown by "Quarter Cubic" pattern with 38,40 MPa.

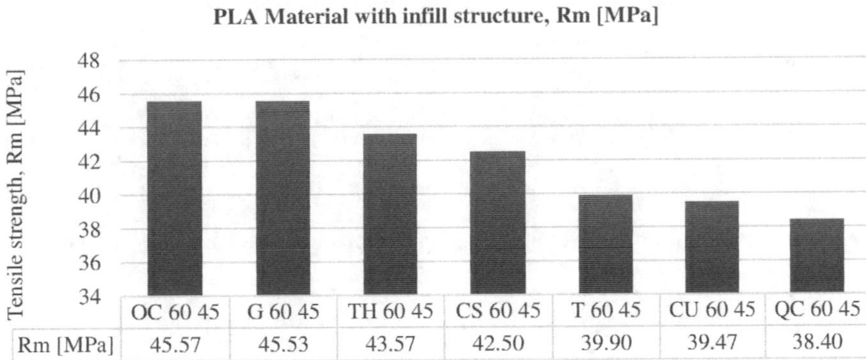


**Figure 13.** Tensile strength of PLA FDM printed material with different infill pattern

Results of elastic modulus for PLA material with infill structure are presented on diagram below (Figure 14). Elastic modulus varies from 2,40 GPa to 2,10 GPa, depending on the infill pattern, where "Octet" pattern is showing highest value, and "Quarter Cubic" pattern the lowest value of elastic modulus.

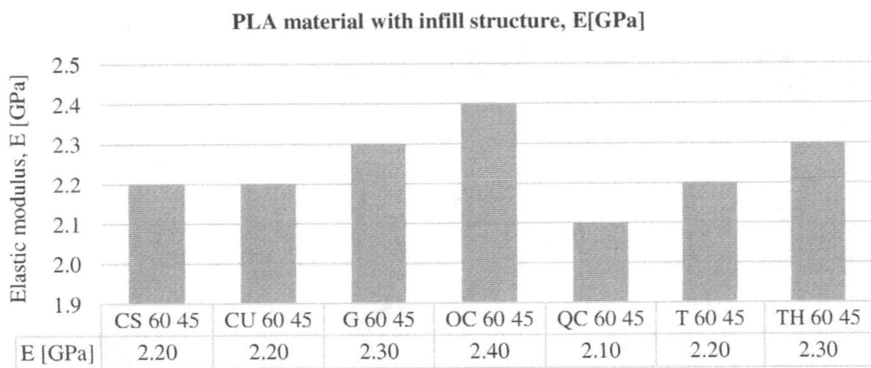


**Figure 14.** Elastic modulus of PLA FDM printed material with different infill pattern

For Tough PLA printed material with infill structure, tensile strength varies from 40,10 MPa to 36,40 MPa, depending on infill pattern (Figure 15). Highest tensile strength was achieved by "Grid" and the lowest by "Quarter Cubic" infill pattern. In general, the results are showing lover tensile strength for Tough PLA compared to PLA material.

Elastic modulus for Tough PLA material with infill structure compared with PLA material is slightly lower. What is noticeable is that for all infill patterns, value of the elastic modulus is the same and amounts to 2,10 GPa, except for "Quarter Cubic" pattern where it is 2,00 GPa. Elastic modulus results for Tough PLA material are presented on diagram below (Figure 16).

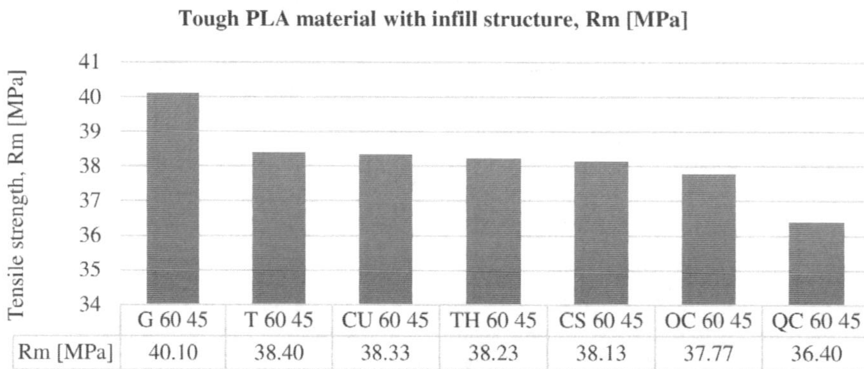


**Figure 15.** Tensile strength of Tough PLA FDM printed material with different infill pattern

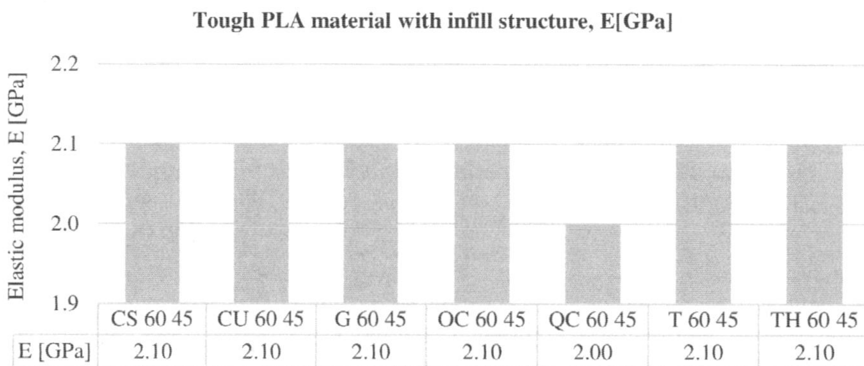


**Figure 16.** Elastic modulus of Tough PLA FDM printed material with different infill pattern

As expected, based on the data from Table 1, PC FDM printed material with infill structure showing highest tensile strength, where values are in range from 50 MPa to 47 MPa. Infill pattern "Octet" is showing highest, and "Cubic Subdivision" the lowest values of tensile strength. Tensile strength results for PC material are presented on diagram below (Figure 17).

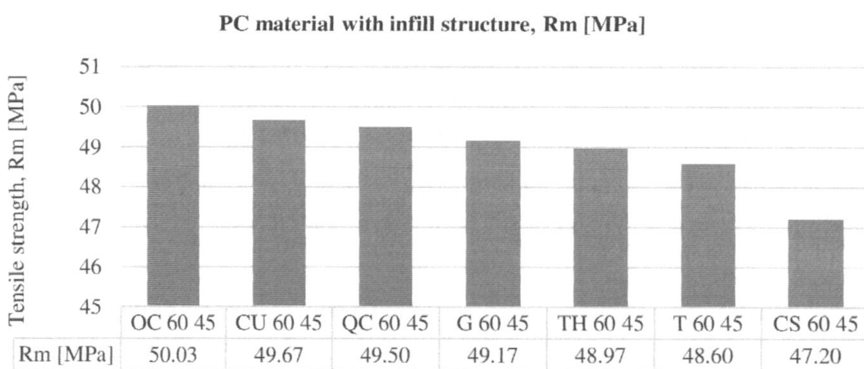


**Figure 17.** Tensile strength of PC FDM printed material with different infill pattern

Elastic modulus results for PC material are presented on Figure 18. Overall, FDM printed PC material with infill structure showed lowest elastic modulus values, compared to PLA and Tough PLA materials.

IOP Conf. Series: Materials Science and Engineering 1208 (2021) 012019 doi:10.1088/1757-899X/1208/1/012019

All infill patterns have 1,90 GPa elastic modulus, while only "Cubic Subdivision" pattern showed a slightly lower value of 1,80 GPa.

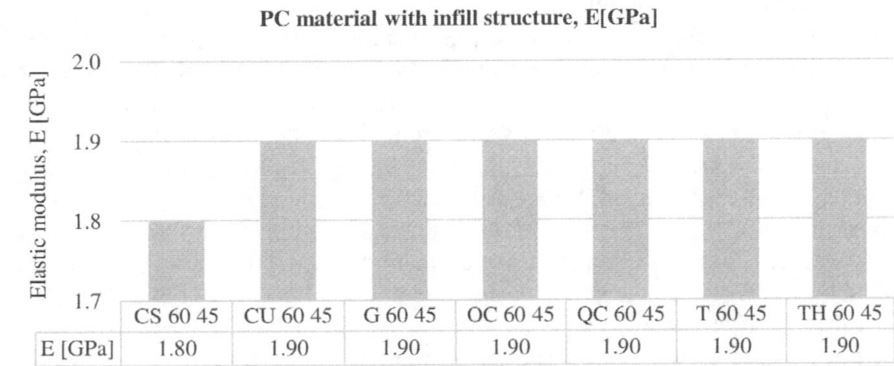


**Figure 18.** Elastic modulus of PC FDM printed material with different infill pattern

From the diagram below (Figure 19), it is evident that when looking at the results for all infill patterns, it can be seen that the highest tensile strength was achieved by PC, than PLA and Tough PLA FDM printed material with infill structure. Also, the highest values of elastic modulus were achieved by PLA, than Tough PLA and PC material (Figure 20).

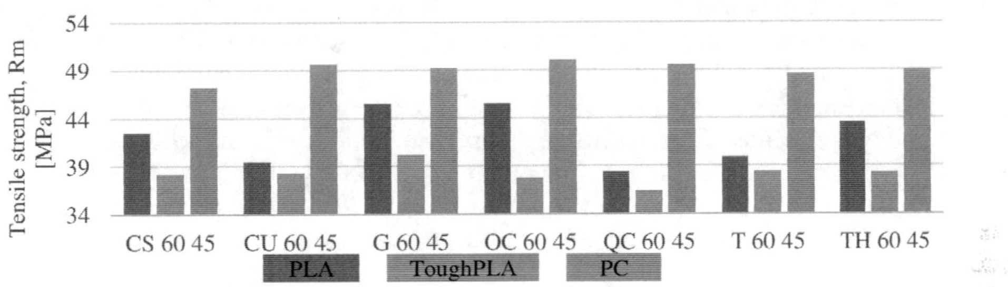


**Figure 19.** Tensile strength of all tested FDM printed materials with different infill pattern

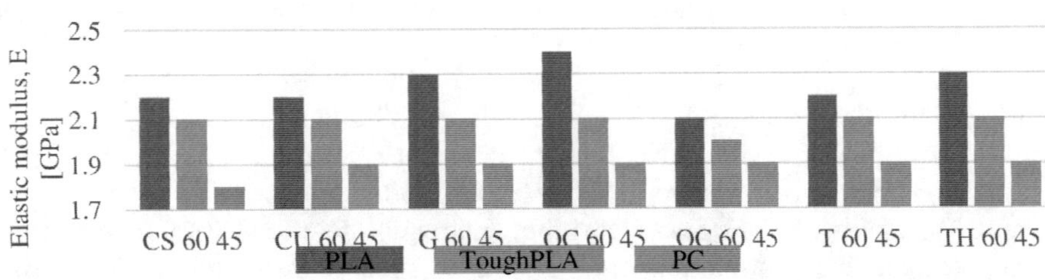


**Figure 20.** Elastic modulus of all tested FDM printed materials with different infill pattern

## 5. Conclusion

The literature review has shown the advantages of materials with infill structure that can be applied to products manufactured by additive technologies such as FDM. Infill design is defined by several factors that affect the mechanical properties of the printed material with infill structure, and one of them is the infill pattern, which through this research has been shown to affect tensile mechanical properties. In this research, influence of infill pattern on PLA, Tough PLA and PC FDM printed materials with infill structure were experimentally analysed. The obtained experimental results were statistically analysed, presented through diagrams, compared and based on those results conclusions are:

- Overall, infill pattern have an influence on tensile mechanical properties such as tensile strength and elastic modulus.
- For PLA material with infill structure tensile strength varies up to 17%, depending on the infill pattern, where maximum value is showed by "Octet" and "Grid" and minimum with "Quarter Cubic" infill pattern. Also, for elastic modulus, the influence of the infill pattern can be seen, where the results varies up to 9%. The "Octet" pattern showed the highest elastic modulus of 2,40 GPa, while "Quarter Cubic" had the lowest value of 2,10 GPa.
- For Tough PLA material with infill structure tensile strength varies up to 10%, depending on the infill pattern, where maximum value is showed by "Grid" and minimum with "Quarter Cubic" infill pattern. For all patterns the elastic modulus was the same with the value of 2,10 GPa, except for "Quarter Cubic" pattern which have up to 5% lower value in the amount of 2,0 GPa.
- For PC material with infill structure tensile strength varies up to 6%, depending on the infill pattern, where maximum value is showed by "Octet", and minimum with "Cubic Subdivision" infill pattern. Similar to Tough PLA material, elastic modulus for PC material was the same for all patterns, except "Cubic Subdivision", which have up to 5% lower elastic modulus.
- In general, when all three materials are analysed together, the highest tensile strength is shown by PC, while the lowest by Tough PLA material. Also, when the elastic modulus is observed, the highest values is shown by PLA, while PC material is showing lowest value of elastic modulus.

Analysis of other infill design parameters such as infill density and infill angle for these seven patterns is recommended for future research. In addition to this, it is necessary to do deeper analysis of all infill design parameters together, in order to determine the most influential factors on the mechanical properties, and also the interactions between them. Also, connect all this with the production time and the amount of material, so that a techno-economic analysis can be done. Of course, in addition to mechanical tensile properties, analyse both compressive and flexural properties, to better understand the behaviour at different loads and to make the most of the advantages of the printed materials with infill structure.

## Acknowledgments

The authors would like to acknowledge the Ultimaker company for supporting this research.

## References

[1] Hsueh M H, Lai C J, Wang S H, Zeng Y S, Hsejh C H, Pan C Y, Huang W C 2021 Effect of printing parameters on the thermal and mechanical properties of 3D-Printed PLA and PETG, using fused deposition modeling. *Polymers* **13** 1758

[2] Valaean C, Marsavina L, Marghitas M, Linul E, Razavi J, Berto F 2020 Effect of manufacturing parameters on tensile properties of FDM printed specimens. *Procedia Structural Integrity* **26** 313-320.

[3] Syrlybayev D, Zharylkassyn B, Seisekulova A, Akhmetov M, Perveen A, Talamona D 2021 Optimisation of strength properties of FDM printed parts-a critical review. *Polymers* **13** 1587

[4] Harpool T D, Alarifi I M, Alshammari B A, Aabid A, Baig M, Malik R A, Mohamed S A, Asmatulu R, EL-Bagory T M A A 2021 Evaluation of the infill design on the tensile response of 3D printed polylactic acid polymer. *Materials* **14** 2195

[5] Hodzic D, Pandzic A, Hajro I, Tasic P 2020 Strength comparison of FDM 3D printed PLA made by different manufacturers. *TEM Journal* **9** 966-970

[6] Pandzic A, Hodzic D, Milovanovic A 2019 Effect of infill type and density on tensile properties of PLA material for FDM process. *Proceedings of the 30th DAAAM International Symposium*

0545-0554
[7] Naveed N 2021 Investigating the material properties and microstructural changes of Fused Filament Fabricated PLA and Tough-PLA parts. *Polymers* **13** 1487
[8] Cabreira V, Santana R M C 2020 Effect of infill pattern in Fused Filament Fabrication (FFF) 3D printing on materials performance. *Materia* **25** 1126
[9] Forster W, Pucklitzsch T, Nickel D 2021 The influence of different infill structures on the mechanical properties in additive manufacturing. *ACC Journal* **27** 7-16
[10] Pandzic A, Hodzic D, Milovanovic A 2019 Influence of material colour on mechanical properties of PLA material in FDM technology. *Proceedings of the 30th DAAAM International Symposium* 0555-0561
[11] Rodriguez A P, Claver J, Camacho A M 2018 The influence of manufacturing parameters on the mechanical behaviour of PLA and ABS pieces manufactured by FDM: A comparative analysis. *Materials* **11** 1333
[12] Cho E E, Hein H H, Lynn Z, Hla S J, Tran T 2019 Investigation on influence of infill pattern and layer thickness on mechanical strength of PLA material in 3D printing technology. *Journal of Engineering and Science Research* **3** 27-37
[13] Oteyaka M O, Aybar K, Oteyaka H C 2021 Effect of infill ratio on the tensile and flexural properties of unreinforced and carbon fiber-reinforced polylactic acid manufactured by fused deposition modelling. *Journal of Materials Engineering and Performance*
[14] Vicente M F, Calle W, Ferrandiz S, Conejero A 2016 Effect of infill parameters on tensile mechanical behavior in desktop 3D printing. *3D Printing and Additive Manufacturing* **3** 183-192
[15] Rismalia M, Hidajat S C, Permana I G R, Hadisujoto B, Muslimin M, Trawan F 2019 Infill pattern and density effects on the tensile properties of 3D printed PLA material. *Journal of Physics: Conference Series* **1402** 044041
[16] Suteja J 2021 Effect of infill pattern, infill density, and infill angle on the printing time and filament length of 3D printing. *Jurnal Rekayasa Mesin* **12** 145-152
[17] Al Hasan N H J 2020 The influence of an infill density, percent on the flexural strength of the 3D. *Key Engineering Materials* **870** 73-80
[18] Yadav P, Sahai A, Sharma R S 2020 Strength and surface characteristics of FDM-based 3D printed PLA parts for multiple infill design patterns. *Journal of The Institution of Engineers (India): Series C* **102** 197-207
[19] Baich L 2016 Impact of infill design on mechanical strength and production cost in material extrusion based additive manufacturing. *Master of Science in Engineering Thesis* Youngstown State University
[20] Harpool T D 2014 Observing the effects of infill shapes on the tensile characteristics of 3D printed plastic parts. *Bachelor of Science Thesis* Wichita State University
[21] Ultimaker 2020 *PLA material technical data sheet*. https://support.ultimaker.com/hc/en-us/articles/360011962720-Ultimaker-PLA-TDS Visited 05.07.2021
[22] Wu N and Zhang H 2017 Mechanical properties and phase morphology of super-tough PLA/PBAT/EMA-GMA multicomponent blends. *Materials Letters* **192** 17-20
[23] Ultimaker 2020 *Tough PLA material technical data sheet*. https://support.ultimaker.com/hc/en-us/articles/360012759599-Ultimaker-Tough-PLA-TDS Visited 05.07.2021
[24] Kannan S and Ramamoorthy M 2020 Mechanical characterization and experimental modal analysis of 3D Printed ABS, PC and PC-ABS materials. *Materials Research Express* **7**, 015341
[25] Ultimaker 2020 *PC material technical data sheet*. https://support.ultimaker.com/hc/en-us/articles/360012061159-Ultimaker-PC-TDS Visited 05.07.2021
[26] Hodzic D, Pandzic A, Hajro I, Tasic P 2020 Strain rate influence on mechanical characteristics of fdm 3d printed materials. *Proceedings of the 31st DAAAM International Symposium* 0168-0175.

# Modeling of Influential Parameters on Bending Strength of Solid Wood Panels Perpendicular to the Grain

**Damir Hodžić, Nedim Hurem and Atif Hodžić**

University of Bihac, Faculty of Technical Engineering Bihac, Bosnia and Herzegovina

E-mail: damir.hodzic@unbi.ba

**Abstract**. The paper presents the results of modeling the bending strength of wood. During the experimental examination and definition of the model, solid wood was taken, where the bending was performed perpendicular to the grain. The experiment was done with thirteen replications and the input values that varied at three levels were wood density and board thickness. The thirteen-repetition experiment also involved four repetitions in the marginal areas, so two more wood densities and two board thicknesses had to be taken. The experimental measurement was performed in the laboratory of the Technical Faculty Bihać. Based on the experimental results, a sufficiently adequate mathematical model of the breaking force of a solid wood panel perpendicular to the grain is obtained.

## 1. Introduction

The paper presents experimental research of the influential parameters of solid wood panel on changes in its bending strength for panels that are width - joined into a structure perpendicular to the grain. Experimental researches were performed according to the standard BAS EN 789 on samples of solid boards for different values of influential parameters of wood density ρ (g/cm3), and thickness of solid board d (mm) which were varied at five levels, where the output parameter was the maximum bending force Fmax.

In wooden constructions, solid wood is often exposed to different types of stresses, and often to complex stresses that are necessary to know before choosing the basic material for making wood products. It is very important to know the physical and mechanical properties of wood and wood-based panel constructions for their wide application in construction and production of wooden furniture [1].

Bending strength of wood means the resistance that a highly elongated wood (wooden beam) supported at both ends or fastened only at one end provides forces that try to bend or break. Bending occurs due to the action of bending moment in the cross section of wood vertical to the plane of the cross section. Bending moments occur as a result of the action of external forces, which can be arbitrarily distributed in relation to the longitudinal axis. Bending strength is of special importance for wooden structures, most often in construction, bridge construction, mining, shipbuilding, etc., as well as in furniture and construction of carpentry structures.

## 2. Preparing for the experiment

Important conditions for performing the experiment are defining the conditions in which the experiment will be performed, with clearly identified parameters of the observed process, ie the choice of parameters that will be included in the modeling, realization of the experiment, optimization of the observed process and other activities. The choice of parameters of a process for analysis, requires prior knowledge of the processing process derived from similar processes [2].

Experimental research related to this work is based on measuring the maximum bending stress force up to the moment of fracture of a solid board, made in five different thicknesses and five types of wood: spruce, poplar, beech, ash and oak previously processed into workpieces and thereafter cut into sample dimensions according to standard BAS EN 789 [3]. Production of workpieces, milling operation of wedge joints, and length and width joining in the format of the board was done in Ltd. "Rosewood" Visoko. Measurement of moisture content, determination of wood density and testing of bending strength up to the maximum fracture force was done at the Faculty of Technical Engineering Bihać.

The experiment used 20 mm thick spruce, 18 mm and 22 mm poplar, 16 mm, 20 mm and 24 mm thick beech, 18 mm and 22 mm thick oak and 20 mm thick ash, of different dimensions in length and width (Figure 1). The final cutting of the samples was done on a two-blade format circular saw in accordance with the plan of the experiment.

**Figure 1.** Panels prepared for bending

The moisture content of the samples was measured with an electric moisture meter, on the surface to be treated. Determination of wood density (bulk density) was done according to the standard BAS EN 13061-2: 2016, where small samples of dimensions 30 x 30 x d mm were taken.

The density of wood (ρ) is determined by the equation:

$$\rho = \frac{m}{V} \tag{1}$$

where:

$m$ - mass of wood in [g]

$V$ - volume of wood in [cm$^3$]

The volume of the samples was determined stereometrically, and the mass of the samples by weighing on a precise analytical balance in the laboratory at the Faculty of Technical Engineering Bihać. The mean value of the calculated wood density according to the equation (1) is: spruce 0,45 [g/cm$^3$], poplar 0,56 [g/cm$^3$], beech 0,65 [g/cm$^3$], oak 0,72 [g/cm$^3$] and ash 0,82 [g/cm$^3$].

The bending strength test was performed in the laboratory at the Faculty of Technical Engineering Bihać on a testing machine SIL-50KNAG, manufactured by Shimadzu (Figure 2), with one working cylinder, located in the upper part of the machine. The working cylinder, acting indirectly, lowers or raises the upper head, while the lower head is stationary. Then, through the computer connected to the testing machine, the bending is measured in four points, and the values of the achieved force on the bending of the test specimens can be read on the diagram

**Figure 2.** Bending of wood panel according to standard BAS EN 789

Test methods for determining the mechanical properties as well as the dimensions, moisture content and wood density of the tested samples are determined according to BAS EN 789: 2004 – Timber Structures – Test methods – Determination of mechanical properties of wood based panels. The tests must be carried out under conditions where temperature T = 20 [°C] at a relative air humidity of 65 [%].

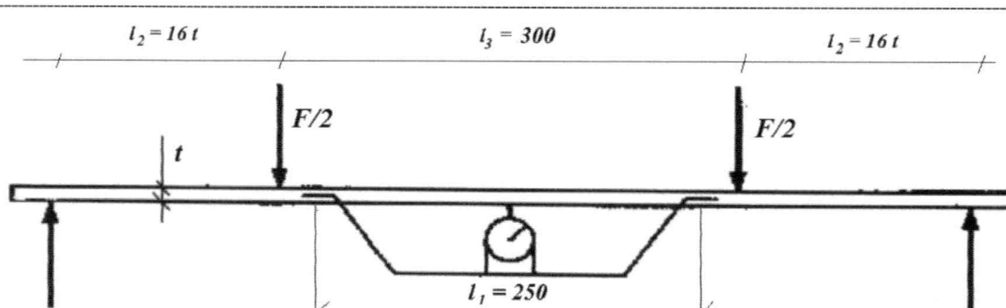


**Figure 3.** Arrangement for bending test according to standard BAS EN 789 [3]

The application of the load shall be as shown in Figure 3 with the load and reaction forces applied by rollers of 30 ± 1 [mm] diameter [3]. The distance between the load points and the supports shall be 16 times the nominal thickness, but not more than 400 [mm] and not less than 240 [mm] with an accuracy of ± 1 [mm].

## 3. Design of experiment

In this research, we selected an experimental design with 13 replicates of which 5 were at the central point. In the case when we have 13 repetitions of the experiment, the number of repetitions is obtained using the equation [2,4]:

$$N = 2^k + 2k + n_c = 2^2 + 2 \cdot 2 + 5 = 13 \quad (2)$$

where:
$n_c$ – central point experiment repetitions,
$k$ – number of factors,
2 – number of factor levels.

Input factors and their values for experiment are shown in table 1.

**Table 1.** Input factors and their values

| Effect | -1,414 | -1 | 0 | 1 | 1,414 |
|---|---|---|---|---|---|
| Factor A – Wood density ρ [kg/m$^3$] | 450 (spruce) | 560 (poplar) | 650 (beech) | 720 (oak) | 820 (ash) |
| Factor B – Thickness [mm] | 16 | 18 | 20 | 22 | 24 |

Design of experiment with 13 repetitions of experiment can be graphically presented, Figure 4. According to that experimental schedule was done and exactly for each replicate of experiment type of wood and thickness was dedicated. For all 13 experiment repetitions maximum force was measured and shown in table 2 [5].

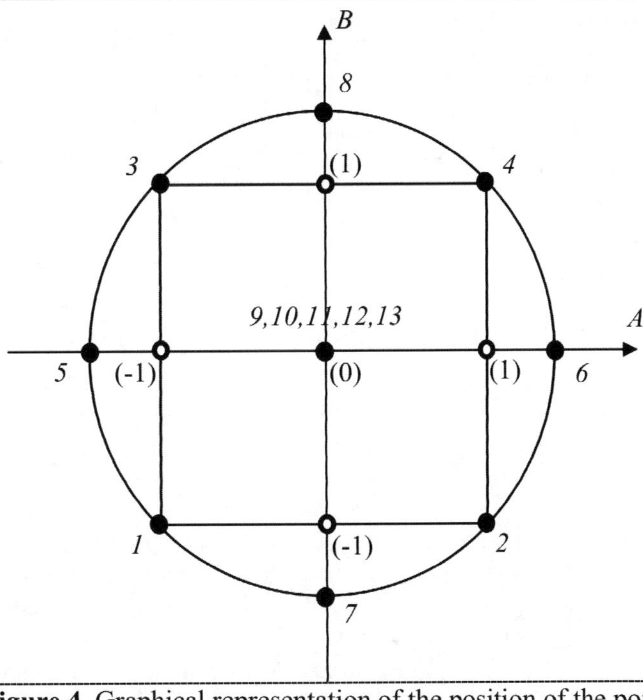


**Figure 4.** Graphical representation of the position of the points of the experiment with 13 repetitions

**Table 2.** Results of experiment

| Sample | Factor A – Density $\rho$ [kg/m$^3$] | Factor B Thickness $b$ [mm] | Force $F_{max}$ [kN] |
|---|---|---|---|
| **P 1** | 560 | 18 | 0,226 |
| **P 2** | 720 | 18 | 0,734 |
| **P 3** | 560 | 22 | 0,63 |
| **P 4** | 720 | 22 | 0,984 |
| **P 5** | 450 | 20 | 0,24 |
| **P 6** | 820 | 20 | 0,912 |
| **P 7** | 650 | 16 | 0,461 |
| **P 8** | 650 | 24 | 1,327 |
| **P 9** | 650 | 20 | 1,021 |
| **P 10** | 650 | 20 | 1,1 |
| **P 11** | 650 | 20 | 1,164 |
| **P 12** | 650 | 20 | 1,077 |
| **P 13** | 650 | 20 | 1,093 |

To determine the significance of the regression coefficients for the linear model, an analysis of variance for a given experiment will be performed. Analysis of variance involves calculating the sum of squares (SS), degrees of freedom (dF) and mean squares (MS) for given input factors, interactions between them, lack of fit and error of the model, Table 3.

**Table 3.** Analysis of Variance (ANOVA) results

| Source of variation | Sum of squares | Degrees of freedom | Mean of squares | Fisher's coefficient | $p$-value | Significance |
|---|---|---|---|---|---|---|
| A | 0,38 | 1 | 0,381 | 3,54 | 0,089 | significant |
| B | 0,43 | 1 | 0,433 | 3,38 | 0,095 | significant |
| Model | 0,81 | 2 | 0,407 | 4,704 | 0,036 | significant |
| Error | 1,60 | 10 | 0,160 | | | |
| Lack of fit | 1,56 | 6 | 0,260 | 2,825 | 0,167 | Not significant |
| Pure error | 0,036 | 4 | 0,009 | | | |
| Total | 2,42 | 12 | 0,201 | | | |

Significance of the factors and model can be determined by calculating Fisher coefficient ($F$) and comparing it with tabular $F$ for significance of 10 % ($\alpha = 0{,}1$), or calculating p-value which can be determined by software R. If p-value is less than 0,1 that means source of variation is significant. In our research both factors are significant.

Equation which can give us maximum force for bending strength of wood panel perpendicular to the grain is:

$$Y = F_{max} = -3{,}228 + 0{,}026\rho + 0{,}116b \tag{3}$$

By including the values for wood density and panel thickness in equation (3), values of the maximum force can be obtained according to the presented model. Table 3 shows the values of the maximum force according to the model and according to the experimental results.

**Table 4.** Maximum force according to the experimental results and according to the model

| Sample | $F_{eksp}$ [kN] | $F_{model}$ [kN] |
|---|---|---|
| **P 1** | 0,226 | 0,316 |
| **P 2** | 0,734 | 0,732 |
| **P 3** | 0,63 | 0,78 |
| **P 4** | 0,984 | 1,196 |
| **P 5** | 0,24 | 0,262 |
| **P 6** | 0,912 | 1,224 |
| **P 7** | 0,461 | 0,318 |
| **P 8** | 1,327 | 1,246 |
| **P 9** | 1,021 | 0,982 |
| **P 10** | 1,1 | 0,982 |
| **P 11** | 1,164 | 0,982 |
| **P 12** | 1,077 | 0,982 |
| **P 13** | 1,093 | 0,982 |

After obtaining the model equation, it is useful to check the homogeneity of the dispersions using the Cochran criterion and the adequacy of the model by determining the multiple regression coefficient. Calculated Cochran criterion was $K_h = 0{,}531$, which is less than table value for Cochran for input factors ($K_t = 0{,}544$). That means dispersions of the model results is homogeneous.

The calculated value of the regression coefficient is $R = 0{,}9786$ and that means that the linear model describes the accuracy of the experimental results with 97.86%, which is an excellent accuracy of the model.

## 4. Discussion of results

For the purposes of this paper and to confirm the obtained statistical results, the Design Expert software package was used. It is a commercial software used for experiment planning and optimization of results, as well as statistical processing and visual display [5, 6].

Using Perturbation Plot diagram, the effects of all input factors at a particular point in the experimental space can be compared. Figure 5 shows a perturbation plot diagram of the input factors of the model for the central point.

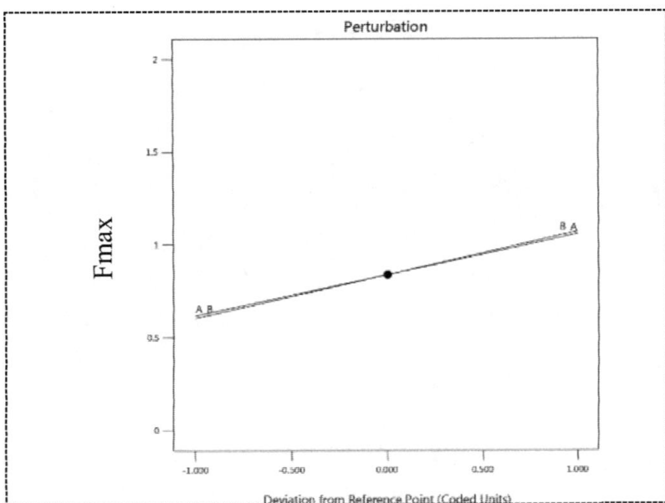


**Figure 5.** *Perturbation plot diagram of the input factors*

The diagrams in Figure 6 show the dependence of the output value of the model, the maximum force on the input factors, the density of the wood and the thickness of the panel. On the input size abscissa, the minimum value is displayed with "-1" and the maximum value of the input size with "+1". It can be seen from the diagram that the value of the maximum force increases relatively proportionally with the increase in the density of the plate or the thickness of the panel.

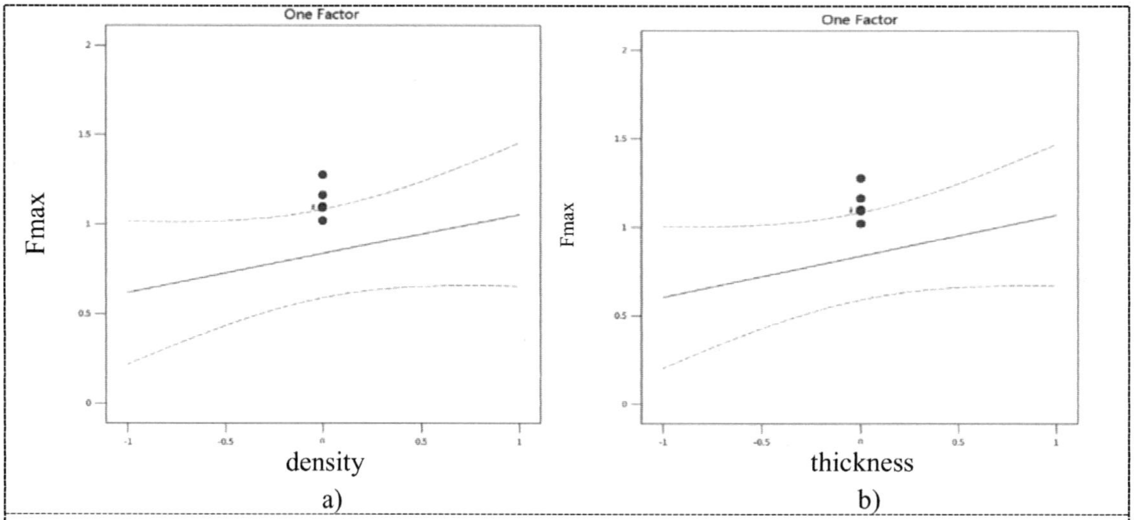


**Figure 6.** Dependence of the output value on the input factors a) density; b) thickness

The dependence of the output value of the model, the maximum force, on the interaction of input factors can be graphically represented by contour two-dimensional diagrams, Figure 7 as well as by three-dimensional display, Figure 8. The diagrams show the values of input factors in coded values from "-1" to "1" and the output value of the maximum force in the values obtained using the model.

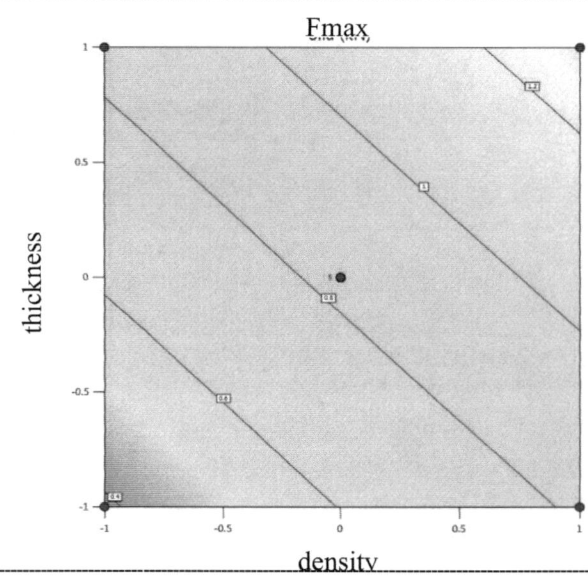


**Figure 7.** Contour two-dimensional diagram - dependence of the output value of the model, the maximum force, on the interaction of input factors

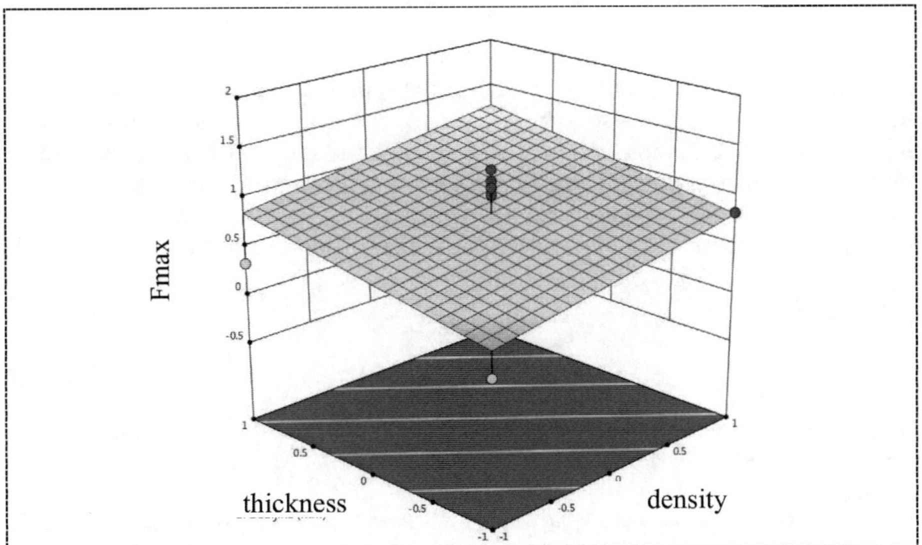


**Figure 8.** Contour three-dimensional diagram - dependence of the output value of the model, the maximum force, on the interaction of input factors

It is possible to graphically show the distribution of experimental and model values of the maximum force, Figure 9.

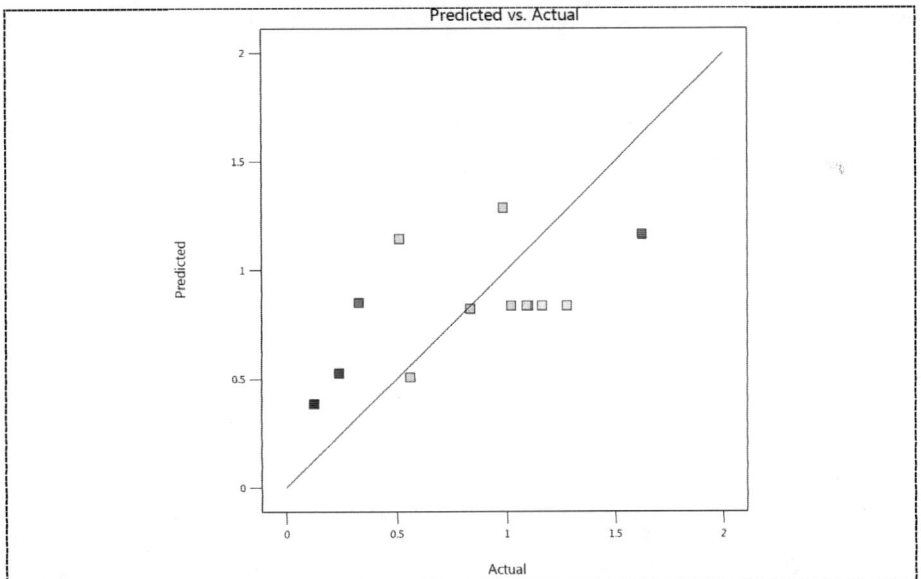


**Figure 9.** Distribution of model values (predicted values) in relation to the experimental ones (actual values)

This diagram helps to detect experimental points in which the output values do not match well with the experimental values. According to the theory, the data points should be evenly distributed in relation to the line, which should be at an angle of 45 °. On the horizontal axis are experimental and on the vertical axis model values and the distribution of model values in relation to the experimental ones is good.

IOP Conf. Series: Materials Science and Engineering 1208 (2021) 012020 doi:10.1088/1757-899X/1208/1/012020

## 5. Conclusion

After the obtained results of the experiment and the model of maximum force, it can be concluded that the obtained model is good enough and adequate for the presented initial conditions, which define the density of wood or the thickness of solid wood panel. The dispersion of the results at the point of repetition of the experiment is homogeneous and the adequacy of the model calculated using the multiple regression coefficient is over 90% and we can conclude that the model is very good and adequate. The value of the maximum force of a solid wood board depends on both the density of the wood and the thickness of the board. As both values increase, so does the magnitude of the maximum force, with the growth increasing as the density of the material increases, which means that this parameter is more influential in the model

## References

[1] Skakic D 2002 *Finalna obrada drveta* (Univerzitet u Beogradu, Šumarski fakultet)

[2] Montgomery D C 2015 *Design and Analysis of Experiment* (Arizone State University, John Wiley and Sons) p 103

[3] BAS EN 789 2005 *Timber Structures – Test methods – Determination of mechanical properties of wood based panels*

[4] Hodzic D and Dzanic A 2020 *Planiranje eksperimenta* (Univerzitet u Bihaću, Tehnički fakultet)

[5] Hurem N and Hodzic D 2021 *Design of experiment and mathematical model of bending strength of solid wood panel* Journal of Mechanical Engineering Mašinstvo, Zenica, Volume 17, Issue 3-4

[6] Hurem N 2020 *Modeling of the influence of wood density and thickness on the strength of solid wood panels*, Master thesis, Univerzitet u Bihaću, Tehnički fakultet

# Computing the heat flux required for warming up of frozen wooden prisms for veneer production in the beginning of their autoclave steaming

**Nencho Deliiski and Dimitar Angelski**

University of Forestry, Faculty of Forest Industry, Sofia, Bulgaria

E-mail: deliiski@netbg.com

**Abstract**. An approach for computing the heat flux required for warming up of frozen wooden prisms in the regimes for their autoclave steaming at limited heat power of the steam generator, depending on the dimensions of the prisms cross section, wood moisture content, and loading level of the autoclave has been suggested. The approach is based on the use of two personal mathematical models: 2D non-linear model of the temperature distribution in subjected to steaming frozen wooden prisms and model of the non-stationary heat balance of autoclaves for steaming wood materials. For numerical solving of the models and practical application of the suggested approach, a software program was prepared in the calculation environment of Visual FORTRAN Professional developed by Microsoft. Using this program computation and research of the non-stationary change of the processing medium temperature and heat fluxes in an autoclave with a diameter of 2.4 m, length of 9.0 m and loading level of 50% at a limited heat power of the steam generator, equal to 500 kW during the initial part of the steaming in it of frozen beech prisms with different moisture content have been carried out. The suggested approach can be used for computing and model based automatic realization of energy efficient optimized regimes for autoclave steaming of different wood materials.

## 1. Introduction

For plasticizing of prismatic wood materials in the production of veneer and plywood the materials are usually subjected to steaming in different types of equipment [1, 3, 10, 12, 14]. The steaming of wood materials under increased pressure of the processing medium in autoclaves is used in many applications due to its higher energy efficiency and lower duration in comparison with the steaming at atmospheric pressure [2, 11, 13].

Information about the duration and energy consumption of steaming wood materials in atmospheric or increased pressure for the cases of unlimited generator power has been given in [4, 5, 6, 7, 12, 14]. For medium and small enterprises, steaming regimes for wood materials at limited power of the steam generator is of considerable interest.

The aim of the present work is to suggest an approach for the computation and research of the processing medium temperature and largest for the whole steaming process heat fluxes in the beginning of regimes for autoclave steaming of frozen wooden prisms for veneer production at limited power of the heat generator, depending on the dimensions of the prisms cross section, wood moisture content, and loading level of the autoclave with prisms subjected to steaming.

IOP Conf. Series: Materials Science and Engineering 1208 (2021) 012021 doi:10.1088/1757-899X/1208/1/012021

## 2. Material and methods

*2.1. Modeling of the 2D temperature distribution in frozen wooden prisms subjected to steaming*
When the length of the prisms, *l*, is larger than their thickness, *d*, at least 4 ÷ 5 times, and simultaneously with this the width, *b*, does not exceed the thickness more than 3 times, for the calculation of the change in the temperature in the prism's cross section, which is equally distant from the frontal sides (i.e. along the coordinates *x* and *y* of this section) during heating and cooling in steaming medium the following 2D mathematical model can be used [2, 5]:

$$c_{\text{w-eff}} \cdot \rho_{\text{w}} \frac{\partial T}{\partial \tau} = \text{div } (\lambda_{\text{w-eff}} \text{grad } T) \tag{1}$$

with an initial condition:

$$T(x, y, 0) = T_{\text{w0}} \tag{2}$$

and following boundary condition:

$$T(x, 0, \tau) = T(0, y, \tau) = T_{\text{m}}(\tau) \tag{3}$$

where $c_{\text{w-eff}}$ is the effective specific heat capacity of the frozen and non-frozen wood during its heating, $\text{J}\cdot\text{kg}^{-1}\cdot\text{K}^{-1}$; $\lambda_{\text{w-eff}}$ – effective thermal conductivity of the frozen and non-frozen wood cross sectional to the fibers, $\text{W}\cdot\text{m}^{-1}\cdot\text{K}^{-1}$; $\rho_{\text{w}}$ – wood density, $\text{kg}\cdot\text{m}^{-3}$; $x$ – coordinate along the prism's thickness of the separate points of the calculation mesh for numerical solving of the model: $0 \leq x \leq d/2$, m; $d$ – thickness of the prism, m; $y$ – coordinate along the prism's width of the separate points of the calculation mesh: $0 \leq y \leq b/2$, m; $b$ – width of the prism, m; $\tau$ – time, s; $T$ – temperature, K; $\text{T}_{\text{w0}}$ – initial average mass temperature of the prism, K; $T_{\text{m}}$ – processing medium temperature, K; $T(x,y,0)$ – temperature of all points in the prism in the beginning of the steaming, K; $T(x,0,\tau)$ – temperature of all points on the surface of the prism parallel to its thickness during the steaming process, K; $T(0,y,\tau)$ – temperature of all points on the surface of the prism parallel to its width during the steaming.

Mathematical descriptions of the effective specific heat capacities of the frozen and non-frozen wood during its heating, $c_{\text{w-eff}}$, and also of the effective thermal conductivities of the wood in different anatomical directions during its heating, $\lambda_{\text{w-eff}}$, have been suggested in [2, 3, 4] based on the experimentally determined in the dissertations of Chudinov [1] and Kanter [9] relations for the change in $c$ and $\lambda$ of frozen and non-frozen wood as a function of the temperature, $t$, and wood moisture content, $u$.

*2.2. Modeling of the heat balance of the steaming autoclave*
The non-stationary heat balance of the autoclave for each moment $n\cdot\Delta\tau$ of the steaming process can be described by the following mathematical model [2]:

$$Q_{\text{ha}}^{n} = Q_{\text{hw}}^{n} + Q_{\text{hf}}^{n} + Q_{\text{hil}}^{n} + Q_{\text{he}}^{n} + Q_{\text{hfv}}^{n} + Q_{\text{hcw}}^{n} \tag{4}$$

where $Q_{\text{ha}}$ is the specific (for 1 $\text{m}^3$ wood) heat energy, which is supplied into the autoclave by the introduced in it water steam, $\text{kWh}\cdot\text{m}^{-3}$; $\text{Q}_{\text{hw}}$ – energy used for heating of the subjected to steaming wood materials, $\text{kWh}\cdot\text{m}^{-3}$; $\text{Q}_{\text{hf}}$ – energy used for heating of the body of the autoclave and of the situated in it metal trolleys for positioning of the wood materials, $\text{kWh}\cdot\text{m}^{-3}$; $\text{Q}_{\text{hil}}$ – energy used for heating of the insulating layer of the autoclave, $\text{kWh}\cdot\text{m}^{-3}$; $\text{Q}_{\text{he}}$ – energy used for covering of the heat emission from the autoclave in the surrounding air, $\text{kWh}\cdot\text{m}^{-3}$; $\text{Q}_{\text{hfv}}$ – energy used for filling in with steam the free (unoccupied by wood materials) part of the working volume of the autoclave, $\text{kWh}\cdot\text{m}^{-3}$; Qhcv – energy, which is accumulated in the gathered in the lower part of the autoclave condense water, $\text{kWh}\cdot\text{m}^{-3}$ ; $n$ – current number of the step along the time coordinate, $\Delta\tau$, with the help of which the solving of the model is carried out: $n = 0, 1, 2, 3, \ldots$

The whole mathematical model (4) with descriptions of all components in it, depending on the influencing factors, has been given in [2, 5].

The specific heat flux $q_{\text{hw}}^{n} = \frac{dQ_{\text{hw}}^{n}}{d\tau}$, which provides the energy required for warming up of 1 m$^3$ wooden prisms for any moment $n\cdot\Delta\tau$ of the steaming process is equal to (in kW·m$^{-3}$)

$$q_{\text{hw}}^{n} = \frac{dQ_{\text{hw}}^{n}}{d\tau} \approx \frac{3600\Delta Q_{\text{hw}}^{n}}{\Delta\tau} \quad (5)$$

The specific heat flux $q_{\text{ha}}^{n} = \frac{dQ_{\text{ha}}^{n}}{d\tau}$, which provides the whole energy $Q_{\text{ha}}^{n}$ during the heating of 1 m$^3$ of prisms for any moment $n\cdot\Delta\tau$ of the steaming process is equal to

$$q_{\text{hw}}^{n} = \frac{dQ_{\text{hw}}^{n}}{d\tau} \approx \frac{3600\Delta Q_{\text{hw}}^{n}}{\Delta\tau} \quad (6)$$

## 3. Results and discussion

For numerical solving of the above considered mathematical models aimed at usage of the suggested approach for the calculation of $T_m$ during the initial part of TTP a software package was prepared in FORTRAN in the calculation environment of Visual Fortran Professional. For the preparation of the models for programming an explicit form of the finite-difference method has been used, which allows for the exclusion of any simplifications in the models [8].

The increase of the processing medium temperature, $T_m$, in the beginning of the steaming regimes is calculated according to the approach given in [6] by taking in mind the available heat power of the steam generator. During simulations the heat power of the steam generator $q_{\text{source}}$ = 500 kW was used. With the help of the software package computations were made for the determination of $T_m$ and also of the 2D non-stationary change of the temperature in 4 representative points of ¼ of the square cross section of beech prisms with thickness $d$ and width $b$ respectively, during their steaming in an autoclave with a diameter $D$ = 2.4 m, length of $L$ = 9.0 m, and working volume $V_a$ = 48 m$^3$ [5, 13].

Simultaneously with this, calculations for determination of the average mass temperature $T_{avg}$ of the prisms, energy end heat flux consumed by the autoclave, $Q_a$ and $q_a$, respectively, and also the heat energy consumption, $Q_w$, and the heat flux, $q_w$, required for the warming up of the prisms themselves, have been carried out according to approaches given in [2, 5].

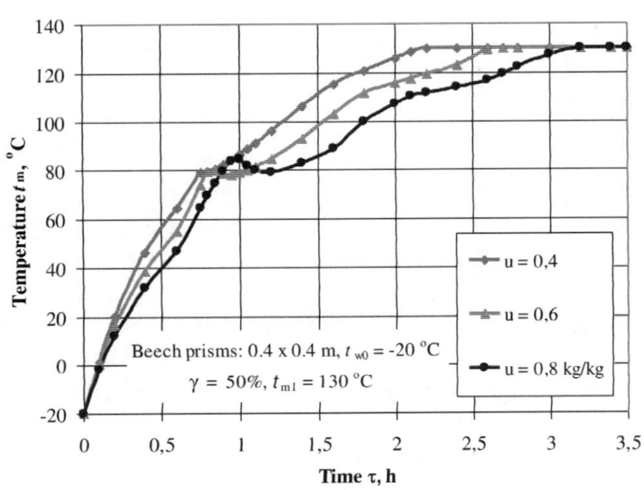


**Figure 1.** Change in tm in the beginning of the steaming regime of frozen beech prisms with $d$ x $b$ = 0.4 x 0.4 m and γ = 50%, depending on $u$

During the solving of the models, the presented in [2, 3] mathematical descriptions of the thermo-physical properties of beech wood with basic density $\rho_b = 560$ kg·m$^{-3}$ and standardized (at 20 °C) fiber saturation point $u_{fsp} = 0.31$ kg·kg$^{-1}$ were used. The initial and the maximal values of the steaming medium temperature were equal to $t_{m0} = -20$ °C and $t_{m1} = 130$ °C respectively. During simulations the initial temperature of the frozen prisms $t_{w0} = -20$ °C and the following values of the factors influencing the duration $\tau_1$ of the initial part of the steaming regimes were set: dimensions of the square cross section of beech prisms with thickness $d$ and width $b$: 0.4 × 0.4 m; moisture content $u$ of the prisms subjected to steaming: 0.4 kg·kg$^{-1}$, 0.6 kg·kg$^{-1}$, and 0.8 kg·kg$^{-1}$; loading level $\gamma$ of the autoclave with filled in prisms for steaming: 0.5 m$^3$·m$^{-3}$ (i.e. $\gamma = 50\%$).

On Figure 1 the calculated change in $t_m$ in the beginning of the autoclave steaming regimes of the studied frozen beech prisms is presented.

The obtained results allow for making the following statements about the change in the duration of the initial part of the regimes for steaming of beech prisms, $\tau_1$, in an autoclave with $D = 2.4$ m and $L = 9.0$ m, depending on the studied influencing factors:

1. By increasing the heating time $\tau$, the temperature of the processing medium $t_m$ increases curvilinear until reaching its largest value $t_{m1}$ after time interval, which depends on the investigated factors. The lack of smoothness of the dependences $t_m = f(\tau)$ is caused by the uneven melting of the frozen free water in the separate points of the wooden prisms. The more points at which the ice melts at a given moment of the steaming process, the more this dependence is deformed.

2. The increase of the prisms moisture content $u$ from 0.4 to 0.8 kg·kg$^{-1}$ at given values of $d$ x $b$ and $\gamma$ causes non-linear increase of $\tau_1$. When $d$ x $b$ = 0.4 x 0.4 m and $\gamma = 50\%$ the duration $\tau_1$ at $t_{m1} = 130$ °C and $q_{source} = 500$ kW is equal as follow: $\tau_1 = 2.2$ h for $u = 0.4$ kg·kg$^{-1}$, $\tau_1 = 2.6$ h for $u = 0.6$ kg·kg$^{-1}$, and $\tau_1 = 3.2$ h for $u = 0.8$ kg·kg$^{-1}$.

The reason for this influence of $u$ is that the larger $u$ means there is a presence of more heat capacity of the wet wood in the autoclave [2, 3, and 4]. Because of this, the larger part of the constant flux $q_a$ is needed for warming up of the prisms with larger $u$, while the remaining smaller part from it leads to a smaller increase in tm during any next step $\Delta\tau$ when solving the mathematical model.

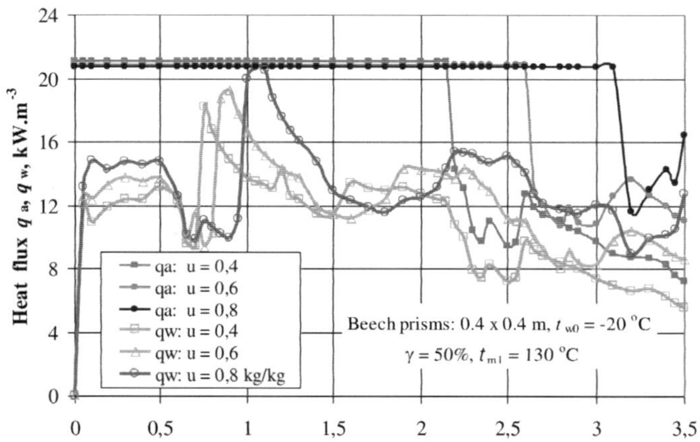


**Figure 2.** Change in $q_a$ and $q_w$ for beech prisms with $d$ x $b$ = 0.4 x 0.4 m at $\gamma = 50\%$, depending on $u$

On Figures 2 the calculated change in the specific heat fluxes $q_a$ and $q_w$ in the beginning of the autoclave steaming regimes of the studied beech prisms for veneer production is presented.

The constant value of $q_a = 20.83$ kW·m$^{-3}$ during the increasing of tm from $t_{m0} = -20$ °C to $t_{m1} = 130$ °C in the beginning of the steaming regimes is determined as follows: By multiplying of the autoclave working volume $V_a = 48.0$ m$^3$ by $\gamma = 50\%$ the volume of the prisms subjected to steaming $V_w = 24.0$ m$^3$ is obtained. After dividing the value of the limited heat power of the steam generator $q_{source} = 500$ kW by $V_w = 24.0$ the shown on Fig. 2 value of the specific heat flux $q_a = 20.83$ kW·m$^{-3}$ is obtained.

The obtained results show that the dependences $q_w = f(\tau)$ reach their maximal values, which are very close to $q_a$, as follow: at $\tau = 0.75$ h for $u = 0.4$ kg·kg$^{-1}$; at $\tau = 0.90$ h for $u = 0.6$ kg·kg$^{-1}$, and at $\tau = 1.05$ h for $u = 0.8$ kg·kg$^{-1}$. These maximum values of $q_w$ are caused by the largest numbers of points in the frozen prisms, in which the ice melts in the specified moments of the steaming regimes.

After reaching of $t_{m1} = 130$ °C at $\tau = \tau_1$ a significant decrease in $q_w$ and $q_a$ is observed. The lack of smoothness of the dependences $q_w = f(\tau)$ and $q_a = f(\tau)$ then is caused by the continuing uneven melting of the frozen free water in the separate points of the prisms.

It can be noted that the difference between fluxes $q_a$ and $q_w$ is equal to the sum of the heat fluxes, which ensure the energy consumptions needed for the last five members in the right side of eq. (4).

## 4. Conclusions

The present paper describes an approach for computing and research of the processing medium temperature, $t_m$, and largest for the whole steaming process heat fluxes, $q_w$ (required for warming up of the wood) and $q_a$ (consumed by the autoclave), in the beginning of regimes for autoclave steaming of frozen prismatic wood materials at limited power of the steam generator, depending on dimensions of the prisms cross section, $d$ x $b$, wood moisture content $u$, and loading level of the autoclave with wood materials subjected to steaming, $\gamma$.

The paper presents and analyses diagrams of the non-stationary change in $t_m$, $q_w$, and $q_a$ in an autoclave with a diameter of 2.4 m, length of 9.0 m and loading level of 50% at a limited heat power of the steam generator, equal to 500 kW during autoclave steaming process of frozen beech prisms with initial temperature of – 20 °C, basic density of 560 kg·m$^{-3}$, cross-section dimensions $d$ x $b = 0.4 \times 0.4$ m, moisture content $u = 0.4$, 0.6, and 0.8 kg·kg$^{-1}$. All diagrams are drawn using the results obtained in the calculation environment of Visual FORTRAN by two personal mathematical models. It has been determined, that the increase of the wood moisture content at given values of $d$ x $b$ and $\gamma$ causes non-linear change in $t_m$ and $q_w$ during the time interval $0 - \tau_1$ in-between tm increases from its initial value of tm = –20 °C to the maximal steaming regime temperature $t_{m1} = 130$ C. The reasons for these changes have been explained in the paper.

The suggested approach and the obtained results can be used for the computation and model based automatic realization of energy efficient optimized regimes for autoclave steaming of different wood materials at limited heat power of the steam generator [7, 8].

**Acknowledgements**

This paper was supported by the Scientific Research Sector at the University of Forestry – Sofia, Bulgaria, under contract № 1079/2020.

## References

[1] Chudinov B.S. 1966 *Theoretical Research of Thermo-Physical Properties and Thermal* Treatment of Wood Dissertation for DSc. Krasnoyarsk SibLTI (in Russian)

[2] Deliiski N 2003 *Modelling and Technologies for Steaming Wood Materials in Autoclaves.* Dissertation for DSc., Sofia University of Forestry, 358 p (in Bulgarian)

[3] Deliiski N 2011 *Transient Heat Conduction in Capillary Porous Bodies.* In Ahsan A. (ed) Convection and Conduction Heat Transfer. InTech Publishing House, Rieka, pp 149-176

[4] Deliiski N. 2013 *Modelling of the Energy Needed for Heating of Capillary Porous Bodies in Frozen and Non-frozen States*, Saarbrücken: Scholars' Press, Germany, 116 p

[5] Deliiski N, Dzurenda L 2010 *Modeling of the Thermal Treatment in the Technologies for Wood Processing*. Sofia, University of Forestry, 299 p (in Bulgarian)

[6] Deliiski N, Dzurenda L, Angelski D, Tumbarkova N. 2018 *An Approach to Computing Regimes for Autoclave Steaming of Prisms for Veneer Production with a Limited Power of the Heat Generator*, Acta Facultatis Xilologiae Zvolen, 60(1), 101-112

[7] Hadjiski M, Deliiski N. 2016 *Advanced Control of the Wood Thermal Treatment Processing*. Cybernetics and Information Technologies, 16(2), pp 179-197

[8] Hadjiski M, Deliiski N, Tumbarkova N 2020 *Intelligent Hybrid Control of Thermal Treatment*

*Processes of Wood*, Proc. of the 16-th Int. Conf. Intelligent Systems, Varna, pp 482-489.
[9] Kanter, K.R. 1955 *Investigation of the Thermal Properties of Wood*. PhD Thesis, Moscow, MLTI (in Russian)
[10] Pervan S 2009 *Technology for Treatment of Wood with Water Steam*. Zagreb, University in Zagreb (in Croatian)
[11] Riehl T, Welling J, A. Frühwald A 2002 *Druckdämpfen von Schnittholz, Arbeitsbericht 01: Institut für Holzphysik*, Hamburg: Bundesforschungsanstalt für Forst- und Holzwirtschaft.
[12] Shubin G.S 1990 *Drying and Thermal Treatment of Wood*. Moscow: Lesnaya Promyshlennost, 337 p (in Russian)
[13] Sokolovski S, Deliiski N, Dzurenda L 2007 *Constructive Dimensioning of Autoclaves for Treatment of Wood Materials under Pressure*. Proc. of the 2nd Int. Sc. Conf. Woodworking techniques, Zalesina, Croatia, pp 117-126
[14] Trebula P, Klement I 2002 *Drying and Hydrothermal Treatment of Wood*. Zvolen: TU in Zvolen, Slovakia, 449 p (in Slovak)

# Spectral Range and Wood Surface Impacts on Equilibrium Moisture Content Estimation in Thermally Modified Beech Wood by FT-NIR Spectroscopy

**Nebojša Todorović**

University of Belgrade, Faculty of Forestry, Department of Wood Science and Technology, Kneza Višeslava 1, 11030 Belgrade, Serbia

E-mail: nebojsa.todorovic@sfb.bg.ac.rs

**Abstract.** Fourier transform near-infrared (FT-NIR) spectroscopy and partial least squares regression (PLS-R) were tested for the possibility of equilibrium moisture content (EMC) prediction in thermally modified beech wood (Fagus moesiaca C.). The samples were modified for 4h at temperatures of 170, 190 and 210 °C. After thermal modification, the samples were kept in a climatic chamber until EMC was reached. FT-NIR spectra (100 scans and 4 cm-1) were collected on the cross-section and radial surfaces at four points. PLS –R models were developed for four spectral regions: the first overtone, the second overtone, the third overtone and the combination band region. Applied thermal treatment caused a decrease of EMC by 42 % at 170 °C, by 53 % at 190 °C, and by 62 % at 210 °C. Principal component analysis (PCA) indicated that there is a difference both between treatments and between wood surfaces. The results of the spectra taken from the radial surface were, in all models, better than the spectra of the cross-section. Related to chemical changes, the first and second overtone region play an important role in the calibrations. The best prediction models for EMC of thermally modified beech wood were obtained from radial surface spectra in the first (Rp2=0.86, RPD=2.69) and second overtone region (Rp2=0.87, RPD=2.70). The obtain results could contribute to the development of predictive models in monitoring of EMC which could significantly improve the quality of industrial production of thermally modified wood.

## 1. Introduction

Thermal modification at temperatures up to 260 °C causes structural and chemical transformations which can significantly modify wood properties. These temperatures degrade the hemicelluloses which are the main factors of hygroscopicity in wood [1]. The lowered hygroscopicity leads to the reduction of equilibrium moisture content (EMC) and the level of this reduction depends on several factors: applied temperature and time, wood species, type of heating medium, etc. [2] reported that this reduction in beech wood averages at 30 % and [3] showed that, by using the same parameters of conditioning, beech wood showed a greater EMC reduction than spruce wood.

Manufacturing processes of thermally modified wood in the current wood industry involve various procedures such as regimes, sorting, moisture content monitoring, scaling and grading of wood products. In order to optimize these processes, it is necessary to have accurate real-time knowledge about thermally modified wood products properties and characteristics. Moisture content is one of the most important properties of wood because it affects the final quality of a thermally modified product. When wood adsorbs or loses water, significant changes in wood mechanical or physical properties occur. The techniques used to monitor moisture content in thermally modified wood must be fast, efficient and inexpensive. Current methods are mostly used in laboratories and are destructive, expensive and time-consuming.

Using an FT-NIR spectroscope, as a non-destructive method, has taken a significant role in determining the quality of wood and wood products. Apart from assessing the physical and mechanical properties [4, 5, 6], the FT-NIR method, along with PLS-R (partial least square – regression) analysis, was researched for predicting moisture content of non-modified and thermally modified wood. [7] reported a very high coefficient of determination (R2=0.99) between the predicted and measured values of moisture content in small samples of spruce wood. [8] however reported a much lower correlation (R2=0.58) while examining moisture content in red oak wood. [9] concluded that wood surface had an important role in assessment of moisture content. [10, 11] confirmed that reducing of spectral range did not significantly affect quality loss in their models. On the other hand, there are studies about the application of FT-NIR spectroscopy in assessment of moisture content in thermally treated wood. [12, 13] reported high values of statistical parameters for certain species of genera Eucalyptus (Eucalyptus globules and Eucalyptus urophylla) and Pinus (Pinus pinaster and Pinus ocarpa). [13] also reported that quality of their model in assessment of MC of kiln-treated wood was much lower compared with wood treated in autoclave.

Considering the importance of thermal modification and the assessment of its properties by using non-destructive methods, and also the fact that beech (Fagus moesiaca C.) is the most important industrial wood species in Balkan peninsula, the focus of this research was the assessment of EMC of thermally modified beech wood by using NIR spectroscope. A secondary goal of this paper was to show the influence of wood surface and of the reduced spectral range on the quality of models. The results could contribute to the development of predictive models in monitoring of moisture content which could significantly improve the quality of industrial production of thermally modified wood.

## 2. Materials and methods

### *2.1. Materials*

For this research, five beech logs were cut from Goč mountain forest in western Serbia. Each log was cut into four radial boards 30 mm thick. In total, 20 boards without visible knots and deformations were cut. The boards were each cut into four samples (one modified and three for thermal treatment). The samples had clearly defined anatomical directions and had no visible defects. They were cut into 20x20x20mm samples used for assessing EMC.

### *2.2. Thermal modification of wood and EMC testing*

Thermal modification was carried out in a laboratory chamber (1 m3, ± 1 °C sensitivity, water vapor atmosphere), where samples were exposed to temperatures of 170, 190, or 210 °C. It took approximately twenty-four hours to heat the samples from room temperature to the treating temperature, after which the temperature was kept constant for four hours [6]. The chosen schedules are often used in industrial thermal modification of beech timber. After the thermal treatment, the samples that did not have visible cracks and defects were used for recording the FT-NIR spectra and moisture content assessment. There were a total of 180 specimens (90 unmodified + 90 thermally modified).

The unmodified and thermally modified samples were conditioned at 23 °C and relative humidity of 50 % until EMC was reached. EMC were determined by oven-drying method after conditioning.

### 2.3. Near-infrared spectroscopy and multivariate analysis

FT-NIR spectra were collected after thermal treatment with a Nicolet Nexus 670 FT-IR spectrometer equipped with a Thermo Nicolet Smart Near-IR UpDrift probe in the wavenumber range from 11000–4000 cm-1 using the default parameters. The FT-NIR spectra were acquired by an integrating sphere scanning an area of about 7 mm in diameter. For each scanning point, 100 scans (4 cm-1) were collected and averaged into a single spectrum. Four spectra on the cross-section and radial longitudinal face of test samples were averaged to one average spectrum per face. The measurements were taken on a planed surface.

PLS-R analysis (Unscrambler 9.7) was applied in the prediction of EMC by FT-NIR spectroscopy. PLS-R calibration was done with two subsets (calibration set and validation set) with a maximum ten latent variables (LV). Samples for calibration and validation sets were divided manually. First, the samples were ranked in ascending order of their dependent variables and every third or fourth sample was taken into the validation set. Secondly, NIR variables were regressed against the values of EMC by cross-validation of five randomly chosen groups. Different data treatments were evaluated for the spectral data: first derivative (1stDer), second derivative (2ndDer), standard normal variate (SNV), normalization (norm.), multiplicative scatter correction (msc) and combinations of all. First derivative was obtained by using a 9-point filter and a second order polynomial as well as the Savitzky-Golay algorithm [14]. For the 2nd derivative, a 15-point filter and a second order polynomial were used. Outlier samples were identified by Student residuals and leverage value plot analyses. Model efficiency was then tested by the validation set. Beside the full FT-NIR region, PLS –R models were developed for four spectral regions: the first overtone, the second overtone, the third overtone and the combination band region [15].

The evaluation of the statistical models was done based on RPD (residual prediction deviation or ratio of performance to deviation), and two groups were created for model classification: preliminary screening,$1.5 \leq RPD \leq 2.0$; and screening, $2.0 < RPD \leq 5.0$; The RPD was introduced by [16] several and is calculated as the ratio of two standard deviations: the standard deviation of the reference data for the validation set and the standard error of prediction (from cross-validation or test set validation).

In this paper, chemometrical properties of wood were examined by principal component analysis (PCA). PCA is one of the dimensional reduction techniques that reduce the difficulties in interpreting high and low-dimensional signals. When the data are distributed in a multidimensional space, it is possible to find an axis that can reduce the dimension most efficiently and to reduce the dimension of that axis. In other words, it is essential to find the optimal axis among several axes, and the exact axis can be obtained by the PCA [17,18].

## 3. Results and discussion

### 3.1. EMC of thermally modified beech

Table 1 shows the average values of EMC of unmodified and thermally modified beech wood, after conditioning, and the basic statistical parameters. Average EMC of unmodified samples was 9.56 % and this research confirmed that thermal modification decreases the hygroscopicity. Applied thermal treatment caused a decrease of EMC by 42 % at 170 °C, by 53 % at 190 °C, and by 62 % at 210 °C. The Tukey HSD test (with 95 % probability) indicated a significant difference ($F=204.3$ $p<0,05$) in EMC between the used temperatures. The decreases and variations of EMC are similar to those reported by [19] who also researched of beech wood.

**Table 1.** Equilibrium moisture content of thermally modified beech wood after conditioning (t=23 °C; φ=50 %)

| Statistical indicators | EMC (%) | | | |
|---|---|---|---|---|
| | Unmodified | 170 °C | 190 °C | 210 °C |
| N | 90 | 30 | 30 | 30 |
| x (%) | 9.56 | 5.56 | 4.52 | 3.68 |
| SD (%) | 0.28 | 0.39 | 0.34 | 0.32 |
| Min (%) | 8.82 | 4.6 | 3.8 | 2.62 |
| Max (%) | 10.09 | 6.64 | 5.42 | 4.54 |

*N – number of samples; x – average value; SD – standard deviation; Min – minimum; Max - maximum*

According to [20], the reason for lower absorption water could be the altered chemical composition after heat treatment. Beside the changes of cellulose, extractives and lignin, high temperature range degrades hemicelluloses in the wood, the main component responsible for its hygroscopic character. This will cause reduction in moisture adsorption capacity, resulting in lower equilibrium moisture content (EMC), and radial and tangential swellings [20]. Other authors reported that reaction of lignin polycondensation [21] or the increase of crystalinity of cellulose chains [22] could also be the cause of lower EMC after thermal modification.

### *3.2. Estimation of EMC by FT-NIR spectroscopy*

Obtained values of coefficient of determination and RPD in the prediction test (Rp2) ranged from 0.75 (RPD=1.80) in the spectra recorded on cross-section surface by using the absorbance of wavelengths in third overtone range, to those recorded in radial surface in the entire NIR range (Rp2=0.89, RPD=2.92) – Table 2.

**Table 2.** Summary statistics of the calibration and prediction models for EMC of thermally modified beech wood samples

| FT-NIR region | Wood surfaces | Pre-treatment | Calibration | | | | Prediction | | | |
|---|---|---|---|---|---|---|---|---|---|---|
| | | | No. of samples | SEC | SECV | $R_c^2$ | No. of samples | SEP | $R_p^2$ | RPD |
| CbR | CS | norm | 51 | 0.48 | 0.53 | 0.8 | 27 | 0.55 | 0.79 | 2.1 |
| | RA | msc | 51 | 0.38 | 0.41 | 0.87 | 28 | 0.45 | 0.85 | 2.57 |
| FoR | CS | msc+1$^{st}$ | 52 | 0.43 | 0.46 | 0.85 | 28 | 0.53 | 0.8 | 2.17 |
| | RA | norm | 51 | 0.35 | 0.38 | 0.89 | 27 | 0.43 | 0.86 | 2.69 |
| SoR | CS | snv | 51 | 0.43 | 0.47 | 0.85 | 27 | 0.5 | 0.85 | 2.32 |
| | RA | raw | 51 | 0.34 | 0.37 | 0.89 | 27 | 0.43 | 0.87 | 2.70 |
| ToR | CS | msc+1$^{st}$ | 51 | 0.55 | 0.6 | 0.76 | 28 | 0.64 | 0.75 | 1.8 |
| | RA | snv | 52 | 0.46 | 0.5 | 0.82 | 27 | 0.56 | 0.8 | 2.05 |
| Full range | CS | 1$^{st}$ | 51 | 0.39 | 0.43 | 0.86 | 28 | 0.46 | 0.85 | 2.52 |
| | RA | 2$^{st}$ | 52 | 0.28 | 0.3 | 0.93 | 27 | 0.39 | 0.89 | 2.92 |

*Key: FoR–first overtone region, SoR–second overtone region, ToR–third overtone region, CbR-combination band region, CS–cross-section, RA– radial section, $Rc^2$-determination coefficient of calibration set, SEC-standard error of calibration, (SECV)-standard error of cross-validation, $Rp^2$ (value calculated to show the ability of the calibration to account for the variation in the validation set)-determination coefficient of validation set, SEP (the measure of the calibrations ability to predict wood properties in samples not used in the calibration set)- standard error of prediction and RPD (the ratio of two standard deviations: the standard deviation of the reference data for the validation set and the standard error of prediction from cross-validation or test set validation) - residual prediction deviation or ratio of performance to deviation*

In both wood parts, the highest statistical parameters were obtained by using the full NIR range. As for individual reduced regions, the best prediction models for EMC on cross-section of thermally modified wood were obtained from spectra in second overtone range (Rp2=0.85, RPD=2.32), while on radial surface it was the first (Rp2=0.86, RPD=2.69) and second overtone range (Rp2=0.87, RPD=2.70–Fig. 1). Obtained values of statistical indicators were mostly in the line of the published values of prediction of EMC of oven treated eucalypt and pine [13]. [12] tested the possibilities of application of NIR in prediction of EMC of eucalypt and pine treated at atmospheric pressure with presence of air. The authors reported values of coefficient of determination in calibration samples for eucalypt-0.94 and for pine-0.93 and these values are significantly higher as compared to the data obtained in this research. On the other hand, statistical indicators obtained with coniferous species were somewhat better considering that they have a more homogeneous structure.

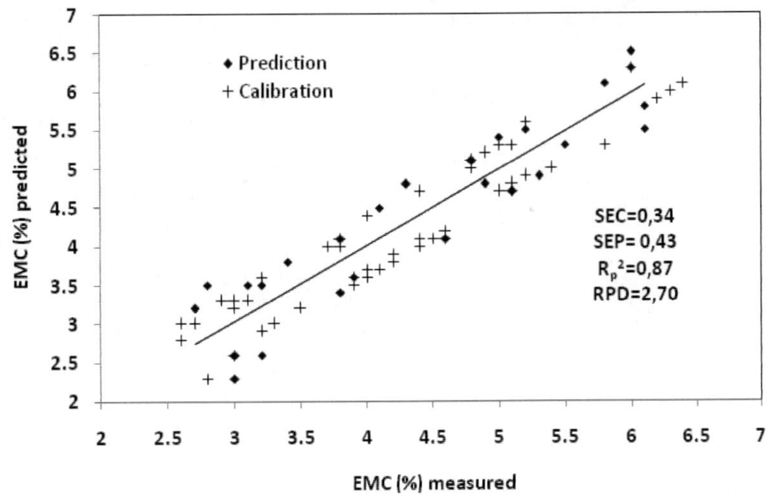


**Figure 1.** Calibration plot for EMC of thermally modified beech wood using NIR variables from second overtone region

Table 2. shows that wood surface affected the quality of models predicting the EMC. Much better parameters in all wavelength regions were obtained by recording the spectra on radial surface. [8] found that models built to estimate moisture content of red oak (Quercus spp.) lumber from spectra collected on cross-section were better than models based on spectra recorded over tangential or radial sections. These authors reported that better prediction models obtained from cross section or longitudinal (radial or tangential) sections could also be due to the distinct anatomical differences within the growth rings. The difference between longitudinal and cross-sections can also be due to surface roughness differences, which relate to particle size differences and influence radiation scattering [15]. This can result in spectral changes such as baseline offset as well as slope and band shift [15]. According to [22], the roughness of the cross-section is higher than the roughness of the tangential and radial section. Also, cross-sections consist of transversal sections of tracheid cells, which have a longitudinal axis that is parallel to the direction of the NIR incident radiation. Therefore, the NIR radiation can travel further into the wood, causing an increase in absorbance [15].

### *3.3. Spectroscopic characterization and principal component analysis*

Showed results of the influence of spectral range on the quality of statistical models were in line of the published results of [9]. Their research reported that the first and second overtone regions were the most informative and that they had a low noise level. Recorded FT-NIR spectra of unmodified and thermally modified beech wood indicate that absorbance increases with temperature (Fig. 2). This is caused by the darker tone of the wood [24]. The biggest change in absorption was found in the second overtone range, on -CH groups at 8300 cm-1 and in first overtone regiona on -OH groups at 7100 i 5150 cm-1. However, the raw spectra do not give enough information about spectral changes relative to applied treatment. Because of that, FT-NIR spectra were analyzed in the second derivative mode (2nd; not shown). The differences in spectral lines and absorption at 5800 and 5865 cm-1 indicate a degradation of carbohydrates and a deacetylation of poliosa. Differences before and after treatment were found on -OH groups in amorphous (7000 cm-1), semi-crystalline (6775 cm-1) and crystalline (6460 cm-1) regions of cellulose. This result is caused by degradation of -OH groups under high temperatures. Differences observed at 5980 cm-1 i 6900 cm-1 show the modification of lignin [15]. The obtained 2nd spectra are in line with the results of [5,6,24] for beech wood.

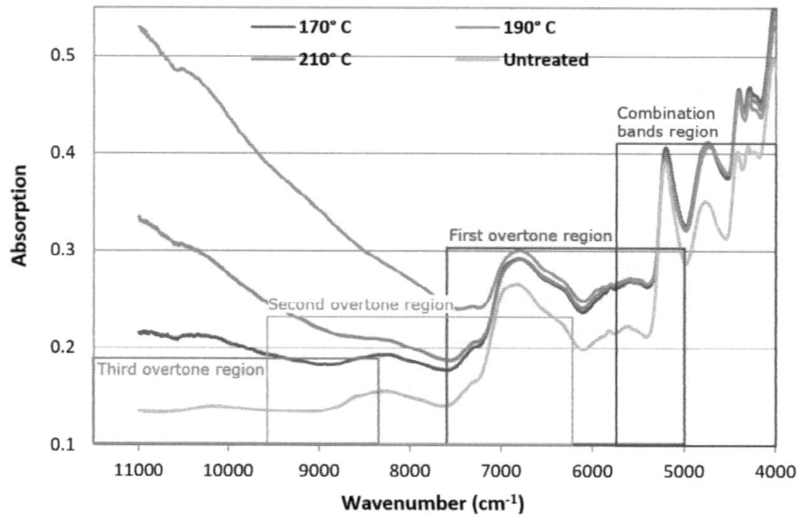


**Figure 2.** Normalized FT-NIR spectra of beech samples from radial surface after thermal modification

In this paper, PCA were carried out with original spectra obtained in the radial and cross–section surfaces of the the thermally modified wood, to carry out a preliminary evaluation of the behavior of the spectra and possible separation of the specimens according to differences temperature and wood surfaces. Three clusters indicated that there is a difference between treatments as a result of different level of chemical changes (Fig. 3). The two main components together account for 100 % of the variability of the analyzed data, 91 % are explained by the main component 1 (PC1) and 9 % are explained by the main component 2 (PC2). Also, there is a difference between the cross section and radial surface at all temperatures, which is a consequence, beside of the anatomical structure, and the different colors of the wood surface [21].

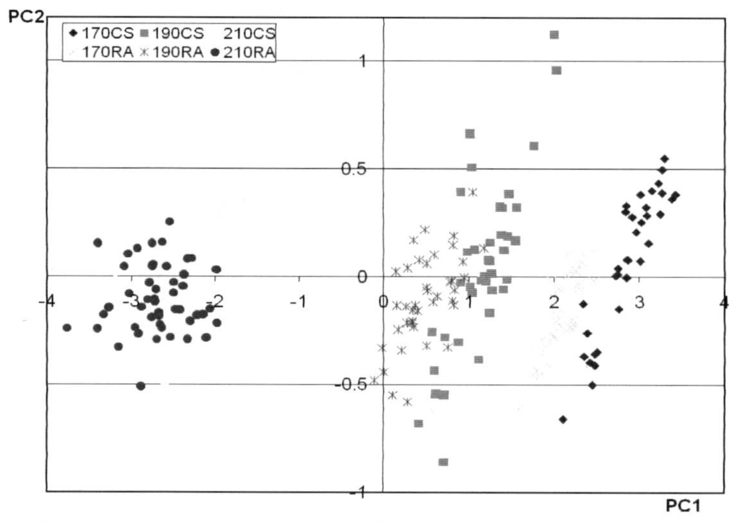


**Figure 3.** PCA score plot of normalized NIR spectra recorded on cross-section (CS) and radial (RA) surface of thermally modified beech samples

## 4. Conclusions

The applied thermal modification reduced hygroscopicity of beech wood for 42 to 62 %. The obtained results show that is possible to predict EMC of thermally modified beech wood by using NIR spectroscopy and PLS regression. The anatomical structure and color of the wood surface lead to different models of FT-NIR spectra taken from cross section and radial surface. Statistical parameters were better in predicting EMC on radial surface. According to PCA analysis, there is a difference between treatments as a result of different level of chemical changes. Beside full FT-NIR range, very good models were also obtained from reduced wavelength ranges in different overtone regions. In general, according to statistical indicators, most models could be used for screening (2.0<RPD<5.0). This is particularly visible in first and second overtone regions – regions with most important chemical compounds susceptible to modification of wood under high temperatures. The possibility of using a reduced ("narrow") spectral range from radial surface enables using a smaller, faster and cheaper spectrometer for monitoring EMC of thermally modified wood.

The obtained results can be useful for the wood industry as it provides accurate estimates of the equilibrium moisture content of thermally modified wood, assisting in the definition of regime parameters, optimizing industrial processes and the consumption of raw material.

**Acknowledgments** The author acknowledge the financial support by the Ministry of Education, Science and Technological Development of the Republic of Serbia (No. 451-02 68/2020/14/2000169).

**References**

[1] Bekhta P and Niemz, P 2003 Effect of high temperature on the change of color, dimensional stability and mechanical properties of spruce wood *Holzforschung* 57, 539-546.

[2] Tjeerdsma B, Boonstra M, Militz H 1998 Thermal modification of non-durable wood species. Part 2. Improved wood properties of thermally treated wood, In: International Research Group Wood Pre., Document no. N°IRG/WP 98- 40124

[3] Kamdem D, Pizzi A and Jermannaud A 2002 Durability of heat-treated wood *Holz als Roh- und Werkstoff* 60 1–6

[4] Hein G, Ricardo P, Campos M, Carolina A, Fernando TP and Tarscisio LJ Gilles C 2009 Near infrared spectroscopy for estimating wood basic density in Eucalyptus urophylla and Eucalyptus grandis Cerne 15(2) 133-141

[5] Bächle H, Zimmer B, Windeisen E and Wegener G 2010 Evaluation of thermally modified beech and spruce wood and their properties by FT-NIR spectroscopy *Wood Sci Technol* 44 421-433

[6] Todorović N, Popović Z and Milić G 2015 Estimation of quality of thermally modified beech wood with red heartwood by FT-NIR spectroscopy *Wood Sci Technol* 49 527–549

[7] Hoffmeyer P and Pedersen JG 1995 Evaluation of density and strength of Norway spruce by near infrared spectroscopy *Holz als Roh Werks* 53 165-170

[8] Defo M, Taylor AM and Bond B 2007 Determination of moisture and density of fresh sawn red oak lumber by near infrared spectroscopy *Forest Prod J* 57(7) 68-72

[9] Leardi R, Seasholtz M and Pell RJ 2002 Variable selection for multivariate calibration using a genetic algorithm: prediction of additive concentrations in polymer films from Fourier transform-infrared spectral data *Analytica Chimica Acta* 461 189-200

[10] Adedipe OE, and B Dawson-Andoh 2008 Prediction of yellow poplar (Liriodendron tulipifera) veneer stiffness and bulk density using near-infrared spectroscopy and multivariate calibration *Journal of Near Infrared Spectroscopy* 16 487-496

[11] Hans G, Leblon B, Stirling R, Nader J, LaRocque A and Cooper AP 2012 Monitoring of moisture content and basic specific gravity in black spruce logs using a handheld MEMS-based near-infrared spectrometer *The Forestry Chronicle* 89(5) 605–618

[12] Zanuncio AJV, Hein PRG, Carvalho AG, Rocha MFV andCarneiro ACO 2018) Determination of heat-treated Eucalyptus and Pinus wood properties using NIR spectroscopy *Journal of Tropical Forest Science* 30(1) 117–125

[13] Esteves B and Pereira H 2008 Quality assessment of heat-treated wood by NIR spectroscopy *Holz Roh Werkst* 66 323-332

[14] Savitzky A and Golay EJ 1964 Smoothing and differentiation of data by simplified least squares procedures *Anal Chem* 36 1627-1639

[15] Schwanninger M, Rodrigues JC and Fackler K 2011 A review of band assignments in near infrared spectra of wood and wood components *J Near Infrared Spectrosc* 19 287-308

[16] Williams P and Norris K 2004 Near-infrared technology in the agricultural and food industries *American Association of Cereal Chemists*, Inc, St. Paul, 296.

[17] Kim KJ and Ji-Ae Ryu Eom TJ 2019 Characteristics of the hest treated wood pack aging materials according to international standards for phytosanitary measures and verifiability of heat treatment *Wood research* 64 (4) 647-658

[18] Matthias S 2006 Approaches to analyse and interpret biological profile data PhD thesis. Faculty of Mathematics and Natural Sciences University of Potsdam Potsdam Germany

[19] Schnabel T, Zimmer B, Petutschnigg A and Schönberger S 2007 An approach to classify thermally modified hardwoods by color *For. Prod J* 57 (9) 105-110

[20] Jämsä S and Viitaniemi P 2001 Heat treatment of wood – Better durability without chemicals In Proceedings of special seminar held in Antibes France

[21] Esteves B, Velez Marques A, Domingos I, and Pereira H 2008 Heat induced color changes of pine (Pinus pinaster) and eucalypt (Eucalyptus globulus) wood *Wood Sci Technol* 42(5) 369-384

[22] Boonstra M, Van Acker J, Tjeerdsma B, and Kegel E 2007 Strength properties of thermally modified softwoods and its relation to polymeric structural wood constituents *Ann Forest Sci* 64 679-690

[23] Tsuchikawa S, Hayashi K and Tsutsumi S 1996 Non-destructive measurement of the subsurface structure of biological material having cellular structure by using near-infrared spectroscopy *Applied Spectroscopy* 50 1117–1124

[24] Windeisen E, Bächle H, Zimmer B and Wegener G 2009 Relations between chemical changes and mechanical properties of thermally treated wood *Holzforschung* 63 773-778

[25] Leblon B, Oluwatosin A, Guillaume H, Ataollah H, Tsuchikawa S, Burger J, Stirling R, Pirouz Z, Groves K, Nader J and LaRocque A 2013 A review of near-infrared spectroscopy for monitoring moisture content and density of solid wood *The forestry chronicle* 89 (5) 595-606

# Researching the change in the weight of split oak firewood due to different drying methods

**Amina Gačo**

University of Ljubljana, Biotechnical Faculty, Slovenia

E-mail: aminagaco5@gmail.com

**Abstract.** In this paper, the influences of stacking methods and drying places of split oak firewood in winter conditions (Quercus cerris) have been experimentally and theoretically investigated, due to the great importance of use in heating with such wood. The research was conducted on a sample of 48 pieces of logs with bark made of freshly cut oak. The logs were obtained by mechanized technology with the help of grafting by a hydraulic splitter and a chainsaw. Each log was measured for a research length of 0.30 m, marked with numbers 1-24. On each log, the weight was measured on a scale of 5 kg with an accuracy of 0.5 grams. Drying of logs was performed in two places. The first drying place was in outdoor conditions, sheltered from the snow and ventilated. The second place was indoors, at approximately constant room temperature. The logs were stacked in a crossed way for better air circulation. They were dried between 15/12/2020 and 15/02/2021, a total of 62 days. The average initial weight of logs dried in the facility was 1130.3 g, in the dried state the average weight was 952.87 g, which represented 84.3% of the original weight. The average initial weight of logs dried in outdoor conditions was 1192.125 g, in the dried state 1076.37 g, which represented 90.29% of the original weight. The average initial water content in the logs dried in the facility was 37.9%, in the dried state it was 27.5%. For logs dried in outdoor conditions, the average initial amount of water was 36.6%, in the dried state the average amount of water was 31.2%.

## 1. Introduction

The most important source of energy for people are liquid fossil fuels. Wood burning has also been successfully used on a global scale. Firewood has also always been an important source of energy in different forms. The most important form is logs which are burned in fireplaces, stoves and boilers (central heating). In addition to logs, wood chips, pellets and wood briquettes are also used. Wood is a clean source of energy that has a very efficient combustion. The amount of releases into the environment is small. The basic property of fuels is calorific value. Wood calorific value is the amount of heat generated during complete combustion. It is expressed in kWh/kg, MJ/kg, kWh/m$^3$ or MJ/m$^3$. Wood of different tree species has different calorific value and burning properties. One cubic meter of fire wood has an energy value of 2,628 kWh, oak 2,884 kWh, beech 3,078 and spruce 2,178 kWh. The calorific value of wood is influenced by the water content or moisture content of the wood, the chemical structure of the wood, the density of the wood, the tree species and parts of the tree and the health condition. Due to the calorific value, the amount of water in the wood is very important. To obtain ideally dried wood, it is important to obtain log technology. Wood drying is one of the most drastic wood processing procedures. Dryer wood has a higher energy value. When burning damp wood, part of the energy is used to evaporate the water. Air-dried wood with a humidity of up to 28%, which has been dried in covered and air-stored warehouses for at least 6 months, is usually used for heating. Proper drying and storage of wood affects the water content in the wood. The duration of drying depends on the shape of the wood, the time of felling due to the water content in the wood, the place of drying (location) and the type of wood. When drying wood, it is important the season when the wood was cut. In spring, the water content in the wood is higher than in winter.

## 2. Research

The aim of the research is to experimentally and theoretically investigate the changes that occur in the middle of natural drying of chopped firewood (*Quercus cerris*), in winter conditions. The aim was to investigate how much split oak firewood lost in mass, what was the initial amount of water and the amount of water lost during natural drying in outdoor conditions and drying indoors. The final goal was to show from the obtained results what changes have occurred in different drying methods. One can present whether drying oak is better in outdoor conditions or indoors during the winter.

### *2.1. Wood drying*

Wood is a porous and anisotropic fibrous material. Wood fibers serve to transport water and nutrients from the roots to the leaves. In the case of wood, there is the moisture content of the wood, which is characterized as the ratio between the weight of water and the weight of absolute dry wood. When drying, the humidity of the wood drops. This causes different deformations in the wood. Each tree being dried has different conditions depending on the presence of water. The wood can be dried to an absolute state by the help of a dryer. Based on the presence of water in wood, it can be divided on fresh wood which contains the most water, semi-dry wood (55%), pre-dried wood (20%) and dry wood (15%), (J. Krpan 1965). By drying the wood, one gets increased dimensional stability, biological resistance, reduce weight and ensure easier processing and use. Due to drying, the wood deforms and cracks. It is noticeable that conifers dry faster, while deciduous trees have tracheae which perform a conducting function and the cell walls are thicker and difficult to be passed through by water. The process of wood drying is complex. Water has two forms; free water in the cells and bound water found in the woody walls of the cells. In the drying process, free water is released first, when the pores are emptied. Drying of wood is divided into natural and technical drying. Natural drying is the placing of wood in the environment, which is the cheapest form of drying. Disadvantages of this drying are high consumption of space, longevity, time depends mainly on the tree species, dependence on climate and air temperature. Technical drying of wood is getting better and began to evolve with the needs of a larger amount of wood. We know two different drying methods based on the mobility of the material. Stationary or chamber wood drying. Based on the temperature, there are low temperature (15 °C - 45 °C), normal (40 °C - 90 °C) and high temperature (90 °C - 130 °C). People are also familiar with the basis of the method, chamber, channel, condensation, vacuum, infrared drying, high frequency and solar drying. In addition to the method of drying the regimes are known. Drying regimes were created on the basis of long-term measurements and the search for optimal drying conditions for wood for individual tree species, thickness and water content in the wood. There are English regimes, American or Madison ones, and Russian drying ones that prescribe gradual drying steps. All newer drying regimes were composed so that the drying sharpness is constant with respect to the tree species and does not change during drying. In addition, there are regimes prescribed by the manufacturers of technological equipment for drying or built into the software for managing the drying process.

## 3. Methodes

For the preparation of the experimental work, an oak assortment from which the logs were made of the desired dimensions was obtained. Before setting up the experiment, the wood was split making 48 logs. The result was 24 logs from one meter of wood. There was a 2-meter piece of wood. It obtained a tiny 48 logs with bark, 0.30 m long, about the same thickness. There were two different drying places, outdoors and indoors. The logs were folded crosswise with24 logs in each group. The outdoor log was placed under the roof protected from rain and snow, and was located on the sunny side of the house. The inner logs were located in the boiler room. All logs were numbered to make it easier to track changes in log weight. Each week, the mass of the wood using a scale was measured. The measurement devices installed in the premises to monitor the temperature and humidity in the environment.

### *3.1. Shape form*

For the oak shape, a quarter of the fine log was selected, which has a triangular shape (Figure 1). Such a form is suitable for home heating which is the most useful. Two sides were without bark but one side with bark. Such a shape was obtained by splitting logs by a splitting machine into four logs splitted twice into 8 logs.

**Figure 1.** The shape of a dried log

### *3.2. Place of drying and stacking of logs*

The logs were dried in two places in the winter (in the outdoor environment and in the building). The outer logs were folded crosswise on the tiles and were covered with a roof (Figure 2, left). There was no rain or snow on the logs. It was positive for drying that it was located on the sunny side of the house (south) and that the wind always blew in this place. In such a position the logs should be well dried. The logs stacked in the building were on the tiles, in a part of the house in front of the room with the boiler (Figure 2, right). In such an environment, there was no draft, but the wood was warm which would affect faster drying. Each log was marked with numbers from 1 to 24.

**Figure 2.** Crossing method of log drying outside and inside

## 4. Results

### *4.1. The mass of firewood*

The log drying was in two different places and the method of drying had an impact on the logs due to the environment. The measurements lasted from 15/12/2020 to 15/02/2021. The average initial weight of logs dried in the facility was 1130.3 g, in the dried state the average weight was 952.87 g and the average lost weight was 84.3%. The average initial weight of logs dried in outdoor conditions was 1192.125 g, in the dried state 1076.37 g, the average lost weight was 90.29%. Table 1 shows the descriptive statistics given on the mass of logs.

IOP Conf. Series: Materials Science and Engineering 1208 (2021) 012023 doi:10.1088/1757-899X/1208/1/012023

**Table 1**. The mass of fire wood.

| Lable | Measuring unit | N | Min. | Median | Max. | Mean | St.deviation |
|---|---|---|---|---|---|---|---|
| 1 | 2 | 3 | 4 | 5 | 6 | 7 | 8 |
| $m_{outdoor}$ | gram | 24 | 2,85 | 12,06 | 16,95 | 10,76 | 3,9 |
| $m_{outdoor}$ | gram | 24 | 3,7 | 11 | 12,98 | 11,92 | 4,1 |
| $m_{indoor}$ | gram | 24 | 6,05 | 11 | 17 | 11,3 | 3,5 |
| $m_{indoor}$ | gram | 24 | 4,5 | 9,3 | 15,6 | 9,5 | 3,2 |

*4.2. Moisture of firewood*

The average initial water content in the logs dried in the facility was 37.9%, in the dried state it was 27.5%. For logs dried in outdoor conditions, the average initial amount of water was 36.6%, in the dried state the average amount of water was 31.2%. Table 2 shows the descriptive statistics given on the moisture content of logs after drying.

**Table 2.** The amount of moisture in the wood of the log

| Lable | Measuring unit | N | Min. | Median | Max. | Mean | St.deviation |
|---|---|---|---|---|---|---|---|
| 1 | 2 | 3 | 4 | 5 | 6 | 7 | 8 |
| $m_{outdoor}$ | % | 24 | 31,2 | 31,7 | 36,6 | 33,36 | 2,14 |
| $m_{indoor}$ | % | 24 | 27,5 | 32,1 | 37,9 | 32,16 | 4,1 |

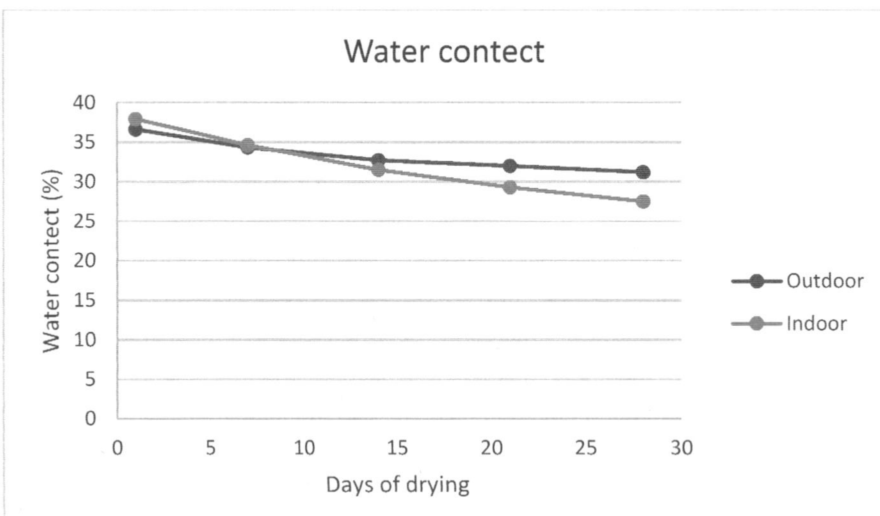


**Figure 3.** Water content of oak logs dried in the outdoor environment and in the building

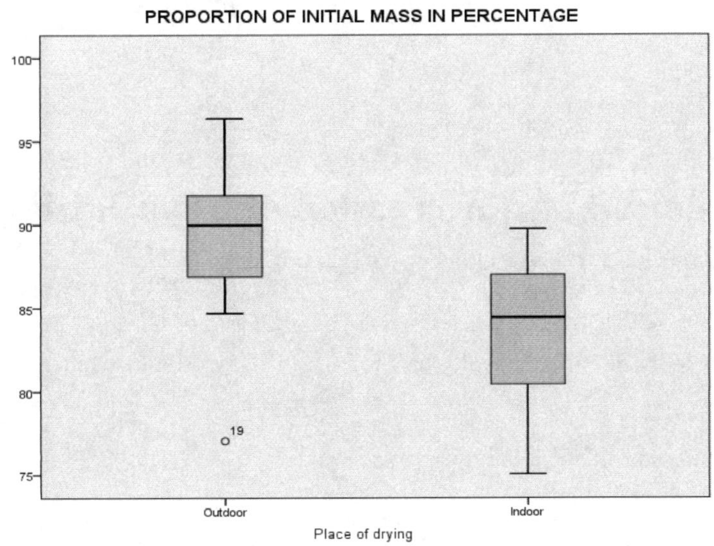


**Figure 4.** Influence of drying site on the mass of oak logs shown in percentage

The results (Figure 4) showed that logs stacked in the facility had 84.3% of their initial weight and logs stacked outside had 90.29% of their initial weight. According to the Mann-Whitney U test (p <0.001), the differences in the drying site were statistically significant. Based on this data, it can be confirmed that the logs in the building dry faster than outside logs.

## 5. Conclusion

The logs dried in the facility actually lost more moisture than the logs dried outside. The outside dried logs were dependent on the weather and due to the very difficult drying of oak logs, the choice of drying logs outside was not a good solution. The logs in the outdoor environment dry slowly and take longer. If one wants to dry the logs as quickly and better as possible, a better solution for drying in a closed facility was confirmed by the results of the research. According to research, logs dried in the building lost water faster, are drier and had a higher calorific value. The weight of logs dried in the building averaged 84.3%, in contrast to logs dried in outdoor conditions where it was 90.2%. During natural drying, there was a drop in moisture in the wood, the average moisture content of logs in the raw state dried outside was 36.6%, while in the dried state the humidity dropped to 31.2%. During the natural drying of logs in the building, the average moisture content of logs in the raw state was 37.9%, while in the dried state the humidity dropped to 27.5%. With this it can be concluded that logs that were cut in the winter with a large amount of water and can be quickly dried and used for burning. In the indoor building the logs can be easily dried to an acceptable humidity and used in heating.

## References

[1] Čuk A 2018 *Energijska in okoljska presoja sušilnice lesa* (Ljubljana: University of Ljubljana) p 104
[2] Ministrstvo za okolje prostor in energijo 2020 *Energetska bilanca Republike Slovenije,* Ljubljana, p 59
[3] Karahasanović A 1992 *Nauka o drvetu* (Sarajevo: Faculty of mechanical engineering) p 425
[4] Klenovšek M 2009 *Porasti družbene odgovornosti in trg biomase* (Ljubljana: Biotechincal faculty) p 65
[5] Novak M 2008 Sušenj lesa (Maribor: Woodwork school ) p 136
[6] Potokar M 2016 Uporaba manj vrednega lesa pri ogrevanju kmečke hiše z lesnimi sekanci (Ljubljana: Biotechnical Faculty, Department of Wood Science and Technology)

# The force required for the creation of wood cuttings

**Hasan Talić**

University of Bihać, Faculty of Technical Engineering, Bosnia and Herzegovina

E-mail: hasan.talic@unbi.ba

**Abstract**. Wood as a material has its own peculiar role during processing due to its characteristics which depend on a number of factors. Therefore, wood-based plate materials tend to encounter the same issues. The creation of the continuous cuttings is conditioned by the strength as it is being cut orthogonally. The cutting force is shown as the sum of the forces for plastic deformation, the force for overcoming the work of the friction force on the front and rear surface of the tool and the force for creating a new surface. Each of the forces is connected to appropriate mechanical features of wood. Examining the mechanical properties of wood, which can be used to calculate the required power to create a new surface, demonstrates the dependence of cutting power on the type of wood, cutting speed, and wood moisture.

## 1. Introduction

There are numerous construction materials used in the world nowadays, some of which pushed out many of the elementary materials. Wood has nevertheless stayed one of the most important materials. As such it is used in production of various products. There is the basic solid wood as well as other materials, products of solid wood with great properties. Solid wood itself and other wood-based materials, most of the time require mechanical processing. The basis for this kind of processing is in the metal processing. Wood is however anisotropic- orthotropic material, which makes its processing more complex.

Mechanical wood processing can be divided in two ways: cutting and plastic deformation. Wood processing by wood shavings separation is a series of technological, physical and economic parameters that vary as well as series of factors that affect those parameters. The simplest way of wood shavings separation processing is orthogonal cutting. This particular cutting method occurs when the tool blade goes through the item being processed, in this case wood, in a manner perpendicular to the direction of movement.

## 2. Wood structure

Wood is one of the natural polymers that has complex structure. In order to thoroughly observe wood structure and understand its influence on wood properties we can divide it into macroscopic and microscopic structure. It is possible to study it with naked eye or magnifying glass (macroscopic structure) or with a microscope (microscopic structure). In the field of forestry for practical purpose we need macroscopic structure knowledge while submicroscopic and microscopic structure expertise is much needed in the study of phenomena and processes such as wood cutting.

Transversal and longitude cuts provide macroscopic view of the trunk (wood). If we take a look at transversal cuts, depending on the type of wood we can see anatomical axis or core, heart, protein, bark as well as growth rings (annual rings) of early and late wood. On longitudinal cuts, radial (if the cut is along the fibers through the axis) or tangential (same cut going outside of axis) we can see same properties in the same or changed shape. Unlike deciduous trees, coniferous trees have simpler microscopic i. e. anatomic structure made of tracheid and parenchymal cells. Deciduous trees, anatomically speaking are considered to have complex structure. Here is an example to illustrate. In deciduous trees tracheid and transitional forms of vascular and vasicentral tracheid elements perform conducting. Libriform and fiber tracheids have the mechanical function while parenchymal cells take on the function of accumulation. [1]

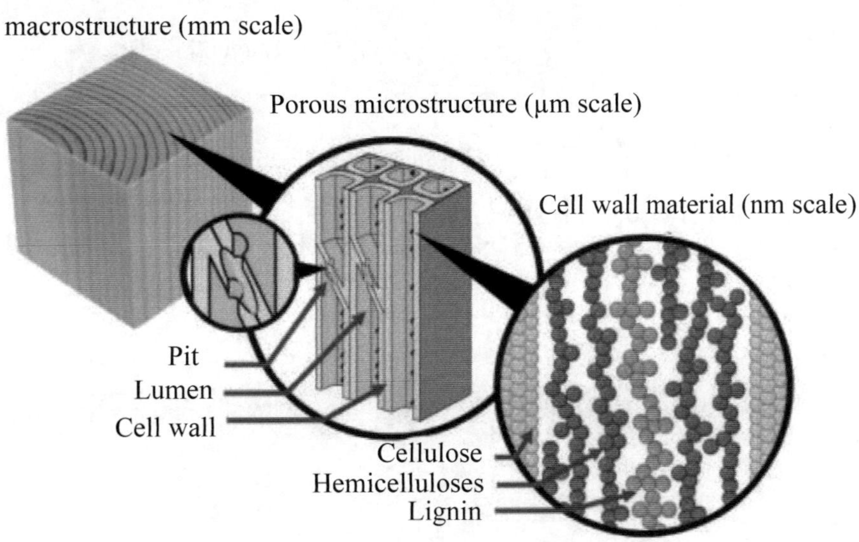


**Figure 1.** Simplified view of wood structure.

Physical properties of wood are directed by anatomic and chemical structure as well as the nature of physical forces that have influence on it. Some of the physical properties to consider when cutting wood orthogonally are moisture, wood porosity and density, change of dimensions (shrinkage and swelling), influence of electricity and heat on wood. While being processed wood is exposed to different kinds of complex straining. Due to the complexity and very specific wood structure, it is quite hard to study them and apply classical estimation methods used with other materials. This is precisely why we need to use conventional and modern methods to study mechanical properties of wood in detail. Some of the frequently studied mechanical properties are elasticity and plasticity, static voltage in wood, (pressure, straining, shearing, chopping, bending), wood hardness, wear resistance, weather influence etc. [1]

Speaking of wood as a material we are obliged to mention other wood-based plates that bare some of the same properties as wood but at the same time have some different properties and conditions for usage, studies show. Fibrous plates, a woodbased material, has numerous advantages compared to solid wood such as more homogenous structure, lesser dimension change under the influence of moisture, possibility to manufacture specific properties, etc. Homogenous structure of these plates is great advantage in the studies and estimates of the material properties, which is in turn applicable to orthogonal cutting which is the focus of this paper. [5]

## 3. Orthogonal wood cutting

Cutting wood is different than cutting metal or other materials. The reason for this is the cutting resistance that changes in wood depending on the cutting planes. This difference occurs because of the

IOP Conf. Series: Materials Science and Engineering 1208 (2021) 012024 doi:10.1088/1757-899X/1208/1/012024

wood structure which is different in different parts of the wood as well as between early and late part of the wood. We differ three basic planes: transversal, radial and tangential plane (Figure 2).

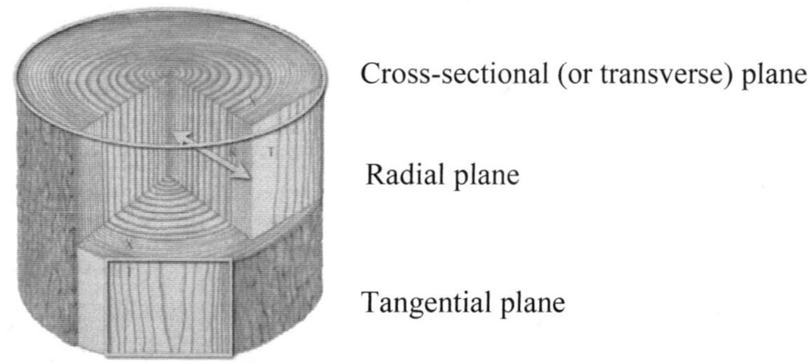


**Figure 2**. Basic wood planes. [4]

Plastic deformation of wood with a tool produces wood shavings. Tool work causes complex strains The plane where the strains are the most intense is called shear plane. Its position is determined by the angle of sheerness (φ). There are three basic methods of cutting defined by wood planes e. i. direction of movement of the tool and they are as follows: transversal (b) longitudinal (c) tangential (a) Figure 3.

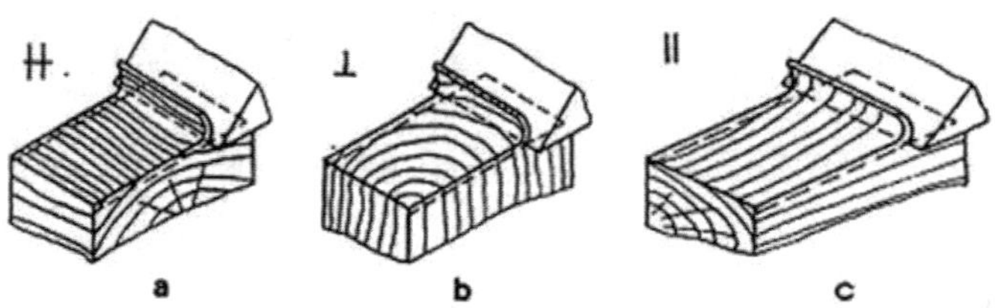


**Figure 3.** Scheme of basic cutting directions (a) tangential (b) longitudinal (c) transversal. [3]

Orthogonal cutting starts with the blade cutting through the wood where the blade is vertically positioned in reference to the direction of the movement otherwise, we are talking about diagonal cut. The orthogonal cut has three components: the tool, the object being cut (wood) and wood shavings. The studies of the orthogonal cut are focused on the mutual influence on each other of tool, wood and wood shavings in the production process. Figure below shows system of orthogonal cut and its respective parts.[2]

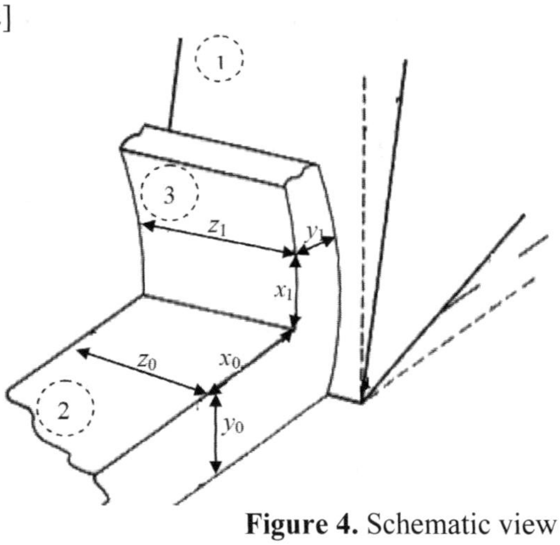


1- Tool (tool blade),
2- Item in process (wood)
3- wood shavings
$x_0$ – length of uncut wood shavings
$y_0$ – height of uncut wood shavings
$z_0$ – width of uncut wood shavings
$x_1$ – length of cut wood shavings
$y_1$ – height of cut wood shavings
$z_1$ – width of cut wood shavings

**Figure 4.** Schematic view of orthogonal cut. [2]

Orthogonal cut can be viewed as a two-dimensional problem where there is equation between cut und uncut sawdust ($z_0 = z_1$). This process can be viewed through three different zones. First zone marked by the distance between points A and B represents length of contact of front part of the tool. Zone marked by the distance from points B to C is the zone where the cut occurs and sawdust appears. Blade point comes in the contact with the wood in point B which is shown in Figure 5.

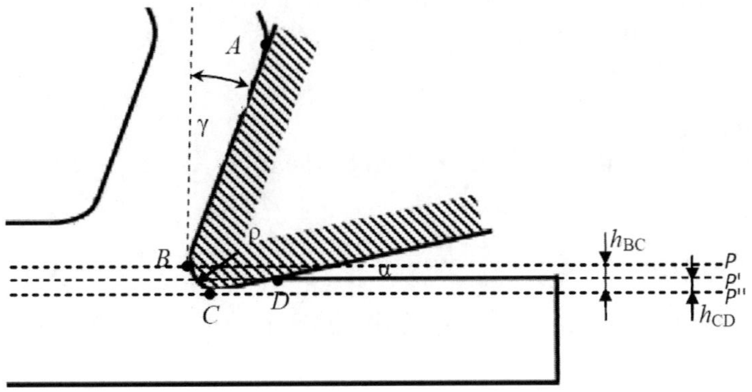


AB – length of contact between front side of the tool and sawdust,
BC – length of contact between wood and curved part of tool's blade,
CD – length of contact between backside of the tool and wood
$h_{BC}$ – wood compression,
$h_{CD}$ –elastic return of wood
γ – front angle of the blade,
α – rear angle of the blade,
ρ – radius of the blade's curve

**Figure 5.** Schematic view of orthogonal cut in two dimensions. [2]

Dynamic of orthogonal cut is combination of tool work, wood and wood shavings. In the zone between points AB front part of the tool's blade causes pressure on the newly formed sawdust. That pressure can be replaced by friction whose impact direction is perpendicular to the front part of the tool. The wood shaving moves in particular speed on the front part of the tool causing the friction. Vector sum of the two forces gives the resulting force on the front part of the tool. [2]

While the tool blade cuts into the wood cutting resistance occurs. When defining degree of material workability, level of weariness of tool blade, and setting the element dimensions, and machine power we need to be familiar with the resistance force. The resulting force in orthogonal cutting process (FR) is shown in the Figure 6.

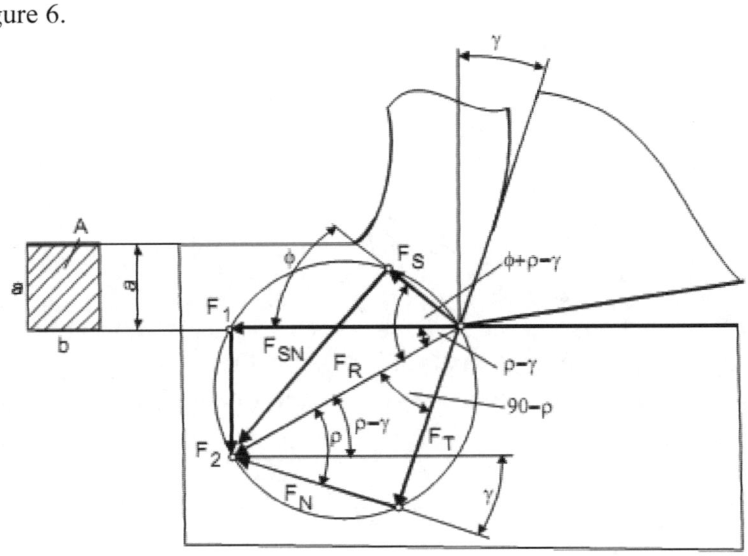


**Figure 6.** Resulting force ($F_R$) in orthogonal cutting system.

- Tangential force ($F_T$) or force of friction that occurs on the front side of the tool surface and Normal force ($F_N$) that occurs in the plane perpendicular to the front part of the tool blade,
- Shear force ($F_S$) and Normal force ($F_{SN}$) that occurs in the plane perpendicular to the shear plane,
- Main force ($F_1$) and penetration force ($F_2$) that affects processed item perpendicularly. [7]

The estimated value is determined based on main cutting force which defines other components as well. This force is in turn mainly determined by material being cut and parameters of the process (cutting depth and length of tool movement). [7]

## 4. Power required for plastic deformation

Power required for plastic deformation of wood shavings is determined by the following equation:

$$P_{pd} = \frac{dW}{dt} = \frac{\bar{K}\cdot\bar{\varepsilon}^{\bar{n}+1}}{\bar{n}+1}\cdot\frac{d_{x0}}{dt}\cdot y_0\cdot z_0 = \frac{\bar{K}\cdot\bar{\varepsilon}^{\bar{n}+1}}{\bar{n}+1}\cdot v_A\cdot y_0\cdot z_0 \tag{1}$$

Formula for power (1) required for plastic deformation is derived from time formula (2) for energy required for plastic deformation.

$$W = \int_V \sigma\cdot d\varepsilon = \int_V \bar{K}\cdot\varepsilon^{\bar{n}}\cdot d\varepsilon = \frac{\bar{K}\cdot\bar{\varepsilon}^{\bar{n}+1}}{\bar{n}+1}\cdot x_0\cdot y_0\cdot z_0 \tag{2}$$

Equation (1) shows connection between power required for plastic deformation of wood shavings ($P_{pd}$) and process parameters ($v_A$, $y_0$ i $z_0$), as well as mechanical properties of the material (wood) K i n. Equivalent deformation (ε) depends on process parameters and mechanical properties of wood, as such it is important demonstrator of plastic deformation of wood shavings. Value of equivalent deformation is most often calculated by Von Mises yield criterion (3):

$$\bar{\varepsilon} = \frac{\sqrt{2}}{3}\cdot\sqrt{\left(\varepsilon_x-\varepsilon_y\right)^2+\left(\varepsilon_y-\varepsilon_z\right)^2+(\varepsilon_z-\varepsilon_x)^2} \tag{3}$$

Further input of main deformation values in the Mises criterion we get the equation for equivalent deformation of wood shavings (4):

$$\bar{\varepsilon} = \frac{2}{3}\cdot\sqrt{ln^2\xi_x - ln\xi_x ln\xi_y + ln^2\xi_y} \tag{4}$$

In equation (4) for calculation of power we need to express compression ratio as well as power and hardening ratio. Adding equation (4) in equation (1) we get equation (5) that links factors of compression and power spent for plastic deformation of wood shavings.

$$P_{p1} = \frac{\bar{K}\cdot\left(\frac{2}{3}\sqrt{ln^2\xi_x - ln\xi_x ln\xi_y + ln^2\xi_y}\right)^{\bar{n}+1}}{\bar{n}+1}\cdot v_A\cdot y_0\cdot z_0 \tag{5}$$

Equation for equivalent value of power factor K and hardening ration n determined according to Hanks' equation (Bodig and Jane, 1982.):

$$\bar{K} = \frac{K_x \cdot K_y}{K_x \cdot sin^2\gamma' + K_y \cdot cos^2\gamma'}$$
$$\bar{n} = \frac{n_x \cdot n_y}{n_x \cdot sin^2\gamma' + n_y \cdot cos^2\gamma'} \quad (6)$$

Total cutting power can be calculated as sum of power needed for plastic deformation of wood shavings and power needed to overcome the friction force on the front and rear part of the tool blade, and power needed to create new surface. [2]

## 5. Conclusion

The aim of this paper is to present different kinds of wood cutting, mainly orthogonal cutting and to present the power required for plastic deformation. Besides this power, we can calculate total cutting power which is the sum of powers: the aforementioned plastic deformation power, power to overcome the force of friction on the front and rear part of the tool blade, and power required to produce new wood shavings. These powers are linked through wood shavings hardening ratio. They share some of the mechanical properties. In order to exactly determine the cutting power required it is necessary to be familiar with mechanical and physical properties of wood that depend on wood structure and conditions of the processing, and their microscopic and macroscopic structure of wood. For instance, moisture, to be exact amount of moisture in wood changes processing parameters as well as the quality of the processed object. We must mention wood-based materials such as fibrous plate compared to solid wood.

## References

[1] Šoškić B M and Popović Z D 2010 Svojstva drveta *University of Beograd* Faculty of Forestry
[2] Radmanović K 1970 Utjecaj mehaničkih svojstava drva na snagu potrebnu za odvajanje čestice pri ortogonalnom rezanju *University of Zagreb* Faculty of Forestry Zagreb
[3] B Crnogaća 2015 Stvaranje strugotine u procesu rezanja drveta *DRVOtehnika* n **47** pp 58-59
[4] George S S 2012 Presentation: What are tree rings? *University of Minnesota*
[5] Thybring E E and Fredriksson M 2021 Wood modification as a tool to understand moisture in *Wood Forests*, **12**(3), p 372
[6] Ćehić M and Omer S E 2018 Pločasti materijali na bazi drveta *University of Bihać*
[7] Tufekčić Dž and Brdarević S 1995 Obrada rezanjem i alatne mašine *University of Tuzla*

IOP Conf. Series: Materials Science and Engineering 1208 (2021) 012025 doi:10.1088/1757-899X/1208/1/012025

# Mechanical properties changes in fir wood (abies sp.), linden wood (tilia sp.), and beech wood (fagus sp.) subjected to various thermal modification process conditions

**Redžo Hasanagić[1], Sauradipta Ganguly[2], Ermin Bajramović[1] and Adem Hasanagić[1]**
[1]Faculty of Technical Engineering, University of Bihać
[2]Forest Research Institute Dehradun, Dehra Dūn, India

E-mail: redzo.hasanagic@unbi.ba

**Abstract**. Wood is one of the most important construction materials in Europe and its use in building applications has increased in the recent decades. To enable even more extensive and reliable use of wood, this article aimed to determine the effect of thermal modification on mechanical properties of fir wood (lat. Abies sp.), linden wood (lat. Tilia sp.), and beech wood (lat. Fagus sp.). The thermal modification was conducted in a laboratory oven at five different temperatures of 170, 180, 195, 210, 220 °C and processed with a different maximum duration of the process of 78, 120, 180, 240, 276 minutes. Mechanical properties of treated wood have shown statistically insignificant fluctuations at lower temperatures compared to control samples. On the other hand, raising the temperature to 210 °C significantly affected the strength of all the species. The results revealed that thermal modification at high temperatures and longer exposure causes a decrease in the maximum force of the three wood species.

## 1. Introduction

Wood is one of the earliest construction materials. The structural use of wood and wood-based composites continues to increase steadily. In fact, new wood-based materials continue to develop and are being successfully introduced into the engineering and construction marketplace [1,2]. Wood is an important natural resource used in many applications, from construction to furniture and different domestic or industrial objects, tools, and artworks. Despite its properties, wood presents some limitations in exterior environment use, such as dimensional instability, susceptibility to fungal decay or weathering, etc. [3]. The main disadvantages of wood, such as poor dimensional stability and biological degradation or deterioration , are mainly due to the nature of the main polymers of the cell wall of wood, especially due to the abundance of free hydroxyl groups (OH), as reported in previous studies [4,5,6]. Though thermal modification has been successful in improving dimensional stability and fungal attack resistance, as stated in the research [7,8,9], it reduces the mechanical properties of wood. The researchers reported that thermal modification of wood at different temperatures and duration decreased the equilibrium moisture content (EMC), increased dimensional stability [10,11], mass loss (ML) [12,13], biological durability [14], and decreased some mechanical properties, as well as the wettability of wood, although in some studies these latter two parameters increased [15]. The diversity of wood species in combination withdifferent thermal processes, cause changes in the physical properties, such as ML, anti-swelling efficiency, and EMC, and produce a varied range of comparable values justifying the success of the process. ML is a determinative factor of the result of heat treatment, i.e. the greater the ML, the more severe the effects on the physical and mechanical properties. In a similar research [16] authors reported that there is a significant relationship between ML and compression strength of wood, whereas [17] observed that there is a significant relationship between ML and equilibrium moisture content of wood [18]. During previous research, a test for optimization of bending and tensile force parameters was performed. The procedure resulted in mathematical models and optimal parameters of the influence of the bending and tensile force process to the maximum fracture force [19, 20, 21]. In line of the above, for the present work, the objectives are to determine the mechanical properties of Abies sp., Tilia sp., and Fagus sp. wood under various temperatures and process durations to determine the mechanical properties of thermally modified samples using a series of standardized four-point static bending and tensile tests on the maximum force.

## 2. Materials and methods

### *2.1. Materials*

The experiment used unmodified samples of beech wood (lat. Fagus sp.), linden wood (lat. Tilia sp.) and fir wood (lat. Abies sp.) for bending and elongation, whose measured average bulk density was 0.366 g/cm3 for fir, 0.472 g/cm$^3$ for linden, and 0.675 g/cm$^3$ for beech samples. In the case of modified samples, the average bulk density was 0.353 g/m$^3$ for fir samples, 0.455 g/m$^3$ for linden samples and 0.655 g/m$^3$ for beech samples. The thermal modification process was performed on rectangular four-point bending samples (Figure 1) with dimensions of 380 x 50 x 20 (mm) and elongation samples with dimensions of 162 x 22 x 18 (mm).

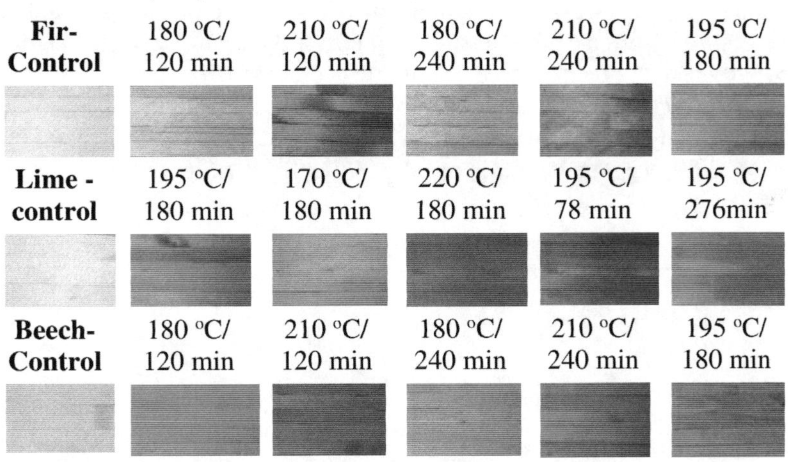


**Figure 1.** Thermally modified samples of fir, linden, and beech wood during different processing processes

### *2.2. Thermal Treatment*

Thermal modification of wood is a process in which the molecular structure of cell wall polymers (cellulose, hemicellulose and lignin) are changed. This can lead to crosslinking, reduction of OH groups and (unwanted) chain cleavage. The reduction of available OH groups leads to limited interaction with water in relation to untreated wood. The structure of the cell wall of wood mainly consists of cellulose, hemicellulose and lignin. All of these components contain free hydroxyl groups. These hydroxyl groups play a key role in the interaction between water and wood. At the same time, these groups present the most reactive space in wood. If wood is exposed to humid conditions, water molecules accumulate between the wood polymers and form hydrogen bonds between the hydroxyl groups and the individual water molecules. This water needs space between the components of the cell wall, which leads to swelling of the wooden piece, and fungi and insects infestation (Figure 2). The wood industry is continuously developing advanced processes, materials and solutions to meet the requirements and increase market competitiveness. One such treatment that occurs is thermal modification of wood.

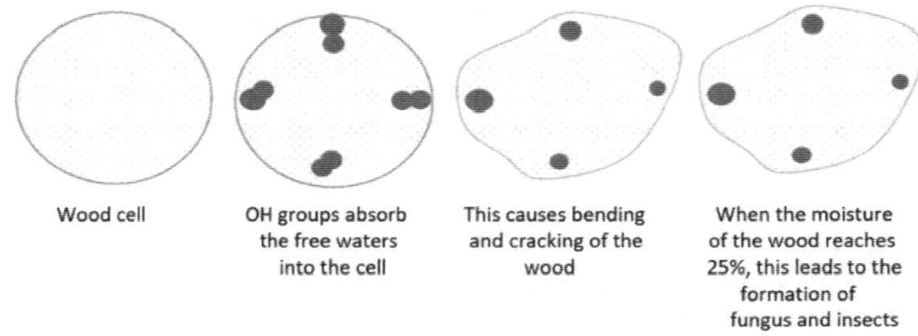


**Figure 2.** Moisture absorption into wood

Thermal modification of wood is performed under controlled conditions at high temperatures. In the process of modification, only heat, water vapor and fresh water are used, without the application of any chemical preparations or additives, which makes it completely environmentally friendly. It is performed at temperatures between 160 °C and 260 °C, in specially designed chambers equipped with a fully automated process control system. The use of high temperature, pressure and water vapor significantly increases the quality of wood and its physical properties. The most important parameters on which the properties of thermal modification of wood depend are the following: type of heating medium, duration of the process, final temperature, wood moisture before temperature, and type of wood. Prior to the thermal modification (TM) process, all samples in this research were marked according to the experimental plan and oven-dried in the drying chamber of the manufacturer "Kambič''. The thermal modification was performed according to a commercial process (Silvapro®, Silvaprodukt, Ljubljana, Slovenia), with an initial vacuum in the first step of the treatment [22,23]. The samples were heated to maximum temperatures ranging between 170 °C - 220 °C and treated with a different maximum duration of the TM process in different time intervals of 78, 120, 180, 240, 276 minutes depending on the wood species

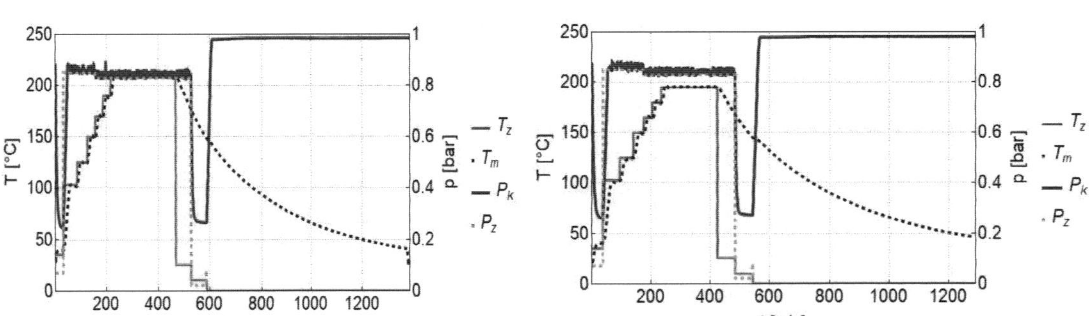


**Figure 3.** Changes of temperature, process time and chamber pressure during TM: a) 210 °C/240 min; b) 195°C/3 min. Tz (defined chamber temperature), Tm (temperature in the central part of the sample), Pk (chamber pressure), Pz (defined chamber pressure)

Figure 3 shows four parameters in the heat treatment process of beech wood samples (lat. Fagus sp.), linden wood samples (lat. Tilia sp.), and fir wood samples (lat. Abies sp.), where the marked curves of temperature and pressure are present in the diagrams.

### 2.3. Bending and tensile testing procedures

Solid wood for various construction applications must be sorted by its strength before exploitation. To make the process as economical as possible, the measurement usually focuses on the most important mechanical properties: density, modulus of elasticity (MOE), and bending and tensile strength. Testing in this paper was carried out in accordance with standard EN 408 (2012) [24]. The tests of the maximum force ($F_{max}$) on bending and elongation of five samples per cycle of thermal modification were performed. The samples were always positioned in such a way that they were subjected to maximum load. The test samples, having a minimum length of approximately 19 times the depth of the section, were simply supported and symmetrically loaded in bending at two points over a span of approximately 18 times the depth. The maximum force was measured within the loading points which is shown in Figure 4 a.

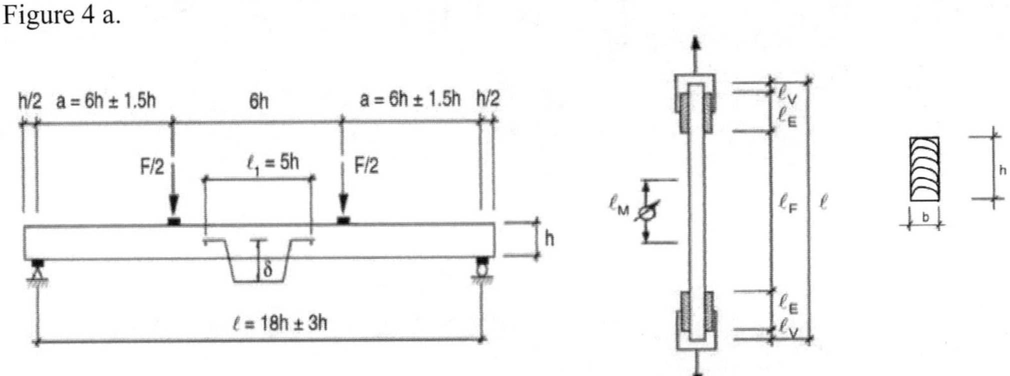


a) b) c)

**Figure 4.** Mechanical tests: a) bending test: 4-point bending procedure, b) tensile test c) dimensions of samples of rectangular cross-section b/h [24]

The tensile test in parallel with the fibers was carried out by EN 408 + A1, where the wood of full cross-section of length 9 is the thickness of the tested samples (Figure 4 b). As mentioned above, the experiments used testing machine type SIL-50KNAG, manufactured by SHIMADZU, in which a working cylinder is located in the upper part of the testing machine. The working cylinder moves down (bending) and up (elongation) depending on the test, while the lower head is stationary. Then, through a computer connected to the SIL-50KNAG type testing machine, the bending is measured at four points so that the values of the force achieved can be read in the diagram.

## 3. Results and discussion

### 3.1. Mass loss (ML) of thermally modified samples

The results have shown a growing trend of ML with increasing thermal modification temperature from 170 °C to 210 °C in bending and tensile samples (Figure 5). The lowest ML were recorded for linden bending and tensile samples at a temperature of 170 °C and duration of 180 minutes while the highest ML were recorded in bending samples of linden wood at a temperature of 220 °C and duration of 180 minutes and tensile samples of beech wood at a temperature of 210 °C for 240 minutes. This indicates that degradation of cell-wall polymers was very low at this temperature. It is expected that mostly hemicellulose and lignin were degraded at this temperature leading to the ML as cellulose tends to have higher resistance to thermal degradation in comparison to hemicellulose and lignin. [25,26,29]. It is noticeable from the diagrams (Figure 5 a and b) that the ML increased with increasing temperature and treatment time. This paper determined that the weight loss was inversely proportional to the increase in temperature and length of thermal treatment which is in line with the findings of previous researchers [13].

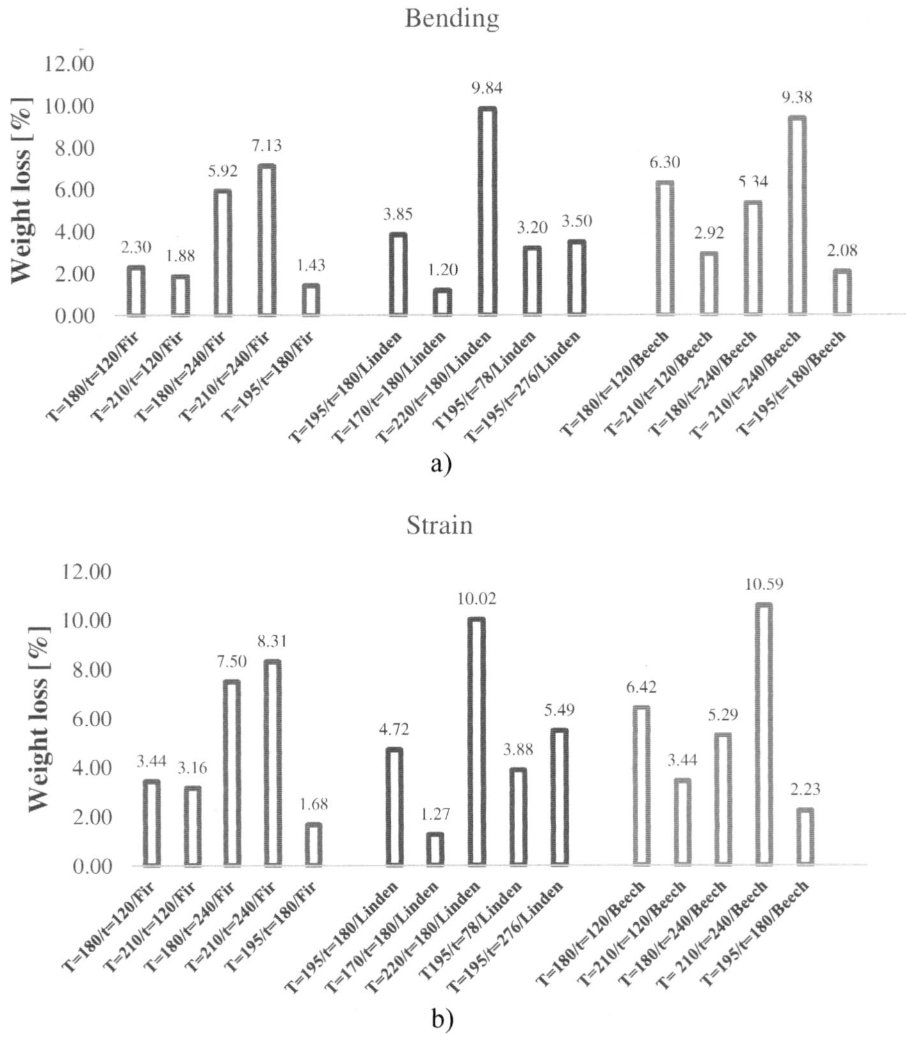


**Figure 5.** Mass loss in fir, linden and beech wood heat-treated from 170 to 210 °C: a) samples for four-point bending; b) tensile samples

### *3.2. Bending strength*

The average maximum four-point bending strength (Fmax) of thermally modified wood showed statistically insignificant fluctuations at low temperatures and shorter duration of the thermal modification process compared to unmodified samples. This is considered to be corroborating evidence of low cell wall polymer degradation at 210 °C/120 minutes for thermally modified fir wood. On the other hand, the increase in temperature and duration time had a significant impact on all properties. There was a significant decrease in maximum force (Fmax), indicating degradation in cell wall polymers, as can be seen in Figures 6 and 7. Thermal modification in this study was performed at five different temperatures and process durations causing a glass transition of lignin in certain wood species, as noted through research in [11,28]. Therefore, as a binder, lignin crosslinking, formation of new bonds and lignin binding points with cellulose and hemicelluloses, which have been thermally developed, can occur [29, 30]. Also for cellulose, the crystallinity remains intact initially but with prolonged exposure the heat degradation becomes dominant and results in strength loss.

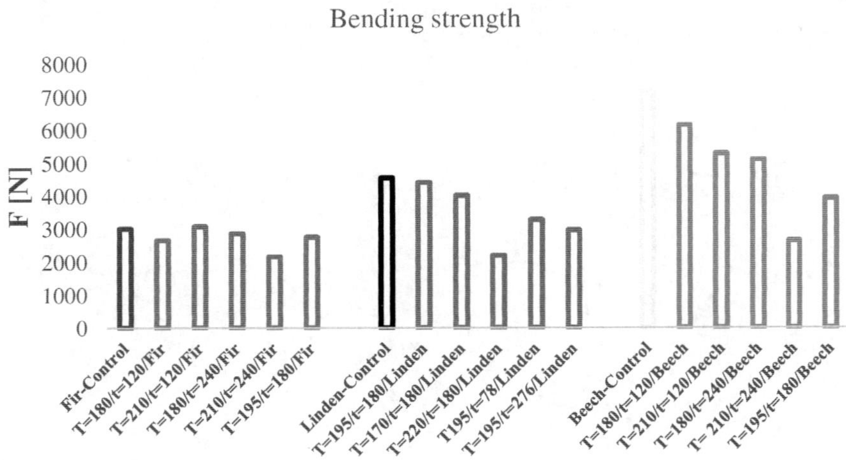


**Figure 6.** Average value of maximum bending strength of thermally modified and unmodified wood

Due to the crosslinking of lignin and binding to cellulose, there is a partial increase in the $F_{max}$ value in fir wood at a temperature of 210 °C for 120 minutes. Although the increase was not statistically significant, there was still an increase of 2.18 % which can be seen in Figures 6 and 7. Crosslinking of lignin has put thermal degradation in the foreground, causing small increases in mechanical properties in fir wood at a temperature of 210 °C during 120 minutes. This indicates that the degradation of the cell wall polymers was very small at this temperature and time duration. At this temperature, hemicellulose and lignin decomposed, whereas cellulose had a higher resistance to thermal decomposition compared to hemicellulose and lignin.

The $F_{max}$ ratio decreases (Figures 6 and 7) in linden wood at a temperature of 195 °C for a process duration of 180 minutes, where there was a slight decrease of 3.35 %. Similar decreases in Fmax were recorded in cycles in the same tree species where at a temperature of 170 °C for 180 minutes there was a decrease of 11.88 %, while at temperature of 195 °C for 78 minutes the decrease was 9.51 %. The largest reductions in bending $F_{max}$ of 51.71 % were recorded in linden wood at a maximum temperature of 220 °C for 180 minutes. There were also larger reductions at 195 °C for 276 minutes where the reduction was 34.68 % compared to unmodified linden samples.

The value of bending $F_{max}$ of unmodified and modified beech samples is shown in Figure 6. The data show a large decrease of $F_{max}$ in thermally modified beech wood, especially at a temperature of 210 °C for 240 minutes with $F_{max}$ decrease of 63.12 %, 26.35 % decrease at a temperature of 210 °C for 120 minutes, 28.97 % decrease at 180 °C for 260 minutes, and about 45.34 % decrease at a temperature of 195 °C for 180 minutes. It can be noticed is that at lower temperatures and time duration there were smaller reductions, which was about 14.53 % at a temperature of 180 °C for 120 minutes compared to unmodified samples (Figure 7).

When it comes to this research, it is important to emphasize that hemicellulose is less thermally stable than cellulose. The data on the exact degrees of temperature for the beginning of the decomposition process are different. Various thermal analyzes of wood show that there is a broad endotherm in the range of 90 to 150 °C which is associated with the evaporation of retained water in the cell wall. The analysis shows that in harder wood species there is a greater loss of bending Fmax than in soft wood species, when monitoring the process of thermal modification under the same conditions, as can be seen in Figure 7 and this is in conformity with the findings of several researchers as mentioned in [13], Hardwood is less thermally stable than softwood, which can be attributed to differences in the content and composition of hemicellulose. Pentosans, which are found in larger quantities in hardwood hemicellulose, are more susceptible to thermal destruction than hexosans. In addition, hardwood generally has a higher proportion of hemicellulose which has a higher content of acetyl groups compared to softer species.

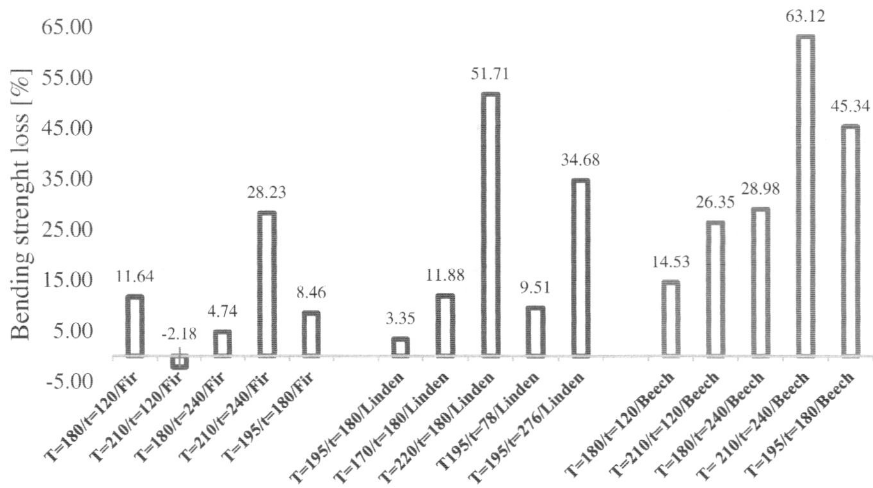


**Figure 7.** Percentage reduction of maximum bending strength

The results have shown a growing trend of decreasing $F_{max}$ as the temperature of thermal modification increased. As previously described, the largest losses of maximum force were recorded in fir wood at a temperature of 210 °C for 240 minutes, in linden wood at 220 °C for 180 minutes and beech wood at 210 °C for 240 minutes. The increase in $F_{max}$ occurred in fir wood at a temperature of 180 °C for 240 minutes .

### *3.3. Tensile strength*

The mean values of tensile $F_{max}$ before and after thermal modification of wood are shown in Figure 8. It is visible that there was a decrease in mechanical tensile properties at elevated temperatures and duration in all cycles. In linden wood, there was the smallest reduction in Fmax of 3.4 %, where the temperature was 195 °C for 180 minutes.

The temperature of 180 °C/120 minutes has shown the smallest decrease of tensile Fmax in fir wood, which was about 23.67 %. With increasing temperature and process duration compared to unmodified wood, a relatively lower Fmax was obtained, as seen in Figure 9. The largest decrease in Fmax of modified samples compared to unmodified fir wood samples (69.31 %) was recorded at a maximum temperature of 210 °C for 120 minutes. In addition, there was a significant reduction of tensile strength Fmax by about 57.18 % at a temperature of 180 °C and duration of 240 minutes. Thus, it can be said that with increasing temperature and duration of the process, there is a decrease of tensile strength Fmax of fir wood, which is best seen in Figure 9.

Diagram 8 shows the results of $F_{max}$ of thermally treated linden wood. It is important to emphasize that the thermal modification carried out at five different levels clearly affects the tensile strength of linden specimens, which is significantly reduced by about 59.56 % at 220 oC for 180 minutes, 54.87 % at 195 °C for 276 minutes, while in the smallest decrease of 3.42 % was recorded at a temperature of 195 °C for 180 minutes.

Heat treatment of beech wood at 180 °C for 120 minutes resulted in Fmax reduction of 53.39 % (Figure 9). Increasing the effective processing temperature to 210 °C for 120 minutes caused a decrease in tensile $F_{max}$ of 48.92 %. In addition, by increasing the process duration while maintaining the temperature at 180 °C for 240 minutes the reduction was as much as 62.16 %, and at 210 °C for

240 minutes it resulted in an even greater reduction of the maximum tensile force by 77.60 %, compared to unmodified samples.

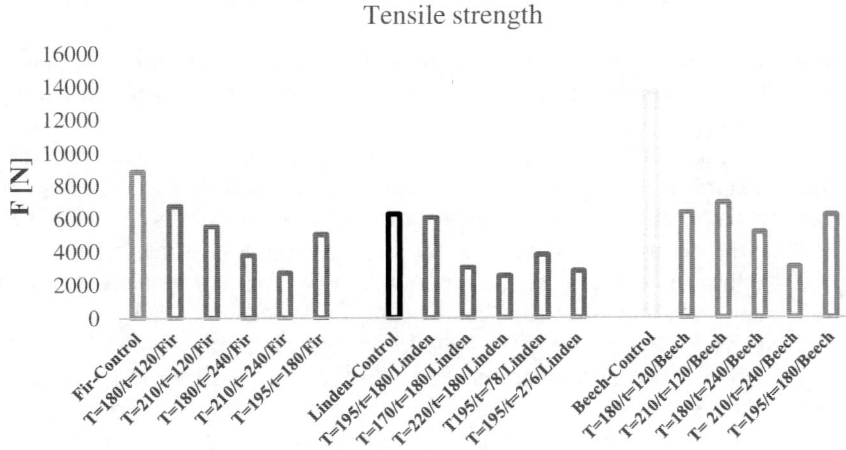


**Figure 8.** Maximum tensile strength of thermally modified and unmodified wood

The results have shown, as with the bending samples, that the lowest tensile $F_{max}$ losses were in the fir wood samples that were thermally modified at 180 °C for 120 minutes where the tensile $F_{max}$ was reduced by about 23 %, and in the linden wood samples that were modified at 195 °C for 180 minutes. The largest reductions in tensile $F_{max}$ were recorded at temperatures of 210 °C for 240 minutes in fir and beech wood, and at a temperature of 220 °C for 180 minutes in linden wood, where the reduction was 59.56 %.

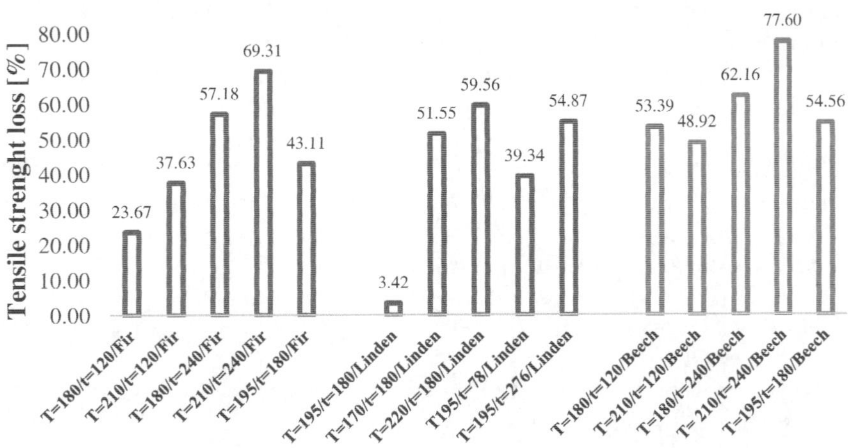


**Figure 9.** Percentage of reduction of maximum tensile strength

Figure 9 shows the percentage decrease in tensile $F_{max}$ of the modified samples for different temperatures, durations, and densities. Based on the analysis of the results, it is clear that there is a decrease in Fmax when processing wood at elevated temperatures. In addition, the loss of tensile strength depends a lot on the maximum duration of the thermal modification process

## 4. Conclusion

Prior to testing the thermally modified wood, it is recommended to perform an analysis of the most influential process parameters (temperature, pressure, process duration, geometry, wood type, etc.).

Every situation is different and the values of the influential parameters are constantly changing. Therefore, making general statements about the mechanical strength of thermally modified wood is not a wise approach. Although, perhaps with a few minor exceptions, the basic knowledge needed to make solid, durable elements still exists.

The possibility of improving the mechanical properties of thermally modified wood is advancing each day. Technology is constantly changing, and a byproduct of seemingly unrelated fields can have a significant effect on solving apparently impossible problems. The development of wood modification and other wood processing processes could influence production methods which, in turn, would result in better dependence on mechanical properties and extension of the life of thermally modified wood.

Based on the analysis of the bending and tensile strength of all materials tested in the experiment, it can be concluded that the duration of the thermal modification process at maximum temperature, wood type and maximum achieved heat treatment temperature are very important parameters that affect the retention of mechanical properties in thermal modification of wood.

Experimental research of the bending and tensile strength determined a higher load, i.e. a higher value of the fracture force at lower temperatures and a shorter duration of the processing. That is, by increasing the temperature and duration of thermal modification, there is a significant reduction in the fracture force of the treated samples, especially in deciduous species. Therefore, heat-treated wood can be used with appropriate heat treatment time and temperature without any loss in strength values in areas where wood elements such as parquet, wooden facades, decorative purposes and where dimensional stability is important.

**References**

[1] Kržišnik D, Lesar B, Thaler N and Humar M 2018 Micro and material climate monitoring in wooden buildings in sub-alpine environments *Construction and Building Materials* **166** 188–95
[2] Kržišnik D, Humar M, Straže A, Šernek M, Šega B, Plavčak D, Lesar B and Grbec S 2020 Durability and mechanical performance of differently treated glulam beams during two years of outdoor exposure *Drvna industrija* **71** 243–52
[3] Popescu C M, Jones D, Kržišnik D and Humar M 2020 Determination of the effectiveness of a combined thermal/chemical wood modification by the use of ft–IR spectroscopy and chemometric methods *Journal of Molecular Structure* **1200** 127-133
[4] Van den Bulcke J, De Windt I, Defoirdt N, De Smet J and Van Acker J 2011 Moisture dynamics and fungal susceptibility of plywood *International Biodeterioration & Biodegradation* **65** 708–16
[5] Schmidt O 2006 Wood and tree fungi, *Springer.*
[6] Reinprecht L. 2016 Wood Deterioration, *Prot. and Maintenance.* Ny, USA: Wiley.
[7] Boonstra M J and Tjeerdsma B 2006 Chemical analysis of heat treated softwoods *Holz als Roh- und Werkstoff* **64** 204–11
[8] Tjeerdsma B F,Stevens M and Militz H 2000 Durability aspects of hydrothermal treated wood. *Inter.Res. Gr. on Wood Preserv.*
[9] Hasanagić R 2019 Thermal modified wood: A critical review, *RIM 2019 – Development and Modernization of Production* Sarajevo **2566-3283** 80-85.
[10] Tiemann H D 1920 The Kiln Drying of Lumber, *JB Lippincott Company*
[11] Phuong L X, Shida S and Saito Y 2007 Effects of heat treatment on brittleness of styrax tonkinensis wood *Journal of Wood Science* **53** 181–6
[12] Hakkou M, Pétrissans M, Zoulalian A and Gérardin P 2005 Investigation of wood wettability changes during heat treatment on the basis of chemical analysis *Polymer Degradation and Stability* **89** 1–5
[13] Esteves B M, Helena M and Pereira H M 2008 Wood modification by heat treatment: A review. *BioResources* **4** 370–404.

[14] Boonstra M 2008 A two-stage thermal modification of wood, Université Henri Poincaré-Nancy 1, Ph.D. dissertation.
[15] Kaboorani A and Englund K R 2010 Water sorption and mechanical performance of preheated wood/ Thermoplastic Composites *Journal of Composite Materials* **45** 1423–33
[16] Gunduz G, Korkut S, Aydemir D and Bekar I 2009 The density, compressive strength and surface hardness of heat-treated hornbeam (Carpinus betulus L.). Maderas. Ciencia y tecnología 11.
[17] Esteves B, Marques A V, Domingos I and Pereira H 2006 Influence of steam heating on the properties of pine (Pinus pinaster) and eucalypt (Eucalyptus globulus) wood, *Wood Science and Technology* **41** 193–207 DOI: 10.1007/s00226-006-0099-0
[18] Brito J O, Garcia J N, Bortoletto G, das Chagas Pessoa AM and Müller da Silva P H 2006 Densidade básica e retratibilidade da madeira de Eucalyptus grandis, submetida a diferentes temperaturas de termorretificação, *Cerne* **12** 182–188.
[19] Hasanagić R 2020 Optimizacija parametara termičke modifikacije konstrukcijskog drveta i njihov uticaj na čvrstoću savijanja i istezanja. Bosna i Hercegovina. Bihać: Univerzitet u Bihaću-Tehnički fakultet. Oktobar
[20] Hasanagić R 2018 Modeling and prediction of fracture force to tighten solid wood elements by genetic programming *Tehnika* **73** 653–7, DOI: 10.5937/tehnika1805653H
[21] Hasanagić R, Hodžić A and Jurković M 2019 Modelling and optimization of tensile break force of solid wood elements lengthened by finger joint *Journal of Adhesion Science and Technology* 1–15, DOI: 10.1080/01694243.2019.1690266
[22] Rep G and Pohleven F 2001 Wood modification-a promising method for wood preservation. *Drvna Industrija* **52** 71–76
[23] Rep G, Pohleven F and Bučar B 2004 Characteristics of thermally modified wood in vacuum, In Proceedings of the IRG Annual Meeting Ljubljana-Slovenia. 9.
[24] BAS EN 408+A1, "Timber structures. Structural timber and glued laminated timber. Determination of some physical and mechanical properties," Institut za standardizaciju BiH, 2013.
[25] Tjeerdsma B F and Militz H 2005 Chemical changes in hydrothermal treated wood: FTIR analysis of combined hydrothermal and dry heat-treated wood, *H.als R.- und Wer.* **63** 102–111. DOI: 10.1007/s00107-004-0532-8
[26] Tjeerdsma B F, Boonstra M, Pizzi A, Tekely P and Militz H 1988 Characterisation of thermally modified wood: molecular reasons for wood performance improvement, *H. als R.-und Wer.***56** 149.
[27] Esteves B, Graça J and Pereira H 2008 Extractive composition and summative chemical analysis of thermally treated eucalypt wood, *Holzforschung* **62** DOI: 10.1515/HF.2008.057
[28] Callum Hill A S 2007 Wood modification: chemical, thermal and other processes. John Wiley & Sons 5.
[29] Boonstra M J, Acker J, Tjeerdsma B F and Kegel E V 2007 Strength properties of thermally modified softwoods and its relation to polymeric structural wood constituents *Annals of Forest Science* **64** 679–90, DOI: 10.1051/forest:2007048
[30] Hasanagić R, Hodžić A and Bjelić A 2019 Mathematical modelling of the surface roughness in orthogonal cutting of the massive wood *Wood Material Science & Engineering* **16** 74–80, DOI: 10.1080/17480272.2019.1608297

**Acknowledgements**
The authors gratefully acknowledge the support of the "Bilateral project between the Republic of Slovenia and Bosnia and Herzegovina", Joint research project within the scientific and technological cooperation between Bosnia and Herzegovina and the Republic of Slovenia and the research program “P4-0015 Wood and lignocellulosic composites” financed by the Slovenian Research Agency (ARRS).

# Influence humidity and heat on tensile shear strength of lap joints made of solid fir/ spruce wood

**Minka Cehic**
University of Bihac, Faculty of Technical Engineering Bihac, Bosnia and Herzegovina

E-mail: minka.cehic@unbi.ba

**Abstract.** Determination of tensile shear strength of lap joints is carried out according to three standards: BAS EN 205:2018 for adhesives used for non–load–bearing structures, and according to BAS EN 302-1:2014 for adhesives used for the manufacture of load-bearing structures and according to EN 14257:2019 for lap joints who are exposed to elevated temperatures. The paper presents the results of tensile shear strength of lap joints made of solid fir/spruce (Abies alba ssp./ Picea abies spp). Tensile shear strength was tested in 4 groups of tests samples. The Group 1 consisted the samples were 7 days in standard atmosphere [20/65].The Group 2 consisted the samples were previously soaked in water at (20 ± 5) °C, then recondition in standard atmosphere [20/65]. The Group 3 consisted the samples were previously 6 h soaked in boiling water 2 h, then soaked in water at (20 ± 5) °C; the samples tested in the wet state. The Group 4 consisted the samples were previously exposed to heat in a preheated fan oven, at (80 ± 2) °C, for (60 ± 2) min. The test results can be applied for gluing windows, doors, stairs, high-frequency gluing, veneering panels, etc.

## 1. Introduction

Climatic conditions are different in each country. There is also a significant difference in atmospheric conditions in the same area depending on the seasons.

Atmospheric conditions such as temperature and humidity as well as other environmental characteristics have a significant impact on the physical, mechanical and rheological properties of solid wood and wood-based materials. The impact of these parameters on the strength and durability of wood and wood-based materials is significant.

For manufacturers of furniture, construction joinery, interior design equipment, but also for all those who produce wooden elements joined with glue, it is important to know the impact of climate on the strength of joints.

In general, by gluing, i.e. using an adhesive joint, certain advantages can be achieved, such as to obtain a material having better mechanical properties (the strength of the adhesive joint is higher than the strength of the material from which the joint is made), the fabricated composite has low weight given the created structure, greater resistance of the glued joint to the action of biotic and abiotic agents and finally gluing is achieved by reducing the cost of production. In contrast, glued joints also have a number of limiting properties. Eg. the glued joints cannot disassemble, making it difficult to check their strength or lifespan or possibly repair them.

The preparation of the contact surfaces of the bonded materials, as well as the method of processing affect the properties of the bonded joint [5].

Some adhesives have a more or less limited life span under the influence of moisture, elevated temperature, or some chemicals. Formaldehyde-based adhesives have better resistance against the effects of water. But their application also caused a certain problem - the spontaneous unwanted release of formaldehyde into surrounding environment.

Therefore, formaldehyde-based adhesives are being replaced by other adhesives that have increased resistance to moisture.

Testing of glued joints has been carried out for a long time. Tested of the influence of material properties, such as wood moisture on the strength of the glued joint, were performed [3], [4]. Based on the test results [9], it was found that a slight increase in relative wood moisture (from 12 to 15%) has a relatively small impact on the speed of the hardening of the bonded joint where the PUR adhesive is used.

Some significant tests are the influence of moisture content on the bond strength and water resistance of glued joints [3], the influence of temperature on the strength and stiffness of glued corner joints of frame structural elements [6], influence of the time between machining and assembly of mortise and tenon joints on tension strength of T-type joints [2], strength of furniture joints constructed with biscuits [11], tensile strength of finger joints type T, strength of furniture joints made using dowels and other. Tested the joints were glued with PVAc, PUR adhesives, UF adhesives and other synthetic adhesives depending on the area of application of the adhesive compound.

The tests were conducted in real time, but also with the use of certain regimes of accelerated aging. Real-time testing takes a long time, often several years.

The tests were performed in real time where the glued joint was exposed to the action of the actual climate (atmosphere) of a certain geographical area, but also with the use of appropriate conditioning sequences and regime of accelerated aging. Real-time testing takes a long time, often several years.

However, the request of today's market for wood products requires that the design and characteristics of the product change rapidly and adapt to the requirements of modern customers. It is therefore resorted to testing the properties of bonded joints that have previously been exposed to the effects of appropriate conditioning sequence according to BAS EN 204: 2014. or accelerated aging treatments. These appropriate conditioning sequence can simulate different climatic conditions.

The aim of this paper is to provide more information on the influence of heat (temperature 80°C) and humidity - cold and hot water on tensile shear strength of lap joints made of solid fir / spruce wood. Prior to the tensile shear test, the test pieces were subjected to appropriate conditioning sequence.

## 2. Experimental study of the influence of humidity and heat on tensile shear strength

Appropriate resources should be provided for each test:

- necessary theoretical knowledge of the desired research - proven information on the desired research issues that can be found in books and similar publications, the results of previous similar research conducted by the author or other researchers and standards and other technical norms necessary for conducting research;
- test material and
- test equipment.

Testing of tensile shear strength by lap joints is performed according to the standards: BAS EN 205:2018 for adhesives used for for non-structural applications, or according to BAS EN 302-1:2014 for adhesives used for the load-bearing timber structures and according to EN 14257:2019 for determination of tensile strength of lap joints at elevated temperature (WATT '91).

We decided to test glued joints for non-structural use, i.e. we used the standards BAS EN 204:2018 and BAS 205:2018, and EN 14257:2019 (WATT '91) to test the tensile shear strength of lap joints at elevated temperature. WATT '91).

Although the standards BAS EN 205:2018 and BAS EN 302-1:2014 are primarily intended for testing the tensile shear strength of lap joints made of beech wood, we wanted to test the shear strength of lap joints made of fir / spruce wood (Abies alba ssp./ Picea abies spp.) with an average density of 0.420 [g/cm$^3$]

The gluing was done with two types of Kleiberit adhesives, namely: Kleiberit 300.0 and Kleiberit 303.0. Both adhesives are PVAc dispersions.

IOP Conf. Series: Materials Science and Engineering 1208 (2021) 012026 doi:10.1088/1757-899X/1208/1/012026

One-component adhesive Kleiberit 300.0 is an industrial adhesive for waterproof joints of durability class D3, which is suitable for gluing windows, doors, stairs, for surface gluing of HPL panels, for veneering, for high-frequency gluing). The technical properties of this adhesive are: density at 20 ° C about 1.10 [g / cm3], pH value 3 ±0.5, viscosity at 20 ° C - Brookfield RVT Sp 6/20 rpm: 12000 ± 3000 mPa·s, the glue is white and the open time of this glue is (bei 20°C: 6-10 minutes.

Kleiberit 303.0 is an industrial adhesive for waterproof gluing with high requirements, durability class D3 and D4. The glue is used for making windows and doors, making stairs, gluing in shipbuilding, surface gluing on wooden boards (e.g. HPL, CPL, etc.), general construction gluing (e.g. pen-groove joint, toothed joint, etc., gluing hard and exotic wood and for high frequency bonding.

If Kleiberit 303.0 is used as a one-component adhesive then it is a waterproof adhesive of durability class D3. The density of this adhesive is 1.10 [g / cm3], the pH value is approximately 3, the viscosity of the adhesive is at 20 ° C: 12,000 ± 2,000 mPa·s (Brookfield RVT spindle 6/20 rpm), the glue is white and the open time is 6-10 min. However, if 5% hardener is added to this adhesive, the durability class of the adhesive changes and becomes D4 and the adhesive is waterproof, with similar technical properties as the adhesive of durability class D3, but in this case the adhesive joint can withstand (be exposed to) and high temperatures 80 ° C).

Tests were performed in accordance with the standard BAS EN 205:2018. The appearance and dimensions of the test pieces are shown in Figure 1.

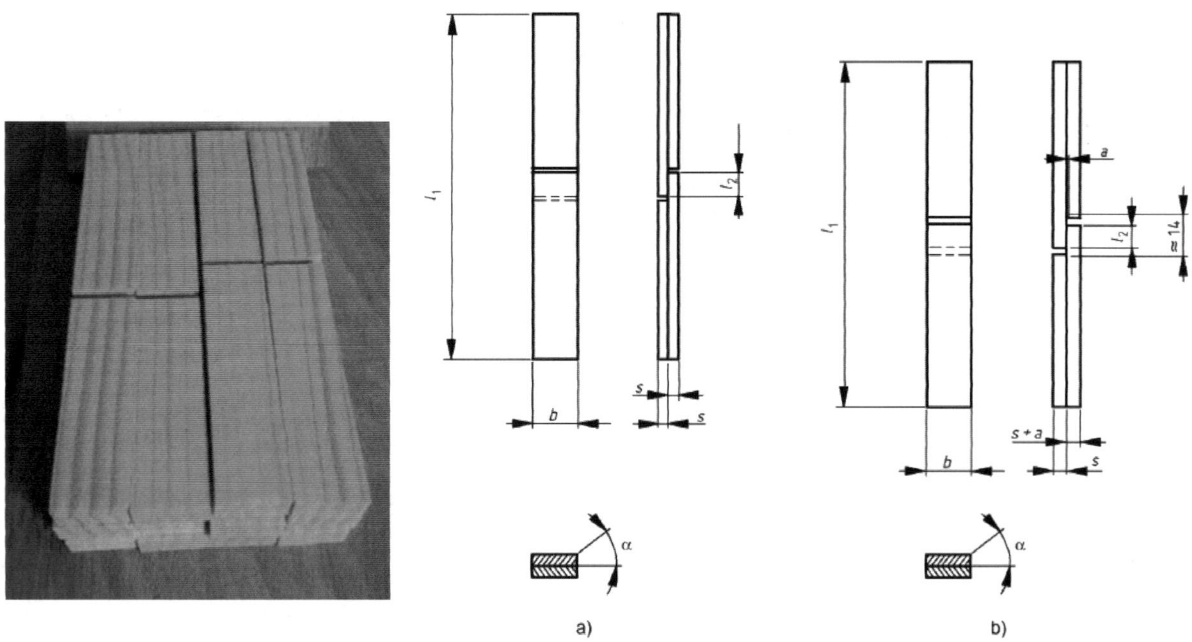


Key
h = 150 ±5 mm total length of test pieces
b = 20,0 ±0,1 mm width of test pieces (width of tested surface)
$l_2$ = 10,0 ± 0,1 mm length of overlap (length of tested surface)
s = 5,0 ± 0,1 mm thickness of the panels
$\alpha$ = 30° to 90° angle between growth ring and surface s to be bonded
a = 1,0 ± 0,1 mm thickness of the thick glue line

**Figure 1**. Lap joint test pieces a) with thin glue line and b) with thick glue line.

IOP Conf. Series: Materials Science and Engineering 1208 (2021) 012026 doi:10.1088/1757-899X/1208/1/012026

After pressing the bonded assemblies, and before cutting the test piece and testing, the assemblies are conditioned in a standard atmosphere (temperature 20 ± 2° C relative humidity 65 ± 5% - standard atmosphere 20/65) for a minimum of 7 days.

After cutting the test pieces from the conditioned glued assemblies, a number of test pieces (minimum 10 test pieces) are subjected to appropriate conditioning treatments - conditioning sequences in accordance with BAS EN 204:2018 ensuring that the test pieces are in horizontal plane and do not touch or support each other, so that they are not affected by stress.

Prior to the tensile shear strength test, a minimum of 10 test pieces for each test were subjected to a specific conditioning treatment - conditioning sequence:

– first group - test pieces are only conditioned in the conditions of standard atmosphere (20/65) for 7 days - T1 conditioning sequence;
– second group - test pieces were conditioned for 7 days in standard atmosphere (20/65), then soaked for 4 days in water at 20 ± 5° C and then conditioned in standard atmosphere [20/65] for 7 days - T4 conditioning sequence;
– third group: test pieces are exposed to the following conditions: 7 days in a standard atmosphere, 6 hours in boiling water and 2 hours in water at a temperature of 20 ± 5° C; - T5 conditioning sequence;
– fourth group: after conditioning in a standard atmosphere [20/65] for 7 days, the test samples were exposed to an elevated temperature of 80 ± 2° C for 60 ± 2 minutes - W conditioning sequence.

Prior to the shear tensile test, at least 10 pieces for each test were subjected to climatic treatment according to the specific treatment - conditioning sequence.

The tensile shear strength test of the lap joints was performed on a SHIMADZU test machine, type SIL-50NAG.

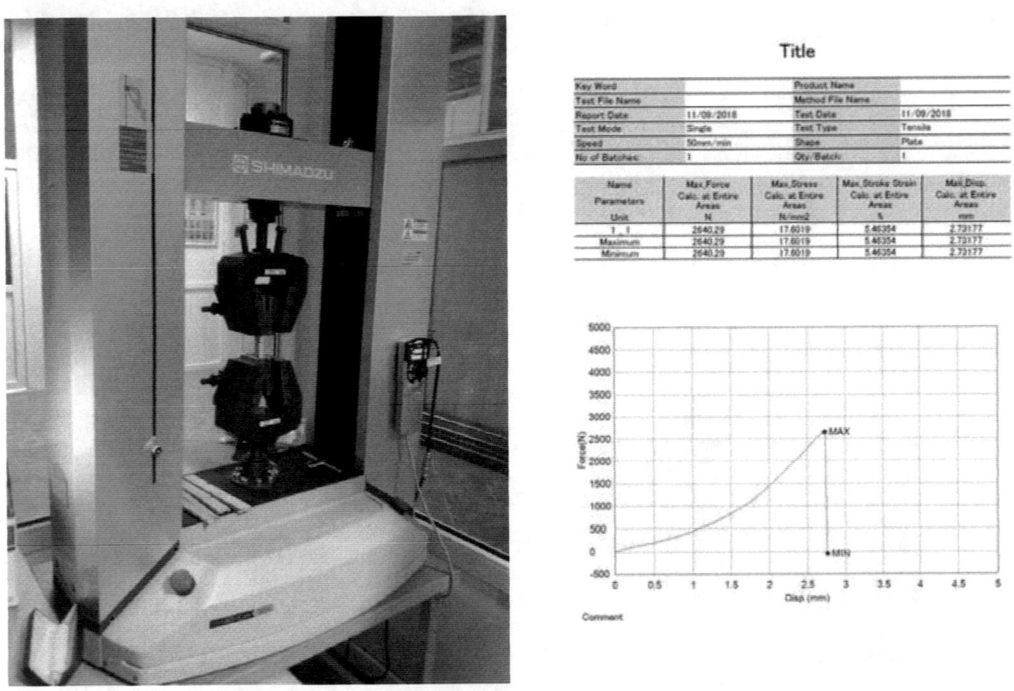


**Figure 2.** Test machine SHIMADZU, type SIL-50NAG and fracture force display on test piece 7.

Calculate the tensile shear strength of lap joints σ t in Newton per square millimetre (N/mm$^2$) using equation:

$$\sigma = \frac{F_{max}}{A} = \frac{F_{max}}{l_2 \cdot b}$$

where
$F_{max}$ = the applied maximum force in Newton (N)
A = the bonded test surface in square millimetres ($mm^2$)
$l_2$ = the length of the bonded test surface in millimetres (mm)
b = the width of the bonded test surface in millimetres (mm).

## 3. Results and discussion

Lap joints that were glued with adhesives that according to EN 204 are in stress group D3 - Kleiberit 300.0 - D3 adhesive and Kleiberit 303.0 - D3 adhesive were subjected to conditioning treatment before the tensile shear strength test (conditioning sequence) T1, T4 and W .

Lap joints made using Kleiberit 303.0 - D4 adhesive before the tensile shear strength test were subjected to conditioning treatments (conditioning sequences) T1, T5 and W.

The results of the fracture force test are shown in Table 1.

**Table 1.** Results of measurement of fracture force and tensile shear strength values of lap joints.

| No. | Type of glue | Sress group | Conditioning sequence | Fracture force [N] | | | Tensile shear strength [N/mm²] |
|---|---|---|---|---|---|---|---|
| | | | | min | max | average | |
| 1. | Kleiberit | | T1 | 1,732.22 | 1,967.10 | 1,848.923 | 9.24461 |
| 2. | 300.0 – PVAc | D3 | T3 | 1,720.04 | 1,891.92 | 1,791.493 | 8.95746 |
| 3. | dispersion | | W | 1,295.11 | 1,459.18 | 1,382.370 | 6.91185 |
| 4. | Kleiberit | | T1 | 1,916.13 | 2,047.55 | 1,981.840 | 9.90920 |
| 5. | 303.0 – PVAc | D3 | T3 | 1,867.10 | 2,021.71 | 1,931.937 | 9.65968 |
| 6. | dispersion | | W | 1,559.73 | 1,727.64 | 1,628.110 | 8.14055 |
| 7. | Kleiberit | | T1 | 2,164.80 | 2,216.71 | 2,190.755 | 10.95377 |
| 8. | 303.0 – PVAc | D4 | W | 1,579.85 | 1,711.15 | 1,648.510 | 8.24550 |
| 9. | dispersion | | T5 | 792.01 | 955.72 | 865.750 | 4.32875 |

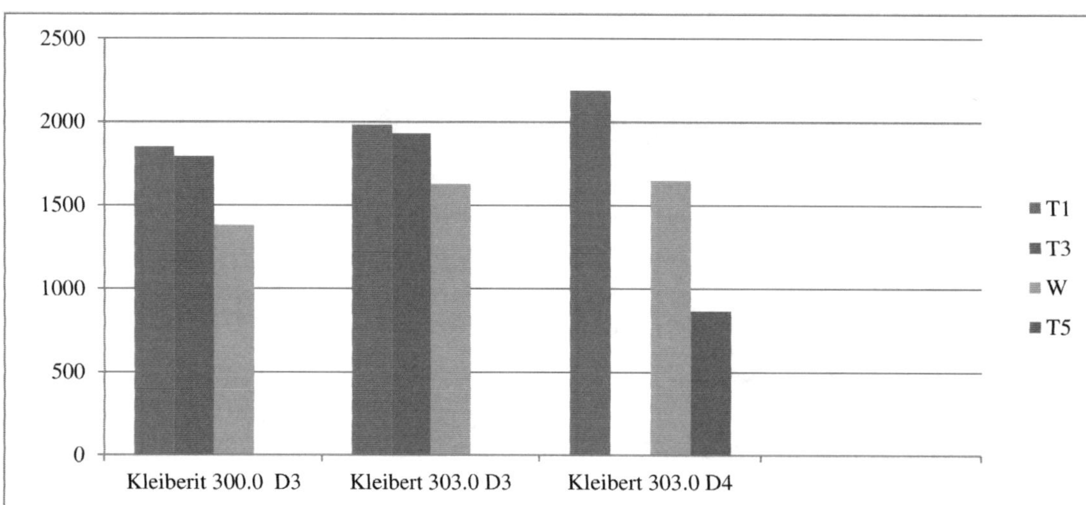


**Figure 3.** Graphical representation of the results of the fracture force test.

Analyzing the results of the experimental examination of the fracture force, it is noticed that the samples glued with Kleiberit 303.0 D4 adhesive had the highest force value, about 18.4% higher force value than in the other two cases (gluing with Kleiberit 300.0 D3 adhesive and Kleiberit 303.0 D3), in case the samples were exposed to conditioning only in accordance with T1 conditioning treatment.

The action of water (cold, and especially the combined action of hot and cold water) caused a decrease in the value of the fracture force.

In the case that the samples after conditioning in a standard climate were also exposed to the action of heat - temperature 80° C (conditioning sequence W) the reduction in the force value was from 17.84% (Kleiberit 303.0 D3) to 25.23% (for Kleiberit 300.0 D3 adhesive).

The combined action of cold and hot water has a significant reduction in the value of the fracture force, which can be seen in the example of test results of tests performed with Kleiberit 303.0 D4 adhesive.

## 4. Conclusion

Experimental testing has shown that the action of elevated temperature and humidity (cold and hot water) negatively affects the value of tensile shear strength.

Even in the case of gluing with adhesives that according to BAS EN 204:2018 belong to the group of adhesives that can be used for external conditions of use, it should be known that there will be a decrease in the value of strength.

This means that when designing and making the glued joint using PVAc dispersions that can be purchased on the Bosnian market, it should be taken into account that the strength value of the glued joint will decrease by about 20 ± 5% depending on environmental conditions and treatment duration.

## References

[1] Bachtiar E V, Clerc G, Brunner A J, Kaliske M and Niemz P 2017 Static and dynamic tensile shear test of glued lap wooden joint with four different types of adhesives *Holzforschung Volume **71 (5)*** p 391-396 DOI https://doi.org/10.1515/hf-2016-0154

[2] Barboutis I and Melissides T 2011 Influence of the time between machining and assembly of mortise and tenon joints on tension strength of T-type joints *Annals of WULS - SGGW. Forestry and Wood Technology 2011;* 73, 23-29

[3] Bomba J, Šedivka P, Böhm M and Devera M 2014 Influence of Moisture Content on the Bond Strength and Water Resistance of Bonded Wood Joints *BioResources* **9(3)** p 5208–5218

[4] Clauss S, Joščak M and Niemz P 2011 Thermal stability of glued wood joints measured by shear tests *Eur. J. Wood Prod.* **69** p 101-111 DOI https://doi.org/10.1007/s00107-010-0411-4

[5] Gaff M and Gašparík M 2013 The development of stresses during the shaping of the surface of aspen wood and their impact on the quality of the surface *BioResources* **8(3)** p 3741-3752

[6] Jivkov V, Marinova A and Petutschnigg A J 2008 Influence of the temperature on the bending strength and stiffness characteristics of glued corner joints of frame structural elements made from solid beech wood *International Conference of Nábytok 2008 (ISBN 978-80-228-1839–1), Faculty of Wood Science and Technology, Technical University in Zvolen, Slovak Republika Volume 2008*

[7] Máchová E, Holouš Z, Langová N and Balážová Ž 2017 The effect of humidity on the shear strength of glued wood based and plastic joints *Acta Facultatis Xylologiae Zvolen,* **60(1)** p 113–120 DOI: 10.17423/afx.2018.60.1.12

[8] Obućina M 2014 *Lijepljenje drveta* Mašinski fakultet Sarajevo ISBN 978-9958-601-46-0

[9] Ren D and Frazier C 2012 Wood/adhesive interactions and the phase morphology of moisture-cure polyurethane wood adhesives *International Journal of Adhesion and Adhesives* Vol. **34 (2012)** p 55-61 DOI https://doi.org/10.1016/j.ijadhadh.2011.12.009

[10] Šedivka P, Bomba J, Böhm M and Boška P 2015 Influence of temperature on the strength of bonded joints *BioResources* **10(3)** p 3999-4010

[11] Vassiliou V and Barboutis I 2008 Strength of furniture joints constructed with biscuits *Proceedings of International Conference Nabytok 2008 (ISBN 978-80-228-1839–1), Faculty of Wood Science and Technology, Technical University in Zvolen, Slovak Republika Volume 2008*
[12] BAS EN 204:2018 *Classification of thermoplastic wood adhesives for non-structural applications*
[13] BAS 205:2018 *Adhesives - Wood adhesives for non-structural applications - Determination of tensile shear strength of lap joints*
[14] EN 14257:2019 (WATT '91) *Adhesives - Wood adhesives - Determination of tensile strength of lap joints at elevated temperature (WATT '91)*

# Analysis of the dimensions of desks in primary schools in Aerodrom Municipality - Skopje

**Elena Jevtoska, Gjorgi Gruevski, Ivana Antovska and Marija Krstev**

Faculty of Design and Technologies of Furniture and Interior – Skopje, Ss. Cyril and Methodius University in Skopje, Macedonia

E-mail: jevtoska@fdtme.ukim.edu.mk

**Abstract**. Educational institutions are facilities where each person spends many years of their life. Desks used in these institutions are designed with specific features of shape, material and size. Long-term use of desks in educational institutions may be the cause of damage to the health of users if they are not carefully designed and dimensioned. In compliance with the prescribed rules and dimensions in the design of desks it is of great importance for proper development of the users that are in the stage of growth and development when using this type of desk. In this paper, the functional dimensions of real samples of desks taken from primary schools in the Municipality of Aerodrom – Skopje will be measured. The testing samples of desks will be designated to the group that they belong to according to the EN 1729-1:2006 standard, and all requested dimensions according to the standard will be measured. The purpose of this work is to show the real condition of desks in primary schools and whether the desks are with the requested dimensions of the EN 1729-1:2006 standard.

## 1. Introduction

The long-term use of office furniture in the educational institutions could be the reason for health worsening of its users if it is not meticulously dimensioned and designed. In order to avoid such issues, the functional measures of the furniture for sitting and working while using the school furniture are prescribed in the EN 1729-1:2006 standard, while the safety conditions that have to meet the satisfactory needs of the school furniture for sitting and working are prescribed in EN 1729-2:2006.

Are the required standards of the school furniture used in the schools respected and to what extent shall be the research of this paper seen through real samples that are found in the primary schools on the territory of the municipality of Aerodrom, Skopje, R. of Macedonia.

The research shall be done only for the primary schools because the users which use them are at the age of 5.5 to 14, i.e., the stage of life when the human body develops the most and changes its height and shape most significantly.

The choice of the municipality of Aerodrom is done for the reason of this being the most inhabited municipality as a territory in the city of Skopje. In its territory of 20 km2, 10.4 m2 are taken as an urban part. In the urban area, there are 8 primary schools in which 6 000 students annually go to school.

During the visit of the primary schools on the territory of the municipality of Aerodrom measures of 5 different types of desks were taken. In terms of what for and on what age of children the desks are used, only one sample is used in first grade, whereas the rest are used in all the other grades from second till ninth. For measuring the samples, the same are numerated from sample 1 to sample 5. The results of

IOP Conf. Series: Materials Science and Engineering 1208 (2021) 012027 doi:10.1088/1757-899X/1208/1/012027

the dimensions from the measurement will be compared to the required dimensions according to the standard.

## 2. Analysis of the THE EN 1729-1:2006 Standard - Furniture – Chairs and desks for educational institutions – Part 1: Functional dimensions

The freedom in creating and making of school furniture is limited by some optimal dimensions, proper stability and the minimum required sustainability which should meet he needs of the school furniture before it is sent for usage. The optimal dimensions that should satisfy the school furniture for sitting and working are prescribed in EN 1729-1:2006 standard. On behalf of this research paper, the requirements of the standard in the part of the desks will be analysed.

In this part of the standard, there are the optimal and compulsory measures which the school furniture for sitting and working has to satisfy for unconditional process of the work and prevention of negative influence towards the growth and development of the final users of this type of furniture.

### *2.1. Required dimensions for desks according to EN 1729-1:2006 Furniture – Chairs and desks for educational institutions – Part 1: Functional dimensions*

The height groups according to which the desks are divided are 8, starting from group 0 to group 7. Every height group despite the numerical designation, is also marked with different colour, so the symbolizing is as follows: (0 group – white colour, 1 group – orange colour, 2 group – violet colour, 3 group – yellow colour, 4 group – green colour, 5 group – green colour, 6 group – blue colour, 7 group - brown colour).

According to EN 1729-1:2006 standard for desks as well as the height of the working surface it is estimated in which group the desk belongs to. According to the group it is necessary to first satisfy the other required dimensions.

- $h_1$ – height of table top
- $t_1$ – depth of table top
- $w_1$- width of top, per person at front edge where students sit
- horizontal distance between front, legs/structure, where students sit, per person (min)

Despite these dimensions, the standard requires dimensions for desks which are used for chairs, where the sitting surface is double rounded. For this type of desks there is also a legroom which for each group is given with minimal dimension.

Since the samples found and taken for measurement are not of this type of desks, these dimensions shall not be analysed at all.

The measured dimensions of the samples will be compared to the allowed range of dimensions which the standard gives specially for each height group of desks.

Despite these dimensions, the desks should satisfy the minimum free legroom as well. This space in the standard is given as a shape with 4 dimensions which limits the space where there must not be any burden that would cause troubles while sitting.

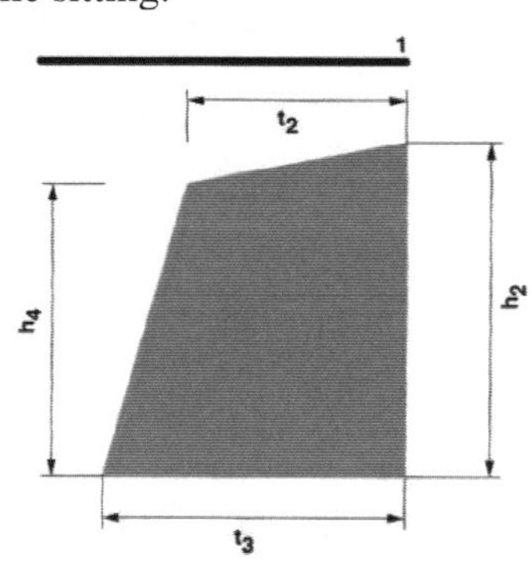


**Figure 1.** Legroom template

IOP Conf. Series: Materials Science and Engineering 1208 (2021) 012027 doi:10.1088/1757-899X/1208/1/012027

**Table 1.** Dimensions and size marks for desks - EN 1729-1:2006

| Size mark | 0 | 1 | 2 | 3 | 4 | 5 | 6 | 7 |
|---|---|---|---|---|---|---|---|---|
| Color cod | □ | ■ | ■ | ■ | ■ | ■ | ■ | ■ |
| Popliteal range (cm) (without shoes) | 200-250 | 250-280 | 280-315 | 315-355 | 355-405 | 405-435 | 435-485 | 485+ |
| Stature range (cm) (without shoes) | 800-950 | 930-1160 | 1080-1210 | 1190-1420 | 1330-1590 | 1460-1765 | 1590-1880 | 1740-2070 |
| $h_1$(mm) - height of top ±10 | 400 | 460 | 530 | 590 | 640 | 710 | 760 | 820 |
| $t_1$(mm) -depth of top (min) | - | 500 | 500 | 500 | 500 | 500 | 500 | 500 |
| $w_1$(mm) – width of top, per person at front edge, where students sit (min) | - | 600 | 600 | 600 | 600 | 600 | 600 | 600 |
| Horizontal distance (mm) between front legs/structure, where students sit, per person (min) | - | 500 | 500 | 500 | 500 | 500 | 500 | 500 |

After determining the basic required dimensions that the desks should satisfy, we move onto measuring the free legroom, and for this type of measurement there are special schemes for each height group separately. The schemes are elaborated according to the required measures for free space which is prescribed by EN 1729-1:2006 standard. The borders of this space are different for each height group and they are presented in Table 2.

**Table 2.** Minimum legroom template dimensions

| Size mark | 0 | 1 | 2 | 3 | 4 | 5 | 6 | 7 |
|---|---|---|---|---|---|---|---|---|
| Color cod | □ | ■ | ■ | ■ | ■ | ■ | ■ | ■ |
| Popliteal range (cm) (without shoes) | 200-250 | 250-280 | 280-315 | 315-355 | 355-405 | 405-435 | 435-485 | 485+ |
| Stature range (cm) (without shoes) | 800-950 | 930-1160 | 1080-1210 | 1190-1420 | 1330-1590 | 1460-1765 | 1590-1880 | 1740-2070 |
| $h_2$ (mm) | 325 | 380 | 440 | 495 | 545 | 610 | 665 | 725 |
| $h_4$ (mm) | 275 | 325 | 375 | 420 | 465 | 520 | 565 | 620 |
| $t_2$ (mm) | 300 | 300 | 300 | 300 | 400 | 400 | 400 | 400 |
| $t_3$ (mm) | 400 | 400 | 400 | 400 | 500 | 500 | 500 | 500 |

## 3. Results

The measured results will be shown separately for each tested sample.

IOP Conf. Series: Materials Science and Engineering 1208 (2021) 012027 doi:10.1088/1757-899X/1208/1/012027

### *3.1. Sample 1*

The sample is a desk used in first grade. Its purpose is for 4 persons sitting simultaneously.

**Figure 2.** Sample 1

**Table 3.** Basic dimensions (sample1)

| Required dimension | Measured dimension | Referent values for height group 2(violet) | Satisfy √ Do not satisfy X |
|---|---|---|---|
| **$h_1$- height of top ±10** | 520mm | 530mm | √ |
| **$t_1$-depth of top (min)** | 1000mm (for two persons) | 500mm | √ |
| **$w_1$ – width of top, per person at front edge, where students sit (min)** | 1000mm (for two persons) | 600mm | X |
| **Horizontal distance between front legs/structure, where students sit, per person (min)** | 855mm | 500mm | √ |

**Table 4.** Minimum legroom (sample 1)

| Required dimension | Measured dimension | Referent values for height group 2(violet) | Satisfy √ Do not satisfy X |
|---|---|---|---|
| **Min. h2** | 855 mm | 440mm | √ |
| **Min. h4** | 855 mm | 375mm | √ |
| **Min. t2** | 855 mm | 300mm | √ |
| **Min. t3** | 855 mm | 400mm | √ |

**Figure 3.** Photos of the measurement of sample 1

*3.2. Sample 2*

The sample is a desk used from second to ninth grade. It is meant for two persons simultaneously.

**Figure 4.** Sample 2

**Table 5.** Basic dimensions (sample 2)

| Required dimension | Measured dimension | Referent values for height group 4(red) | Satisfy √ Do not satisfy X |
|---|---|---|---|
| **$h_1$- height of top ±10** | 670mm | 640mm | X (+20 mm) |
| **$t_1$-depth of top (min)** | 500mm | 500mm | √ |
| **$w_1$ – width of top, per person at front edge, where students sit (min)** | 800-600mm | 600mm | √ |
| **Horizontal distance between front legs/structure, where students sit, per person (min)** | 700mm | 500mm | √ |

**Table 6.** Minimal legroom (sample 2)

| Required dimension | Measured dimension | Referent values for height group 4(red) | Satisfy √ Do not satisfy X |
|---|---|---|---|
| **Min. h2** | 510 mm | 545mm | X |
| **Min. h4** | 510 mm | 465mm | √ |
| **Min. t2** | 420 mm | 400mm | √ |
| **Min. t3** | 440 mm | 500mm | X |

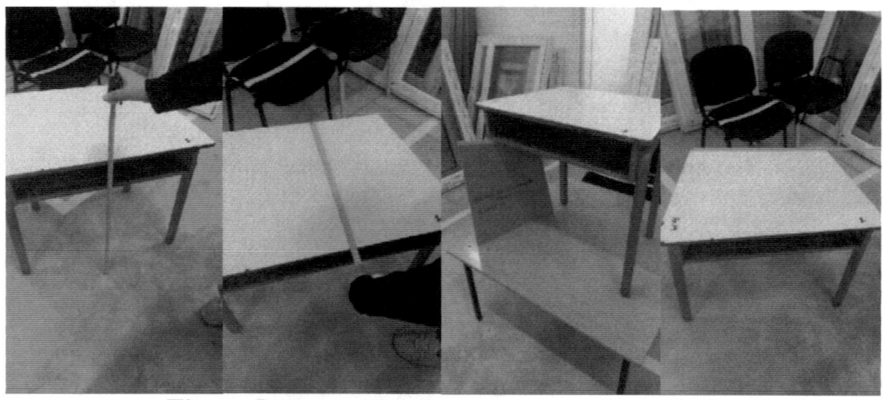

**Figure 5.** Photos of the measurement of sample 2

### *3.3. Sample 3*

The sample is a desk used from second to ninth grade. It is meant for two persons simultaneously.

**Figure 6.** Sample 3

**Table 7.** Basic dimensions (sample 3)

| **Required dimension** | Measured dimension | Referent values for height group 4(red) | Satisfy √ Do not satisfy X |
|---|---|---|---|
| **$h_1$- height of top ±10** | 670mm | 640mm | X (+20 mm) |
| **$t_1$-depth of top (min)** | 500mm | 500mm | √ |
| **$w_1$ – width of top, per person at front edge, where students sit (min)** | 1300mm (for two persons | 600mm | √ |
| **Horizontal distance between front legs/structure, where students sit, per person (min)** | 1230mm(for two persons) | 500mm | √ |

**Table 8.** Minimal legroom (sample 3)

| **Required dimension** | Measured dimension | Referent values for height group 4(red) | Satisfy √ Do not satisfy X |
|---|---|---|---|
| **Min. h2** | 630 mm | 545mm | √ |
| **Min. h4** | 460 mm | 465mm | X |
| **Min. t2** | 200 mm | 400mm | X |
| **Min. t3** | 450 mm | 500mm | X |

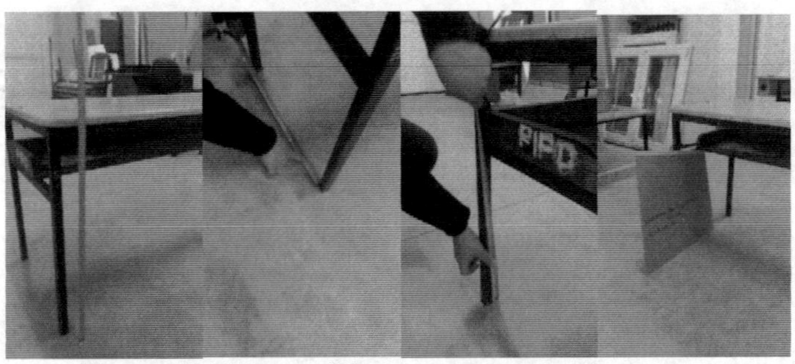

**Figure 7.** Photos of the measurement of sample 3

*3.4. Sample 4*

The sample is a desk used from second to ninth grade. It is meant for two persons simultaneously.

**Figure 8.** Sample 4

**Table 9.** Basic dimensions (sample 4)

| Required dimension | Measured dimension | Referent values for height group 5(green) | Satisfy √ Do not satisfy X |
|---|---|---|---|
| **$h_1$- height of top ±10** | 720mm | 710mm | √ |
| **$t_1$-depth of top (min)** | 475mm | 500mm | √ |
| **$w_1$ – width of top, per person at front edge, where students sit (min)** | 1200mm (for two persons) | 600mm | √ |
| **Horizontal distance between front legs/structure, where students sit, per person (min)** | 1090mm (for two persons) | 500mm | √ |

**Table 10.** Minimal legroom (sample 4)

| Required dimension | Measured dimension | Referent values for height group 5(green) | Satisfy √ Do not satisfy X |
|---|---|---|---|
| **Min. h2** | 690mm | 610mm | √ |
| **Min. h4** | 490mm | 520mm | X |
| **Min. t2** | 100mm | 400mm | X |
| **Min. t3** | 400 mm | 500mm | X |

**Figure 9.** Photos of the measurement of sample 4

### 3.5. *Sample 5*

The sample is a desk used from second to ninth grade. It is meant for two persons simultaneously.

**Figure 10.** Sample 5

**Table 11.** Basic dimensions (sample 5)

| **Required dimension** | Measured dimension | Referent values for height group 6(blue) | Satisfy √ Do not satisfy X |
|---|---|---|---|
| **$h_1$- height of top ±10** | 760mm | 760mm | √ |
| **$t_1$-depth of top (min)** | 640mm | 500mm | √ |
| **$w_1$ – width of top, per person at front edge, where students sit (min)** | 1360mm (for two persons) | 600mm | √ |
| **Horizontal distance between front legs/structure, where students sit, per person (min)** | 1260mm (for two persons) | 500mm | √ |

**Table 12.** Minimal space for legs (sample 5)

| **Required dimension** | Measured dimension | Referent values for height group 6(blue) | Satisfy √ Do not satisfy X |
|---|---|---|---|
| **Min. h2** | 745mm | 665mm | √ |
| **Min. h4** | 560mm | 565mm | X |
| **Min. t2** | 150mm | 400mm | X |
| **Min. t3** | 360mm | 500mm | X |

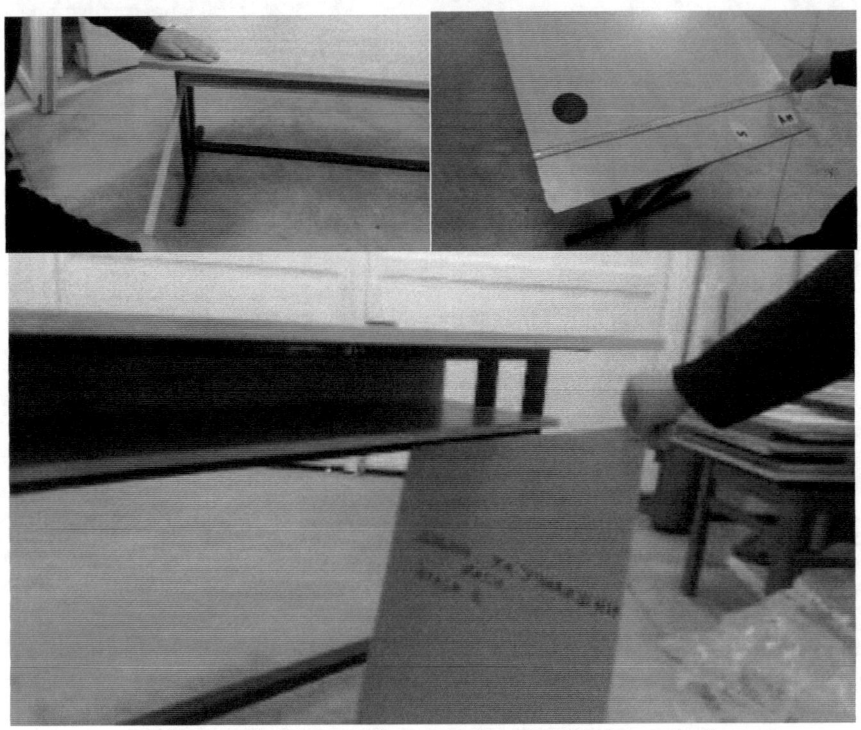

**Figure 11.** Photos of the measurement of sample 5

## 4. Conclusion

In accordance with the gained results, the conclusion is that none of the taken different samples of desks thoroughly meets the required dimensions of the EN 1729-1:2006 standard. All the measured samples have deviations from the prescribed dimensions.

Such results lead to the conclusion that while choosing and purchasing school furniture, i.e., desks there is no control whether they meet the requirements of the obligatory standards.

The biggest and most common deviations are connected to the design of the desks as well. In four of the samples, there is a provided space for leaving objects in the lower zone. In all four samples this element impairs the space for the legs. This element is not most suitable for leaving the objects and it also has a great negative impact concerning sitting on the desks.

**Table 13**. Deviations of the measured dimensions related to the written

| Sample No. | Height group | Deviations |
|---|---|---|
| Sample 1 | 2 violet | There is a minimal deviation in the minimal width per person, because the desk is meant for 4 persons simultaneously. |
| Sample 2 | 4 red | The free space for legroom is impaired with an object for delay of the down part of the desk. |
| Sample 3 | 4 red | The free space for legroom is impaired with an object for delay of the down part of the desk. |
| Sample 4 | 5 green | The free space for legroom is impaired with an object for delay of the down part of the desk. |
| Sample 5 | 6 blue | The free space for legroom is impaired with an object for delay of the down part of the desk. |

## References

[1] Bralic I 2017 *Sekularne promene rasta i razvoja*. (Pediatr Croat 52) pp 25-35

[2] Mehanovic M 2017 *Istraživanje prilagođenosti sjedišta školske stolice antropometrijskim veličinama učenika osnovnih škola*. (11th International Scientific Conference on Production Engineering DEVELOPMENT AND MODERNIZATION OF PRODUCTION) (RIM 2017)

[3] Mehanovic M 2013 *Utjecaj antropometrijskih karakteristika i položaja sjedišta školske stolice na opterećenje kralješnice učenika*. (Proceedings of 5th International Ergonomics Conference, Ergonomics 2013. , Zadar, Croatian Society Of Ergonomics) pp 263-270.

[4] Panero J, Zelnik M. 1990 *Antropološke mere i enterijer, Zbirka preporuka za standarde u projektovanju*, IRO Građevinska knjiga (Beograd).

[5] Domljan D, Grbac I, Bogner A 2004 *Uloga dizajna u procesu razvoja školskog namještaja* (Drvna industrija 55(2)) pp 77-90

# Determination of the bending strength of corner joint – metal corner brace, for corps furniture made of refined panels of particleboards

**Ivana Antovska, Marija Krstev, Elena Jevtoska and Gjorgi Gruevski**

Faculty of Design and Technologies of Furniture and Interior – Skopje, Ss. Cyril and Methodius University in Skopje, Macedonia

E-mail: ivana_antovska@yahoo.com

**Abstract**. This paper presents data about the bending strength of outside corner joint – metal corner brace, for corps furniture made of refined panels of particleboard with a width of 25 mm. The research was conducted on two kinds of samples. The two kinds of sample objects are divided in two groups according to the way of composition as follows:
group 1: the first are obtained by extraction i.e. cutting up an already constructed element – bedside table which was subjected to a test method for determining strength, durability and stability- furniture for indoor use EN 1730: 2000;
group 2: the second sample objects were not subjected to any tests and were composed of two parts of particle board with given dimensions in accordance with the normative and are connected with the chosen connective object.

## 1. Introduction

One of the most studied kinds of strength, which affects the construction of corpus furniture, is the bending strength.

There are a number of studies, which have been carried out most often for particleboards with a thickness of 16 and 18 mm. In this paper, the bending strength of sample objects made of particleboards with a thickness of 25 mm will be experimentally determined.

The very first studies carried out on constructive compositions made of panel elements, were burdened with bending strengths. Those studies were carried out in different countries with very similar methods. In Bulgaria the studies carried out by Kjucukov, Josifov and Zivkov are noted. Analogous studies were carried out in the USA by Eckelman and Zhang. In Germany Albin, Konjer, Kuhne, Bechmann and Hassler published studies on this subject. In Finland Liiri and Haawisto performed analogous surveys. In Macedonia studies of this type were carried out by T Gruevski and N Simakoski.

These studies showed results about the bending strength of corpus furniture. Next are the studies of the strength of stretching which were carried out in Germany, Bulgaria as well as Macedonia.

Finally, there were surveys carried out on constructive compositions made of plate elements burdened with shear strength. Gj Gruevski carried one of these surveys out in Macedonia.

IOP Conf. Series: Materials Science and Engineering 1208 (2021) 012028 doi:10.1088/1757-899X/1208/1/012028

## 2. Purpose and method of examination

Some of the objectives of this study are the following:

- Determination of the specific moments of bending of outside corner joint, dismountable composition of refined panels of particleboard with a thickness of 25 mm joined with a metal corner brace (fixed with screws for pates);
- Determination of the specific moments of bending of outside corner joint, demountable compositions of refined panels of particle board with a thickness of 25 mm taken from an object- a bedside table which was previously subjected to a test method EN 1730, joined with a metal corner brace;
- Comparison of the obtained results of the specific moments of bending of a sample object with a joint element a metal corner brace (fixed with screws for pates) cut off a ready-made element a bedside table and an extra made sample object with the same constructive joint.

### *2.1. Type, shape and dimensions of the sample objects and test fixtures*

The object bedside table, made of refined panels of particle board with a width of 25 mm, with dimensions 550/550/450mm, and a front (door) made of MDF wood fiber board with a thickness of 18 mm and dimensions 500/500/18, on which small trestles are fixed, which raise it from the surface about 100mm.

Four bedside tables object have been tested. On each connective side there are two connective joints. Prior to their delivery to the laboratory, they were joined and delivered as whole objects. Before testing, each of the objects was in an acclimatization room for a period of 7 days at a temperature of 15° to 25 C°. No defects in the product noticed before the testing. These objects were subjected to a test method for determining firmness, durability and stability - furniture for indoor use EN 1730: 2000.

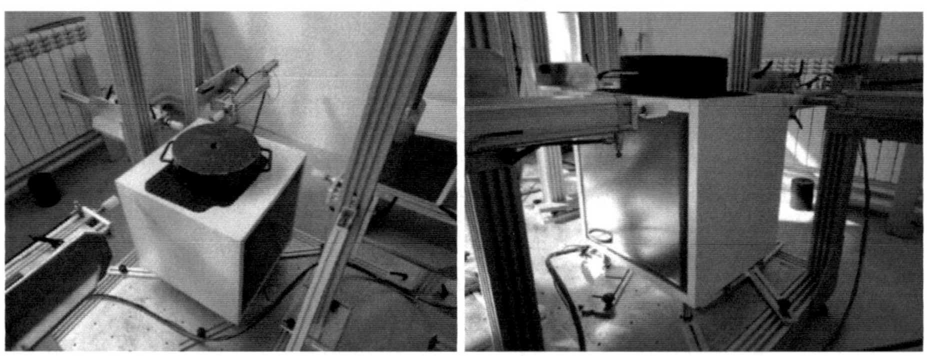

**Figure 1**. Samples for the examination of outside corner, dismountable compositions of refined panels of particle board with a thickness of 25 mm.

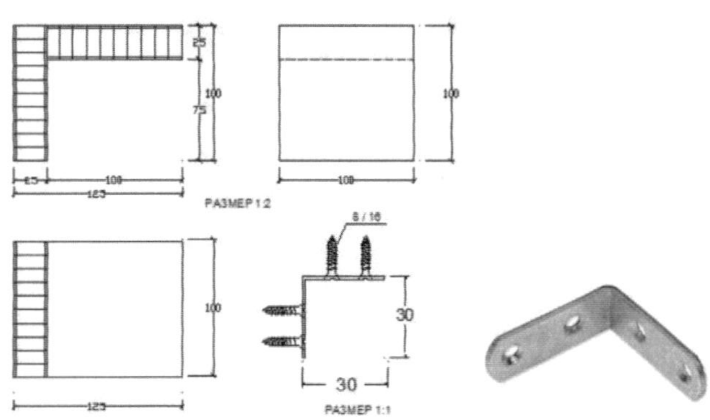


**Figure 2**. Dimension of the test fixtures with metal corner brace.

IOP Conf. Series: Materials Science and Engineering 1208 (2021) 012028 doi:10.1088/1757-899X/1208/1/012028

*2.2. Equipment and devices for examining the bending load of sample objects*

A machine was used for the examination of sample objects that was installed in the laboratory at the Faculty of design and technology of furniture and interior – Skopje, SCHENKE TREBLE, of German origin and with a digital read-out and high accuracy.

The sample objects are placed on moving trolleys, which are used when examining outside corner joints when bending load is applied by opening and closing the sides of the sample object.

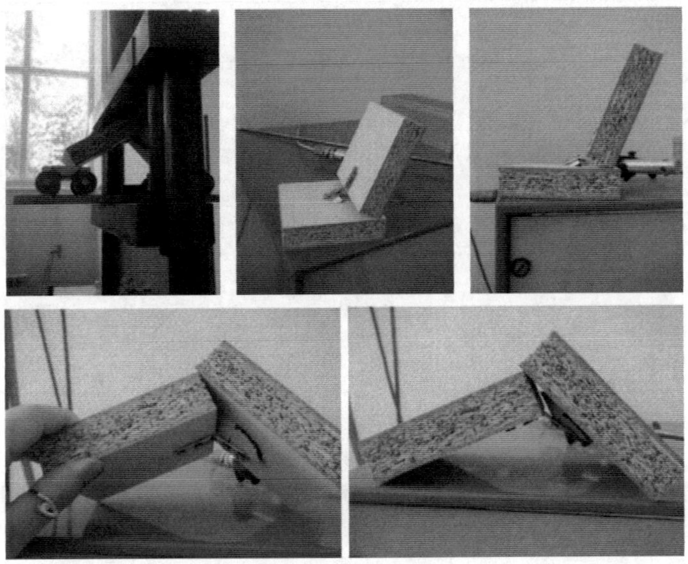

**Figure 3**. Object for the examination of outside corner joints when bending load is applied by opening the sides of the sample object.

**Figure 4**. Object for the examination of outside corner joints when bending load is applies by closing the sides of the sample object.

### 2.3. Results and discussion

The obtained results for the calculated values of the criteria for the bending strength, in each of the groups of sample objects are show in tables.

The processing of data from the conducted research is made according to the usual methods of variation statistics. The medium arithmetic values are calculated[$x_{sr}$], the standard deviation [S], the coefficient of variation [V] and their errors[$fx_{sr}$ , $fS, fV$].

**Table 1.** Results from the bending load with opening the sides of the sample object.

| | $x_{sr}$ [Nm] | S [Nm] | V [%] | $fx_{sr}$ [Nm] | p [%] | fS [%] | fV [%] | n [Number] |
|---|---|---|---|---|---|---|---|---|
| **Group 1** | 36,5 | 5,39 | 14,77 | 3,81 | 10,44 | 0,98 | 2,75 | 15 |
| **Group 2** | 38,94 | 2,96 | 7,61 | 1,96 | 5,04 | 0,541 | 1,40 | 15 |

**Table 2.** Results from the bending load with closing the sides of the sample object.

| | $x_{sr}$ [Nm] | S [Nm] | V [%] | $fx_{sr}$ [Nm] | p [%] | fS [%] | fV [%] | n [Number] |
|---|---|---|---|---|---|---|---|---|
| **Group 1** | 10,75 | 1.13 | 10,58 | 2.73 | 25,42 | 0,207 | 1.95 | 15 |
| **Group 2** | 10,58 | 1,14 | 10,81 | 2,79 | 26,38 | 0,21 | 1,99 | 15 |

The analysis and discussion of the obtained results if firstly done on the strength that affects the sample objects when opening the sides, and then on the strength which affects the sample objects when is closing the sides. The results are presented in tables and graphics. In addition to this paper there are photos of the sample objects after the examination. Finally, all the results are compared so that the conclusions can be made.

## 3. Conclusion

The objective results we obtained from this examination when applying the load in bending, give us an opportunity to determine the strength characteristics of the connective objects that are used for the construction of furniture made of refined panels of particleboard with a thickness of 25 mm.

Some of the conclusions are as follows:

1. The strength of bending with the opening of the sides is significantly greater in comparison with the strength of bending with closing the sides of the sample object;
2. Simulated exploitation, test method EN 1730:2000, does not have a significant impact on the bending strength;
3. Factor that has an impact on the diversity of bending strengths is the human factor when mounting and screwing clutches.

## References

[1] Кючуков Г 1995 *Йакосни показатели на основните видови шарнирани ъединения чрез панти на конструктви елементи на мебели на обликовани и декоративно фолийо плочи от дървени частици с дебелна 16 мм* (София).

[2] Кючуков Г, Кючуков Б, Живков В и Маринова А 1999 *Нормативи за специфичните разрушавацщи сили при натоварване на опьн на освните видове неразглобяеми ьглови сьединенийа на конструктивни елменти от плочи от дьрвесни частици с дебелна 16mm* (София).

[3] Груевски Ѓ 2004 Одредување на јакоста на смолкнување на крајни и средни составни на плочести елементи– Магистерски труд, Шумарски факултет (Скопје).

[4] Груевски Т 1992 *Конструирање на плочите од иверка со средна густина во конструкциите на меебелот и испитување на јакосните показатели на конструктивните елементи.* Шумарски факултет (Скопје).

[5] Груевски Т, Кјучуков Г и Симакоски Н 2001 *Користење на питомиот костен во производството на мебел и плочи од иверки, Испитување на јакост на фиксни аголни состави оптоварени со сили на свивање*. Шумарски факултет (Скопје).

[6] Груевски Т и Симакоски Н 2002 *Елементи на дрвните конструкции*. Шумарски факултет (Скопје).

[7] Eckelman Carl A 2003 *Textbook of product engineering and strength design of furniture*. Purdue University (Indiana, West Lafayette).

[8] Eckelman Carl A and Zhang J L 1993 *Rational design of multi-dowel corner joints in case construction*. (USA: Forest Product Journal 43), pp 52-58

IOP Conf. Series: Materials Science and Engineering 1208 (2021) 012029 doi:10.1088/1757-899X/1208/1/012029

# Influence of furniture exploitation on the bending strength of final mountable-dismountable corner joints

**Marija Krstev, Ivana Antovska, Elena Jevtoska and Gjorgi Gruevski**

Faculty of Design and Technologies of Furniture and Interior – Skopje, Ss.Cyril and Methodius University in Skopje, Macedonia

E-mail: ilievska_11@yahoo.com

**Abstract**. The aim of the researches presented in the paper is to show how the exploitation of furniture affect the strength of its connectors, more precisely the bending strength of the final corner dismountable joints for designed furniture – office desk. For studying this influence, 2 office desks were made (from chosen material OSB panels), joined by dismountable connectors, and also from the same material and the same type of connectors, reference test fixtures were made and their bending strength was the value of comparison. Office desks were tested according to European Standard for determining the stability of the structure (EN 527-3:2003 E) and that was simulated exploration of furniture. After that, the final corner compositions were cut off, and its bending strength was compared with the reference test fixtures strength. The research results showed that the exploiting of dismountable furniture has not a significant impact on their compositions strength.

## 1. Introduction

The quality of furniture and its durability, primarily depends on the material from which it is developed, the type of couplings that are used to merge its constituent elements, the method of processing during the production, as well as the way in which it is used from the buyer.

The furniture has to be solid and firm enough, to withstand its own weight and also the weight of the load it carries. During furniture exploitation, the most common issue that occurs is the bending strength. Bending strength is also one of the most studied kind of strength which affects the construction of cabinet furniture. In this paper, chosen material are oriented strand boards (OSB) with 18 mm thickness, produced by Kronospan Bulgaria EOOD (OSB/2 EN300: High load-bearing boards for indoor use). The advantage of this material is its ability to withstand the load in the direction of the orientation of wooden strands, and the basic negative characteristic is the expressed inhomogeneity due to the size of the chip. The used couplings are mountable-dismountable, eccentric clutch Minifix 15 with clamp and binding element produced by Hafele (Figure 1), which allows us to dismantle the furniture to its constituents.

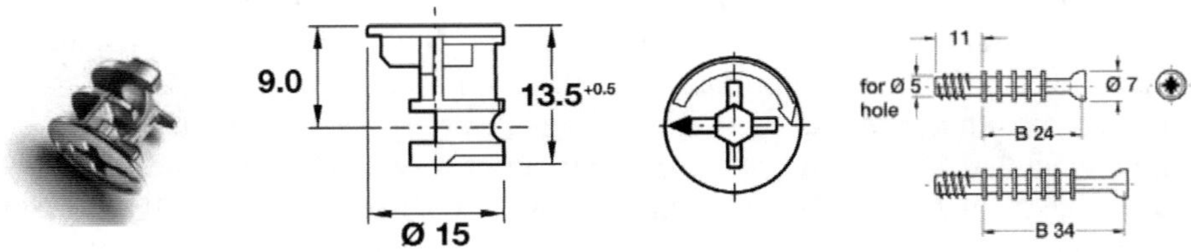


**Figure 1.** Minifix 15 with clamp and binding element.

Through this research, we need to give the basic direction and recommendation to users, whether the cabinet furniture, which is big and heavy, should dismantle it before relocating or transporting to another location, and respond to the question of whether such dismantling and re-installation, will change joints strength, specifically, the strength of bending. In that direction it is necessary to solve the following tasks:

- Determination the stability of the finished product - office desk, according to the European Standard for Office Furniture EN 527-1: 2000, EN 527-2: 2000 (E), EN 527-3: 2003 (E), made of oriented strand boards, assembled with dismountable connectors, which actually simulates the exploitation of furniture;
- Determination the moment of bending of the final corner joints cut off from office desks that passed the stability test according to the European Standard;
- Determination the moment of bending of the final corner joints, made from oriented strand boards, taken as reference test joints.

## 2. Methods of the experimental work

In order to conduct the survey, two office desks were made from the selected material, with dimensions 1200/738/550 mm, assembled with dismountable couplings.

The office desks, such as finished products, subjected to the test method cycles according to the European Standard for Office Furniture, which simulated the exploitation of furniture (Figure 2 and Figure 3) and we enabled the angular compositions that will later cut off from these desks be already burdened with certain stretching. The examination is performed in the furniture testing laboratory at the Faculty of Design and Technologies of Furniture and Interior - Skopje, on the multifunctional test field from the German manufacturer Weinmann.

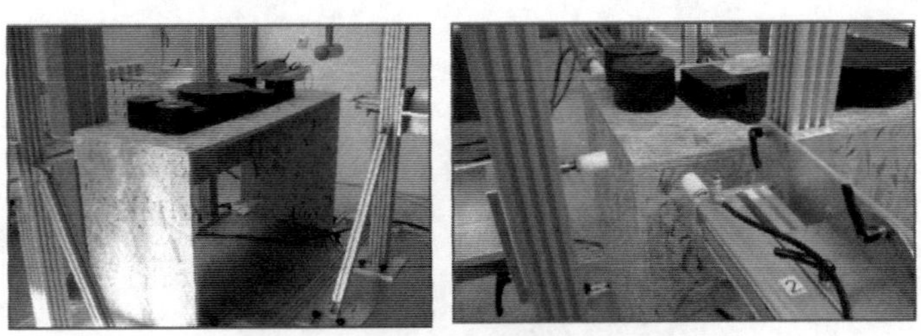

**Figure 2.** Testing stability under load with force in a horizontal direction.

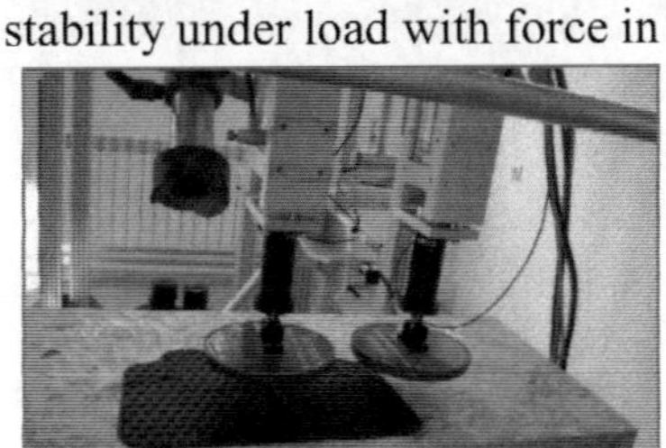

**Figure 3.** Application of vertical short-term loads during the testing process.

Strains in the corner compositions of furniture, occur as a result of the exploitation, or internal forces during relocating or transporting the furniture in mounted condition. The analysis showed that occurs bending strains, which cause the deformations of the furniture. Under the action of these forces, some of the corner compositions are loaded with bending forces by opening the shoulders, and some with their closure. Given that, the test method is moving in two directions: with load with bending forces at opening (test objects of Group I) and when closing (test objects of Group II) the shoulders of the composition (Figure 4) .

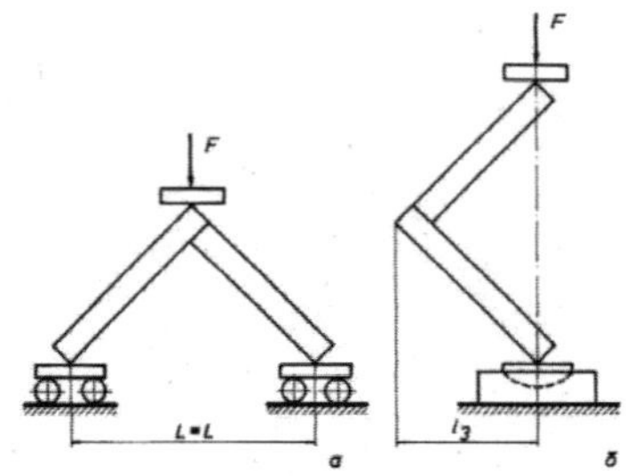

**Figure 4**. Test scheme for testing corner joints
a) - when load opens the shoulders of the test samples
b) - when load closes the shoulders of the test samples.

The type, shape and dimensions of objects for testing are displayed on the image (Figure 5). In this paper, two types of test objects are tested: the one is reference, specially made for the purpose and others are cut off from the office desks we used when examining the durability according to the European Standard. In this way, we have two values to compare: bending strength in angular compositions of new furniture, unused, and already exploited desks, dismantled, and re-mounted in order to further use. The testing of samples is carried out after five days of conditioning at temperature t = 15 - 30 ° C and humidity 45 to 70% and then measure the dimensions.

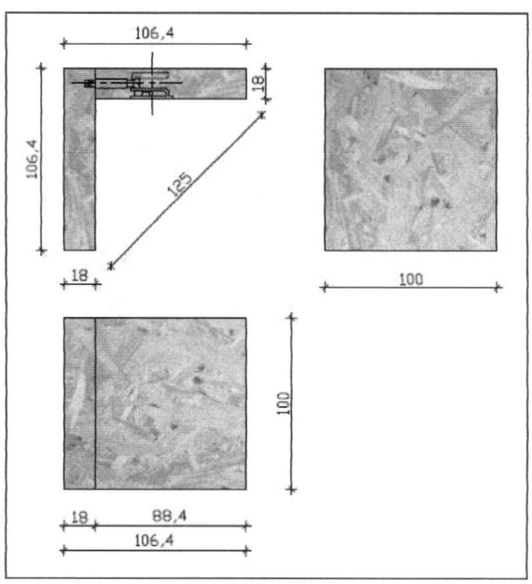


**Figure 5**. Front and cross-sectional view of final corner joint coupled with eccentric clutch.

**Figure 6.** Examination of corner joints when bending loaded by opening the sides of the sample object.

**Figure 7.** Corner joints when bending loaded by closing the sides of the sample object.

The necessary apparatus for carrying out the test of the test fixtures includes a universal test machine, which ensures the accuracy of the test of 1% of the force of breaking, moving trolley and a device for measurement of linear dimensions with an accuracy of 0.1 mm. The examination is carried out according to the Figure 4. The test objects are burdened with a smooth force until breaking. The breaking time is 60 ± 30 seconds from the start of load.

### *2.1. Results and discussion*

The processing of data from the conducted research is made according to the usual methods of variation statistics.

The medium arithmetic values are calculated[$x_{sr}$], the standard deviation [s], the coefficient of variation [V] and their errors[$fx_{sr}$ , $fs, fv$]. The test subjects had been divided into two groups with three models, with following marks description:

- *Group I* - Examination with bending forces when opening shoulders:

Model A – test objects merged with eccentric clutch Type A (minifix 15 with clamp and binding element): A / 1 - reference test objects; A / 2 – test objects at 2500 cycles; A / 3 – test objects targeting 5000 cycles.

**Table 1**. Results from the examination Group I objects.

| Model | Xsr [Nm] | fXsr [Nm] | S [Nm] | fS [Nm] | V [%] | fV [%] | P [%] | n [Number] |
|---|---|---|---|---|---|---|---|---|
| A/1 | 6,41 | 0,61 | 1,49 | 0,43 | 23,24 | 7,06 | 9,49 | 6 |
| A/2 | 6,17 | 1,02 | 2,05 | 0,72 | 33,17 | 12,95 | 16,59 | 4 |
| A/3 | 6,33 | 0,69 | 1,38 | 0,49 | 21,85 | 8,09 | 10,93 | 4 |

The results of the examination of the test objects of Group 1 show small, insignificant variations of the values of the bending, which shows that the swinging of the desks according to the European

Standard does not have a significant impact on the strength of the exploited corner composition compared to the reference.

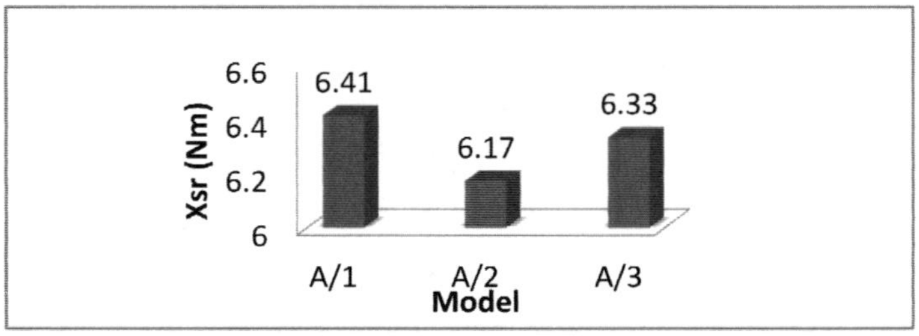


**Figure 8.** Results from the examination Group I / Model A (Examination with bending forces when opening shoulders).

- *Group II* - Examination with bending forces when closing shoulders:

Model A – test objects merged with eccentric clutch Type A (minifix 15 with clamp and binding element): A / 1 - reference test objects; A / 2 – test objects at 2500 cycles; A / 3 – test objects targeting 5000 cycles.

**Table 2.** Results from the examination Group II objects.

| Model | Xsr [Nm] | fXsr [Nm] | S [Nm] | fS [Nm] | V [%] | fV [%] | P [%] | n [Number] |
|---|---|---|---|---|---|---|---|---|
| A/1 | 6,21 | 0,77 | 1,89 | 0,55 | 30,45 | 9,57 | 12,43 | 6 |
| A/2 | 3,47 | 0,35 | 0,7 | 0,25 | 20,16 | 7,41 | 10,08 | 4 |
| A/3 | 6,02 | 1,5 | 3 | 1,06 | 49,85 | 21,57 | 24,93 | 4 |

As can be noted from the data in Table 2, the swinging of the desks according to the European Standard has no significant impact on the moment of bending strength in the corner compositions, regardless of the model.

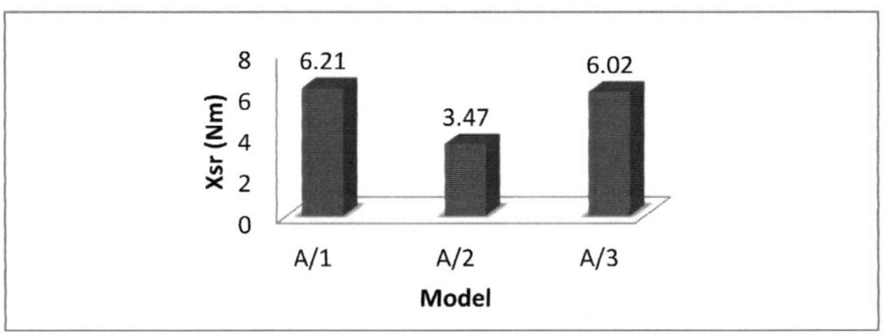


**Figure 9.** Results from the examination Group II / Model A (Examination with bending forces when closing shoulders).

## 3. Conclusion

The results of the examination of the bending strength of the dismountable final corner compositions made of OSB panels, when loading with bending forces and analysis of the data obtained before and after simulated exploitation, provide the following conclusions:

1. Simulated exploitation does not have a significant impact on the strength of bending on corner composition, which means that the furniture that has the ability to dismantle, can be used in the full of its strength, again.
2. The previous conclusion is formed on the basis of the technical characteristics of the coupling element that provides an elastic, not a rigid connection and who succeeds to receive and withstand the strain in furniture.
3. Impact on the results obtained is the interconnection of the eccentric clutch with the type of material (OSB plate). The different values and deviations in a minimum and maximum for the specific moment of bending are the result of the incompatibility of the OSB plate as a material for making furniture due to its non-proficiency and volatility in quality and the purpose of the eccentric joints for material with a finer and equated structure.
4. Another factor that has an impact on the diversity of bending strengths, represents the human factor present when mounting and screwing clutches.
5. The practical application of the results of this paper is in the recommendation to keep an eye on the selection of materials and shackles and to be used appropriately for the purpose for the piece of furniture to fully respond to the function and purpose for which it is provided.

## References

[1] Кючуков Г 1995 *Йакосни показатели на основните видови шарнирани ъединения чрез панти на конструктви елементи на мебели на обликовани и декоративно фолийо плочи от дървени частици с дебелна 16 мм* (София).

[2] Кючуков Г, Кючуков Б, Живков В и Маринова А 1999 *Нормативи за специфичните разрушававщти сили при натоварване на опън на осовните видове неразглобяеми ъглови съединенийа на конструктивни елменти от плочи от дървесни частици с дебелна 16mm* (София).

[3] Груевски Ѓ 2004 Одредување на јакоста на смолкнување на крајни и средни состави на плочести елементи– Магистерски труд, Шумарски факултет (Скопје).

[4] Груевски Т 1992 *Конструирање на плочите од иверка со средна густина во конструкциите на меебелот и испитување на јакосните показатели на конструктивните елементи*. Шумарски факултет (Скопје).

[5] Груевски Т, Кјучуков Г и Симакоски Н 2001 *Користење на питомиот костен во производството на мебел и плочи од иверки, Испитување на јакост на фиксни аголни состави оптоварени со сили на свивање*. Шумарски факултет (Скопје).

[6] Груевски Т и Симакоски Н 2002 *Елементи на дрвните консрукции*, Шумарски факултет (Скопје).

IOP Conf. Series: Materials Science and Engineering 1208 (2021) 012030 doi:10.1088/1757-899X/1208/1/012030

# Influence of accelerated aging on bending strength of particleboards

**Hasan Talić and Minka Ćehić**

University of Bihać, Faculty of Technical Engineering, Bosnia and Herzegovina

E-mail: hasan.talic@unbi.ba

**Abstract**. Particleboards are used in the manufacture of furniture, as well as the furnishing of structures and prefabricated homes. Their quality dictates where they can be used and how long they can last. The boards' characteristics deteriorate as they are used. To examine the change - degradation of the properties of the boards are exposed to the effects of the external climate (rain, snow, wind, sun) or regimes of accelerated aging. For this purpose, a degradation test of the properties of wood-based panels was performed, which aims to create a model that will describe the change in bending strength of the mentioned panels. The paper presents the results of bending strength and thickness swelling tests of 16 mm thick particleboards that were exposed to accelerated aging regimes, and the basic model of property degradation.

## 1. Introduction

For years, the most used wood-based panels are particleboards. There are many reasons for this such as large raw materials base, raw material price, low-cost adhesive, for example, urea formaldehyde adhesive, favorable properties for the inside usage as well for the construction. All these qualities make the particleboards cost less which makes them more favorable compared to other wood-based panels.

Particleboards are defined as boards made of lignocellulosic materials (mostly wood) in the shape of small pieces – wood particles combined with synthetic or other types of adhesives made into a compact panel under high temperatures and pressure. There are many criteria for particleboard classification. They can be classified according to the type of raw material, based on the position of particles in the panel, according to the dimensions of particles, according to look of the cut and its structure, according to the panel processing and structure, according to the volume, according to the type of usage, according conditions of usage, according to the type of adhesive used and formaldehyde emission [1]. Although, they have somewhat different qualities to other types of panels (particleboard and fibrous boards) particleboards have found wide range of usage, primarily in furniture manufacture as well as, interior design as wall coating, etc. (Figure 1.). Place of usage determines which type of panel to use (based on the standard BAS EN 312: 2012). There are following classes of panels: P1 – particleboards of general purpose in dry conditions, P2 –boards for interior space design and furniture and usage in dry conditions, P3 –non-bearing boards for humid conditions, P4 – bearing particleboards for dry conditions, P5 – bearing particleboards for humid conditions P6 –bearing particleboards for heavy load conditions in dry conditions, P7 – bearing particleboards for heavy load conditions in humid conditions.

**Figure 1.** Raw particleboard and examples of usage for furniture and wall coating [6].

Every construction type has specific conditions that a material, in this case particleboard needs to comply with. Particleboard can be used raw but, in most cases, it is coated with materials made for this purpose such as veneer, melamine foil, sieve coating, paint and varnish, impregnated paper, textile and different composite materials. All the wood-based plates used in construction throughout Europe should comply with standard BAS EN 13 986 (Wood-based panels for use in construction – Characteristics, evaluation of conformity and marking). This standard determines properties of wood-based panels for construction purposes and methods of determination and testing of properties of mentioned panels. Based on place of usage, chosen panels are tested for some or all of the properties, i. e. weather they are used in dry or humid conditions, or in the external weather conditions.

## 2. Testing of particleboard qualities

Panel qualities play a great role when deciding which type of panel is best suitable for a specific area of usage. Every space has specific conditions that panels should be compatible with in order to be used. As time goes by the panel qualities change based on the conditions of the space where the panel is used. This is why, aside from basic quality controls, we need to estimate conditions that will affect panel properties while in use and test the panel in such conditions to see its behaviour under the influence of the environment (change of humidity, temperature, etc.) To see the results faster, panels undergo tests in different conditions in a laboratory. This way panel properties deteriorate faster. These conditions are then compared to real conditions where the panel will be applied so that lasting period and resistance of panels in given conditions can be determined.

### *2.1. Theoretical study of testing of particleboards*

There are many studies done on the influence of climate and wood usage conditions and wood-based materials. The aim of these studies is to show how much climate and usage conditions affect wood materials over time as well as estimating influence of external factors on wood materials and level of change of those materials after years of exposure. Another aim is to understand difference between simulated and real external conditions. The ultimate goal being building a model that shows different influence of weather conditions on wood material. That being said, it is crucial to know the climate of the area where the material will be used. I. Teodorescua, D. Tapusia, R. Erbasua (2016), did a study of this matter, e. i. how the change in temperature and humidity change wood and wood material properties [11]. An analysis of temperature and humidity results recorded over a course of a few years in different parts of the day was done. These studies were based on different variables of wood such as loss of mass, resistance, influence of humidity, the lifespan of a tree and its shape. This study was done in France where they studied the influence of weather changes on wood materials. For temperature estimate minimum and maximum temperatures are recorded which gives insight in the daily temperature trend. Mechanical and physical properties of wood elements in wood construction are changed under the influence of humidity, temperature and bioactivities so the goal is building a model to estimate deterioration of wood for elements exposed to weather conditions. According to this the lifespan of the material depends on relative air humidity combined with different temperatures and time of exposure.

Aside from the aforementioned researchers, Y. Kojima and S. Suzuki (2010) made great contribution in this field [8]. Great deal of their research was focused on wood and wood materials deterioration under the influence of external weather conditions and comparison of lab induced aging with real aging of wood after years of exposure. Wood panels are used in construction more than ever. That is why we need durable wood e. i. we need to calculate how long could wood panels could be used in real life conditions, while keeping necessary properties given by law. Sekino and Suzuki (2010) presented the findings of their ten years long study of different types of panels (particleboard, OSB plate, MDF – fibrous plate, plywood plate etc). There are studies from Europe and USA that compared results with those gotten in Japan with regards to influence of climate properties on exposure conditions.

Methods used in studies of wood durability can be long term and short term. Long term studies examine exposure in outside conditions in long time frame including conditions from the past. These tests have many flaws such as long term and heavy damages and their results differ depending on the testing location. Short term studies analyse change of mechanical properties under lab conditions using dipping in water, boiling, steaming, freezing and drying methods. Sekino and Suzuki (2010) tested link between accelerated aging and exposure in the city of Shizuoka using bending properties. Eight different wood-based plates underwent five different accelerated aging treatments as well as five-year long test of external exposure. For every accelerated aging treatment bending strength decreased as cycles number increased. Bending strength after six cycles of treatments for JIS-B, APA D-1 and ASTM almost completely matched [9]. Five-year long test of exposure in open air showed that the bending strength of all the plates decreased with time. Bending strength of particleboard and OSB plate were less than 30 % and 10 % after five-year long exposure. Change of strength after five years of external exposure was equal to the change after six cycles of ASTM treatments [8][10].

## 3. Experimental studies of particleboards

The aim of this paper is to find out how will a raw particleboard be affected e. i. how will its properties change and deteriorate under the influence of different climate and load. Influence of climate on panels properties is shown through sample treatments. The treatments applied were exposing samples to different changes in temperatures and humidity in specific time intervals. These treatments were done in three cycles. Changes in sample panels were noticeable after every cycle visually as well after measurements of parameters (humidity, thickness, density and fracture force).

### *3.1. Standards and other resources for experimental studies*

Resources can be material (samples, tools, instruments, etc) as well as immaterial. Knowledge, as part of resources of experimental studies, means skills and experience of a person who conducts the experiment and relies on standards and guidelines for a given experiment. The experiment presented in this paper applied the following standards: BAS EN 326 – Sampling, cutting and inspection – Part 1: Sampling and cutting of test pieces and expression of test results, BAS EN 321 – Cyclic tests in humid conditions, BAS EN 322 – Determination of moisture content, BAS EN 325 – Determination of dimensions of test pieces, BAS EN 310 – Determination of modulus of elasticity in bending and of bending strength.

Due to variability in and between panels it is inevitable to test a number of panels [n], and a number of samples [m], cut out of one panel, to get reliable results. Minimum number of samples for testing [m] is: sample for density testing [6], for bending test [6], for humidity testing [4], treatment depending on test method requirements. When cutting minimum distance between two samples in the same test has to be 100mm, except for test of quality of glued joints. All samples cut from one panel have to have the identification number of the panel and serial number of the sample. They must have straight edges; no damage and they should be perpendicular on the panel. Every other panel property of a sample should be tested according to the in-effect standards, for samples cut from a whole panel. This testing requires two types of samples: BAS EN 310 for bending test; sample had following dimensions: sample length L, L= 20 t + 50 mm, sample width b, b= 50 mm and sample thickness t = 16 mm (Figure 2.a.). Humidity and density testing samples have following dimensions 50 mm x 50 mm x panel thickness (16) mm, and they are shown in (Figure 2.b) [2].

**Figure 2.** Samples of particleboards.

16 mm particleboard used here is a three-layer non-bearing panel suitable for indoor spaces and dry conditions. It has great dimensional stability and rigidity. It is used for furniture, partition walls, floor bases, sandwich panels. etc. Basic properties of this panel are shown in the table below (Table 1).

**Table 1.** Basic properties of particleboard – 16 mm.

| PARAMETAR | Classes of panels |
|---|---|
| Panel type (standard BAS EN 312) | P2 |
| Fire resistance | D-s1, d0 |
| Formaldehyde emission (EN ISO 12460-5) | Class E1 |

Classes of panel shown in the table are gotten from the supplier (Babilon doo. Tuzla) i. e. panel manufacturer (Kronospan). Aside from mentioned standards and particleboard samples other equipment was used in this study, such as digital scale, dryer, tensimeter - SHIMADZU, movable scale, micrometre, hygrometer, freezer, computer with suitable software and other lab equipment.

### *3.2. Particleboard testing*

It is necessary to identify process parameters that will be included in the experiment. It is possible to identify them through an experiment where a certain number of parameters is chosen from the total and defined as independent variables. The rest is viewed as constants (although these parameters can be independent variables). Maximum fracture force of wood-based panel (particleboard) depends on many factors such as: plate density [$\rho_p$], plate thickness [$d_p$], plate type, glue type, humidity [$w_p$], modulus of elasticity [$E_p$], plate structure- based on plate type (type of wood, thickness/size of veneer/chips, raw material quality, etc.), manufacturing technology (pressure and temperature of pressing, and other properties of manufacturing equipment). In this study, considering the aim of this research independent variables were plate type and conditions of exposure.

Testing of change in humidity is based on standard BAS EN 322. Humidity values change shows as change in mass of samples scaling of which occurs in precisely determined intervals. Samples are dried in a dryer on temperature 103 ± 2°C until constant mass. Data acquired in this way is used to calculate panel humidity. Mass measurement is accurate up to 0,01 g. Number and method of sample cutting is based on standard BAS EN 326-1. One panel gives four samples provided that none of the samples is lighter than 20,0 g. As needed clean the samples of dust and wood particles in order to get results as accurate as possible [14]. Calculating density value of tested wood-based panels is done on panels of following measurements 50 mm x 50 mm x panel thickness in mm. Panel density is the arithmetic mean value of densities of all samples cut from one panel. Sample needs to be tempered until constant mass in the following conditions: air humidity 65 ± 5% and temperature 20 ± 2°C. If the results in two consecutive tests do not differ more than 0,1 % from the test tube mass, constant mass is achieved. When measuring e. i. scaling accuracy has to be 0,01 g. Panel density [ρ] is calculated as arithmetic mean value of densities of all the sample panels made from one panel and it is expressed in [kg/m$^3$] [5].

Sample treatment is done according to standard BAS EN 321. This experiment has three cycles. Every cycle consists of the following: immersing sample into the water, freezing and drying samples in high temperatures. Sample immersion in water was done in controlled conditions. Water temperature was 20±1°C, pH value 7 ±1. Samples were put in plastic containers, separated from each other, in depth of 25±5 mm. Duration of the immersion in water was 70±1 hours. Samples are then taken out of the water, dried and put to freeze at temperature -12°C to -25°C. It is important to have the distance between the samples be at least 15mm and that upside/downside are the same i freezer as they were in the water. Freezing duration was 24±1 hours. Next step is drying of frozen samples in a dryer at a temperature of 70°C. Samples are positioned properly with defined distance and in accordance with dryer capacity. Downside/upside are the same as in the freezer. Drying duration was 70±1 hours. This concludes the first cycle (immersion, freezing, drying) Second and third cycle are identical to the first one.

Testing samples were cut and processed according to standard BAS EN 326. Those samples were put in tensimeter - SHIMADZU where the force is increased until sample destruction (fracture) in 90 ± 30 seconds interval. Sample is put on lower supporters properly spaced apart (20 values of sample thickness). Properly positioned sample should be in the middle of the supporters with 25mm of distance from the edges on each side. Control unit (computer) had Trapezium software where all the parameters were adjusted before the test start. While testing all the study- relevant data is being recorded such as fraction force, time, elastic deformation, etc. Sample thickness was measured before and after the testing. They were measure before testing and after each cycle which is why we can calculate the level of swelling after each treatment.

### *3.3. Results of experimental properties testing*

Density calculations are done according to standard BAS EN 323. Eight samples were measured and scaled and then mean density values were calculated. Sample humidity was measured three times and the values are shown in the table below. Swelling values was calculated in this experiment by measuring thickness before and after treatment. Based on this data results comparison was made and swelling value for specific samples was determined (Table 2).

**Table 2.** Parameters of measurement while testing (density, humidity, thickness of sample).

| Testing | Raw particleboard panel– thickness 16 mm | | |
|---|---|---|---|
| | Untreated | 1. cycle | 3. cycle |
| Panel density (mean value ρ) [kg/m$^3$] | 657,682 | 409,820 | 370,010 |
| Sample humidity [%] | 8 ± 2 | 7 ± 2 | 6 ± 2 |
| Sample thickness [mm] | 16,17 | 24,01 | 26,56 |

Fracture force testing was done on tensimeter (SHIMADZU), where maximum fracture force was determined in three points. As in previous tests it was measured on untreated samples, and then after first and third cycle.

**Table 3.** Maximum value of fracture force – mean values.

| Panel type | Sample No. | Maximum fracture force- mean value $F_{max}$ [N] | | |
|---|---|---|---|---|
| | | Untreated | 1. cycle | 3. cycle |
| Particleboard | 6 | 282,013 | 30,341 | 11,521 |

Raw particleboard thickness 16 mm

Based on maximum fracture force values mean values were calculated for every type of panel. They were compared with mean values after treatment e. i. after first and third cycle (Table 3).

## 4. Model of changes in properties of particleboards

Optimal choice of particleboard for a specific area of use is determined by the algorithm for material choice, that defines mechanical, physical, rheological, functional, technical, economic, ecological, social and other requirements. When used (furniture, partition walls, ceiling and roof construction) particleboards are exposed to different weather conditions and load, that changes over time. Choice algorithm can be understood through three units: input data, request verification and output data. Input data includes the choice of wood-based material and defining basic requirements determined by structure type. It, the structure type: bearing or non-bearing defines stability of shapes and dimensions, strength, heath, sound and humidity protection requirements, resistance to climate changes e. i. minimum and maximum values predetermined and many other properties specific to a certain type of structure. Request verification means that any suggested material will be checked for each and every requirement. If it passes all the checks it will proceed to the further controls. If it does not comply with all the specifications, it will be discarded. We will be at the beginning checking different material. This procedure will be repeated until the required conditions and properties are met [2].

Analysing results gotten in the experiment on the 16mm thick particleboard, we can notice change of values of some parameters after 3 cycles of treatment e. i. change of mean value of density of particleboard depending on testing cycle as well as humidity value of the panel before every cycle. (Figure 3). Change of panel density is notable after every cycle, that is after first and third cycle compared to untreated panel. Change is much greater after first cycle compared to the other two cycles. Decrease in density is cca. 38% after first cycle compared to the untreated panel while the difference between first and third cycle is cca. 10%, which practically means that the biggest change in density value happens in the first cycle while the other cycle cause much less of a change.

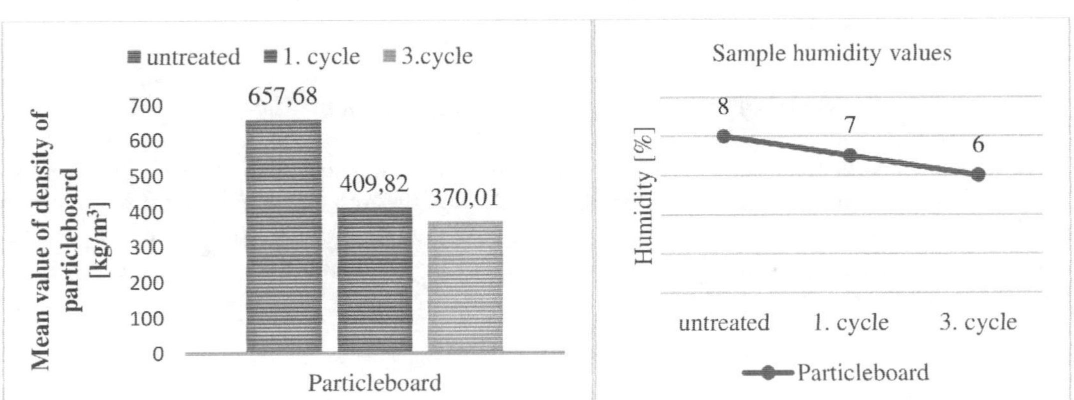


**Figure 3.** Mean values of density and humidity of particleboard (sample).

Thickness swelling of particleboard is one of the parameters noted after first and third cycle that showed similar tendencies as change in density. After first cycle, panel thickness increased by cca. 32% but after the second and third cycle it increased by only 7%. Relation between density and thickness swelling is shown in the figure below (Figure 4).

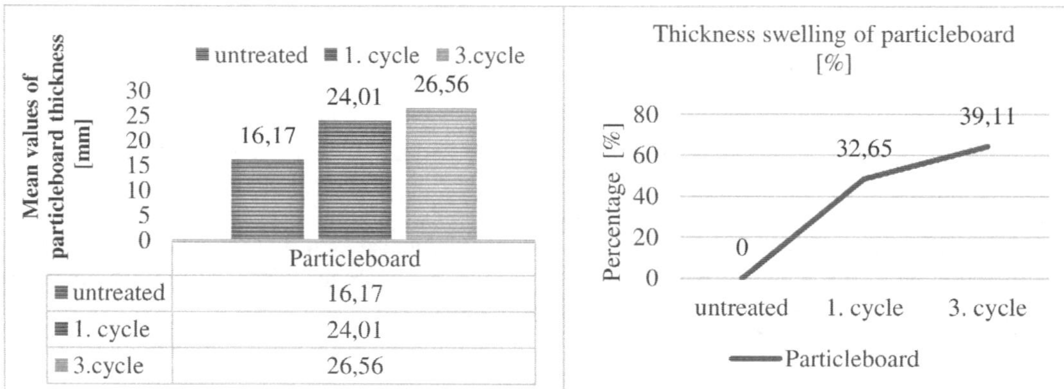


**Figure 4.** Mean values of panel thickness by cycles expressed in percentage.

Change of value in modulus of rapture is shown in (Figure 5), where we can see change after first and third cycle. Aside from that we can see minimum, maximum and mean values of fracture force for untreated and treated (1st and 3rd cycle) particleboards. Important finding is that the particleboard, after just first, cycle. lost strength almost completely. The value of force decreased to cca. 10 %. After additional two cycles it decreased even more, lost its stability and strength. As such it was practically useless.

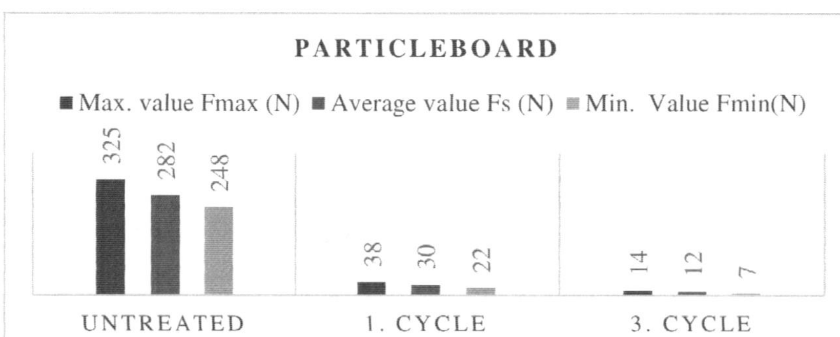


**Figure 5.** Values of force of fracture on a particleboard in different treatments.

Hook's law is used in most cases to calculate strength. It is a linear correlation between straining and deformation. A classical approach to the problem of strength says that the fracture occurs when a combination of factors such as staining, deformation, temperature and other parameters (that describe condition of the material and its specific properties) reaches critical value which can be expressed through the following equation:

$$\sigma = f(\sigma_0, \varepsilon, T, W, C_n) \qquad (1)$$

IOP Conf. Series: Materials Science and Engineering 1208 (2021) 012030 doi:10.1088/1757-899X/1208/1/012030

where $C_n$– represents parameters of macro real volume of the body, determined through an experiment. Change of the environmental conditions (change of temperature and relative humidity as well as duration of influence of climate parameters and how fast these influence change) influences the changes in temperature and humidity of the panel; the value of strength which is different from the maximum strength value that material can have in ideal conditions.

That is why to describe this model phenomenological equation for material conditions is more suitable and can be generally expressed as follows:

$$\sigma = f(\sigma_0, T, W, t) \quad (2)$$

Studies of influence of temperature on strength and strength [14], kinetic measurements of strength of particleboards in a long-term period [15], studies of influence of temperature and humidity on values of model of elasticity and module of rigidity [16], as well as our own experimental studies have shown that change of strength is not in fact linear and can be shown as Arrhenius equation:

$$\sigma_{W,T} = \sigma_0 \, e^{\alpha T_e} \, e^{\beta W_e} \, e^{\gamma W_e/T_e} \, e^{\delta N} \quad (3)$$

Where:

$\sigma_{W,T}$ – real (life) strength of wood in real humidity and temperature,
$\sigma_0$ – wood strength when humidity value is 0 % and temperature is 0°C
$T_e$ – effective temperature,
$W_e$ – effective humidity,
α, β, γ, δ, - constant values:

- α - influence of temperature on value of the material strength
- β – influence of humidity on the value of the material strength
- γ - change of material strength properties in combined influence of humidity and temperature
- δ – changes of strength caused by cyclical changes of humidity and temperature

N – number of cyclical changes of humidity and temperature.

$e^{\delta N}$ can exist as part of the γ value when the change happens in the real life (external) conditions or as a separate value in accelerated aging treatments depending on the number of cycles.

This model can present influence of humidity and temperature on values of panel strength, that is change in strength due to influence of climate where the boars is exposed. Coefficients (α, β, γ, δ) differ for different types of wood and its properties such as density, humidity, wood quality (wood fallacy). Influence of these factors manifests differently during temperature and humidity change in testing conditions and as such gives different results.

## 5. Conclusion

It is necessary to know the properties of material so that we can predict its changes over the course of time so that the material would meet necessary construction requirements. Particleboard is widely used as it has good qualities. This paper presented experimental testing of raw particleboards to study the influence of weather conditions in which the panel is used on the panel qualities. This is show through simulated conditions of aging of panels in three cycles. panels were measured before any treatments and then after first and third cycle. Based on these measures results were compared (density, humidity, thickness swelling, and maximum fracture force while bending) where we can see (degree of) changes after every testing cycle. panel evidently deteriorates after first cycle, although, it continues to do so in the next two cycles to a much lesser degree.

IOP Conf. Series: Materials Science and Engineering 1208 (2021) 012030 doi:10.1088/1757-899X/1208/1/012030

After analysing results of particleboard testing, we can see changes of variables of certain parameters during treatment:

- Change of panel density occurs after every cycle,
- Change is greater after the first cycle (cca. 38%) compared to the change in the next two treatment cycles (cca. 10%),
- Change of thickness swelling is similar to the change of density. After the first cycle the swelling increased approximately 32%. After the following two cycle it increased by approximately 7%,
- Modulus of rapture significantly decreased after first treatment cycle. The panel lost its strength almost completely (value of the force decreased to approximately 10% of the original value) After the other two treatment cycles strength and stability of the panel additionally decreased and as such the panel became unusable.

A model was made based on this data: Function of dependency of parameters such as temperature, humidity, testing duration and number of testing cycles. This shows the influence of climate conditions of the area on the qualities of panel. Thus, we need to consider it when choosing a panel. There is an (aforementioned) algorithm that helps choose panels and includes all the material and environment parameters and then suggest a panel suitable for any given requirements.

**References**

[1] Ćehić M and Omer S E 2018 Pločasti materijali na bazi drveta *University of Bihać*

[2] Talić H 2020 Izbor materijala na bazi drveta za izradu montažnih objekata *University of Bihać*

[3] BAS EN 326-1 2002 Wood-based panels – Sampling, cutting and inspection- Part 1: Samping and cutting of the pieces and expression of test result

[4] BAS EN 310 2001 *Wood-based panels – Determination of modulus of elasticity in bending and of bending strength*

[5] BAS EN 323 1993 *Wood-based panels – Determination of density*

[6] Frischeis 2021 Frischeis oplemenjena iverica. (n.d.)

[7] APA 2012 The engineered wood association: Panel design specification *Tacoma, Washington*

[8] Kojima Y and Suzuki S 2010 Evaluating the durability of wood-based panels using internal bond strength results from accelerated aging treatments Journal of Wood Science, **57**(1), pp 7–13.

[9] Mcnatt J and Link C 1989 Analysis of ASTM D 1037 Accelerated-aging test: Semantic scholar *Forest products journal*

[10] Kajita H, Mukudai J and Yano H 1991 Durability evaluation of particleboards by accelerated aging tests *Wood Science and Technology*

[11] Ioana T, Daniela Ţ and Ruxandra E 2016 Influence of the climatic changes on wood structures behaviour Sustainable *Solutions for Energy and Environment Bucharest Romania*

[12] Kulman S and Boiko L 2016 Non-linear effects in the reaction of wood composite materials during the thermal, humidity and power loads *Annals of Warsaw University of Life Sciences - SGGW Forestry and Wood Technology* 95: pp 159-165

[13] Sergiy K 2017 Method for predicting the bending ultimate strength (MOR) and modulus of elasticity in bending (MOE) composite materials based on wood *Patent 114560* – UA

[14] Kulman S, Boiko L and Antsyferova A 1970 Bending strength (modulus of rupture) and modulus of elasticity of MDF different density at various temperature: Semantic scholar *Annals of Warsaw University of Life Sciences - SGGW Forestry and Wood Technology* 91: pp 101-106

[15] Luidmyla B, Sergiy K and I G 2013 Durability particleboards in furniture *Osvita Ukrainy* Ukraine p 210

[16] Kulman S 2019 The effect of temperature and moisture changes on modulus of elasticity and modulus of rupture of Particleboard *Acta facultatis xylologiae Zvolen*

# Improving Comb Aliasing Rejection Using Sharpening of Modified Comb

**Gordana Jovanovic Dolecek**

Institute INAOE, E. Erro 1, Tonantzintla, Puebla, 72840, Mexico

E-mail: gordana@ieee.org

**Abstract**. This paper presents an efficient method to improve the comb aliasing rejection in a comb decimation filter without increasing its passband droop. This problem is important since aliasing and comb passband droop may deteriorate the decimated signal. We propose here to apply sharpening of the modified comb in the second stage of a two-stage comb structure. The modified comb is obtained by decreasing the middle coefficient of the impulse response of the cascade of two combs by 1/2. The sharpening polynomial with the first order tangencies is used here. As a result, the comb folding bands, where the aliasing occur, become wider and with an increased attenuation in comparison with the original comb filter. However, this improvement in the folding bands did not result in an increased passband droop. The compensator from literature is used to further decrease passband droop. The method is illustrated with examples and compared with the original comb and the methods proposed in literature for increasing aliasing rejection.

## 1. Introduction

This paper treats the problem of the increasing aliasing rejection in comb decimation filters. It is well known that the comb filter is the most popular decimation filter, due to its simplicity, since all its coefficients are equal to unity [1].

The system function of the comb filter is given as:

$$H(z) = \frac{1-z^{-M}}{M(1-z^{-1})}, \tag{1}$$

where $M$ is the decimation factor.

Comb filter attenuates aliasing which occurs in the folding bands, (bands around the comb zeros). However, this attenuation is not enough in many applications, and should be increased [2].

The simplest method to increase the aliasing rejection is to use the cascade of $K$ combs (1). $K$ is called the order of the comb filter. The system function of the comb filter of order $K$ is given as:

$$H(z)^K = \left[\frac{1-z^{-M}}{M(1-z^{-1})}\right]^K. \tag{2}$$

The increase of $K$ increases aliasing rejection, and also increases the comb passband droop, which may deteriorate the decimated signal. This problem is illustrated in Figure 1, where the overall comb magnitude responses, and the passband zooms, for two values of $K$, are presented.

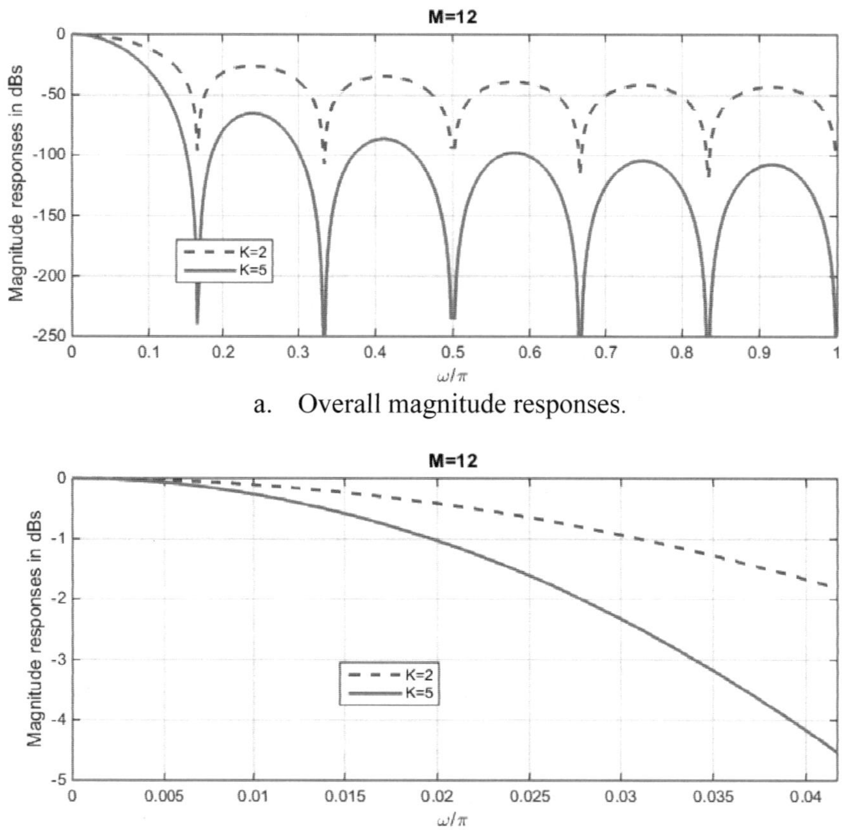


a. Overall magnitude responses.

b. Passband zoom.

**Figure 1**. Magnitude responses of two combs of different orders.

Different methods have been proposed to improve comb aliasing rejection. The rotation of comb zeros, requiring two multipliers, was proposed in [3-4]. Additionally, a recursive term is introduced with the perfect cancellation of poles and zeros on the unit circle for the case, in which the coefficients are presented in an infinite precision. However, when the coefficients are presented in a finite precision, the cancellation of poles and zeros can be lost, leading to instability.

Different methods have been proposed to get the improvement of the aliasing rejection without introducing multipliers and possible instability. The authors in [5-7] used combs of different lengths, while the authors in [8-10] introduced additional zeros into comb folding bands using simple multiplierless filters. The sharpening technique using Kaiser sharpening polynomial was introduced in [11]. Recently, different sharpening polynomials were proposed in [12-14], and the designs of the corresponding compensators to compensate for the passband droop, were also elaborated.

As a difference to works in [11-14], where the sharpening of comb filters was applied, we consider here the sharpening of a modified comb at the second stage. As a result, we get an improved aliasing rejection and a decreased passband droop in comparison with the equivalent comb. In that way, a simple comb compensator can be used to compensate for the passband droop.

The rest of the paper is organized as follows. Next section briefly presents the modified comb and the sharpening technique. The proposed filter is introduced in Section 3, and compared with the equivalent comb filter. The passband compensation is also addressed. The comparisons of the proposed method with the methods recently proposed in literature are presented in Section 4. Finally, the conclusions are given in Section 5.

## 2. Modified comb and sharpened comb

In this section we present brief review of the modified comb and the sharpened comb from literature.

### *2.1. Modified comb [8]*

The modified comb was introduced in [8-9]. This filter is obtained by subtracting the ½ from the middle coefficient of the impulse response of the cascade of two comb filters. The system function of the proposed comb filter can be presented using the modified comb as [8]:

$$G(z) = H^{K-2}(z)H_m(z), \tag{3}$$

where $K$ is the order of the equivalent comb filter and the modified comb is equal to,

$$H_m(z) = \left[\frac{1-z^{-M}}{M(1-z^{-1})}\right]^2 - z^{-M}\left(M - \frac{1}{2}\right). \tag{4}$$

Next example illustrates the benefit of the filter (3).

**Example 1**: We consider $M$=14 and $K$=4. Figure 2 compares magnitude responses of the equivalent comb filter of the order $K$, and the filter (3).

Note that filter (3) has wider folding bands than the equivalent comb filter, thus providing better aliasing attenuation.

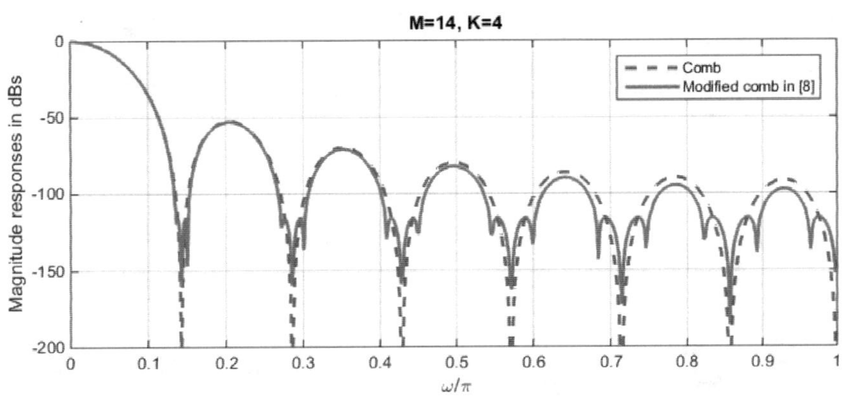


**Figure 2.** Magnitude responses of filter comb and filter given in [8].

### *2.2. Sharpened comb [11]*

We will use the sharpening polynomial Sh{$x$}=$3x^2$-$2x^3$, introduced in [11] where $x$ is a comb filter of order $K$, and Sh{$x$} means sharpening of $x$.

The sharpening polynomial improves the comb magnitude characteristic in both, passband and stopband, as shown in next example.

**Example 2:** We choose $M$=15 and $K$=1. Figure 3 contrasts the magnitude responses of comb and the sharpened comb.

The zoom in the passband is also shown in order to demonstrate that the sharpening improves also the passband of the original comb.

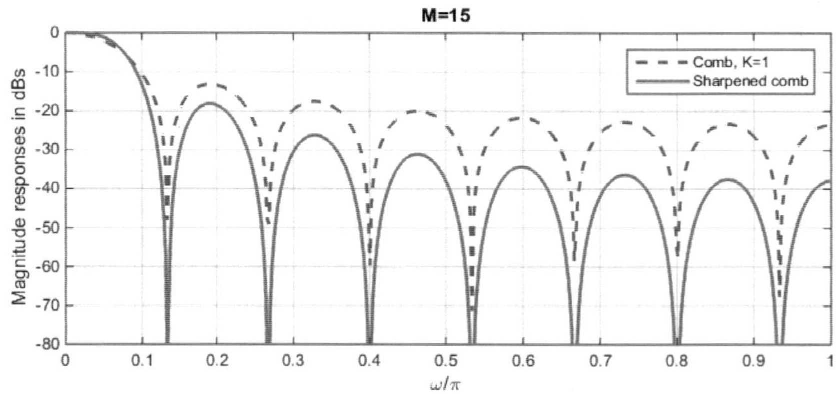


a. Overall magnitude responses.

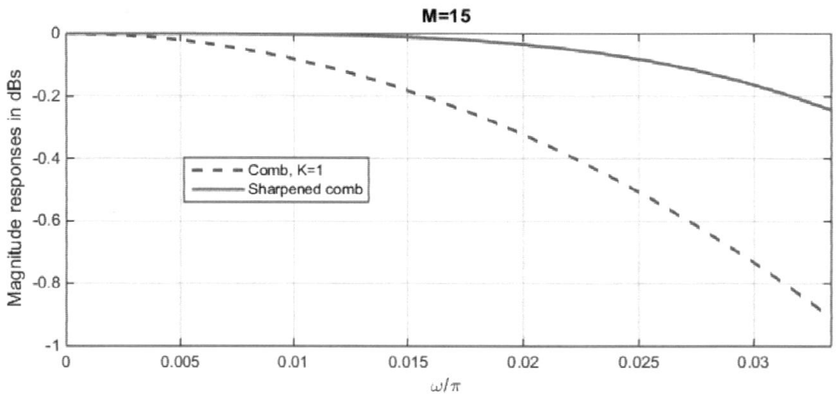


b. Passband zoom.

**Figure 3**. Magnitude responses of comb and sharpened comb.

## 3. Proposed filter

In this section we first present improving of aliasing rejection. Next we present the improving the passband characteristic of the proposed filter.

### *3.1. Improving aliasing rejection*

We propose to sharpen the modified comb at the second stage instead of to sharpen comb itself. The system function of the proposed filter is given as:

$$H_{sh}(z) = H_1^K(z)H_2^{K-3}(z^{M_1})Sh\{H_{2_m}(z^{M_1})\}, \tag{5}$$

where $H_1(z)$ and $H_2(z)$ are the system functions of combs at first and the second stages, respectively, and $H_{2m}(z)$ is the system function of the modified comb $H_2(z)$.

Note that the system functions of all filters at the second stage are expanded by $M_1$ since the overall system function is presented at high input rate.

The method is illustrated in the following example.

**Example 3**: We consider $M$=16 and $K$=4. Figure 4 contrasts magnitude responses of the equivalent comb and the proposed filter.

Note that the proposed filter has wider folding bands, thus providing a better aliasing rejection than the equivalent comb filter of order K. Additionally, the proposed filter exhibits lesser passband droop.

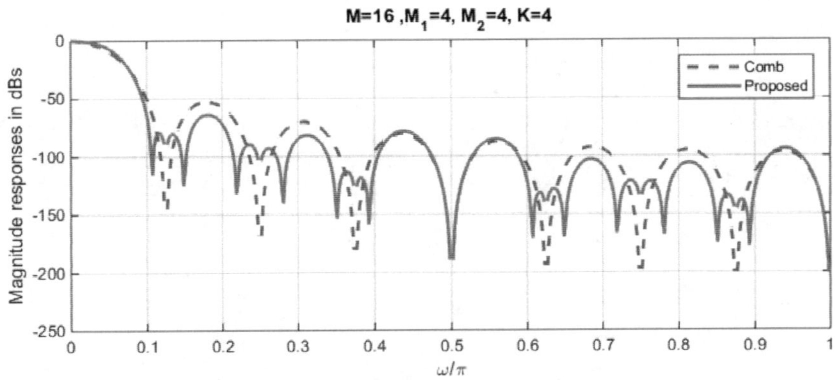


a. Overall magnitude responses.

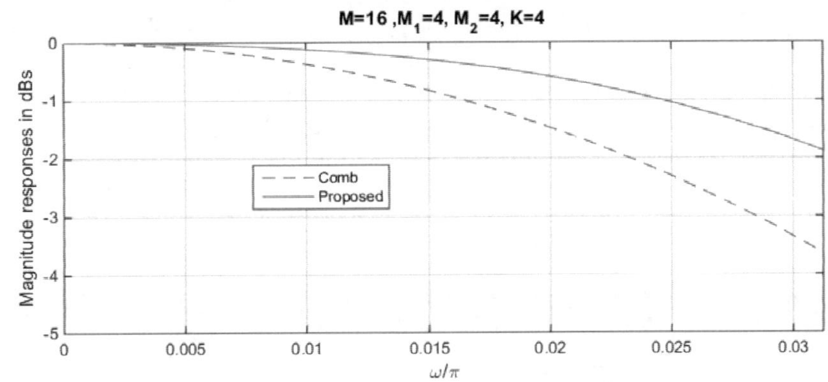


b. Passband zoom.

**Figure 4.** Magnitude responses of the proposed sharpened filter and the equivalent comb filter.

*3.2. Improving passband*

In order to decrease the passband droop in the proposed sharpened filter (5), we applied the compensator $C(z)$ at low rate. We propose here to adapt the compensator, proposed for the comb compensation in [15]. The magnitude characteristic of this compensator is given as:

$$C(e^{j\omega}) = \left[1 + A\sin^4\left(\frac{\omega M}{2}\right)\right] \times \left[1 + A\sin^2\left(\frac{\omega M}{2}\right)\right], \tag{6}$$

where the parameters $A$ and $B$ are obtained using MATLAB simulation, using the similar procedure as described in [15] for a comb filter. Instead of comb filter used in [15], we use here the sharpened modified comb (5).

Table I shows the obtained results for parameters $A$ and $B$, and the absolute values of the maximum passband deviations $\delta$ in dBs, of the compensated filter for a given values of $K$. The passband edge is equal to $\omega_p=\pi/5$.

It is worth to mention that the parameters in Table 1 do not depend on the decimation factor $M$. The total number of compensator adders is $N_A=9$, for all cases in Table 1.

Using (5) we get the system function of the proposed sharpened compensated filter as,

$$H_p(z) = H_{sh}(z)C(z^M), \tag{7}$$

where $H_{sh}(z)$ is given in (5) and $C(z)$ is the system function of the compensator.

Observe that the system function of the compensator $C(z)$ is expanded by $M$ since the compensator works at low rate, i.e. after the decimation by $M$.

**Table 1.** Compensator parameters and values of the absolute deviation δ in dBs

| $K$ | $A$ | $B$ | $\delta$ in dBs |
|---|---|---|---|
| 3 | $2^{-3}$ | $2^{-4}$ | 0.026 |
| 4 | $2^{-3}$ | $2^{-2}$ | 0.037 |
| 5 | $2^{0}$ | $2^{-2}$ | 0.035 |
| 6 | $2^{0}$ | $2^{-1}$ | 0.028 |

Next example illustrates the compensation of the proposed filter from Example 3, ($M$=16, $K$=4).

**Example 4**: The compensator parameters are obtained from Table 1, for $K$=4. The magnitude responses of the proposed sharpened, and the proposed compensated sharpened filters are shown in Figure 5. The passband zoom is also shown to present the passband improvement. The passband deviation in the compensated filter is equal to 0.027 dBs. Note that the compensator does not affect the aliasing rejection.

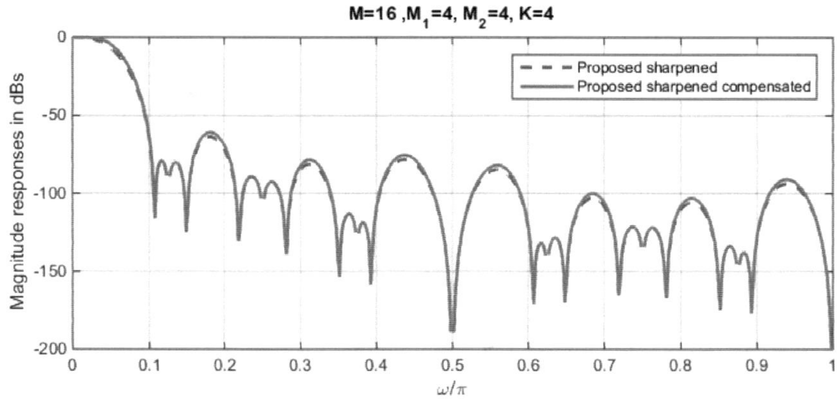


a. Overall magnitude responses.

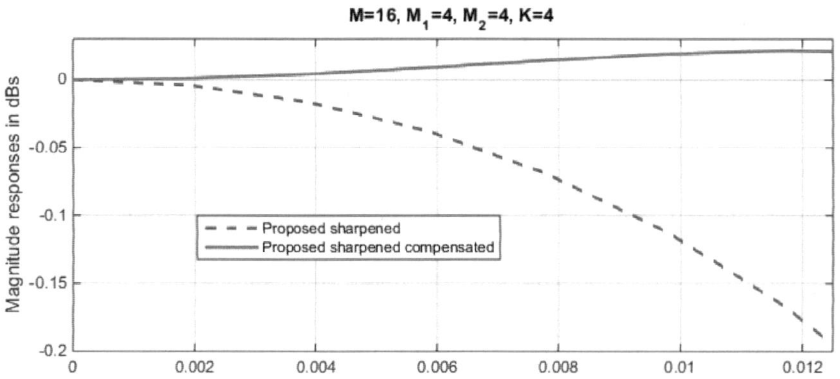


b. Passband zoom.

**Figure 5**. Magnitude responses of proposed sharpened and sharpened compensated filters.

## 4. Comparisons

In this section we present the comparison of the proposed method, with the method in [8] and the sharpening methods recently proposed in literature.

### *4.1 Comparison with method in [8]*

Since the proposed method is the modification of the method in [8], our first comparison is with the method in [8]. We will compare only alias rejection since in method [8] is not presented the compensation.

We chose $M$=20, $M_1$=4, $M_2$=5, and $K$=5. The magnitude responses are compared in Figure 6. We can observe that the proposed method provides much better aliasing reduction.

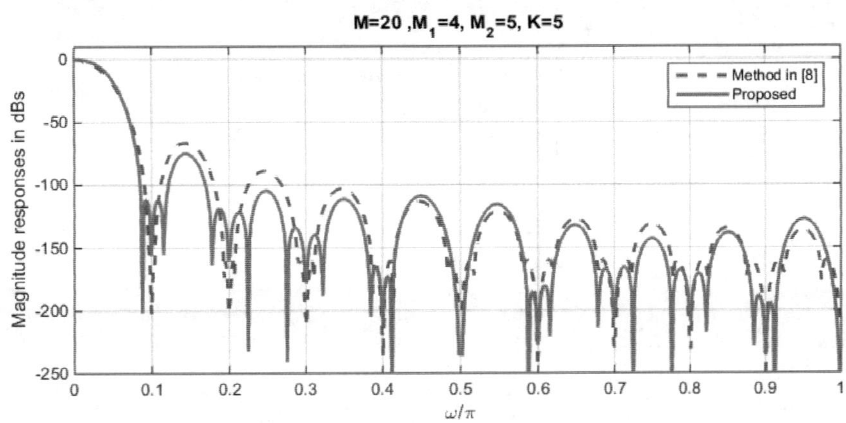


**Figure 6**. Comparison with method in [8].

### *4.2. Comparison with method in [13]*

The compensator design for different sharpening polynomials was proposed in [13]. The obtained results are shown in table for $M$=32. For a sake of comparison, we consider the sharpening polynomial $p(x)=2^{-8}x^2-2^{-3}x^4+x^6$, where $x$ is a comb filter of order 1. In the proposed design we have $M_1$=4, $M_2$=8, and $K$=4. The compensator parameters are obtained from Table 1.

The magnitude responses are shown in Figure 7. We can observe that the proposed method provides better magnitude characteristic in both, the passband and the stopband. Additionally, the proposed method has lesser complexity since the sharpening is performed at lower rate.

### *4.3. Comparison with method in [14]*

A two-stage structure, in which the sharpening is applied at the second stage, was proposed in [14]. The coefficients of the sharpening polynomial are obtained using linear programming optimization. The example of design, presented in table in [14] for $M$=32, is used here for the comparison. Like in [14], in the proposed method $M_1$=4, $M_2$=8. In the proposed method, $K$=3 and the parameters of compensator are obtained from Table 1 for $K$=3.

The magnitude responses are contrasted in Figure 8.

The proposed method has better aliasing rejection, while method in [14] has slightly better passband compensation.

Moreover, the advantage of the proposed method is the design simplicity, since the design parameters do not depend on $M$, and the compensator parameters are given in table for any value of $M$.

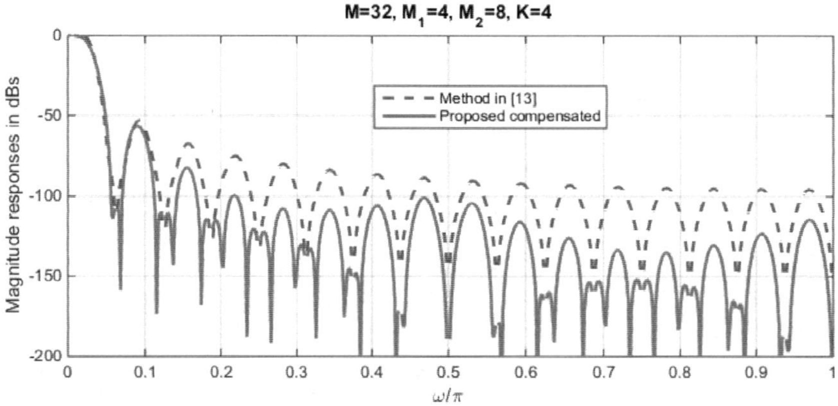


*a.* Overall magnitude responses.

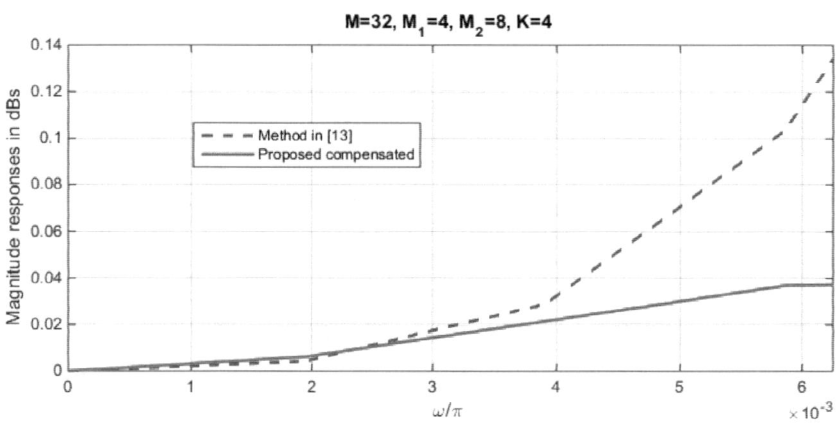


*b.* Passband zoom.

**Figure 7**. Comparison with method in [13].

## 5. Conclusions

This paper presents a simple method to increase aliasing rejection and improve passband characteristic in the comb decimation filter. The proposed filter has two stages decimated by $M_1$ and $M_2$, respectively, where the overall decimation factor is equal to $M= M_1 M_2$.

In the first stage is the comb filter, while in the second stage is the cascade of comb filters, and the sharpened modified comb filter.

The sharpening methods recently proposed in literature, require new parameter calculations for each new value of $M$. In contrast, in the method proposed here, the parameters of the proposed sharpened filter do not depend on $M$.

Similarly, the compensator parameters depend only on the comb parameter $K$, and are provided in table. The presented examples show that the proposed method provides better aliasing rejection and better or similar passband characteristics in comparison with methods from literature. The complexity is decreased by realizing the sharpening at the second stage. In future work we plan to design a new compensator in order to get a better compensation in the passband.

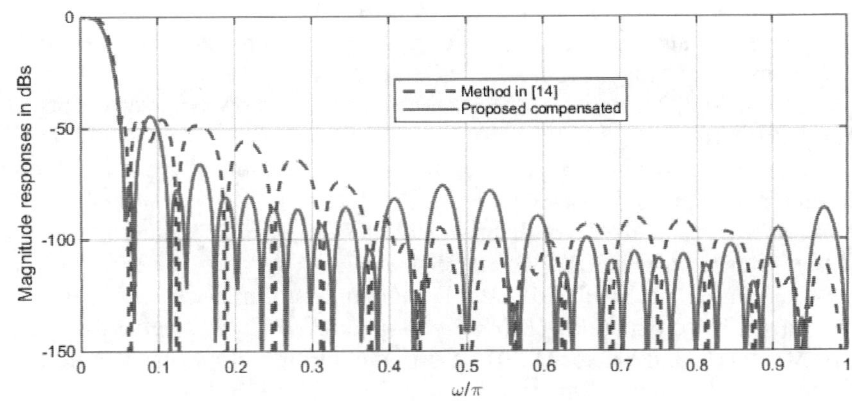


*a.* Overall magnitude responses.

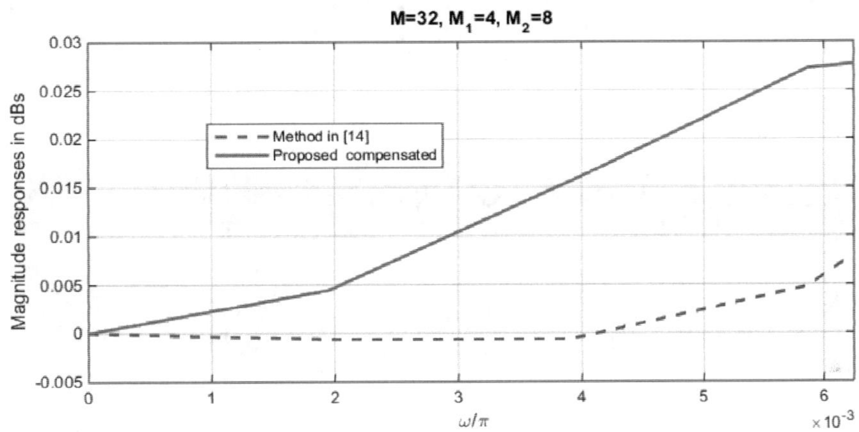


*b.* Passband zoom.

**Figure 8**. Comparison with method in [14].

## References

[1] Jovanovic Dolecek, G (Editor) 2001 *Multirate systems: Design and applications* (Hershey: Idea Group Publishing).

[2] Harris, F 2004 *Multirate signal processing for communications systems* (New Jersey: Prentice Hall).

[3] Presti, L L 2006 Efficient modified-sinc filters for Sigma-delta A/D converters. *IEEE Trans. on Circ. and Syst. II: Analog and Dig. Sig. Proc.* **47**, 1204–13.

[4] Laddomada, M 2007 Generalized comb decimation filter for Sigma-delta A/D converters: Analysis and design *IEEE Trans. on Circ. Des. I: Reg. papers* **94**, 994-1005.

[5] Stošić, BP and Pavlović, VD 2016 Design of new selective CIC filter functions with passband-droop compensation. *El. Lett.* **52,** 115-7.

[6] Milic, DS and Pavlovic,VD 2016 Comparison of classical CIC and a new class of stopband-improved CIC filters formed by cascading non-identical comb sections *Facta Univ. Series: Electr. and Energ.* **29**, 61-76.

[7] Stosic, BP, Milic, D and Pavlovic, VD 2015 New CIC filter architecture: Design, parametric analysis and some comparisons *IETE J. of Res.* **61**, 244-50.
[8] Jovanovic Dolecek, G and Dolecek, L 2016 Exploiting features of symmetric polynomials for improved comb filter design Proc. *IEEE SPA 2016*, 26-9.
[9] Harris, F and Jovanovic Dolecek, G 2016 Bifurcate repeated stop band zeros of CIC filters Proc. *IEEE DSP 2016,* 1-4.
[10] Jovanovic Dolecek, G 2017 Improving comb alias rejection by exploiting certain class of palindromic polynomials. Proc. *ECCTD 2017* 1-4.
[11] Kwentus, AJ and Jiang Z 1997 Application of filter sharpening to cascaded integrator comb decimation filter *IEEE Trans. Sig. Proc.* **25** 457-67.
[12] Molnar, G, Dudarin, A andVucic, M 2018 Design and multiplierless realization of maximally flat sharpened CIC compensator *IEEE Trans. on Circ. and Syst. II: Express Briefs* **65** 51-5.
[13] Dudarin, A, Molnar, G and Vucic, M 2018 Optimum multiplierless compensators for sharpened cascaded-integrator-comb decimation filters *Ellec. Lett.* **54**, 971-72.
[14] Gautam, D, Khare, K and Shrivastava, BP 2021 A novel approach for optimal design of sample rate conversion using linear optimization techniques *IEEE Acc.* **9** 44436-41.
[15] Jovanovic Dolecek G, Garcia Baez R, Molina Salgado G and de la Rosa, JM 2017 Novel multiplierless wideband comb compensator with high compensation capability *Circ. Syst. and Sig. Proc.* **36** 2031–49.

# Application of genetic algorithm in solving the travelling Salesman problem

**Fatka Kulenović and Azra Hošić**

University of Bihać, Faculty of Technical Engineering, Bosnia and Herzegovina

E-mail: kulenovic.fatka@gmail.com

**Abstract**. The Travelling Salesman Problem is categorized as NP-complete problems called combinatorial optimization problems. For the growing number of cities it is unsolvable with the use of exact methods in a reasonable time. Genetic algorithms are evolutionary techniques used for optimization purposes according to survival of the fittest idea. These methods do not ensure optimal solutions, however they give good approximation usually in time. Studies have shown that the proposed genetic algorithm can find a shorter route in real time, compared with the existing manipulator model of path selection. The genetic algorithm depends on the selection criteria, crosses, and mutation operators described in detail in this paper. Possible settings of the genetic algorithm are listed and described, as well as the influence of mutation and crossing operators on the efficiency of the genetic algorithm. The optimization results are presented graphically in the MATLAB software package for different cases, after which a comparison of the efficiency of the genetic algorithm with respect to the given parameters is performed.

## 1. Introduction

The travelling salesman problem is one of the best known problems of combinatorial optimization with the goal of minimization. It represents a logistical problem of everyday life whose solution is found in graph theory. The mathematical model of travelling salesman problem comes down to searching for the Hamiltonian cycle f least weight in a weight graph. The weight of a given edge can represent the amount of cost, distance, time or any other measure that characterizes that edge. There is a well-known set of cities to visit and their mutual distances. The optimal travel route must be found so that each city is visited only once, and then returned to the starting city that represents the starting point and with the shortest possible distance. The more cities to visit, the more complicated this problem becomes. Since exact methods do not provide an optimal solution to a given problem, heuristic methods are used that can find solutions very close to the optimum in a reasonable time. The aim of this paper is to define the methodology and apply it to solve the travelling salesman problem. The complexity of the problem itself is reflected in the set of partial tasks that need to be solved in order to obtain the most rational solution. In other words, the aim of the paper is to determine the shortest distances between network nodes, determine locations and determine optimal routes. Solving this problem is a major challenge, partly because of its extreme difficulty and partly because of its connection to interesting practical issues. Confirmation of the popularity and applicability of The Travelling Salesman Problem is found in many papers [1], [2] and [3].

## 2. Mathematical formulation of the travelling Salesman problem

There are numerous formulations of TSP in the literature whose basic form is most often formulated as follows [4]:

1. The travelling salesman has a given set of cities, each of which he must visit exactly once and return to the starting point. The question is in what order he should tour the cities so that the total length of the journey is minimal.
2. Find the Hamiltonian cycle of minimum weight in the weight graph.

To better understand the theory of the travelling salesman problem, it is necessary to get acquainted with the basic definitions of graphs [4].
Graph definitions are given:

1. A graph is an ordered triple $G = (V, E, \varphi)$, comprising:
   - $V = V(G)$ is a nonempty set whose elements are called vertices,
   - $E = E(G)$ is a set disjoint with V whose elements are called edges and,
   - $\varphi$ is a function that associates a pair $u, v \in V$ with each edge $e$ of $E$.
2. Graph is in short marked as $G = (V, E)$ or just $G$.
3. Expressed in the terminology of graph theory, let $V$ be a set of $n + 1$ vertices (locations) $V = \{0, 1, 2, \ldots, n\}$, the problem is to find a route (cycle) that begins and ends at vertex 0, and passes through all vertices from 1 to n exactly once, so that the total distance is minimal.
4. The Hamiltonian path (cycle) in graph $G$ is the path (cycle) that contains all the vertices of the graph.
5. In other words, the travelling salesman problem is finding the Hamiltonian cycle of minimum length.
6. A graph containing a Hamiltonian cycle is called a Hamiltonian graph.
7. In complete graphs with $n$ verticesthe number of Hamiltonian cycles is $(n - 1)!$, while in undirected graphs $\frac{(n-1)!}{2}$, since the two routes are identical, only inverse, and grows exponentially with the number of vertices.

If the variables$x_{ij}$are considered, which are defined as follows [5]:

$$x_{ij} = \begin{cases} 1, & if\ branch\ (i,j) belongs\ to\ the\ optimal\ route\ of\ the\ travelling\ salesman \\ 0, & otherwise \end{cases}$$

The travelling salesman problem can be formulated as follows [5]:

Minimize

$$\sum_{i=1}^{m}\sum_{j=1}^{n} d_{ij}$$

under restrictions

$$\sum_{j=1}^{m} x_{ij} = 1 \qquad i = 1, 2, \dots, m$$

$$\sum_{i=1}^{m} x_{ij} = 1 \qquad j = 1,2, \dots, m$$

$$\sum_{i \in S, j \in S} x_{ij} \leq |S| - 1 \qquad S \subset \{1, 2, \dots, m\}$$

$$x_{ij} = 0 \ \ or \ \ 1 \qquad \forall i, j$$

where $S$ is any subset of cities $1,2, \dots, m$. The first two restrictions take care of the movement of travelling salesman and indicate that only one branch can come out of each node visited by travelling salesman, ie that only one branch can come to each node in the route of the travelling salesman. The third constraint prevents the creation of cycles that do not represent the complete route of the travelling salesman [5]. Methods of solving the travelling salesman problem can be divided into four main groups: exact, approximation, heuristic and metaheuristic.

## 3. Optimization route of the travelling Salesman problem using genetic algorithm

The solution of the travelling salesman problem using a genetic algorithm was done in the MATLAB software package. Using a graphical user interface, two basic tasks of creating an application are integrated: setting up visual components and programming the operation of the application, which allows a quick transition between visual design on the screen and code development in the MATLAB editor. An application has been created that calculates the shortest route and selects a route to tour the cities of Bosnia and Herzegovina, so that each city hevisits only once and returns to the source point. For a detailed description of the problem, UML (Unified Modelling Language) modelling was used, the main goal of which is to present a detailed procedure, as well as the functioning of genetic algorithms and how to solve problems by applying them [6]. With the development of an application to solve this problem, it was determined that one of the most important components for the successful operation of this algorithm is the selection of the correct genetic operators and parameters that will determine the behaviour of these operators. If set correctly, the algorithm gives excellent results, but if the parameters are set incorrectly the algorithm will tend to a local optimum, which is closer or farther from the true optimum, depending on the parameters. The initial view of the application is given in Figure 1.

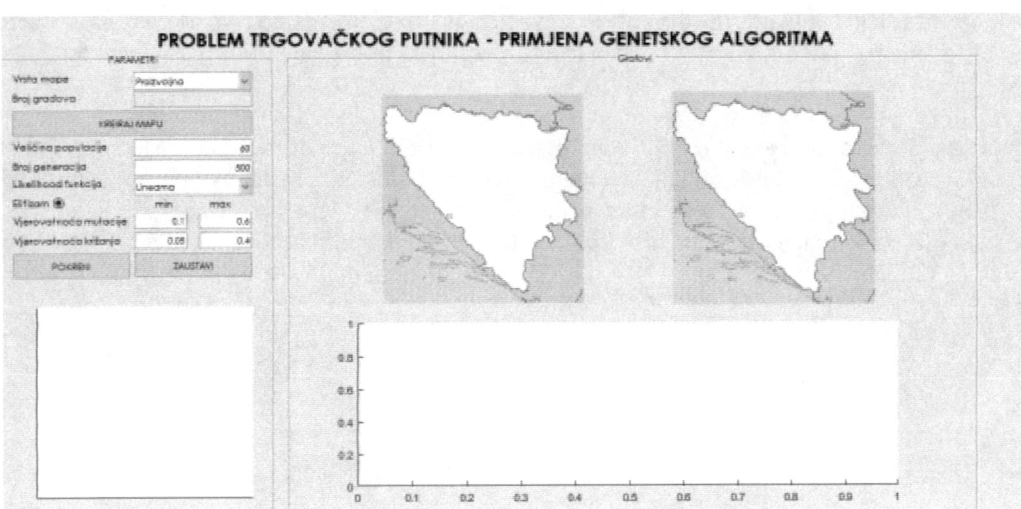


**Figure 1.** Initial view after starting the application.

In order to create a network of cities to visit, the user must select the type of map to create. The program offers the ability to select two types of maps that are implemented using a drop-down menu: an arbitrary map and a map created by randomly selecting positions. The performance of the algorithm depends on the parameters. Regardless of the genetic algorithm, it has the following parameters: population size, number of generations (number of iterations, repetitions) and probability of mutation and probability of crossing. One of the earlier studies on parameter values was performed by De Jong, and according to his results, the best population size is between 50 and 100 individuals, the probability of crossing is about 0.6, and he determined the probability of mutation as 0.001 per bit in Figure 2.

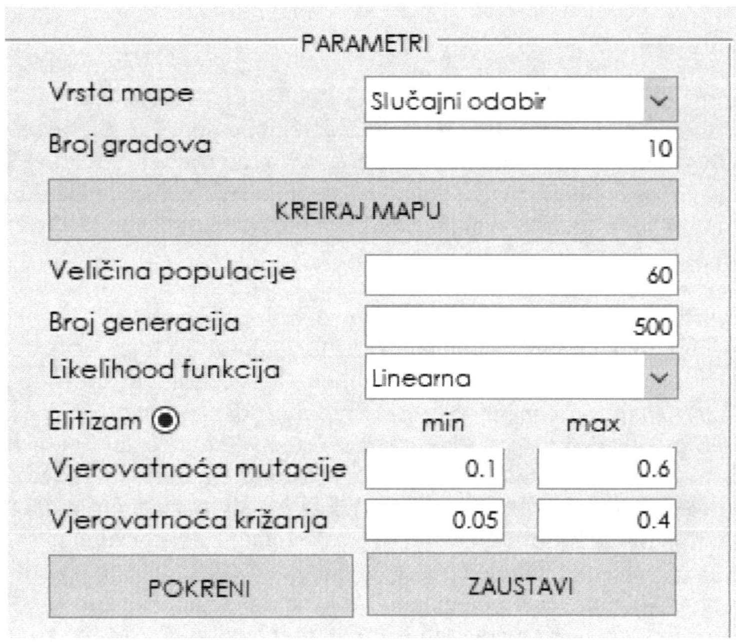


**Figure 2.** Display of genetic algorithm parameters.

The probability function measures the goodness of a statistical model for given parameter values. In mathematics, computing and especially in graph theory, the distance matrix is a square matrix (two-dimensional array) that contains the distances between pairs, or elements of a set. If there are $N$ elements, the size of the matrix will be $N x N$. The data can be displayed in graphical form as a heat map.

As it can be seen in Figure 3, a distance matrix was created based on a randomly formed map of random cities. There are several types of heat maps. It is possible to define a heat map as a map of five colors: red, green, blue, yellow and black. The user has the ability to change the properties of the map, where using different functions will change the appearance of the map, create edges on the map, interpolate colors with each other and the like. Testing of the genetic algorithm was performed on the example of a randomly created map of 30 cities. The initial parameters are programmatically defined in Figure 4.

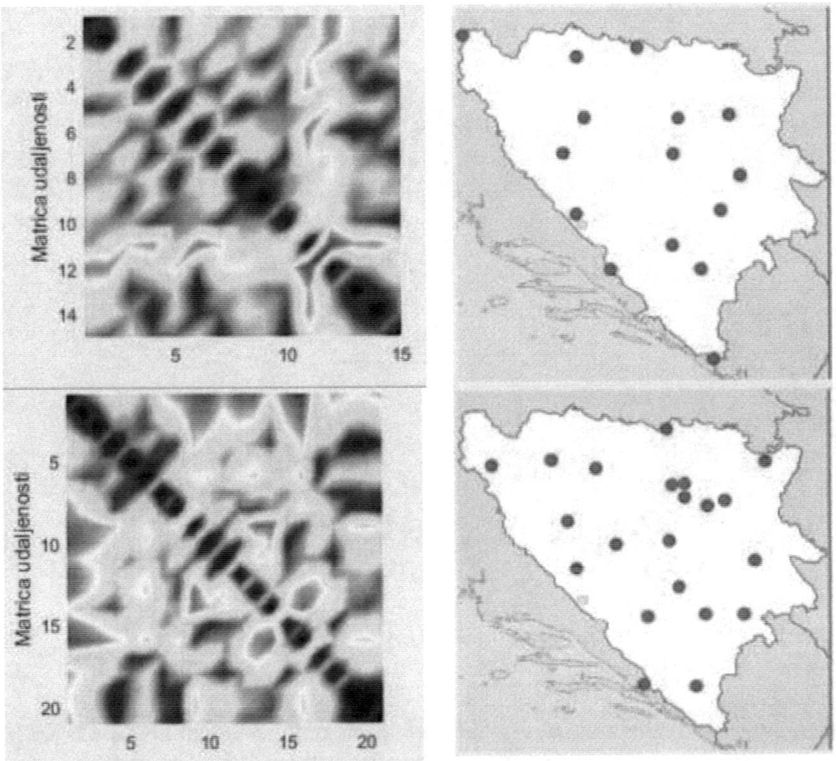


**Figure 3.** Example of heat maps based on two created maps.

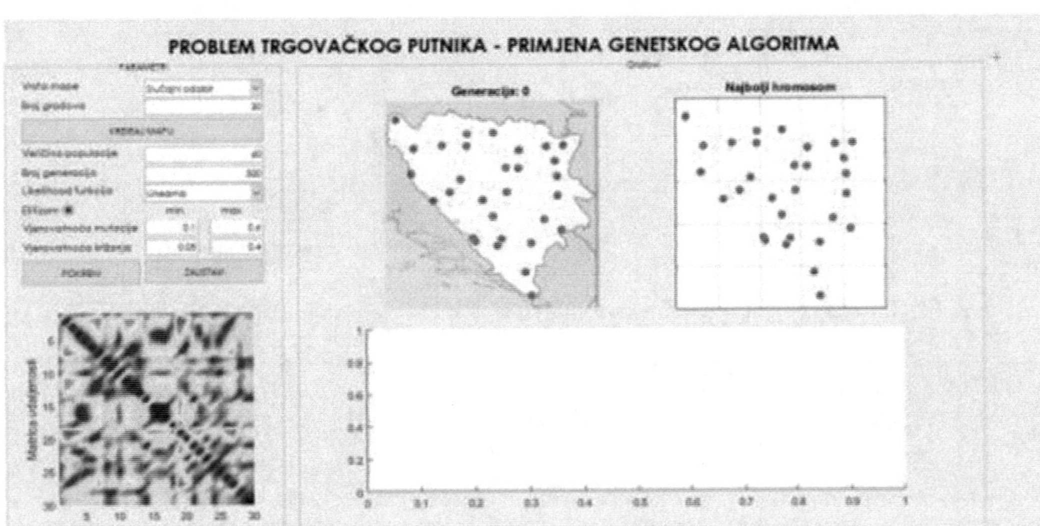


**Figure 4.** Creating randomly selected 30 cities for path optimization using a genetic algorithm.

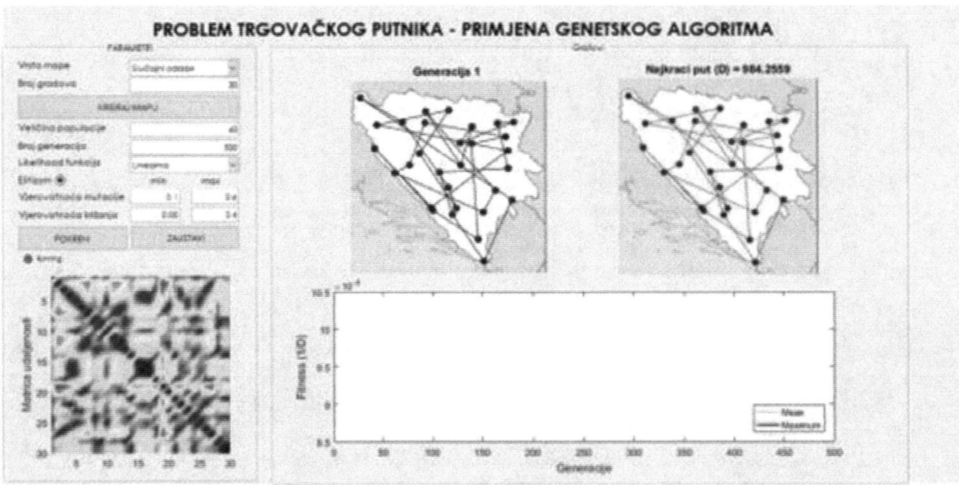


**Figure 5.** Changing the path of the travelling salesman through iterations for a randomly created map of 30 cities.

By running the program, the initial locations are drawn with mutual paths that will change through iterations all the way to a certain number of given generations, trying to reach the ideal path as shown in Figure 5. Figure 6 shows the completed process of optimizing the route of a commercial traveller on the example of a randomly created map of 30 cities whose parameters are population size: 60 and number of generations: 300. The minimum distance between the given locations is 543.1525. The obtained results do not represent the optimal path and visually do not represent the best solution, therefore the parameters will be changed.

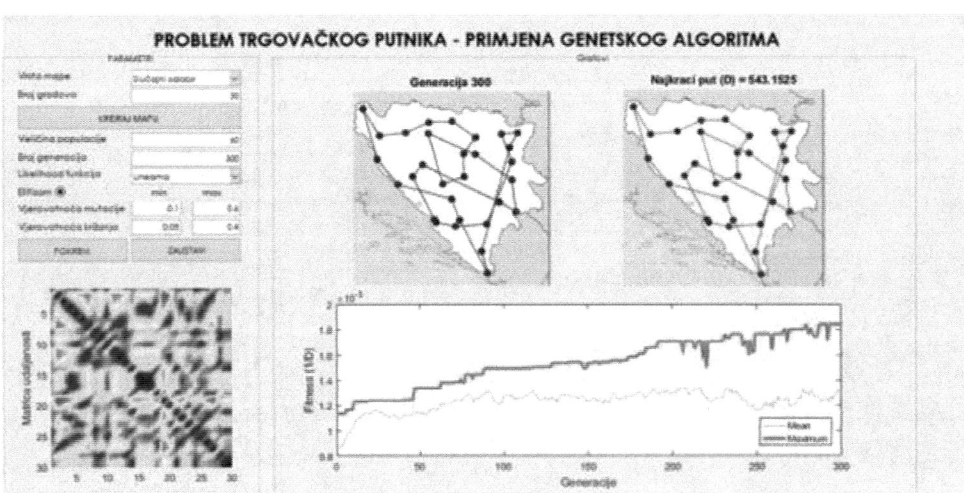


**Figure 6.** Completed process of optimizing the route of the travelling salesman on the example of a randomly created map of 30 cities.

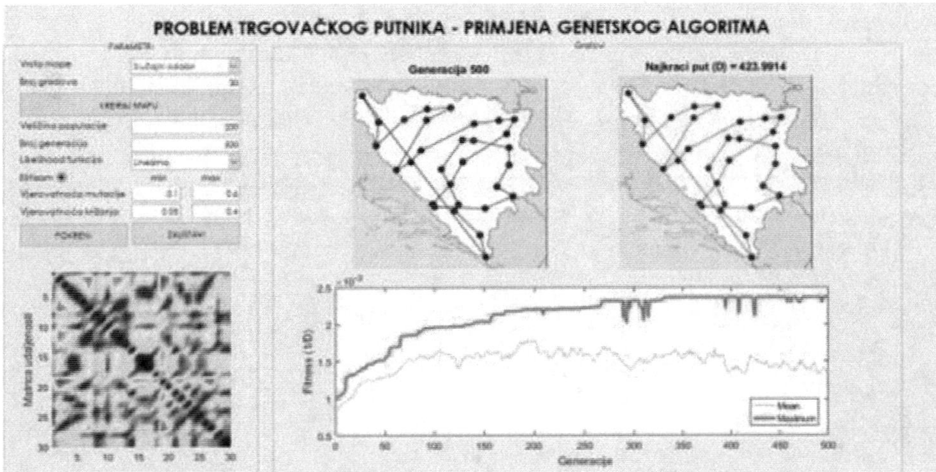


**Figure 7.** Changing path optimization parameters on the example of a randomly created map.

By adjusting the parameters of population size to 200 and the number of generations to 500, the final solution is obtained as in Figure 7 and in comparison with the previously set parameters, a significant reduction in the distance between the points is visible.

## 4. Conclusion

This paper presents a practical implementation of an application for optimizing the path of a travelling salesman using a genetic algorithm in the software package MATLAB. The travelling salesman problem and the parameters used in the realization are described, and the cases of use in the work of the application whose solutions are graphically presented. Genetic algorithms represent an easy approach to complex large space search problems. Using models of learning and evolution, they enable modelling of the desired properties of the system, and automatic finding of solutions. In the described optimized cases, it is shown that abstract models of the genetic algorithm can be applied to real problems from different areas. The implementation of GA involves the use of various parameters: population size, number of generations (number of iterations, repetitions) and the probability of mutation and the probability of crossing. Genetic algorithms have various advantages and disadvantages that make them more or less applicable to various problems. The disadvantage of the genetic algorithm is that there is no single combination of parameters that is best for all problems, or at least for different instances of the same problem, but in each case we have to adjust them experimentally [7]. The achieved results, although they show satisfactory solutions, do not always ensure finding the optimal solution. As the commercial traveler problem is an NP-complete problem, and the genetic algorithm has proven to be suitable for solving this problem, it is obvious that the genetic algorithm is suitable for solving related problems from the class of NP-complete problems, provided that the problem can be represented according to the genetic algorithm.

## References

[1] Matai R, Singh S and Mittal M L 2010 Traveling salesman problem: An overview of applications, formulations, and solution approaches *IntechOpen*

[2] Iqbal Ali, A., & Kennington, J. L. (1986). The asymmetric M-travelling salesmen problem: A Duality based branch-and-bound algorithm. *Discrete Applied Mathematics*, *13*(2-3), 259–276. https://doi.org/10.1016/0166-218x(86)90087-9

[3] Applegate, D. L. (2007). *The traveling salesman problem: A computational story*. Princeton Univ. Press.

[4] Golemac, A., & Vojković, T. (1970, January 1). *Kako pomoći trgovačkom Putniku Velga*

*Bosančić*. Semantic Scholar. Retrieved October 13, 2021.
[5] Teodorović Dušan. (2007). *Transportne mreže*. Univerzitet u Beogradu, Saobračajni fakultet.
[6] Azra Hošić, (2021). *Optimizacija putanje trgovačkog putnika primjenom genetskog algoritma u programskom paketu Matlab*, diplomski rad, Univerzitet u Bihaću, Tehnički fakultet. (in Bosnian)
[7] Fatka Kulenović, (2014). *Metode matematičkog programiranja za rješavanje lokacijskih i alokacijskih problema*, Doktorska disertacija, Univerzitet u Sarajevu, PMF. (in Bosnian)

IOP Conf. Series: Materials Science and Engineering 1208 (2021) 012033 doi:10.1088/1757-899X/1208/1/012033

# Fire prediction with logistic regression on territory of Bosnia and Herzegovina

**Mursel Musabašić, Denis Mušić and Elmir Babović**

"Džemal Bijedić" University of Mostar, Faculty of Information Technologies

{mursel.musabasic}{denis}{elmir.babovic}@edu.fit.ba

**Abstract.** The Canadian Fire Weather Index system [1] has been used worldwide by many countries as classic approach in fire prediction. It represents system that account for the effects of fuel moisture and weather conditions on fire behaviour. It numerical outputs are based on calculation of four meteorological elements: air temperature, relative humidity, wind speed and precipitation in last 24h. In this paper meteorological data in combination with Canadian Fire Weather Index system (CFWI) components is used as input to predict fire occurrence using logistic regression model. As logistic regression is a supervised machine learning method it's based on user input in the form of dataset. Dataset is collected using NASA GES DISC Giovanni web-based application in the form of daily area-averaged time series in period of 31.7.2010 to 31.7.2020, it's analysed and pre-processed before it is used as input for logit model. CFWI components values are not imported but calculated in run-time based on pre-processed meteorological data. As a result of this research windows application was developed to assist fire managers and all those involved in studying the fire behaviour.

## 1. Introduction

Fire can be a process caused by a natural sequence of events or it can occur as a human-caused activity. Fire also represents one of the key elements in forest renewal and succession. Beside it's natural behaviours, fire is global issue which affects, not only ecological and economical environments, but also affects all living things. How great the danger of fire is can be shown through recent events in the world. During 2019 – 2020 Australian bushfires burnt over 12 million hectares, destroying thousands of homes, and killing at least 30 people [2]. There were over hundreds of millions wild animals killed. In California 2020 fire season there were almost ten thousand fires detected resulting in at least 30 deaths and over four million hectares burnt [3]. Thousands of buildings were destroyed. Economists estimated that both countries suffered cost over hundreds of million US dollars.

Although smaller in territorial size, Bosnia and Herzegovina has problems with fires during it's fire seasons. In period of 2010 to 2020, European Forest Fire Information System (EFFIS) established by Joint Research Centre (JRC) and Directorate General for Environment of the European Commission EC) [4] reported at least 970 number of fires resulting in 380.000 hectares of burnt area (about 7.5% of country size).

Of the total burnt area, over 30 thousand hectares fall on agriculture areas, about 145 thousand hectares of forest areas, and over 100 thousand on other natural land and artificial surfaces. Worst fire seasons were recorded in 2012, 2017, 2019 and 2020 with 767 number of fires and about 305 thousand hectares of burnt area.

**Table 1.** Last decade fire seasons statistics for Bosnia & Herzegovina

| Year | Burnt areas (ha) | Number of fires |
|---|---|---|
| 2010 | 3692 | 10 |
| 2011 | 23676 | 41 |
| 2012 | 92809 | 147 |
| 2013 | 3616 | 10 |
| 2014 | 3203 | 5 |
| 2015 | 15318 | 26 |
| 2016 | 34538 | 89 |
| 2017 | 83449 | 146 |
| 2018 | 3155 | 12 |
| 2019 | 28839 | 134 |
| 2020 | 100107 | 340 |
| Total | 392402 | 960 |

Fire season of 2020 recorded the greatest number of fires and areas burnt. A reason for this number of fires can be found in the situation where the population left urban areas and went to rural areas thus escaping possible corona-virus infection. From an economic aspect, the fires that occurred in the last decade have cost Bosnia and Herzegovina almost 60 million US dollars.

Even 25 years after the war country is struggling with formal strategic plans in fire prevention. In addition to the fact that firefighters use old equipment, there are almost no indications that the early warning system will improve soon [5]. The aim of this study is to show that it is possible to provide at least a step towards better fire risk management with the aid of supervised machine learning methods.

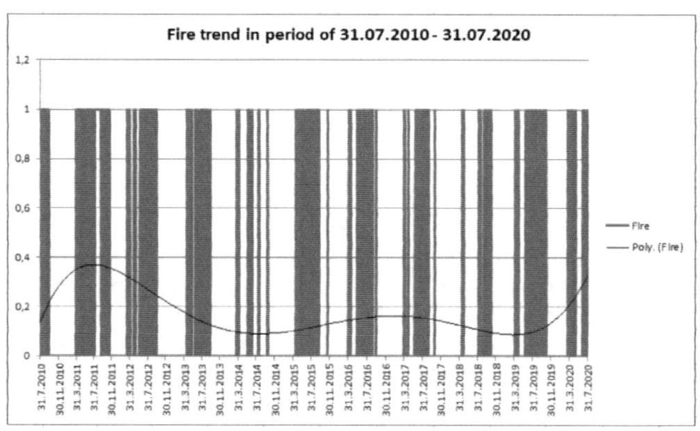


**Figure 1.** Fire trend shows faster grow during 2020.

## 2. Data and methods

### *2.1. Data preparation*

Dataset for logistic regression model consists of three datasets acquired through NASA GESDISC Giovanni web application. Air temperature (C°) and relative humidity (%) are collected from AIRS3STD dataset [6]. Precipitation (mm) is collected from GPM_3IMERGDL_06 dataset [7] and wind speed (m/s) from M2T1NXFLX_5_12_4 dataset [8]. Air temperature, relative humidity and precipitation raw values are in daily temporal resolution as time series, area averaged. Wind speed data is based on hourly temporal resolution, and it is subsequently converted to daily temporal resolution including unit of measure conversion from m/s to km/h. Time period of dataset ranges from 31.7.2010 to 31.7.2020 counting in 3654 rows where every row corresponds to single day. There are two phase of data preparation. The first step is done using Microsoft Excel, but any other spreadsheet application can be used, and second through application. During data loading Canadian Fire Weather Index components are calculated. As many other countries Federal Hydrometeorological Institute of B&H also use CFWI system [9] to calculate six components through WRF-NMM model created by NOAA's National Weather Service. First three components (FFMC, DMC and DC) are so-called „fuel moisture codes“ and other three components (ISI, BUI and FWI) are called „fire behaviour indices“ (Wagner 1987). Every code and index is represented by numerical value. All six components can contain own scale of relative values, but they are structured in way to indicate potential fire burning behaviour. With higher values of fuel moisture codes, the lower are the values of indices and vice versa.

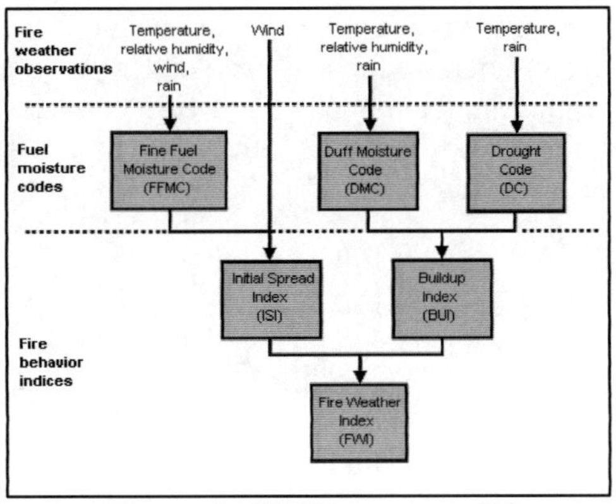


**Figure 2.** CFWI structure [1]

FWI (Fire weather index) is the result based on previous components calculation outputs. As per EFFIS, FWI output value is divided in six fire danger classes as shown in Table 2.

**Table 2.** Fire danger classes per EFFIS

| Fire danger classes | FWI value ranges |
|---|---|
| Very low | < 5.2 |
| Low | 5.2 – 11.2 |
| Moderate | 11.2 – 21.3 |
| High | 21.3 - 38 |
| Very high | 38 - 50 |
| Extreme | > 50 |

In final step of data preparation output values for dependent variable „Fire“ are calculated based on following condition: if final value of FWI is over or equal 11.2 it will generate 1 as fire present indication else 0 as there is no fire present.

*2.2. Model training*

By default, application is designed to split prepared dataset in ratio 80:20 within which 80% of data is used as training set and 20% of data is used for model testing. There is option for users to input custom trainset size in range 0.0 to 1.0. Besides splitting data in train and test dataset's, there is an option for user to get insight of correlations between independent variables. Before training user needs to select appropriate columns as input for a logistic model.

*2.3. Logistic regression*

Logistic regression - is a method coming from classic statistics. It is used to model binary dependent variables. With logistic regression we can predict if something is true or false, right or wrong, is someone ill or not, 1 or 0, etc. Logistic regression assumes existence of linear relationship between dependent variable and independent variables [10]. This relationship is expressed by equation as shown below:

$$\log\left(\frac{p}{1-p}\right) = \beta_0 + \beta_1 X_1 + \cdots + \beta_m X_m$$

$$p - probability\ of\ event\ being\ true \tag{1}$$
$$\beta_0 - bias$$
$$\beta_{1...m} - beta\ coefficients$$
$$X_{1...m} - independent\ variables, predictors$$

Left of equation is called log of odds and its main step in linear transformation from non-linear form. On the right side we have beta coefficients which are also called weights and are in direct association with independent variables. Weights are real numbers and give importance to independent variables in classification decision and determine how predictors influence on classification decision if is something true or false. This linear form is written as:

$$z = \left(\sum_{i=1}^{m} w_i x_i\right) + b$$
$$w_i - weights \tag{2}$$
$$x_i - predictors$$
$$b - bias$$

To be able to solve linear combination of predictors and weights (2) we need a function which is called sigmoid or logistic function:

$$y = \sigma(z) = 1/(1 + e\hat{}(-z)) \tag{3}$$

Logistic function outputs values in range from 0 to 1. The sigmoid function results in effect that if one is divided by a big number, it will get closer to zero, and if it's divided by number a little bit bigger than one it will get closer to one. Sigmoid function is used to map predictions values to probabilities values.

### 2.4. Calculating the parameters

To find best parameters for our model we use maximum likelihood estimation. First, we determine log-likelihood function then we find our weights. As sigmoid function (3) outputs our values as probabilities we can interpret them as Bernoullies variables, Y = Ber(p), where:

$$p = \sigma(\omega^T x) \tag{4}$$

Given Bernoulli probability function for single point:

$$P(Y = y|X = x) = p^y (1-p)^{(1-y)} \tag{5}$$

we can write likelihood function for complete data as:

$$L(\omega) = \prod_{i=1}^{n} P(Y = y_i|X = x_i) = \prod_{i=1}^{n} p^{y_i}(1-p)^{(1-y_i)}, p = \sigma(\omega^T x_i) \tag{6}$$

The reason we are multiplying all the data points is that we are assuming they are independent and not related (0/1, fire/no fire, etc). So, we need to maximize our likelihood function and to avoid multiplication form we take log of likelihood function.

(7)

$$LL(\omega) = \sum_{i=1}^{n} y_i \, log(p) + (1 - y_i) \, log(1 - p), p = \sigma(\omega^T x_i)$$

As we want to maximize our log-likelihood we take derivative of our function and move in direction of gradient using gradient ascent algorithm.

$$\frac{\partial LL(\omega)}{\partial(\omega)} = (y - p)x, p = \sigma(\omega^T x) \tag{8}$$

Taking in account derivative of LL our update weight rule becomes like this:

$$\omega_{i+1} = \omega_i + \frac{\partial LL(\omega)}{\partial(\omega)} \alpha \tag{9}$$
$$\alpha = \text{learning rate, scaling size}$$

To practically solve our problem and as it's more convenient, update weight rule is calculated through our software application iteratively including the creation of complete logit model.

*2.5. Fire prediction software application*

Specifically, for this study we have created software application which can be used by people who are involved in wildfires research area. In this early version of software application, users can choose multiple options in fire analysis/prediction, such as correlation, maximum number of iterations, step size/learning rate, regularization penalty (only L2), data split, data shuffle (Fisher-Yates method) by iteration and data normalization using Min-Max scaler.

Besides training process setup application results contains values for: cost, McFadden pseudo-R2, accuracy on train/test data, sensitivity on train/test data, confusion matrix, AIC and BIC model selection methods and final regression equation.

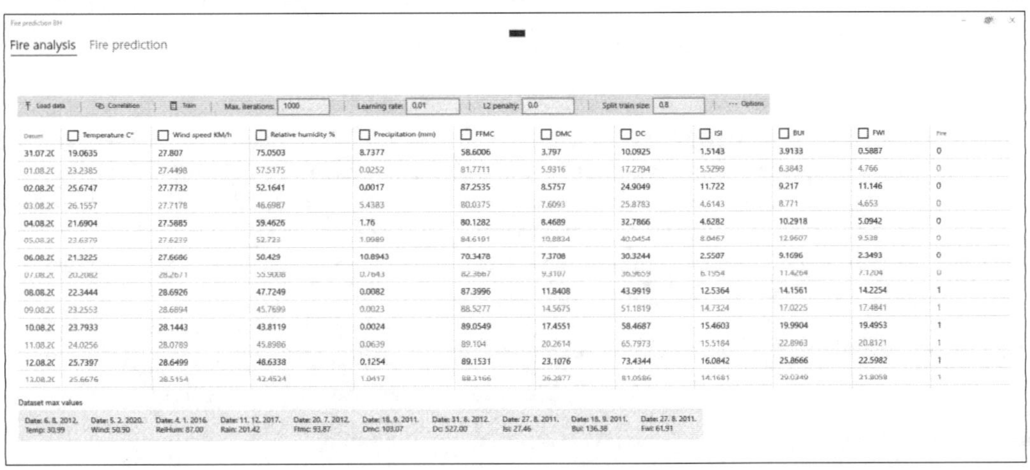


**Figure 3.** Application main screen

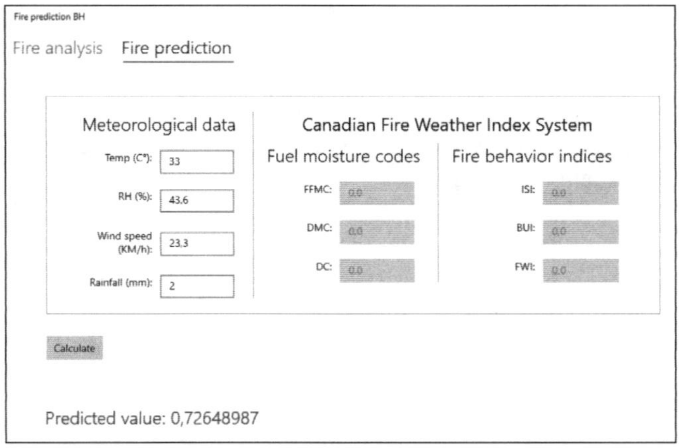


**Figure 4.** Fire prediction input

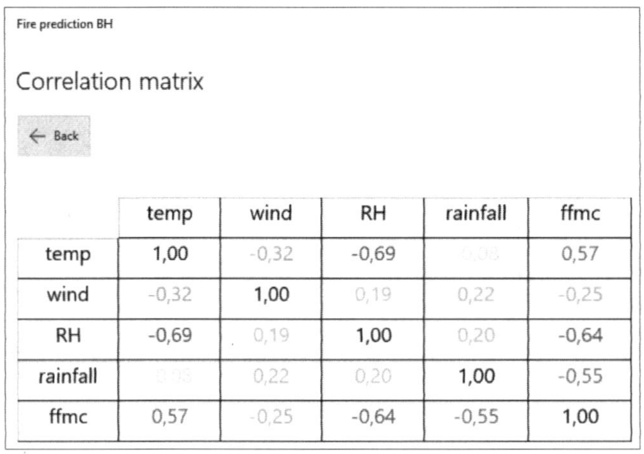


| | temp | wind | RH | rainfall | ffmc |
|---|---|---|---|---|---|
| temp | 1,00 | -0,32 | -0,69 | [illegible] | 0,57 |
| wind | -0,32 | 1,00 | 0,19 | 0,22 | -0,25 |
| RH | -0,69 | 0,19 | 1,00 | 0,20 | -0,64 |
| rainfall | [illegible] | 0,22 | 0,20 | 1,00 | -0,55 |
| ffmc | 0,57 | -0,25 | -0,64 | -0,55 | 1,00 |

**Figure 5.** Correlation matrix window

## 3. Results

Three models were compared during model training. In first model (model A) meteorological data is combined with fuel moisture code components (FFMC, DMC and DC), second model (model B) is combination of meteorological data and fire behaviour indices (BUI, ISI and FWI) and the third model (model C) includes only meteorological data (temp, RH, wind, and rainfall). As rainfall values deviate in higher percent then others variable values data needed to be normalized before training process. Following table shows prediction power of tested models:

**Table 3.** Model training results

| | Model A | Model B | Model C |
|---|---|---|---|
| Cost | 0,1567 | 0,0813 | 0,1969 |
| McFadden $R^2$ | 0,6691 | 0,8283 | 0,5841 |
| Accuracy on trainset | 0,93 | 0,97 | 0,92 |
| Accuracy on testset | 0,62 | 0,86 | 0,88 |
| Sensitivity / Specificity (test) | 0,63 / 0,94 | 0,70 / 0,95 | 0,58 / 0,94 |
| AIC | 929,78 | 489,17 | 1158,93 |
| BIC | 971,64 | 531,03 | 1182,85 |

Looking at the Table 3. we can conclude that model B have most desirable results. Compared to other models it's loss function returned smallest error value. With largest pseudo-R2 value and by sensitivity and specificity percentages which are derived from confusion matrix, shows more accurate

IOP Conf. Series: Materials Science and Engineering 1208 (2021) 012033 doi:10.1088/1757-899X/1208/1/012033

prediction compared to model A and C and therefore the best-chosen logistic regression model. AIC and BIC selection methods returned lowest values for model B and thus confirm the above statements. These results tell the fire manager that fire behaviour indices (including wind as weather element) have most significant impact on the occurrence of fire and its behaviour.

## 4. Conclusion

„All models are wrong, but some are useful“. This citation is attributed to the British statistician George Box. It means that models will always be wrong because of the complexity of the nature of the things being modelled, but they can be useful. Although machine learning methods can be very helpful, they still require a larger amount of data to create models as accurately as possible.

This means that countries (especially Bosnia & Herzegovina) must in some way establish a stable and reliable network of many meteorological stations where measurements will be made constantly, and the stored data will be used for various types of model evolution.

Despite the above, the model created as part of the research, which is, as stated in the research, based on average satellite data for 4 basic meteorological elements, the evaluation of the model still showed almost 87% accuracy. This percentage is not so small, if we consider missing variables which are almost always taken into account when evaluating fires, such as distances from populated areas, vegetation index, elevation, forest types, etc.

## References

[1] Canada N R Canadian wildland fire information system: Canadian forest fire weather index (FWI) system [Internet]. Canadian Wildland Fire Information System | Canadian Forest Fire Weather Index (FWI) System [cited 2020Oct12] https://cwfis.cfs.nrcan.gc.ca/background/summary/fwi

[2] Elisabeth P L Y 2020 The 2019-2020 Australian bushfires IDMC 2020 [cited 2020Oct15]. https://www.internal-displacement.org/publications/the-2019-2020-australian-bushfires-from-temporary-evacuation-to-longer-term

[3] California Department of Forestry and Fire Protection (CAL FIRE) 2020 incident archive Cal Fire Department of Forestry and Fire Protection [cited 2020Oct17] https://www.fire.ca.gov/incidents/2020/

[4] European Forest Fire Information System Gwis statistics [cited 2020Oct5] https://effis.jrc.ec.europa.eu/apps/gwis.statistics.portal

[5] Fire risk management: *UNDP in Bosnia and Herzegovina* [Internet] UNDP 2020 [cited 2020Oct10] https://www.ba.undp.org/content/bosnia_and_herzegovina/en/home/climate-and-disaster-resilience/FireRisk.html

[6] AIRS S T and Teixeira J 2013 AIRS/Aqua L3 Daily Standard Physical Retrieval (AIRS-only) 1 degree x 1 degree V006 (Greenbelt: Goddard Earth Sciences Data and Information Services Centre - GES DISC) [cited 2020Sep10] doi:10.5067/Aqua/AIRS/DATA303

[7] Huffman G J, Stocker E F, Bolvin D T, Nelkin E J and Jackson T 2019 GPM IMERG Late Precipitation L3 1 day 0.1 degree x 0.1 degree V06 (Greenbelt: Goddard Earth Sciences Data and Information Services Centre - GES DISC) [cited 2020Sep12] doi:10.5067/GPM/IMERGDL/DAY/06

[8] Global Modelling and Assimilation Office 2015 MERRA-2 tavg1_2d_flx_Nx: 2d, 1-Hourly, Time-Averaged, Single-Level, Assimilation, Surface Flux Diagnostics V5.12.4 (Greenbelt: Goddard Earth Sciences Data and Information Services Centre - GES DISC) [cited 2020Oct6] doi:10.5067/7MCPBJ41Y0K6

[9] Federal Hydrometeorological Institute of B&H Prognoze Index opasnosti od požara – FWI [cited 2020Oct20] http://www.fhmzbih.gov.ba/latinica/AGRO/pozar.php

[10] Sampson M G 2018 *Statistical Analysis in JASP – A Students Guide* Logistic regression 2nd Edition JASP v0.9.1 p 70

IOP Conf. Series: Materials Science and Engineering 1208 (2021) 012034 doi:10.1088/1757-899X/1208/1/012034

# Remote control and monitoring in smart homes using voice commands

**Azra Vojić, Edin Mujčić and Una Drakulić**

University of Bihać, Faculty of Technical Engineering Bihać, Irfana Ljubijankića bb

E-mail: una.94@hotmail.com

**Abstract**. With the development of modern technology, smartphones have become a necessity for most people. Among other uses, mobile phones are increasingly being used in smart home systems. In smart homes, mobile phones are used to remotely control and monitor various systems such as simply turning on/off lights and household appliances, various monitoring systems, etc. Nowadays, sending coded messages or pressing application buttons is increasingly being avoided in process of developing smart systems. More and more frequently is used voice commands. The system which uses voice commands for remote control and monitoring smart home is described in this paper. In the implemented system, the user is able, using specific voice commands to remotely control the operation of various appliances in his home. An Android application was designed to control the implemented system. Using the designed Android application, the user controls the desired home devices with specific voice commands. Also, on the designed Android application are buttons that the user can use, in case the user's voice is not recognized in the implemented system. For experimental work analysis, the model of the home is made with lights and different home appliances inside. The results of the experimental work analysis of the implemented system show this system is very simple to use and very efficient. Also, the latest technology for remote control and monitor smart systems is applied in the proposed smart home system.

## 1. Introduction

Home automation is the automatic or semi-automatic control and monitoring of household appliances and home features like doors, gates, windows, and etc. [1]. Home automation is used to provide convenience for users to remotely control and monitor the appliances and provides better use of electricity. The increased use of PCs, smart devices, and wireless technology makes it easy for a user to remotely control and monitor automated systems. A lot of research has been done for smart homes. Some of the solutions are using the Internet, wireless technology, others are using Bluetooth and GSM technology for controlling the home appliances [2]. Controlling the device or home appliances through the voice command represents the smart, modern and user-friendly system [3]. There are various home automation solutions based on voice control on the market today, which to a greater or lesser extent suit the end-user. The most known voice command systems are Amazon's Alexa and Apple's Siri, Google Home, Home Assistant, HAL, OpenHAB, Crestron, Amazon Echo, Castle OS, etc. These systems are becoming more popular way of voice control and monitor different systems. Voice-control home automation systems offer people a more comfortable lifestyle and simplify ordinary tasks [4]. According to the major companies involved in voice recognition research, the voice will be the main interface between humans and machines in the near future [2]. Such systems are especially suitable for people with disabilities and the elderly. Therefore, a home automation system with voice control has been designed so that users can perform certain tasks using voice over their smartphones. The entire system is controlled via smartphone application. In this paper is designed and implemented the smart home system which uses voice commands for remote control and monitor smart home. This smart home system enables the user to use his voice to remotely control the operation of different appliances in his home. For voice recognition of specific commands is created the android application. Using the created android application, the user controls the desired home devices with specific voice commands. Also, on the created android application are buttons that the user can use, in case the user's voice is not recognized. For experimental work analysis, the model of the home is made with indoor and outdoor lights and different home appliances placed inside.

## 2. Smart homes using voice commands

Home automation is becoming popular due to its numerous benefits. Home automation refers to the control of home appliances by local networking or by remote control [3]. A voice-controlled home automation system is a specific automation system. Using Natural Language Processing and the accessible hardware in most smartphones enables to use for controlling electrical equipment [4]. The first speech recognition system, named Audrey, was created by Bell Laboratories in 1952. That system was understanding only ten digits spoken by particular people. Ten years later, IBM developed and demonstrated their Shoebox Machine. The device recognized and responded to 16 different spoken words, including all ten digits “0” to “9” as well as calculating commands such as “plus” or “minus”. In the 1970s is developed the Hidden Markov model. In this model, speech recognition started using a statistical method measuring the probability of unknown sounds. Now, the potential to recognize an unlimited number of words became imminent due to the method allowing the number of understandable words to go up to a few thousand. These choices of observation distribution in each state of the model allow accurate modeling of virtually unlimited types of data. The first mass accessible voice command system was launched by Apple Inc. as they released the virtual assistant named Siri in 2011 [4]. There are several commercial and research projects on smart homes and voice recognition systems. One of them is an integrated platform for home security, monitoring, and automation SMA from uControl. This system is a 7-inch touchscreen that can wirelessly be connected to security alarms and other home appliances. The home automation through this system requires holding and interacting with a large panel which constraints the physical movements of the user. Another popular commercially available system for home automation is Home Automated Living HAL. This software uses the power of an existing PC to control the home. It provides a speech command interface. A big advantage of this system is it can send commands all over the house using the existing highway of electrical wires inside the walls of the house. No new wires mean HAL is easy and inexpensive to install. However, most of these products sold in the market are heavily priced and often require a significant home makeover [5].

## 3. Design and implementation of the remote control and monitoring in smart homes using voice commands

In this part of the paper, a system for remote control and monitoring in smart homes using voice commands is designed. For experimental work analysis, the system is implemented in a home model. For control and monitoring the smart home, the android application is created. MIT App Inventor is used to create the android application. MIT App Inventor is a software environment that allows the creation of fully functional applications for smartphones [6]. First is created a graphical appearance for the android application and enabled communication with the smart home system. After that is programmed the created application for recognize voice commands. It is also programmed that the system can return information to the user when it performs a given task. The *SpeechRecognizer* and *TextToSpeech* blocks are used for that. The SpeechRecognizer component is used to listen, recognize voices and convert them to text. The *TextToSpeech* component pronounces the default text aloud and confirms to the user that it will perform the given task. The voice color and pitch of the pronunciation can be adjusted. In case of not recognizing voice commands, the user is enabled to press the button next to the command on the android application to perform desired tasks. The appearance of the created android application is shown in Figure 1.

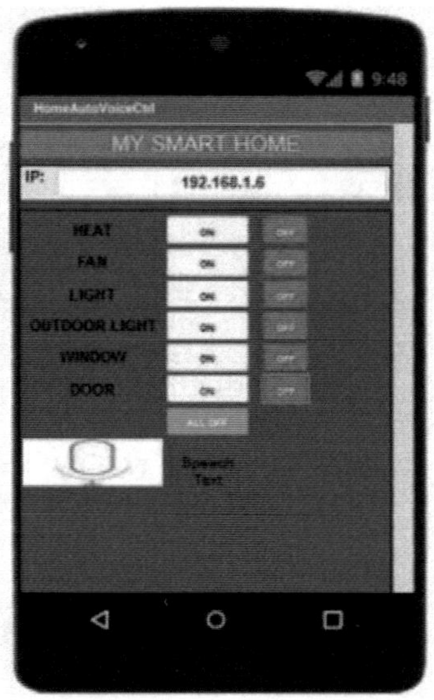


**Figure 1.** The appearance of the created android application

The android application sends the recognized voice command converted to text wirelessly (WiFi network) to the nodeMcu ESP8266 module. This module performs the tasks based on the received voice command. In the proposed smart home system, the user, using voice commands, is able to control: heating, indoor and outdoor lighting, open and close doors and windows. Expansion of this system can be done in a simple way by adding desired elements to control. Figure 2 shows a functional block diagram for remote control and monitoring in smart homes using voice commands system.

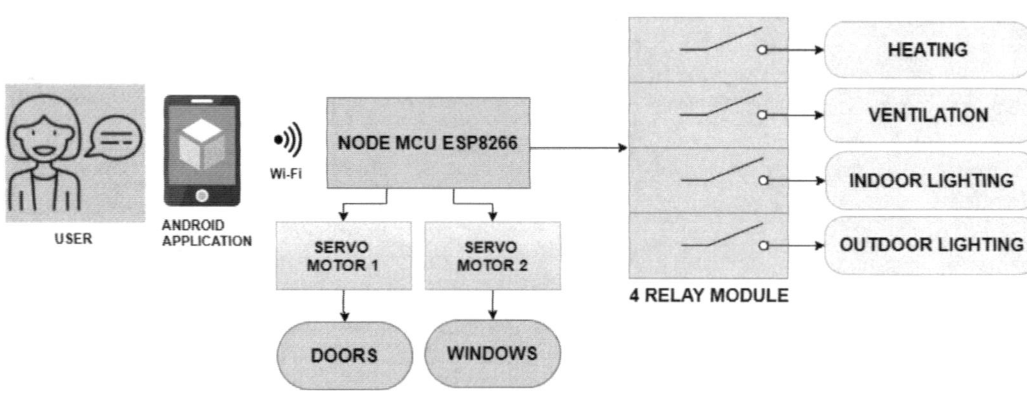


**Figure 2.** The functional block diagram of the remote control and monitoring in smart homes using voice commands system

4 Relay Module contains EL817 optocoupler for signal isolation and amplification with NodeMCU and four relays. Working voltage is 5 V, maximum load AC 250 V / 10 A, and DC 30 V /

10 A. From Figure 2 can see that NodeMCU esp8266 acts on individual elements of the smart home based on the received voice commands. Servo motors are used as actuators for opening and closing doors and windows. The relays are used for switching on and off individual elements of the smart home model. For the experimental work analysis of the remote control and monitoring in smart homes using voice commands the model of the home is made. The made home model is size 80x60 cm, and is shown in Figure 3.

**Figure 3.** The model of the smart home (a) appearance of the smart home model, (b) the hardware part of the smart home model

## 4. Experimental work analysis of the remote control and monitoring in smart homes using voice commands

For the first experimental work analysis it tested the voice command for turning the heater on and off (see Figure 4).

**Figure 4.** The experimental work analysis of the heater

The required data of the remote control and monitoring system of the heating control subsystem is the value of the current room temperature from the temperature measuring sensor. The exact value of the temperature given by the sensor inside the smart home can be read on the LCD display, and thus determine whether it is necessary to turn the heating on or off. The voice command "HEAT ON" is used on the mobile phone. When the voice command is spoken to the android application, after one second we can notice that the heating has been turned on in the model of the home. When we want to turn off the heating, we do it with the voice command "HEAT OFF". In case the system does not recognize the voice, the user can turn on / off the heating manually on the Android application. For the second experimental work analysis we tested the voice command for turning the fen on and off (see Figure 5).

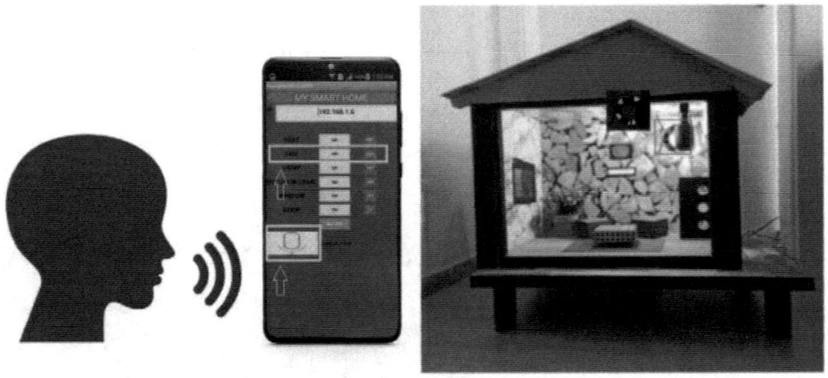

**Figure 5.** The experimental work analysis of the fen

Voice commands for controlling the fen are 'FAN ON' and 'FAN OFF'. Also, on the Android application are buttons for manually controlling the fan. For the third experimental work analysis we tested the voice command for turning the outdoor lights on and off (see Figure 6).

**Figure 6.** The experimental work analysis of the indoor lighting

Voice commands for controlling the indoor lighting are 'LIGHT ON' and 'LIGHT OFF'. Also, on the Android application are buttons for manually controlling the indoor lights. Also, the outdoor lights are able to control this way using voice commands' OUTDOOR LIGHT ON' and 'OUTDOOR LIGHT OFF' (see Figure 7).

**Figure 7.** The experimental work analysis of the outdoor lighting

After experimental analysis and detailed testing of the smart home system which is controlled using voice commands, we can conclude that the system is harmonized, functional, and user-friendly. Also, according to the needs of different users, this system has the possibility of connecting additional subsystems for voice controlling, without ruining the operation of the implemented ones.

## 5. Conclusion

In this paper, the system for remote control and monitoring of smart homes using voice commands is designed. The designed system uses smartphones that have become irreplaceable in the modern world, and voice commands that represent the future for human-machine communication. An android application is created. The use of the created android application is very simple. It is enough to open the android application and press the button "Speech text" and say the desired command. In case of not recognizing the voice command or staying in places where it is not appropriate to speak, the user can press the button on the application. These buttons are placed on the android application, so the user can perform desired tasks if not able to use his voice. The smart home model is made for experimental proof of the proper work of the designed smart home system. Application in the real smart home does not require any additional system upgrades because relays and higher power servomotors are used at the output. Experimental work analysis of the designed smart home system is performed. Based on the results of the performed experimental work analysis can be concluded that the designed smart home system works correctly. Of the 100 random voice commands, all were recognized and tasks were performed correctly. Based on the above, can be concluded that the designed smart home system works well and can be used for voice control of desired elements in a real smart homes.

## References

[1] Arham S, Roshan K, Latika C, Sheha N and Pushpa U 2015 Voice Controlled Smart Home System *International Journal of Electrical, Electronics and Data Communication* (vol 3, Issue 4)

[2] Amrutha S, Aravind S, Ansu M, Swathy S and Rajasree R P 2015 Voice Controlled Smart Home *International Journal of Emerging Technology and Advanced Engineering* (vol 5, Issue 1)

[3] Škrgić M, Drakulić U and Mujčić E 2018 Smart Home System - Remote Monitoring and Control Using Mobile Phones *International symposium on information and communication technologies ISICT (BHAAAS)*

[4] Ch.Venkateswara R, Anirvan S, Debasish T and Klanidhi K 2016 Smart home wquipped with voice recognition *International Journal of Industrial Elektronics and Electrical Engineering* (vol 4, Issue 4)

[5] Mujčić E, Drakulić U and Škrgić M 2016 Advertising LED system using PIC18F4550 microcontroller and LED lighting *International symposium on information and communication technologies ISICT (BHAAAS)*
[6] Patton E W, Tissenbaum M and Harunani F 2019 MIT app inventor: Objectives, design, and development *Computational Thinking Education* 31–49

# A web application for remote control of ROS robot based on WebSocket protocol and Django development environment

**Zinaid Kapić, Aladin Crnkić, Edin Mujčić and Jasna Hamzabegović**
University of Bihać, Faculty of Technical Engineering, Bosnia and Herzegovina

E-mail: zinaid.kapic@unbi.ba

**Abstract**. The development of teleoperation systems, robots, or any physical part of the system can be costly and if something goes wrong it can lead to development overdue. Precisely for these reasons, engineers and scientists today resort to the development of simulated systems before the construction of a real system. Robot Operating System (ROS) is one of the most popular solutions for robot development, manipulation, and simulation. In this paper, we present a web application for remote control of a ROS robot. The robot is controlled via a web application that is used as a virtual Joystick. Also, in this paper, an experimental work analysis of the projected system is performed. Further research possibilities include upgrading the presented web interface, adding certain motion autonomy sensors, or integrating some path planning algorithms.

## 1. Introduction

Teleoperations are present in all spheres of human activity. Many endeavors of human civilization are inconceivable without the presence of teleoperations, ie remote control. Exploring impassable remote places, doing hard work, reducing the physical distance of people, or some everyday activities involve some form of teleoperation [1-3]. The choice of media for teleoperations is extremely important for the development of the teleoperating system, and the Internet is certainly one of the most popular ways of transmitting information [4]. Teleoperations over the Internet also depend on the choice of protocol for transmitting the information. The protocol used in this paper is the popular WebSocket protocol, which has recently been imposed as a standard in working with the Internet. The speed and possibility of full-duplex communication are especially emphasized [5]. The WebSocket protocol is mainly used in applications that include real-time chat applications, live broadcasting, and in recent works, it found its applicability in teleoperation systems and is used for ROS communication [6].

The development of teleoperation systems, robots, or any physical part of the system can be expensive and sometimes time-consuming. Precisely for these reasons, scientists today resort to the development of simulated systems before the construction of a real system. In this way, the system and its environment can be modeled sufficiently to examine the validity of the physical construction of that system. The necessity for this kind of development produced several robotics middleware platforms, such as OROCOS [7], Player [8], YARP [9], ROS [10], etc.

ROS (Robot Operating System) is a meta-operating system with a huge library and support for the development of robots and similar systems. It is this software that in recent years has become synonymous with the development of robots, and is increasingly advancing and gaining in importance

[11]. Since its beginning in 2007, ROS has found application in almost every robotic field, from mining, healthcare, home automation, and others [12-16].

It enables the creation of sufficiently good simulations of various systems, including teleoperations, before the realization of the system itself. It is based on the C ++ and Python programming languages, and the code created within this system can be transferred directly to the physical robot [11]. Through this paper, we have developed an application for direct teleoperation based on the master-slave system. The master system is a joystick internet application that controls the movement of the ROS simulated robot. The communication between a web application and a ROS robot is based on the WebSocket protocol. The application is implemented using Python programming language and the Django development environment.

In the next chapters, we present an implementation process with a brief explanation of all parts used in developing an application and connecting it with a simulation. After that, we display an experimental analysis of our application. Finally, we present a conclusion and possible upgrades of this application.

## 2. Implementation of the proposed system

The scheme of the implemented system is presented in figure 1. As is seen in the figure, the web application is using the internet as a medium for data transfer. Protocol for establishing communication and sending data is a WebSocket protocol. After establishing a successful connection with the ROS server, the user can send commands consisted of Twist structure messages.

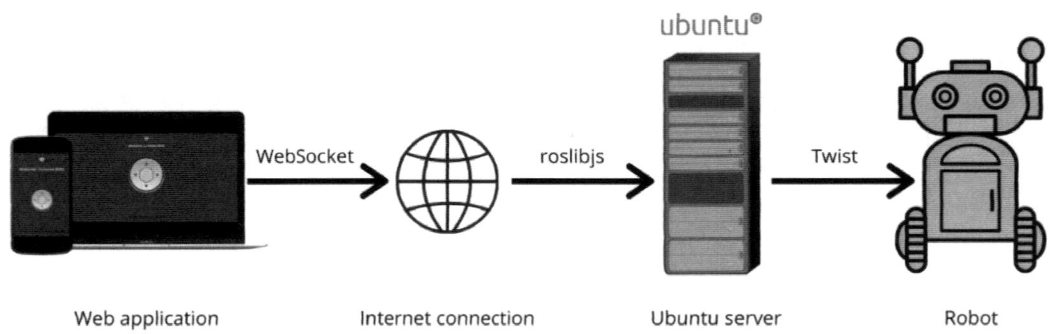


**Figure 1.** Scheme of the implemented system

Messages that are sent consist of two ROS structures, named geometry-msgs/Vector3 that are consisted of linear and angular float64 data. Twist messages are sent from a web application with roslibjs library [17].

The web application is implemented with a Django development environment. Django is a popular framework that uses Python programming language [18]. We have used Django because of the power that Python provides. Python is also a language used by ROS for developing robot functionality. By using Django as a web development tool, we achieve consistency, clean code, and limit a need for knowledge of multiple programming languages.

Besides that, Python exploits many libraries such as OpenCV, Numpy, Panadas, gmapping, PCL, and others [19]. These libraries are useful when developing simple or advanced web, mobile, or other applications integrated with ROS. There have been previous works that integrate Robot Operating System with Jupyter and Conda ecosystems [20], or even ReactJS [21]. Besides controlling mobile robots [22], ROS can be useful for working with 3D data, with applications that include rotations [23-24], and path planning algorithms [25].

IOP Conf. Series: Materials Science and Engineering 1208 (2021) 012035 doi:10.1088/1757-899X/1208/1/012035

The design of a web application is presented in figure 2. Application is built as a virtual joystick that is accessible through the browser and is responsive for multiple platforms, from desktop devices to mobile phones.

**Figure 2.** Application design

Virtual joystick sends commands for specific movements to a simulated ROS robot. In our case, we use the standard TurtleSim [26] robot for explaining purposes. Commands sent from a web application to TurtleSim simulation are Twist messages consisted of data for a specific movement. For example, forward movement is achievable with a message presented in figure 3.

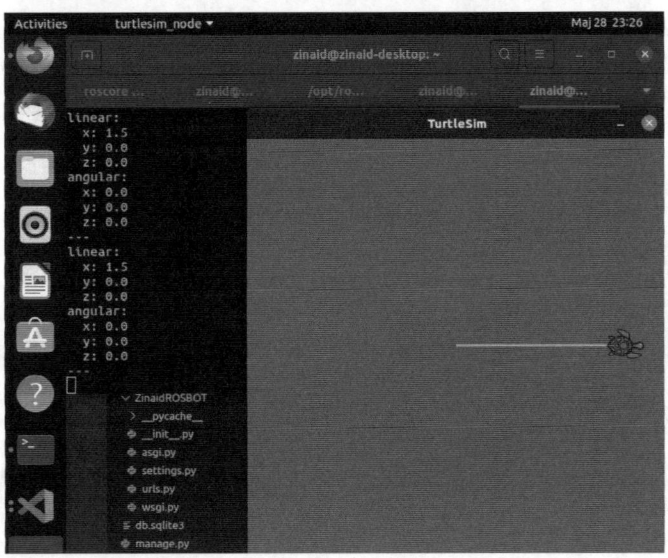


**Figure 3.** Forward movement

Forward movement, as is seen in Figure 3, can be achieved by sending a linear value of $x = 1.5$ where 1.5 represents a movement's distance. The greater the value, the greater the movement. Reducing a value we reduce a distance of a movement and achieve finer-grained movements. Any other movement, linear or angular, is achieved in the same way.

## 3. Simulations

The movement of a TurtleSim is presented in figure 4. Sending specific values of a Twist message to TurtleSim results in different movement paths.

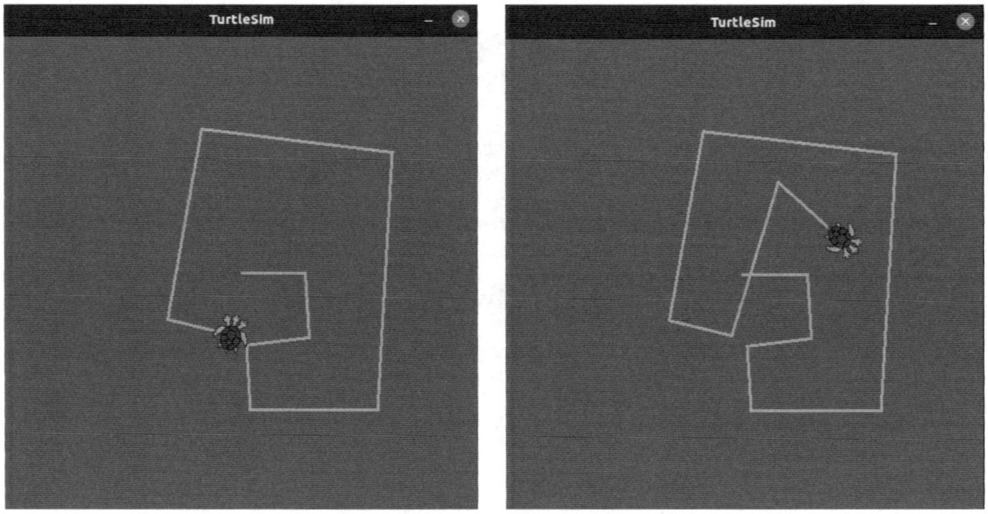


**Figure 4.** TurtleSim movement

The strongest part of a Robot Operating System is its 3D visualizations. ROS visualization of a 3D TurtleSim is presented in figure 5. This simulation is created with a Gazebo library that is integrated into a ROS ecosystem [27].

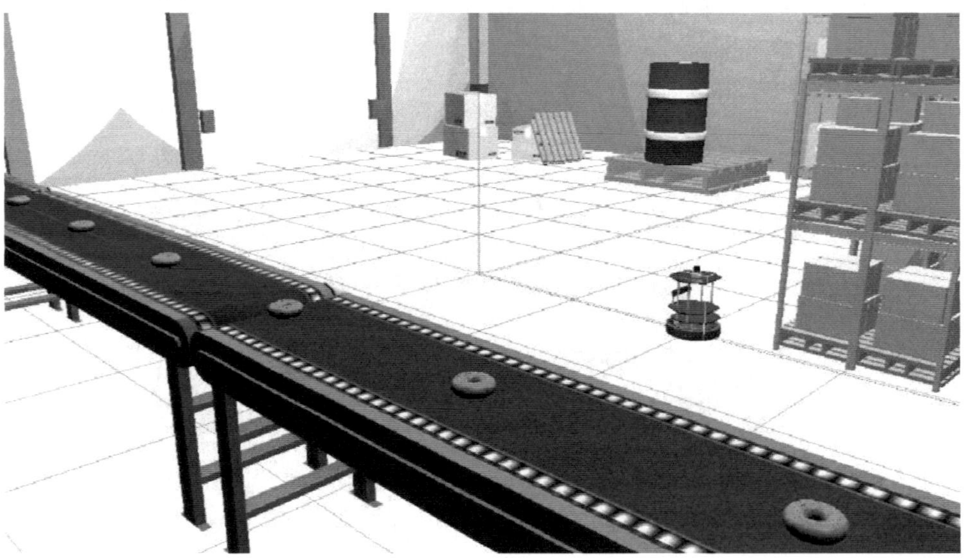

**Figure 5.** Gazebo simulation

In the abovementioned figure, we present movements of a controlled TurtleSim in a 3D environment of a production factory. With these simulations, engineers can test the entire process of working with robots and their interactions with the environment.

## 4. Conclusion

Teleoperations, and thus teleoperated robots, play an important role in the modern world. The possibility for simulations before the construction of a real system plays an important upgrade and provides beneficial time and money savings to engineers and scientists. Robot Operating System is currently the most popular tool for studying robotics, developing robot systems, and simulations.

In this paper, we have presented an application for internet communication with robots based on WebSocket protocol. Through experimental analysis of our application, we have concluded that WebSocket's exceptional speed and full-duplex communication is an advantage for teleoperation system development. A combination of web developing tools such as the Django framework, which exploits the strength of a Python programming language, with ROS software can provide fast testing and simulation development, before the construction of a real system.

Possible upgrades of this application are graphical interface improvements, integration of controllers for motion control, or adding sensors and path planning algorithms into a system for motion automation. In future work, we will also address the time delay problems caused by the internet connection.

## References

[1] Tomic T, Schmid K, Lutz P, Domel A, Kassecker M, Mair E, Grixa I, Ruess F, Suppa M and Burschka D 2012 Toward a fully Autonomous uav: Research platform for indoor and OUTDOOR Urban search and rescue *IEEE Robotics & Automation Magazine* **19** 46–56 doi: 10.1109/mra.2012.2206473

[2] Su H, Yang C, Ferrigno G and De Momi E 2019 Improved Human–Robot collaborative control of Redundant robot for Teleoperated minimally invasive surgery *IEEE Robotics and Automation Letters* **4** 1447–53 doi: 10.1109/lra.2019.2897145

[3] Drakulić U and Mujčić E 2019 Remote monitoring and control system for greenhouse based on IoT *Advanced Technologies, Systems, and Applications IV - Proceedings of the International Symposium on Innovative and Interdisciplinary Applications of Advanced Technologies (IAT 2019)* 481–95 doi: 10.1007/978-3-030-24986-1_38

[4] Pala M and Sincak P 2012 Towards the assisted teleoperation systems *2012 IV International Congress on Ultra Modern Telecommunications and Control Systems* doi: 10.1109/icumt.2012.6459715

[5] Srinivasan L, Scharnagl J and Schilling K 2013 Analysis of websockets as the new age protocol for remote robot tele-operation *IFAC Proceedings Volumes* **46** 83–8 doi: 10.3182/20131111-3-kr-2043.00032

[6] Sorrentino A, Cavallo F and Fiorini L 2020 A plug and play transparent communication layer for cloud robotics architectures *Robotics* **9** 17 doi: 10.3390/robotics9010017

[7] Bruyninckx H 2001 Open robot control software: The OROCOS PROJECT *Proceedings 2001 ICRA. IEEE International Conference on Robotics and Automation (Cat. No.01CH37164)* doi: 10.1109/robot.2001.933002

[8] Vaughan R T and Gerkey B P 2007 Reusable robot software and the PLAYER/STAGE PROJECT *Springer Tracts in Advanced Robotics* 267–89

[9] Fitzpatrick P, Ceseracciu E, Domenichelli D, Paikan A, Metta G and Natale L 2014 A middle way for robotics middleware *Journal of Software Engineering for Robotics* **5** 42–9

[10] Quigley M, Conley K, Gerkey B, Faust J, Foote T, Leibs J, Wheeler R and Ng A Y 2009 ROS: An open-source robot operating system *ICRA workshop on open source software*

[11] Koubaa A 2015 ROS As a Service: Web Services for Robot Operating System *Journal of Software Engineering for Robotics* **6**

[12] Araújo A, Portugal D, Couceiro M S and Rocha R P 2014 Integrating arduino-based educational mobile robots in ros *Journal of Intelligent & Robotic Systems* **77** 281–98 doi: 10.1007/s10846-013-0007-4

[13] Lleras N O, Brennan S, Murphy D, Klena M J, Garvey P and Sommer H 2016 Development of

an open-source tractor driving simulator for tractor stability tests *Journal of Agricultural Safety and Health* **22** 227–46 doi: 10.13031/jash.22.11774

[14] Bouchier P 2013 Embedded ros [ros topics] *IEEE Robotics & Automation Magazine* **20** 17–9 doi: 10.1109/mra.2013.2255491

[15] Badger J, Gooding D, Ensley K, Hambuchen K and Thackston A 2016 Ros in space: A case study on robonaut 2 *Studies in Computational Intelligence* 343–73 doi: 10.1007/978-3-319-26054-9_13

[16] Kehoe B, Patil S, Abbeel P and Goldberg K 2015 A survey of research on cloud robotics and automation *IEEE Transactions on Automation Science and Engineering* **12** 398–409 doi: 10.1109/tase.2014.2376492

[17] Tatarian K, Pereira S, Couceiro M S and Portugal D 2018 Tailoring a ros educational programming language architecture *Robotics in Education* 217–29 doi: 10.1007/978-3-319-97085-1_22

[18] Marani F 2018 Introduction to django *Practical Django 2 and Channels 2* 1–22 doi: 10.1007/978-1-4842-4099-1_1

[19] Millman K J and Aivazis M 2011 Python for scientists and engineers *Computing in Science & Engineering* **13** 9–12 doi: 10.1109/mcse.2011.36

[20] Fischer T, Vollprecht W, Traversaro S, Yen S, Herrero C and Milford M 2021 RoboStack: Using the Robot operating system alongside the conda and Jupyter data Science Ecosystems *arXiv.org* arXiv:2104.12910

[21] Ivanov A, Zakiev A, Tsoy T and Hsia K-H 2021 Online monitoring and visualization with ros and reactjs *2021 International Siberian Conference on Control and Communications (SIBCON)* doi: 10.1109/sibcon50419.2021.9438890

[22] Alisher K, Alexander K and Alexandr B 2015 Control of the mobile robots with Ros in Robotics Courses *Procedia Engineering* **100** 1475–84 doi: 10.1016/j.proeng.2015.01.519

[23] Kapic Z, Crnkic A, Jacimovic V and Mijajlovic N 2021 A new dynamical model for solving rotation averaging problem *2021 20th International Symposium INFOTEH-JAHORINA (INFOTEH)* doi: 10.1109/infoteh51037.2021.9400663

[24] Kapic Z and Crnkic A 2021 Interpolating rotations with non-Abelian Kuramoto model on the 3-sphere *Advanced Technologies, Systems, and Applications VI - Proceedings of the International Symposium on Innovative and Interdisciplinary Applications of Advanced Technologies (IAT 2021)* doi: 10.1007/978-3-030-90055-7_48

[25] Galli M, Barber R, Garrido S and Moreno L 2017 Path planning USING Matlab-ROS integration applied to mobile robots *2017 IEEE International Conference on Autonomous Robot Systems and Competitions (ICARSC)* doi: 10.1109/icarsc.2017.7964059

[26] Joseph L 2018 Programming with ros *Robot Operating System for Absolute Beginners* 171–236 doi: 10.1007/978-1-4842-3405-1_5

[27] Takaya K, Asai T, Kroumov V and Smarandache F 2016 Simulation environment for mobile robots testing using ros and gazebo *2016 20th International Conference on System Theory, Control and Computing (ICSTCC)* doi: 10.1109/icstcc.2016.7790647

# Challenges and opportunities for semiconductor and electronic design automation industry in post-Covid-19 years

**Galia I Marinova[1] and Aida K Bitri[1,2]**

[1]Technical University of Sofia, 8, Boulevard „St. Kliment Ohrisdki", Sofia-1000, Bulgaria
[2]Aleksander Moisiu University of Durres, L1. St. Currilave, Durres-2000, Albania

E-mail: gim@tu-sofia.bg

**Abstract.** The coronavirus pandemic found the semiconductor industry and the chip production supply chain ecosystem unprepared. Companies and main actors in the sector could not read the signs. The decision-makers suffered to deal with the challenges in time and take the right actions. The bullwhip effect caused by the COVID-19 destabilized the operations and some of the experts say that these problems might last and on the other side, this might open doors to innovative solutions that might change the game. The global shutdowns, the misread of the demand for electronics, underestimating customers' demand for the automotive sector, and the Internet of Things in general, were some of the main problems causing chaos in the industry. The paper studies the state-of-the-art and the solutions offered by the semiconductor industry and by the initiatives that Europe, the USA, and especially China, took to make companies and their countries take the most out of this situation.

## 1. Introduction

One of the problems caused by the COVID-19 long shutdown was the "chip shortage" problem. From a business and strategy point of view, this anomaly had a significant impact on businesses related closely to the semiconductor industry and chip production ecosystem. Automakers were the ones most impacted by losing billions of dollars in earnings during 2020 [1]. The automotive industry could not have in time the necessary supply to deliver its main products. The chip shortage caused a ripple effect for the entire network.

Electronic chips are used in everything, in electronic systems and electronic cars. This shortage is one of the first big problems for the Electric Vehicle industry [1]. Due to the high demand for electronic-based products during this period, the demand for semiconductor chips increased more and more. Foundries and chip designers found themselves under pressure to keep up with filling the needs in a short time in a middle of a crisis.

Companies are still trying to figure out how to recover from the lack of supply and how to fulfill the demand. Managers and decision-makers are divided when it comes to how all this will impact the current balances in the industry and how long will be needed to overcome the barriers and challenges. The EDA providers, once offering online and free options products tended during the pandemic to change the approach with pressure on users for payments with different schemes. Their revenues increased during the pandemic time [2]. Some companies planned to invest in opening new plants to cover the needs of the production, others are still trying to develop monitoring and controlling activities to keep in control the distribution of chips and the demands.

Different causes contributed to the expansion of this situation. On the other hand, different analysts suggest that this situation might change the game for the electronic-supported industries.

*1.1 Global shutdowns*

Due to global shutdowns and people staying at home, consumers increased their demand for electronic-based products. Companies in charge of producing those electronics increased their request for more semiconductor chips from the suppliers to fulfill their needs. Due to restrictions, some plants were closed, and the production of some parts was stopped. This increase in demand overwhelmed the foundries and manufacturers [1].

*1.2 The demand for electronics increased*

According to Gartner, the sale of PCs rose to 275 million units during the difficult year of 2020 [3]. Despite the Apple increase of orders for its product and especially the iPhone, still, the company said it did not have enough supply for its new electronic products [3].

Manufacturers of consumer electronics and automakers pushed the chip designers to provide them with a bigger amount of complex chips. These companies caused foundries an extra wave of demands putting them in a difficult position to fulfill the request [6]. A shortage of chips was the start of 'chaos' for companies working directly with semiconductor chips in their electronic products as shown in Figure 1.

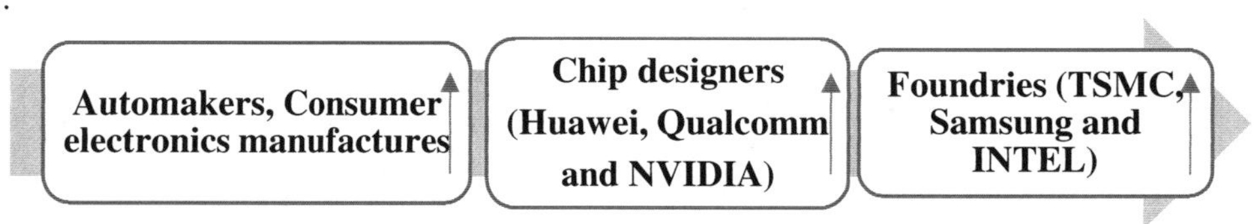


**Figure 1.** Increase of demand for semiconductor chips.

*1.3 Automakers underestimated customers' demand for cars and decreased production*

The chip shortage situation exposed the priority of automakers in the foundries' demand plan fulfillment. Consumer Electronics and the Internet of Things (IoT) are the "volume" companies demanding the biggest amount of semiconductor chips. Automakers discovered that they were after these companies and that is why the foundries were focused on the production of consumer electronics semiconductor chips. On the other hand, there was an increase in demand for automotive cars and the lack of chips caused problems. In 2020, only 3% of Taiwan Semiconductor Manufacturing Company's (TSMC) sales were from automotive chips, compared to 48% for smartphones [3].

The global market of green cars has already been struggling with technological challenges required to keep up with the future of the automotive industry and mobility and with the initiatives to regulate the market through regulators. The automotive industry requires electronic chips more than ever and more than any other industry pushed by the latest technological advancements. In 2017, electronics made up more than 40% of the automotive total cost [3]. Despite the low sales of these cars and mobility options, the automotive industry is one of the near future industries that will change the game. The advanced functionalities of these cars will request more advances and personalized chips. Manufacturers and companies are under continuous pressure to lead the high demand for these chips in time. There is an imbalance of supply and demand causing shortage and latency in the process [4]. COVID-19's unpredictable situation enhanced further the supply and demand problems between customers and chip manufacturers.

Volkswagen revealed that was difficult for the company to build 100 000 cars due to the 2020 chip shortage, Tesla was also affected by this situation declaring "having 'some of the most difficult supply challenges we've ever experienced". Jaguar Land Rover and Daimler companies took some measures to reschedule production activities in their plants due to lack of chips [5].

IOP Conf. Series: Materials Science and Engineering 1208 (2021) 012036 doi:10.1088/1757-899X/1208/1/012036

## 2. Reaction to Chip Shortage

According to Goldman Sachs, this problem might last till the end of 2021 and according to Acer, this crisis might last until 2022/2023 [7][8].

There are different approaches to deal with crises. Companies can move production in-house or can invest in the new expansion to deal with the increased demand. Investing in in-house production might not be the solution since the problem comes from the lack of inputs and not from the process of manufacturing. Some semiconductor manufacturers have announced an investment in new foundries and factories hoping to increase the supply. These factories might take two years to build and during this time a lot of things might change, and the chip shortage problem might ease [4].

Fast reaction without a proper analysis might unbalance the industry and especially in companies operating in the value chain processes. Semiconductor leaders should manage this situation properly otherwise they might create a similar situation soon. Some suggest that the market requires a controller to monitor the distribution of the chips based on the company's needs and ensure that the chip allocation will allow all companies and chip customers to survive and grow. Companies should focus more on developing and cultivating strong and stable relations with their co-operators, suppliers, partners, competitors, and even their personnel. Customers of chips should choose smart and gain trust to work with suppliers that know and can align with their business's needs [9][10].

Tan Hock, CEO of Broadcom said that their company will follow a strategy to ensure that all their customers get the product, despite some latencies [4]. Proper distribution of chips and prioritizing demands might be a good solution to problems similar to this. TSMC announced that they will focus on fulfilling demands from the automotive industry, one of the industries that had the biggest damages during this year [4]. Tower Semiconductor Ltd. declared that they could speed up the chip production process and sort their lines in a way that the customers that have top priority to be supplied with the required products [11].

### *2.1. The Electronic Design Automation industry reaction to COVID-19*

Despite problems in the semiconductor ecosystem caused due to COVID-19, the Electronic Design Automation (EDA) sector is experiencing increased revenues and profits. In general, the software development industry managed to grow its profit during this difficult period and EDA did not stay behind. According to Walden C. Rhines, CEO of Mentor Graphics, now known as Siemens EDA, he believes that one of the reasons why the industry received increased revenues, was due to long-term and stable investments in research during the years [2].

An interesting trend to be seen is the increased demand for EDA tools during long shutdowns and at the same time the rise for cloud EDA. The market of EDA cloud is expected to grow and to receive special attention during this time [12][13]. Taking advantage of the cloud can help companies to provide the tools to designers 24/7 and at lower costs [13]. Another reason why U.S EDA companies seem to have secure and sustainable growth is that due to this trade war between China and the USA, the USA government is investing in a system-level design that will help produce advanced and high level of security chips [14].

### *2.2. The reaction of countries to COVID-19 chip shortage*

Leader countries in high technologies are trying to build policies on how to be independent to chip manufacturers that are mainly covered by South Korea and Taiwan, around 75% of the chip production according to US-based semiconductor industry association [15]. USA government and European politicians desire those chips to be made and developed locally, some of them fear the reunion of China with Taiwan, especially since the high-tech chips coming from China increase [15] [16].

Amid chip shortage and fears that China is investing in being independent and leading the industry, the US government released a plan to boost the production and development of the semiconductor industry further and propose an investment of 50 billion dollars on infrastructure. This technological war between the two countries has made governments of both countries develop plans to increase their presence and to develop advanced tools to make their nations independent and secure. Security concerns that come with advanced technologies are one of the reasons why Biden is taking this seriously [17].

The third entering this war for the chips is the European Union. Angela Merkel, the German Chancellor, declared that Germany in cooperation with European companies is investing in the chip industry [17]. There are around 22 European countries that have joined forces to work together on developing a local semiconductor ecosystem, according to Thierry Breton, the chief of European industry [18]. The EU is focusing on local production of chips and becoming more independent from the USA and Taiwan and other east countries [19]. There is still no clear how things will go but different scenarios exist and one of them is the rise of China as the leading country in the industry. The following sections give an overview of the situation in China regarding semiconductor investments and development plans.

### *2.3. The rise of China in the semiconductor ecosystem*

Trade and technological war between the USA and China uncovered some of the weaknesses of China's semiconductor ecosystem, especially in the electronic design automation industry subsector. The government and the Chinese companies started to think and look for local opportunities and policies to support, use and grow their presence in the industry [20].

The chip shortage advanced the current problem since China is rising demand for electric and new advanced cars and the lack of know-how, manufacturers, and the tension in the region might be a big problem for it [21]. On the other hand, these problems go beyond the COVID-19 chip problem. (Yuan Chengyin, general manager of the National New Energy Vehicle Technology Innovation Center). According to him, China will be more vulnerable to global impact and dynamics in the chip supply chain and ecosystem. Investing and growing in this area seems like a challenge for China but at the same time, a new country that will invest in technological know-how in this area might cause distort the current stable market of the semiconductor ecosystem [20] [21].

In the last years, the Chinese government, being aware of entering too late in the market, is investing in the semiconductor sector by developing policies, realizing concrete plans on chip production, and supporting new initiatives in the business field [22]. The history of China during the years and the Chinese government policies to support the advancements in high technologies in the nation are listed in Table 1 (Adopted by [20]).

**Table 1.** Chinese government policies to support the advancements in high technologies in periods.

| Years | Event |
|---|---|
| 1990 to 2000 | State planning focus |
| 1991 | Development of the Eighth Five-Year Plan (FYP) document where the "the main task" of the state was the industry development. |
| 2000 to 2013 | First China-based foundries. Semiconductor Manufacturing International Corporation (SMIC) |
| 2014 | - Release of the National Integrated Circuit Industry Development Guidelines<br>- National IC Investment Fund establishment |
| 2015 | The Made in China 2025 (MIC2025)4 initiative. |
| 2018 | Section 301 report by the U.S. Trade Representative (USTR) on China's trade policies in March 2018, the start of the trade war between China and the USA. Chinese officials increased their focus on making China less dependent on companies currently leading the market and preparing the domestic market for its resources and local supply chain development. |

The case of building a chip production ecosystem in China will require a lot of knowledge and capital, mature technology, and a sustainable business ecosystem to enhance development. New companies in China must deal with established companies that already lead the market and have an advantage in terms of covering most of the market, economies of scale, brand recognition, and patent quality [23].

On the other hand, the idea of everything developed and made by China seems like an "aspiration" difficult to be achieved, especially since most of the companies operating in China desire to go beyond the domestic market towards global markets. Table 2 (Adopted by [20]) shows the current situation in China of the companies and market share it covers.

**Table 2.** Current situation in China of the companies and market share it covers.

| Semiconductor industry subsectors | China's current place |
|---|---|
| Foundries | 2 companies (SMIC and Hua Hong) |
| IDMs (integrated device manufacturers) | 0 companies |
| Semiconductor Design companies | China is the second after the USA, surpassing Taiwan. |
| Electronic design automation industry | China is far beyond when it comes to EDA tools developers. It is the weakest point of the semiconductor supply chain in China together with the manufacturing materials and supplies |
| OSAT (outsourced assembly and test) | Tongfu Microelectronics, JCET, and Tianshui Huatian Technology Chinese companies trying to gain more share in the market which is led by the Taiwan firms. |

Currently, China is encouraging foreign companies to invest and bring their technologies, talents, and research centers to China. There are specific policies, such as tax, tariff, and materials offered by the Chinese government for the following years to support the development in this area. On the other side, many members of the USA Congress and policymakers are concerned about the potential of China to take the lead in the semiconductor industry based on their good attraction policies [20].

## 3. Conclusions

The chip shortage problem exposed the weaknesses in the chip production ecosystem and chip supply chain. More than lack of materials, or a shutdown caused by COVID-19, the problem was a lack of proper decision making. Leaders of companies failed to read properly the demands, trends and to allocate the resources according to the needs. This led to an unstable situation that will require some years to come back to normality. On the other hand, it seems that it opened new possibilities and opportunities to change the game. New companies might take the lead. EU, the USA, and China are fighting to shift the production of chips locally and to control the market. It is early to predict where this is going but the industry might experience some fundamental changes. The currently established organizations might have to restructure the market and develop new strategies to keep their place in the industry.

Additionally, academia in the EU and USA should increase the courses and studies in Computer-aided design of electronic chips to avoid the loss of know-how in this area and to be ready for further challenges.

## Acknowledgments

This study is realized and partly supported by the CEEPUS network CIII-BG-1103-06-2122, especially in the part Joint doctoral program within CEEPUS.

## References

[1] Pizzemento A 2021 The 2021 Semiconductor Chip Shortage: What, Why, and What's Next?

IOP Conf. Series: Materials Science and Engineering 1208 (2021) 012036 doi:10.1088/1757-899X/1208/1/012036


[2] Blyler J 2020 Semiconductor chip developers will face COVID-19 challenges and new tech
[3] Leswing K 2021 Why there's a chip shortage that's hurting everything from the PlayStation 5 to the Chevy Malibu
[4] Deloitte 2019 Semiconductors – the Next Wave Opportunities and winning strategies for semiconductor companies
[5] Rivero N 2021 The global semiconductor shortage can be explained by the bullwhip effect
[6] Moldrich C 2021 Global chip shortage: everything you need to know
[7] Shein E 2021 The global chip shortage: What caused it, how long will it last?
[8] Abigail Ng 2021 Goldman Sachs says disruptions from the chip shortage should diminish in the second half of 2021
[9] Hopkins J 2021 Yes, the global microchip shortage is COVID's fault. No, it won't end any time soon
[10] Sweney M 2021 Global shortage in computer chips 'reaches crisis point'
[11] Shead S 2021 The global chip shortage is starting to have major real-world consequences
[12] Jeong E Strumpf D 2021 Why the Chip Shortage Is So Hard to Overcome
[13] Global Industry Analysts Inc 2021 Cloud EDA - Global Market Trajectory & Analytics report
[14] Market research 2021 Electronic Design Automation (EDA) Market: Global Industry Analysis 2015 - 2019 and Opportunity Assessment 2020 – 2030
[15] Leopold G 2020 EDA industry maintains healthy growth
[16] Josephs J 2021 Cisco says computer chip shortage to last six months
[17] Leary A and Ziobro P 2021 Biden Calls for $50 Billion to Boost U.S. Chip Industry
[18] Su-bin L Dong-hyun K Shin-young P and Jin-won K 2021 Quest for semicon autonomy pushes countries to brink of chip cold war
[19] Nienaber M 2021 More EU member states join forces to support local chip production - Breton
[20] Kelly É 2021 EU promises 'significant' funds to boost local semiconductor production
[21] Paul T 2021 The Future of China's Semiconductor Industry American Affairs Vol 5.1 p 90–113.
[22] BloombergQuint 2021 China's Car-Chip Shortage Could Persist for as Long as a Decade
[23] Jingjing M 2021 Rapid growth seen in IC industry
[24] Paszak P 2020 Chinese Ambitions in the Semiconductor Industry

IOP Conf. Series: Materials Science and Engineering 1208 (2021) 012037 doi:10.1088/1757-899X/1208/1/012037

# Interpolation on the special orthogonal group with high-dimensional Kuramoto model

**Aladin Crnkić and Zinaid Kapić**

University of Bihać, Faculty of Technical Engineering, Bosnia and Herzegovina

E-mail: aladin.crnkic@unbi.ba

**Abstract**. The construction of smooth interpolation trajectories in different non-Euclidean spaces finds application in robotics, computer graphics, and many other engineering fields. This paper proposes a method for generating interpolation trajectories on the special orthogonal group SO(3), called the rotation group. Our method is based on a high-dimensional generalization of the Kuramoto model which is a well-known mathematical description of self-organization in large populations of coupled oscillators. We present the method through several simulations and visualize each simulation as trajectories on unit spheres $\mathbb{S}^2$. In addition, we applied our method to the specific problem of object rotation interpolation.

## 1. Introduction

3D rotations find usage in many different applications in robotics, computer vision, computer graphics, etc. Different representations of 3D rotations, notably rotational matrices, Euler angles and unit quaternions, are used for solving different engineering problems [1]. For our purpose in this paper, we use rotational SO(3) matrices and Euler angles as its parameterization and explain it in the second section in more detail. The most common problems when dealing with rotation data are rotation averaging and interpolation. Rotation averaging is extensively studied in many papers and finds different applications in the field of structure-from-motion [1-2].

Construction of smooth interpolation curves in non-Euclidean spaces finds application in many scientific fields, for example, robotics, computer animation, and computer graphics [1,3]. Problems in robotics such as motion control, obstacle avoidance, and object approach are in constant need of solving rotation interpolation problems [4]. Besides that, it is also applied in computer animations and computer graphics for different use-cases such as smooth interpolation between camera frames [5]. This paper focuses on rotation interpolation between two rotations, where we introduce a new model for rotation interpolation on the group manifold SO(3).

Different approaches are used for interpolation on Lie groups SO(3) and $S^3$. The most common rotation interpolation algorithm on group $S^3$ is a famous Spherical Linear Interpolation algorithm (SLERP) that uses quaternionic representation and gives the shortest and the most direct path between two rotations [6]. Interpolation on unit sphere $S^3$ based on the high-dimensional Kuramoto model is introduced and explained in the paper [7]. For better understanding, it is important to emphasize that a result of the multiplication between two unit quaternions is always a unit quaternion. This characteristic of unit quaternions allows a spherical interpolation between rotations. On the other hand, a result of the multiplication between two rotational SO(3) matrices is not always a new rotational matrix. This characteristic of rotational matrices produces an anomaly in creating an interpolation curve because the point in the interpolation trajectory does not always preserve a unit sphere [1]. Common algorithms for rotation interpolation with SO(3) matrices and Euler angles are a so called Linear Matrix Interpolation (LinMat) and Linear Euler Interpolation (LinEul), respectively [8]. These algorithms are subjected to the abovementioned anomaly, and there are several papers that solve this problem by producing spherical interpolation of rotations parameterized with rotational matrices or Euler angles [9-10]. In our paper, we present two novel models, one for rotation interpolation with rotational matrices and other with Euler angles. These models always preserve SO(3) group and give the shortest and the most direct spherical path between two rotations.

Model used in this paper is based on one generalization of a famous Kuramoto model which is the most adequate for studying collective behaviour and self-organization in large populations of coupled oscilators [11]. Through the last decades, Kuramoto model has been subjected to various generalizations on higher dimensions. One of the most studied generalization has been introduced by Lohe and is known as the non-Abelian Kuramoto model, or just Lohe model [12]. Non-Abelian Kuramoto model can be used for clustering of static and stream data [13-14], coordination and consensus on groups $S^3$ and SO(3) [15], rotation averaging [16], and other applications in different scientific and engineering fields [17].

In the next section we explain parametarizations used in this paper and mapping between them. In the section 3 we describe our dynamical model and present our method through an algorithm. Simulations are implemented using programming environment Wolfram Mathematica and illustrated in the third section. Finally, in the last section, we draw some conclusion and present an outlook on future research.

## 2. Mapping between special orthogonal group SO(3) and Euler angles

As it is mentioned above, different parameterizations can be used for representing rotations. In this paper we use rotational SO(3) matrices and Euler angles. SO(3) is the group of all rotations about the origin in three dimensional Euclidean space. This group is represented by 3 × 3 rotation matrices with determinant 1, as it follows [1]:

$$SO(3) := \{R \in R^{3x3} | R^T R = RR^T = I_3, \det(R) = +1\}. \tag{1}$$

In (1) $I_3$ represents identity 3 × 3 matrix, $\det(\cdot)$ represents determinant of the matrix, and $R^T$ is a transpose of the matrix $R$. Besides SO(3) matrices, interesting representation of rotations are so called Euler angles [6]. These angles are easy to visualize and understand. Euler angles are consisted of roll, pitch and yaw, as it is visualized in figure 1.

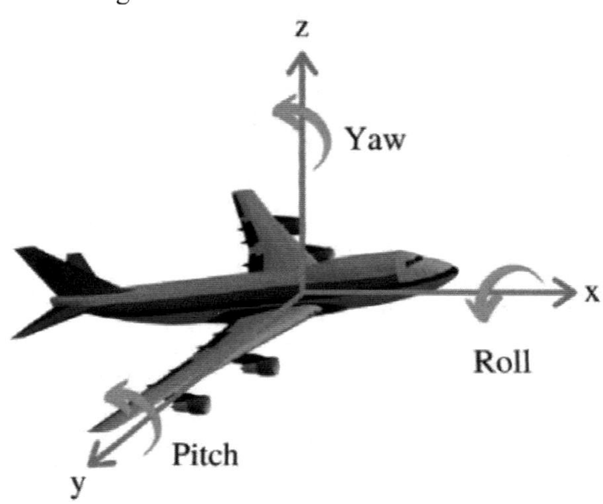


**Figure 1.** Euler angles as parameterization of rotations

Roll angle represents rotation about the $x$-axis and is denoted as $\psi$, pitch angle refers to rotation about $y$-axis and is denoted as $\theta$ and yaw angle is a rotation about $z$-axis and is denoted as $\phi$. It is possible to map Euler angles to SO(3) matrices, and vice-versa. Let us assume that a rotation about axis $x$ by an angle $\psi$ is represented as [18]:

$$R_{x,\psi} = \begin{bmatrix} 1 & 0 & 0 \\ 0 & cos\psi & -sin\psi \\ 0 & sin\psi & cos\psi \end{bmatrix}.$$

With the same analogy we represent a rotation about axis $y$ by an angle $\theta$, and a rotation about axis $z$ by an angle $\phi$:

$$R_{y,\theta} = \begin{bmatrix} cos\theta & 0 & sin\theta \\ 0 & 1 & 0 \\ -sin\theta & 0 & cos\theta \end{bmatrix} \text{ and } R_{z,\phi} = \begin{bmatrix} cos\phi & -sin\phi & 0 \\ sin\phi & cos\phi & 0 \\ 0 & 0 & 1 \end{bmatrix}.$$

A rotational matrix can be obtained by performing a rolling about $x$ axis by an angle $\psi$, pitching about $y$ by an angle $\theta$ and yawing about $z$ by an angle $\phi$. In this case, $R_{(\phi,\theta,\psi)} = R_{z,\phi}R_{y,\theta}R_{x,\psi}$, yielding:

$$R_{(\phi,\theta,\psi)} = \begin{bmatrix} cos\theta cos\phi & -cos\psi sin\phi + sin\psi sin\theta cos\phi & sin\psi sin\phi + cos\psi sin\theta cos\phi \\ cos\theta sin\phi & cos\psi cos\phi + sin\psi sin\theta sin\phi & -sin\psi cos\phi + cos\psi sin\theta sin\phi \\ -sin\theta & sin\psi cos\theta & cos\psi cos\theta \end{bmatrix}. \quad (2)$$

Obtaining Euler angles from a given rotational matrix from (2) one has that roll, pitch and yaw angle are, respectively [18]:

$$\psi = \arctan\left(\frac{R_{(3,2)}}{R_{(3,3)}}\right), \theta = -\arcsin(R_{31}), \phi = \arctan\left(\frac{R_{(2,1)}}{R_{(1,1)}}\right).$$

Every rotation representation has its own preferred application, advantages, and disadvantages. Rotation matrices are used in computer graphics and computer vision, while Euler angles are mainly used in robotics [1]. The advantages of rotation matrices are their well-known mathematical background and the ability to represent all other transformations such as translation, scaling, and projection. On the other hand, matrix representation is redundant because it uses 9 parameters for rotation representation, while Euler angles use only 3. The biggest drawback for the usage of Euler angles is a Gimbal lock problem. Gimbal lock is a loss of one degree of rotational freedom [8]. Quaternions, that are not subject of research in this paper, are a better solution for interpolation purposes as it is seen in paper [7], but Euler angles are also used because they are simpler to understand and visualize, especially for those unfamiliar with quaternion algebra.

## 3. Algorithm

In this section we will introduce our method for interpolation on the rotation group SO(3). As we have said our method is based on the non-Abelian Kuramoto model that is defined as:

$$i\dot{U}_j U_j^* = H_j - \frac{iK}{2N}\sum_{i=1}^{N}(U_j U_i^* - U_i U_j^*),\ j = 1, \dots, N. \quad (3)$$

In the abovementioned equation, $N$ is a number of generalized oscillators, $U_j(n) \in U(n)$ is unitary matrix representig the state of generalized oscillator $j$, $H_j(n) \in \mathfrak{u}(n)$ represents intrinsic frequency of generalized oscillator $j$. The matrix $U_i^*$ is conjugate transpose matrix of the matrix $U_i$. Finaly, $K$ is a strength of a global coupling.

Firstly, let us suppose that the rotations are given with SO(3) rotational matrices, where $R_S$ represents a starting rotation and $R_E$ represents the final one.

Consider the following matrix-valued ordinary differential equation which can be easily obtained from the system (3):

$$R'(t) = R_E - R(t)R_E^T R(t) \quad (4)$$

with initial condition $R(0) = R_S$. In above equation, ${R_E}^T$ corresponds to a transposed matrix of the matrix $R_E$. It is important to emphasize that (4) defines dynamics on the group SO(3), meaning that

$R(t) \in SO(3)$ for all $t > 0$. Also, notice that the solution of the differential equation (4) is an interpolating smooth curve $R(t)$ that connects the starting rotation $R_S$ ($R(0) = R_S$) and the final one $R_E$ ($R_E = \lim_{t\to\infty} R(t)$).

Analogously, we can adapt the method for interpolation between two rotations parameterized with Euler angles. Inserting (2) into (4), we obtain a following system of differential equations:

$$
\begin{aligned}
\psi'(t) = &- \cos(\psi_E)\,[\cos(\psi(t)) \sin\theta_E \sin(\phi_E - \phi(t)) \\
&+ \cos(\phi_E - \phi(t))\, sin\psi(t)] \\
&+ sin\phi_E sin\psi_B [cos\psi(t) sin\phi(t) \\
&- cos\phi(t) sin\theta_E sin\psi(t)] \\
&+ cos\theta_E sin\psi_E [cos\phi(t) cos\psi(t) \\
&+ sin\theta_E sin\phi(t) sin\psi(t)] \\
&+ cos\theta_E [(sec\theta(t) \sin(\psi_E - \psi(t)) \\
&+ \sin(\phi_E - \phi(t))\, tan\theta(t)] \\[1em]
\phi'(t) = &\left[\cos(\psi_E - \psi(t) + cos\theta_E sec\theta(t)\right] \sin(\phi_E - \phi(t)] \\
&- \cos(\phi_E - \phi(t))\, sin\theta_E \sin(\psi_E - \psi(t))\big] \\
&+ cos\theta_E \sin(\psi_E - \psi(t))\, tan\theta(t) \\[1em]
\theta'(t) = &\frac{1}{2}\cos\theta(t)\left[\cos(\phi_E + \psi_E - \phi(t) - \psi(t))\,(-1 + sin\theta_E)\right] \\
&+ 2sin\theta_E + \cos\,(\phi_E - \psi_E - \phi(t) + \psi(t))(1 + sin\theta_E)] \\
&- cos\theta_E[\cos(\phi_E - \phi(t)) + \cos(\psi_E - \psi(t))\, sin\theta(t)].
\end{aligned}
\tag{5}
$$

The abovementioned system corresponds to non degenerate case where $\theta \neq \frac{\pi}{2}$ and $\theta \neq -\frac{\pi}{2}$. For a degenerate case of $\theta = \frac{\pi}{2}$ or $\theta = -\frac{\pi}{2}$ system enters in a so called Gimbal lock.

For a better presentation, we will explain our methods through algorithms. Algorithm for rotation interpolation with rotations given with SO(3) rotational matrices is as follows:

1: Enter $R_S, R_E$
2: Choose tolerance $\varepsilon$, step $\delta$, and define $T=0$
3: Solve (4) with $R(0) = R_S$
4: **loop**
5:     **if** $\|R_E - R(T)\| < \varepsilon$ **then**
6:         **return** $R(t)$ for $t \in [0, T]$
7:     **else**
8:         $T=T+\delta$
9:     **end if**
10: **end loop**

Algorithm for rotation interpolation with rotations given with Euler angles is equivalent to the abovementioned algorithm for rotation interpolation with rotations given as SO(3) rotational matrices, but the process is performed by solving a system of differential equations (5) and the initial and final positions are given with Euler angles $\psi$, $\theta$, and $\phi$.

## 4. Simulations

In this section we present and illustrate simulation results of our method. It is well known that is difficult to visualize points or curves on SO(3). Therefore, we have visualized three different curves on unit spheres $S^2$, obtained with $R(t) \cdot v_i \in S^2$ for $i = 1,2,3$, where $v_1 = [1\ 0\ 0]^T$, $v_2 = [0\ 1\ 0]^T$, $v_3 = [0\ 0\ 1]^T$ are unit vectors and $R(t) \in SO(3)$ for all $t > 0$.

For the first simulation we have used initial and final rotation (represented by SO(3) rotational matrices) given as:

$$R_S = \begin{bmatrix} -1 & 0 & 0 \\ 0 & -1 & 0 \\ 0 & 0 & 1 \end{bmatrix} \text{ and } R_E = \begin{bmatrix} 0 & 0 & 1 \\ 1 & 0 & 0 \\ 0 & 1 & 0 \end{bmatrix}.$$

Interpolation curve from a starting rotation $R_s$ to the final rotation $R_E$ is presented in figure 2. It is obtained using the algorithm with model (4), and by setting initial condition $R(0) = R_S$, tolerance $\varepsilon = 10^{-5}$ and step $\delta = 0.01$. Final point is reached at the moment $T = 11.66$ with a set tolerance $\varepsilon$.

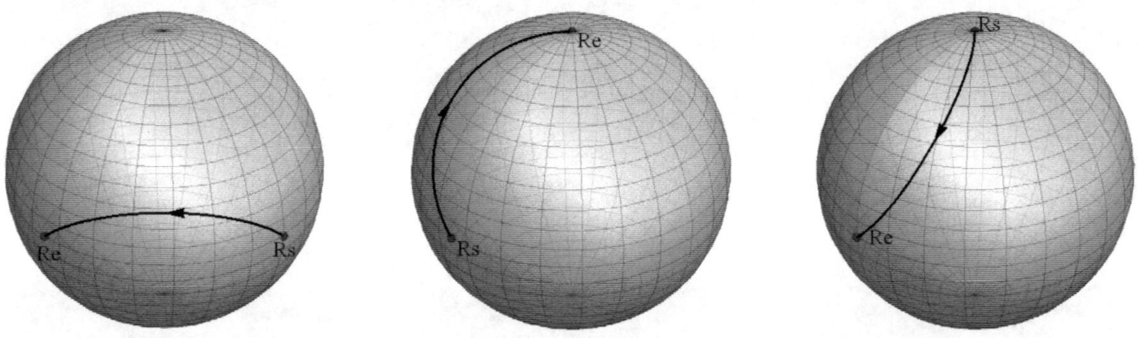


a) b) c)

**Figure 2.** Interpolation curve from a starting rotation $R_s$ to the final rotation $R_E$: a) $R(t) \cdot v_1$ for all $t \in [0, T]$; b) $R(t) \cdot v_2$ for all $t \in [0, T]$; c) $R(t) \cdot v_3$ for all $t \in [0, T]$.

For the second simulation we will use Euler parametarization of rotations. Initial and final rotation are given as Euler angles. Let us assume that the initial rotation is given as: $\phi_S = 2\pi/3$, $\theta_S = 0$, $\psi_S = 0$, and final rotation as: $\phi_E = 0$, $\theta_E = 0$, $\psi_E = \pi/2$.

In this case we use the algorithm with model (5). Initial conditions for this simulation are as follows: $\psi(0) = \psi_S$, $\theta(0) = \theta_S$, and $\phi(0) = \phi_S$, tolerance is $\varepsilon = 10^{-5}$ and step is $\delta = 0.01$. At the moment $T = 13.1$ the stopping criteria in the algorithms step 4 is satisfied.

For illustration of this simulation result we have simulated an objects interpolation from a starting postion $P_s$ to the final position $P_F$ shown in figure 3.

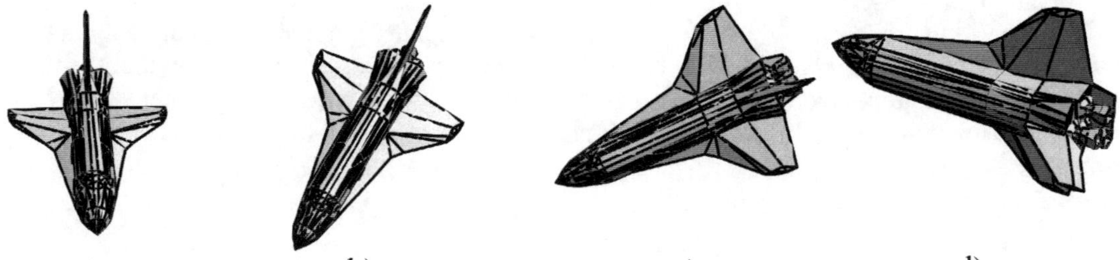

a) b) c) d)

**Figure 3.** Object positions at specific moments: a) $t = 0$ ($P_S$); b) $t = 0.4$ ($P_1$); c) $t = 1.2$ ($P_2$); d) $T = 13.1$ ($P_E$).

Table 1 shows numeric values of rotations presented using Euler parameterization and their equivalent SO(3) matrices at specific positions given in figure 3.

**Table 1.** Numerical values of $SO(3)$ rotational matrices and Euler angles at specific positions.

| Position | SO(3) matrix | Euler angles |
|---|---|---|
| $P_S$ | $\begin{bmatrix} -0.5 & -0.866 & 0 \\ 0.866 & -0.5 & 0 \\ 0 & 0 & 1 \end{bmatrix}$ | $\phi = 2\pi/3$, $\theta = 0$, $\psi = 0$. |
| $P_1$ | $\begin{bmatrix} 0.0026 & -0.934 & 0.358 \\ 0.934 & -0.126 & -0.335 \\ 0.358 & 0.335 & 0.871 \end{bmatrix}$ | $\phi = 1.5681\pi$, $\theta = -0.3660\pi$, $\psi = 0.3670\pi$. |
| $P_2$ | $\begin{bmatrix} 0.727 & -0.558 & 0.400 \\ 0.558 & 0.140 & -0.818 \\ 0.400 & 0.818 & 0.413 \end{bmatrix}$ | $\phi = 0.6544\pi$, $\theta = -0.4118\pi$, $\psi = 1.1032\pi$. |
| $P_E$ | $\begin{bmatrix} 1 & 0 & 0 \\ 0 & 0 & -1 \\ 0 & 1 & 0 \end{bmatrix}$ | $\phi = 0$, $\theta = 0$, $\psi = \pi/2$. |

Notice that our method produces a spherical interpolation curve. That means that at each moment, model (4) preserves the SO(3) group. For better illustration, we have also provided a video that showcases object's movement from starting to final position.[1]

## 5. Conclusion and outlook

In this paper we have introduced a novel model for rotation interpolation between two points using rotational matrices SO(3) and Euler angles as parameterizations. This method is based on high-dimensional generalization of Kuramoto model and offers a most direct and shortest way from initial to final rotation. Simulation results are visualized as a trajectory on a unit sphere and for a better insight we have provided one short video illustration of objects movement.

Approach presented in this paper offers universal solution for rotation interpolation with any given parameterization. This method finds spherical interpolation curve for rotations in SO(3) form or Euler angles, like paper [9] that uses quaternions for these purposes. We find it interesting to integrate all rotation parameterizations into a model that utilizes high-dimensional Kuramoto model for interpolation purposes.

The algorithm proposed in this paper can be used in computer graphics for creating animations and robotics for motion control and robotic hand movement. It is important to mention that the model can be used only for interpolation between two given rotations. Therefore, we find it interesting to upgrade this model for multiple rotations interpolation. This open question will be addressed in our future work.

## References

[1] Hartley R, Trumpf J, Dai Y and Li H 2013 Rotation averaging *International Journal of Computer Vision* **103** 267–305 doi: 10.1007/s11263-012-0601-0

[2] Purkait P, Chin T-J and Reid I 2020 NeuRoRA: Neural Robust Rotation Averaging *Computer Vision – ECCV 2020* 137–54 doi: 10.1007/978-3-030-58586-0_9

[1] Supplementary video is available at: https://www.youtube.com/watch?v=bP6C7ZlCC18.

[3] Zhao X, Liu Y and Xie Z 2021 The application of acceleration-level quaternion interpolation in the visual-servoing process *2021 13th International Conference on Computer and Automation Engineering (ICCAE)* doi: 10.1109/iccae51876.2021.9426103

[4] Legnani G, Fassi I, Tasora A and Fusai D 2021 A practical algorithm for smooth interpolation between different angular positions *Mechanism and Machine Theory* **162** 104341 doi: 10.1016/j.mechmachtheory.2021.104341

[5] Haarbach A, Birdal T and Ilic S 2018 Survey of higher order rigid body motion interpolation methods for keyframe animation and continuous-time trajectory estimation *2018 International Conference on 3D Vision (3DV)* doi: 10.1109/3DV.2018.00051

[6] Shoemake K 1985 Animating rotation with quaternion curves Proceedings of the 12th annual conference on Computer graphics and interactive techniques - SIGGRAPH '85 doi: 10.1145/325334.325242

[7] Kapic Z and Crnkic A 2021 Interpolating rotations with non-Abelian Kuramoto model on the 3-sphere *Advanced Technologies, Systems, and Applications VI - Proceedings of the International Symposium on Innovative and Interdisciplinary Applications of Advanced Technologies (IAT 2021)* doi: 10.1007/978-3-030-90055-7_48

[8] Dam E, Koch M and Lillholm M 1998 Quaternions, interpolation and animation *CiteSeerX*

[9] Shingel T 2008 Interpolation in special orthogonal groups *IMA Journal of Numerical Analysis* **29** 731–45 doi: 10.1093/imanum/drn033

[10] Wu J, Liu M, Ding J and Deng M 2019 Robust rotation Interpolation based on So(n) geodesic distance *Lecture Notes in Computer Science* 126–32 doi: 10.1007/978-3-030-34995-0_12

[11] Kuramoto Y 1975 Self-entrainment of a population of coupled non-linear oscillators *International Symposium on Mathematical Problems in Theoretical Physics* 420–2 doi: 10.1007/bfb0013365

[12] Lohe M A 2009 Non-Abelian Kuramoto models and synchronization *Journal of Physics A: Mathematical and Theoretical* **42** 395101 doi: 10.1088/1751- 8113/42/39/395101

[13] Crnkić A and Jaćimović V 2018 Data clustering based on quantum synchronization *Natural Computing* **18** 907–11 doi: 10.1007/s11047-018-9720-z

[14] Crnkić A, Ivanović I, Jaćimović V and Mijajlović N 2020 Swarms on the 3-sphere for online clustering of multivariate time series and data streams *Future Generation Computer Systems* **112** 11–7 doi: 10.1016/j.future.2020.05.018

[15] Crnkić A, Jaćimović M, Jaćimović V and Mijajlović N 2020 Consensus and coordination on groups SO(3) and S3 over constant and state-dependent communication graphs *Automatika* **62** 76–83 doi: 10.1080/00051144.2020.1863544

[16] Kapic Z, Crnkic A, Jacimovic V and Mijajlovic N 2021 A new dynamical model for solving rotation averaging problem *2021 20th International Symposium INFOTEH-JAHORINA (INFOTEH)* doi: 10.1109/INFOTEH51037.2021.9400663

[17] Ha S-Y and Park H 2020 A dynamical systems approach for the shape matching of polytopes along rigid-body motions *arXiv.org* arXiv:2006.11533

[18] Hashim H A 2019 Special orthogonal Group SO(3), Euler Angles, Angle-axis, Rodriguez vector and Unit-quaternion: Overview, mapping and challenges *arXiv.org* arXiv:1909.06669

# Design and implementation of fuzzy control system for egg incubator based on IoT technology

**Edin Mujčić and Una Drakulić**

University of Bihać, Faculty of Technical Engineering, Irfana Ljubijankića bb

E-mail: una.94@hotmail.com

**Abstract**. Nowadays, automatic systems are using in more spheres of industry, and in this way, human intervention is avoided and used as minimally as possible. In the chicken poultry industry, the use of mother hens is transferring to automatic egg incubating systems. Such systems are helpful for the farmers to incubate the eggs automatically without the need for human intervention. These systems work by keeping the physical quantities, temperature and humidity, at the optimal level. In that way, the fetuses inside eggs are growing without the presence of the mother hen. The egg incubating systems not only improve poultry production considerably but also help in the regularity of income making, enabling the farmers to be able to get transition into possible rural entrepreneurship. This paper describes the design and implementation of a fuzzy control system for egg incubating based on IoT technology. The microcontroller is programmed to work as a fuzzy logic control system for controlling microclimate conditions in the egg incubator to keep the conditions for different eggs type optimal. Informations from the temperature and humidity sensors are sent wirelessly to the cloud. Also, the implemented egg incubating system enables automatic tracking of the remaining days until hatching chickens. In this way, remote monitoring, from any location, of microclimatic conditions inside the egg incubator is enabled. For the experimental work analysis of the implemented egg incubating system, the egg incubator is made. Based on the results of the experimental work analysis can be seen that the egg incubating system works well and that it helps with improving poultry production.

## 1. Introduction

The principles of artificial egg incubation were established centuries ago. Based on historical records, Paniago (2005) and van den Sluis (2011) mention that, in ancient Egypt, eggs were incubated in mud-brick buildings, called an "incubation house" [1]. In the upper part of the egg incubator were shelves for burning, straw, dung, or charcoal to provide heat. Vents in the roof allowed the smoke from the fires to leave and provided light. Humidity was controlled by placing moistened jute on eggs. The first mechanical incubator was built in 1749 (Reamur), and the first commercial incubator was built in 1881 (Hearson) [1]. In the last few years, artificial egg incubation systems have experienced a technological, economic, and social revolution [2]. In this paper is described the design and implementation of an egg incubator based on artificial intelligence and IoT technology. The fuzzy controller has been designed to regulate the temperature inside the incubator, which must be constant regardless of any disturbances that may occur. Also, all important microclimatic conditions inside the incubator are sent to the cloud (ThingSpeak platform) further to the Android application for a 24-hour insight into the proper operation of the incubator.

## 2. Required microclimate conditions for egg incubator

Different types of poultry require different egg incubation conditions as well as the duration of incubation. This is given in Table 1 [3].

**Table 1.** Incubation period and the condition of the incubator for different types of poultry.

| Species | Incubation period/days | Temperature in °C | Humidity in % |
|---|---|---|---|
| Chicken | 21 | 37.5-37.7 | 90 |
| Turkey | 28 | 37.2 | 90 |
| Duck | 28 | 37.2 | 90 |
| Goose | 23-28 | 37.2 | 90 |
| Pheasant | 23-28 | 37.7 | 92 |
| Peafowl | 28-30 | 37.2 | 90 |

By the time the eggs hatch, the humidity is around 35%, and when they hatch, the humidity increases to about 90%, or vinegar is added to soften the eggshell.

### *2.1. Temperature*

Eggs have four types of heat transfer: conduction, radiation, convection and evaporation [2]. Temperature control in the egg incubator is very important for successful incubation. Large temperature fluctuations in the egg incubator lead to a reduction in the number of hatched chicks. The transfer function in egg incubators is assumed to be:

$$\frac{\theta_1}{\theta_2}(s) = \frac{1}{\tau s + 1} \tag{1}$$

Where: $\theta_1$ - max temperature in egg incubator (in our case 37.5 °C)
$\theta_2$- start temperature in egg incubator before turning on system
$\tau$ - time needed for temperature rise to max in egg incubator

### *2.2. Humidity and air ventilation*

During incubation, egg water content is absorbed by the embryo and fetus from embryonic annexes, where a small portion is lost to the external environment [3]. In addition, metabolic water is also produced inside the eggs by embryo lipid metabolism, and particularly by fetal lipid metabolism during the last week of incubation, accounting for 8 to 13% of the fetal water content [4]. Low air relative humidity during incubation may cause excessive egg water loss, resulting in embryo dehydration and death or the hatching of small and dehydrated chicks. If the air relative humidity is too high, the incubation period is shortened [5]. There is a consensus among studies that egg water loss should be approximately 12-14% of the initial egg weight [6].

### *2.3. Egg position and egg turning*

The effects of egg turning and changing egg position influence on gas exchange and heat transfer between the eggs and the external environment, egg water loss, adhesion of the embryo and nutrient availability. The egg turning process is most important during the first week of incubation because of the long distance between the embryo and the eggshell. The position of the eggs in the egg incubator is critical for the formation of the air chamber inside eggs. If the eggs are in wrong position in the egg incubator results are embryonal oxygen deficit and delayed metabolism. Wrong positioning of the eggs may be related to slightly more rounded shape of the eggs, impairing the formation of the air chamber, and consequently, internal pipping [7] [9]. Figure 1 shows the candling method of chicken eggs. In Figure 1 can be seen the difference between healthy fertilized egg and unfertilized egg.

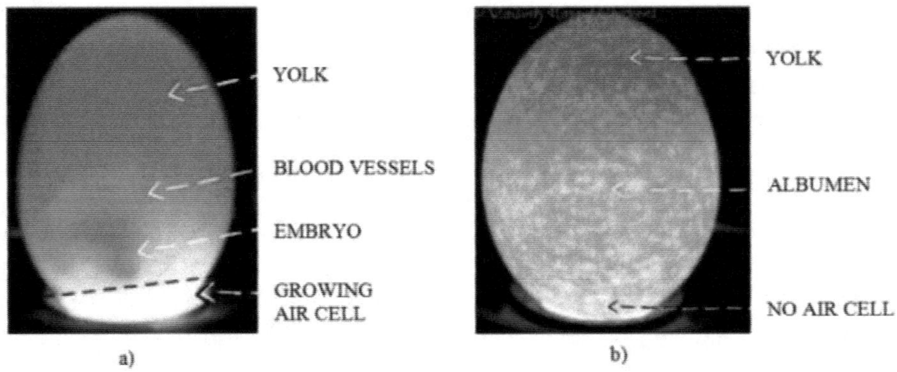


**Figure 1.** Candling method of chicken eggs: a) healthy fertilized eggs, b) unfertilized eggs

## 3. Design and implementation of fuzzy control system for egg incubator based on IoT technology

In this part of the paper, design and implementation of fuzzy control system for egg incubator based on IoT technology is described. The functional block diagram of the egg incubator system is shown in Figure 2.

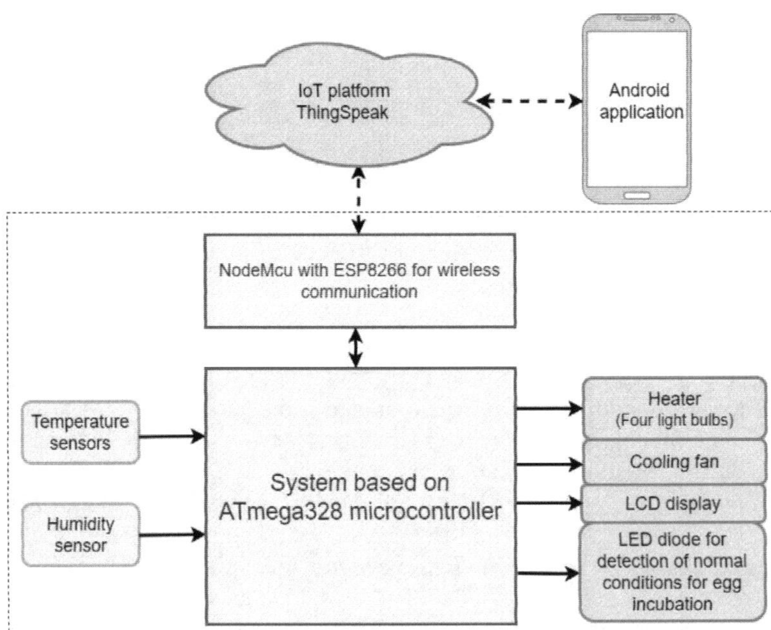


**Figure 2.** The functional block diagram of designed fuzzy control system for egg incubator based on IoT technology

This system consists of the sensors for collecting input information, the control system based on the ATMega 328 microcontroller, the actuators for maintaining micro climatic conditions inside the egg incubator and the nodemcu esp8266 module for sending and receiving the desired information to/from the cloud and android application [8]. Two thermistors with a value of 10 kΩ are used to measure the temperature, and sensor DHT22 is used to measure the humidity. The sensor manufacturer guarantees the accuracy of the sensor within the defined limits in datasheets. Providing enough humidity can be done in two ways. The first way is to place a container of water inside the egg incubator and due to the higher temperature (37.7 ℃) water will evaporate and the humidity will increase. Another way is to place additional heater for the water and thus increase the humidity. In this paper is used the first way. If the humidity inside egg incubator is higher than the desired humidity set by the user, the cooling fan is turn on, and in this way the humidity will stay in desired range. The cooling fan also is turn on to periodically ventilate the egg incubator. The ventilate period is chosen to be every 30 minutes. If the humidity exceeds 35%, the cooling fan is turned on until the humidity value returns in limits, i.e. 35%. Ventilation is periodically turned on every 30 minutes for air renewal and stays on for 3 minutes after which is automatically turn off. The Figure 2 shows that for cloud, the ThingSpeak IoT platform is used. Also, for easier access to data, an android application is designed. The temperature and the humidity read from the sensors and the desired temperature and humidity set form user are sending to the cloud. For the temperature control the fuzzy controller is projected. In the next part of the paper is described the designed of fuzzy logic controller for controlling the temperature inside egg incubator based on experiential knowledge.

### *3.1. Designing the fuzzy logic controller*

The input data in the fuzzy controller is: an error between the desired temperature, ie. the value of the temperature received from the user and the real temperature, and the error rate change. Those values are crisp and need to be converted to fuzzy values i.e. perform fuzzification. That is shown in Figure 3. Trapezoidal and triangular functions were used for fuzzification of input variables, and triangular functions were used for output variable. In Figure 3, abbreviations were used, the meaning of which is given in Table 2.

**Table 2.** The membership functions.

| Name | Meaning |
|---|---|
| NH | Negative High |
| NL | Negative Low |
| Z | Zero |
| PL | Positive Low |
| PH | Positive High |
| N | Negative |
| P | Positive |
| VL | Very Low |
| LOW | Low |
| MED | Medium |
| HIGH | High |
| VH | Very High |

IOP Conf. Series: Materials Science and Engineering 1208 (2021) 012038 doi:10.1088/1757-899X/1208/1/012038

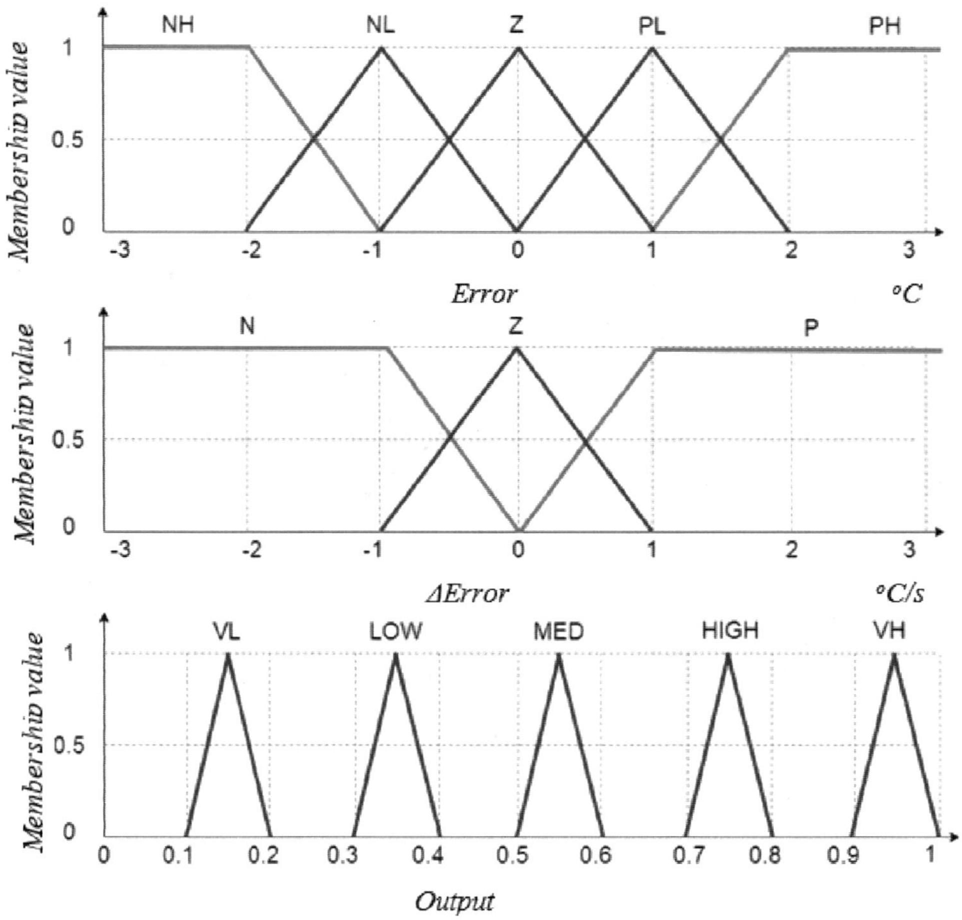


**Figure 3.** The membership functions and range for input and output variables

After creating the membership functions, the rules for Mamdani type fuzzy logic controller are created based on experience. These rules are listed in Table 3.

**Table 3.** The rules for fuzzy controller.

| ΔError \ Error | PH | PL | Z | NL | NH |
|---|---|---|---|---|---|
| P | HIGH | MED | MED | LOW | VL |
| ZE | VH | HIGH | MED | LOW | VL |
| N | VH | HIGH | MED | MED | LOW |

After defining the rules need to choose the method of defuzzification. In our case, was selected centroid method of defuzzification because it gave the best results. These processes are graphically shown in the Figure 4.

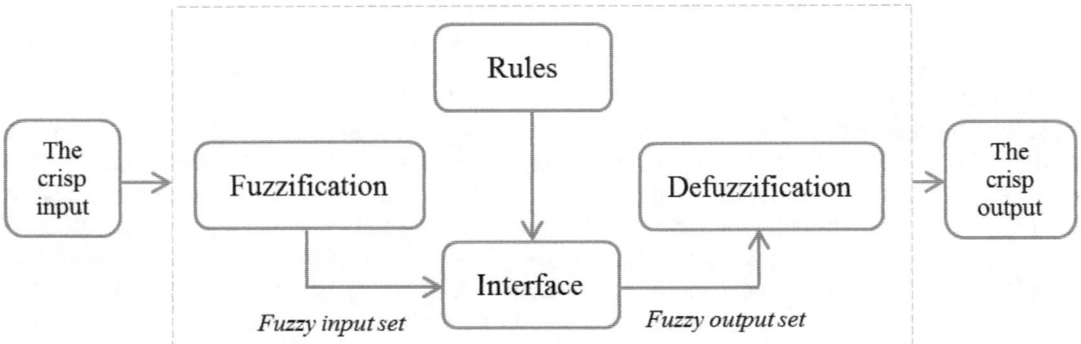


**Figure 4.** The block diagram of designed fuzzy controller

It is now necessary to apply the designed fuzzy controller to control the temperature inside the egg incubator. In Figure 5 is shown block diagram system for control temperature using designed fuzzy controller.

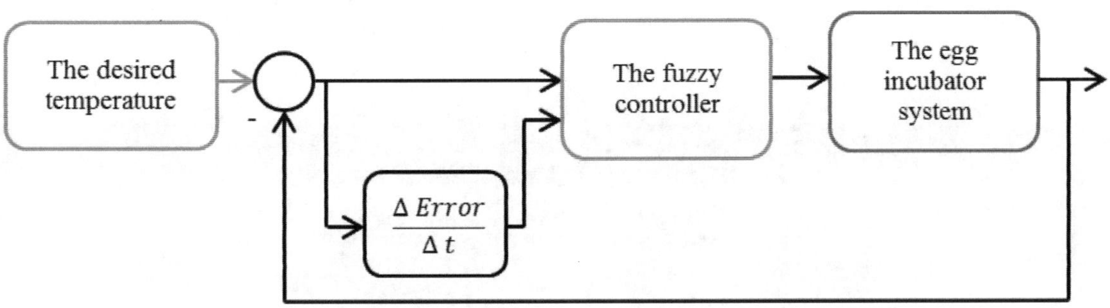


**Figure 5.** The functional block diagram system for control temperature using designed fuzzy controller

In Figure 5, desired temperature is temperature received from user using android application. The temperature read from the temperature sensor returns from the output. The desired temperature and the real temperature are compared and their difference is sent in the fazzy controller. Also, the error rise / decrease rate is sent to the fuzzy controller as a second input. From the fuzzy controller the output crisp value is send as a PWM signal to the MOSFET transistor which is used as an amplifier. The amplified PWM signal is send to four 10-watt tungsten bulbs.

Values $Error$ and ΔError are:

$$Error(i) = T_{des}(i) - T_{real}(i) \tag{2}$$

$$\Delta\, Error(i) = \frac{\mathrm{Error(i)-Error(i-1)}}{\mathrm{t_i - t_{i-1}}} \tag{3}$$

Where is: $T_{des}(i)$ - the desired temperature in the i-*th* step,
$T_{real}(i)$ - the real temperature in the i-*th* step,
$Error(i)$ - temperature difference in the i-*th* step,
$\mathrm{Error(i-1)}$ - temperature difference in the (i-1)-*th* step
$t_i$ - current time i-*th* step,
$\mathrm{t_{i-1}}$ - time in (i-1) –*th* step.

If the $T_{real}$ is lower than $T_{des}$ then $Error$ is the positive value. Else if $T_{real}$ is higher than $T_{des}$ then $Error$ is the negative value. If temperature rises then $\Delta Error$ is positive, else if temperature decreases then $\Delta Error$ is negative.

*3.2. Implementation of fuzzy control system for egg incubator based on IoT technology*

After designing the fuzzy controller, it is necessary to connect all system components based on Figure 2. After connecting all the components, the system shown in Figure 6 is obtained. The power supply to the system is 12 V. This supply voltage value is selected so that the system could operate smoothly in the case of a 220 V mains supply failure. Nodemcu ESP8266 is connected used serial communication with an ATMega 328 microcontroller and connected to 5 V power supply. In our case, an IRFZ44 MOSFET transistor with an additional heatsink is used. A voltage regulator 7805 is used to obtain a stabilized voltage of 5 V required for the working the ATMega328 microcontroller and ESP8266. After connecting all the components, the ATMega328 microcontroller is programmed to perform the function of a fuzzy controller and the ESP8266 microcontroller to send / receive data to / from the cloud using wireless communication.

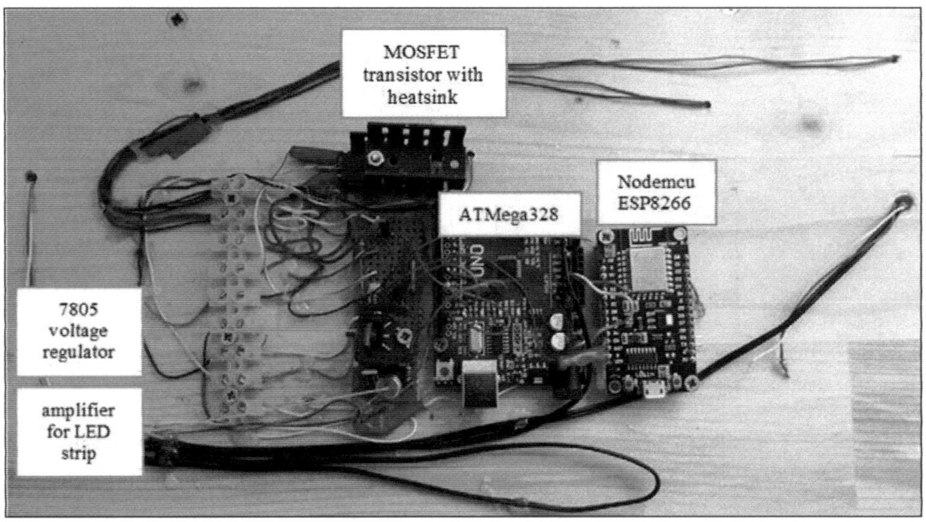


**Figure 6.** Appearance of system for maintaining microclimatic conditions inside the egg incubator

For the implementation of the system, egg incubator size 26 x 36 x 20 cm is made of wood. Inside the egg incubator are placed four 10-watt tungsten bulbs as heater, LED strip, cooling fan, water container and stand for maximum 12 eggs. On the right side of egg incubator is placed display on which is showing temperature and humidity in real time. On the back side of egg incubator is placed hardware part of the system shown in Figure 6. In Figure 7 is shown appearance of the finished egg incubator system.

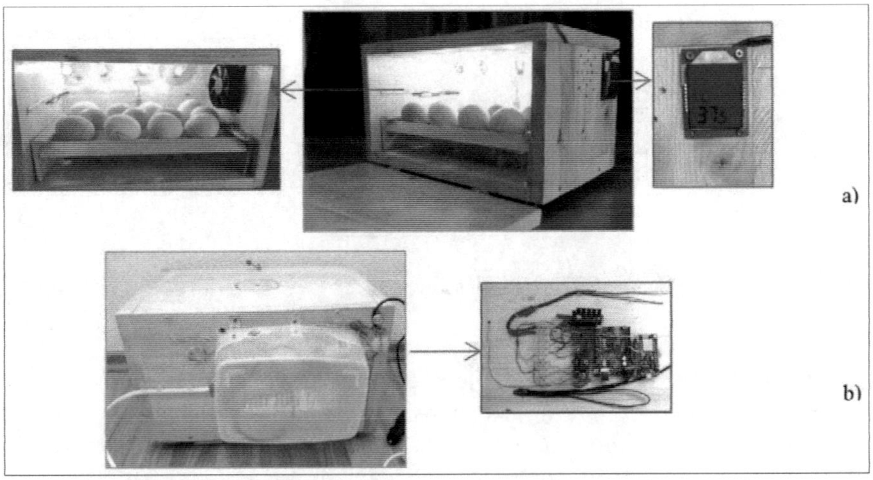


**Figure 7.** The final appearance of the egg incubator system, (a) the front side of the egg incubator, (b) the back side of the egg incubator

### *3.3. Designing android application*

The design of the android application is done using the MIT App Inventor. Using free online software MIT App Inventor enables creating applications by dragging and dropping components into a design view and using a visual language blocks to program application behaviour [10].

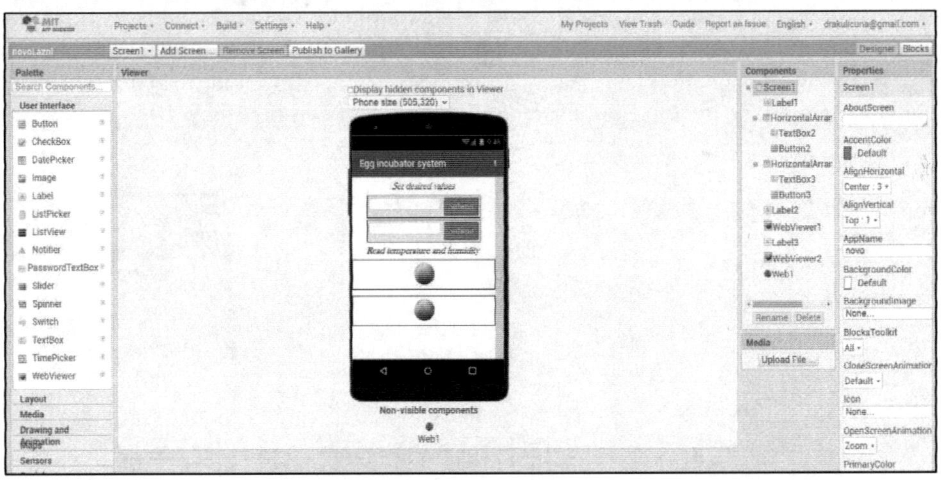


**Figure 8.** The appearance of the designed android application in MIT App Inventor

The appearance of the designed android application in MIT App Invertor is shown in Figure 8.

### *3.4. ACloud*

The designed android application sends the desired data (temperature and humidity) from user to cloud. This data are forwarding from the cloud to the created system for controlling the microclimate conditions in the egg incubator. From created system for control microclimatic conditions inside egg incubator, real temperature and humidity are sending to cloud and forwarded to android application. This is shown in Figure 9. As cloud is used ThingSpeak platform. ThingSpeak is a free platform used to collect, visualize and analyze data in the cloud.

**Figure 9.** The communication block diagram

## 4. Results of experimental work analysis

After turning on the egg incubator system, the temperature inside the egg incubator rise very quickly to the desired temperature send from designed android application. After that temperature stay in the desired limits as well as the humidity. In our case sending data from the egg incubator system to the cloud is every 90 s, which is enough for such systems. The microcontroller reads from the sensor every 2 seconds, which is quite enough for the correct operation of this system. The power supply of the egg incubator system is V_B=12 V. The total electricity consumption of the designed egg incubator system from putting eggs to hatching of chickens is:

$$E = P * t \approx 40.48\ W * 21\ day = 40.48\ W * 504\ h = \ 20402\ Wh = 20.402\ kWh.$$

The ventilator with a DC motor is used for ventilation and to reduce the humidity in case it increase above limits. For LCD display on egg incubator is used another termistor. This LCD display is using to visually confirm the temperature inside the egg incubator. In this way, the possibility of incorrect temperature reading via the main sensor is avoided. The ThingSpeak platform (cloud) is using to store, analysing and monitoring data. The android application is designed to access this data. The ThingSpeak channel and the designed android application is shown in Figure 10.

a) b)

**Figure 10.** (a) The ThingSpeak channel, (b) The android application

In Figure 10 (a) is shown the created ThingSpeak channel where in *Field1* is *Temperature* read from the sensor, in *Field2* is *Humidity* read from the sensor, in *Field3* is *Desired temperature* and in *Field4* is *Desired humidity*. In Figure 8 (b) is shown designed android application where at the top are text boxes for user to enter desired temperature and desired humidity values for the egg incubator system. Below those values in andriod aplication are shown read values the temperature and humidity from sensors. After analyze the data collected in ThingSpeak channel, is obtained that the average value of the humidity is 34.878 %, and the avarage value of the temperature in the egg incubator is 37.4568 °C at set desired values humidity 35 %, and temperature 37.5 °C. The average values for humidity and temperature are obtained on 2000 samples of received data. Time needed to collect 2000 samples on ThingSpeak cloud was 3 000 minutes or 50 hours. Based on results obtained from performed experimental analyze can be conclude that the designed fuzzy controller works very well and maintains the temperature within the desired limits. In our case under ±0.1 °C. Humidity is maintained within the desired limit, in our case below 35 %.

## 5. Conclusion

In this paper, the fuzzy controller is proposed to control the temperature inside the egg incubator. The IoT system is designed to collect data from the egg incubator system and set the desired values. The android application is designed for easier access and assignment of data by users. An experimental work analysis of the designed system is performed, based on which it can be concluded that the designed system works properly.

## References

[1] Boleli I C, Morita V S, Matos Jr J B, Thimotheo M and Almeida V R 2016 Poultry egg incubation: Integrating and optimizing production efficiency *Revista Brasileira de Ciência Avícola* **18** 1–16

[2] Šabić Z, Drakulić U and Mujčić E 2020 The Smart Greenhouse System Based on The Mobile Network and IoT *International Symposium on Innovative and Interdisciplinary Applications of Advanced Technologies* (IAT) (BHAAAS)

[3] Lourens A, van den Brand H, Meijerhof R and Kemp B 2005 Effect of eggshell temperature during incubation on embryo development, hatchability, and posthatch development *Poultry Science* **84** 914–20

[4] van der Pol C W, van Roovert-Reijrink I A M, Maatjens C M, van den Brand H and Molenaar R 2013 Effect of relative humidity during incubation at a set eggshell temperature and brooding temperature posthatch on embryonic mortality and chick quality *Poultry Science* **92** 2145–55

[5] Adegbulugbe T A, Atere A O and Fasanmi O G 2013 Development of an Automatic Electric Egg *Interantional Journal of Scientific & Engineering Research* (vol 4, Issue 9)

[6] Bhosale P, Jagriti, Gillurkar H, Baraparte V, Ramteke P, Burange R 2018 Development of Smart Egg Incubator System Using Arduino *International Conference on Emanations in Modern Engineering Science & Management (ICEMESM)*

[7] Gutierrez S, Contreras G, Ponce H, Cardona M, Amadi H and Enriquez-Zarate J 2019 Development of hen eggs smart incubator for hatching system based on internet of things *2019 IEEE 39th Central America and Panama Convention (CONCAPAN XXXIX)*

[8] Mujčić E and Drakulić U 2019 Monitoring and Control of the Greenhouse Using Remote Control Device and IoT *International Journal of New Innovations in Engineering and Technology* (vol 12, Issue 3)

[9] Mushtag Z, Tayyaba S and Ashraf M W 2014 Liquid level controlling by fuzzy logic technique *International Journal of Innovation and Scientific Research* (vol. 12), pp. 372-379

[10] Škrgić M, Mujčić E and Drakulić U 2018 Smart Home System-Remote Monitoring and Control Using Mobile Phone *International Symposium on Innovative and Interdisciplinary Applications of Advanced Technologies (IAT) (BHAAAS)*

# Model for predicting heart failure using Random Forest and Logistic Regression algorithms

**Vedran Grgić, Denis Mušić, Elmir Babović**

"Džemal Bijedić" University of Mostar, Faculty of Information Technologies

{vedran.grgic}{denis}{elmir.babovic}@edu.fit.ba

**Abstract.** The paper analyzes the cardiovascular parameters of patients with heart disease. The aim of this study was to predict death in a patient with cardiovascular disease based on 12 parameters, using Random Forest and Logistic Regression algorithms. Parameters were tuned for both algorithms to determine the best settings. The most significant factors in the process predicted were found using the FEATURE SELECTION method of both algorithms. By comparative analysis of the obtained results, the highest accuracy of 90% was obtained using the Random Forest Algorithm.

## 1. Introduction

Heart failure is a serious condition in which the amount of blood expelled from the heart every minute is insufficient to meet the body's normal needs for oxygen and nutrients. Heart failure can affect the right, left, or both sides of the heart. Due to the reduced ability of the heart as a pump, there may be an amount of blood in other parts of the body, and it is mistakenly believed that the term heart failure means that the heart has stopped, the term heart failure means that the heart has weakened and lost its ability to work. Heart failure occurs in all age groups but is statistically more common in the elderly with pre-existing conditions. Most often, the condition worsens with age, but with heart disease you can live undisturbed for years if you follow the recommended lifestyle habits. In the United States, about 550,000 new patients develop heart failure each year, and more than half of those who develop heart failure die within 5 years of diagnosis [18][27].

Heart failure is a global pandemic affecting at least 26 million people worldwide and is increasing in prevalence. Problems and mortality caused by heart failure will be dramatically associated with an aging population. Cardiovascular diseases are the leading cause of death in the world. More people die annually from cardiovascular diseases than from any other disease [17].

Most cardiovascular diseases can be prevented, it is necessary to monitor and eliminate risk factors. It is very important that people who have heart disease or extremely high values of the risk factors, one or more, such as hypertension, diabetes, hyperlipidaemia, etc. discover these facts in time, and undergo adequate treatment. Treatment at such an early stage can be only with drugs or only with changes in life habits (elimination of risk factors). When it comes to prevention, there are two ways to prevent cardiovascular diseases, population-wide, and individual.

Examples of comprehensive intervention against cardiovascular diseases can be seen in taxation of goods that cause illness or risk factors, tobacco, certain foods, salts, sugars, etc. Then it is necessary to animate people to healthy living habits, such as walking, playing sports. Find ways to reduce the distribution and consumption of alcohol. It is very important to implement all the above examples in education and teach children healthy living habits and in a passive way to influence the reduction of the number of patients. At the individual level, for the prevention of first heart attacks and strokes, individual health-care interventions need to be targeted to those at high total cardiovascular risk or those with single risk factor levels above traditional thresholds, such as hypertension and hypercholesterolemia.

Efforts should be made in individual prevention and early recognition of disease in which the most significant support is expected from the application of modern technologies and algorithms. The Machine Learning area achieves the most significant results in data processing, grouping, pattern recognition and prediction making. Considering previously mentioned results, which imply some form of learning, applications of Machine Learning within medicine area is suitable for: Disease Diagnosis (key for prevention), Personalized Treatment/Behavioural Modification, Drug Discovery/Manufacturing, Clinical Trial Research, Radiology and Radiotherapy, Smart Electronic Health Records, etc.

The algorithms used within Machine Learning are most divided into two groups: supervised and unsupervised. Supervised machine learning refers to techniques in which a model is trained on a range of inputs (or features) which are associated with a known outcome. In medicine, this might represent training a model to relate a person's characteristics to a certain outcome. Once the algorithm is successfully trained, it will be capable of making outcome predictions when applied to new data. The supervised algorithms are used to solve problems with classification and regression [5][7][9][16][20][21][24][26][30].

Model used in this paper produces discrete values, a patient died or not, so we used classification algorithms. In practice, classification algorithms return the probability of a class. This paper provides an example of a classification algorithm in which a heart failure is predicted. A model that returns a prediction of a continuous value is known as a regression algorithm.
This paper describes the process of training and validating an algorithm to predict mortality based on heart disease parameters [35], using Random Forest and Logistic Regression algorithms.

### *1.1. Overview of research area*

Considering the importance of the area, a significant number of research addresses this topic. Emphasizing the significance of prevention, some of the previous research has focused on early diagnosis in order to prevent heart damage and lethal outcomes. Research [32] proposed model which uses weighted voting of Logistic Regression, Random Forest, K Nearest Neighbour, Gaussian Naive Bayes, and Artificial Neural Network. They used 6 different metrics (Accuracy, Precision, Sensitivity, Specificity, F Measure, and Error) to increase the prediction accuracy and precision. Authors of research [33] investigated different ML-based models to predict HF (heart failure) readmission or death. They reported improvement in performance by using the MLP (Multi-Layer Perceptron) and concluded that the proposed approach is superior to other ML and regression techniques for the prediction of 30 day HF readmission or death. In the context of our research, it is important to mention the paper [34] which used Logistic Regression and Random Forest Algorithm to classify hearth disease. By comparing the accuracy of the aforementioned algorithms, they concluded that Logistic Regression generates slightly better results on used data set.

### *1.2. Random Forest*

Random Forest is a supervised machine learning algorithm, used mostly for classification but can also be used for regression. Classification is method when we have predicting of outcome, based on input data. We can have several predictors X's, and the one response Y. We can make prediction of death based on input parameters. Two possible model outcomes (death or life) we can present with 1 – death, 0 – life. Classification is used to understand relationship between predictors and response [6][8][19][29]. Random Forest Algorithm creates decision trees which are used for training with sample data, prediction and selection of the best solution by voting. More trees mean more robust forest.

Steps of Random Forest Algorithm [13]:

- Step 1 – selection of random samples from a given dataset.

IOP Conf. Series: Materials Science and Engineering 1208 (2021) 012039 doi:10.1088/1757-899X/1208/1/012039

- Step 2 – algorithm will construct a decision tree for every sample. Then it will get the prediction result from every decision tree.
- Step 3 – voting will be performed for every predicted result.
- Step 4 – select the most voted prediction result as the final prediction result

*1.3. Logistic Regression*
Logistic Regression is supervised machine learning algorithm or technique "borrowed" from mathematical statistics. Logistic Regression is used for binary classification problems which enables us to get the answer, will patient live or die. Logistic Regression in one of the most important analytic tools in science. In machine learning is commonly used for classification problems and it has a very close relationship with neural networks [14][25]. It provides mechanisms to train classifier that can make decision about class based on input data. For this step of classification, we use Sigmoid function and with this function we mapped real values into values between 0 and 1. With Sigmoid function, given threshold value, the algorithm determines which class the input parameter belongs to 0 or 1. Everything below threshold value is one class (0), and everything above threshold value is another class (1). For inputs (x) we have linear model, which measure relationship between dependent and independent variables and estimating probabilities by using logistic function. Using Sigmoid function, every real number is map into value between 0 and 1, but it is never exactly 0 or 1, so we have threshold classifier which classifies given number to class 0 or 1.

## 2. Application of algorithms
For implementation of this project, we used Python and Jupyter Notebook application [2][4][10][11][15][31]. Our dataset contains 299 rows of patients and 13 columns (features) which provide clinical, body and lifestyle data. From a logical perspective, we can split the data into categorical and continues variables.

Categorical Variables:
- Anaemia - Decrease of red blood cells - 0 corresponds to false, 1 to true
- High blood pressure - If the patient has high blood pressure - 0 corresponds to false, 1 to true
- Diabetes - If the patient has diabetes - 0 corresponds to false, 1 to true
- Sex - If 0 the patient was a woman and if 1 the patient was a man
- Smoking - If the patient smokes - 0 corresponds to false, 1 to true
- DEATH_EVENT - If the patient died during the follow-up period - 0 corresponds to false, 1 to true

Continues Variables:
- Platelets - Thrombocytes in the blood - range (25.01k - 850.00k)
- Ejection fraction - % of blood leaving the heart with each heartbeat- range (14 - 80)
- Creatinine phosphokinase (CPK) - Level of CPK enzyme in the blood - range (23 - 7861)
- Serum creatinine - Level of creatinine in the blood - range (0.5 - 9.4)
- Serum sodium - Level of sodium in the blood - range (114 - 148)
- Age - Age of the patient - range (40-95) years
- Time - Duration of the follow-up period - range (4-285) days

Feature of interest is DEATH_EVENT, we are going to predict will patient die or live, based on other features. First of all, we have to split data in test and training sets for both algorithms. We are taking 30% of data for training. Then we used sklearn pre-processing feature to remove the mean and scale to unit variance - this leads to better prediction. The sklearn is Python library which contains a lot of efficient tools for machine learning and statistical modelling including classification, regression, clustering and dimensionality reduction.

IOP Conf. Series: Materials Science and Engineering 1208 (2021) 012039 doi:10.1088/1757-899X/1208/1/012039

*2.1. Random Forest*

Random Forest Algorithm have many attributes, but we will consider only the following: n_estimators - the number of trees in the forest., criterion {"gini", "entropy"}, max_depth - the maximum depth of the tree, min_samples_split - the minimum number of samples required to split an internal node.

After applying Random Forest Algorithm on the dataset, we have prediction accuracy of death event with 87% accuracy.

| | precision | recall | f1-score | support |
|---|---|---|---|---|
| 0 | 0.86 | 0.97 | 0.91 | 62 |
| 1 | 0.90 | 0.64 | 0.75 | 28 |
| accuracy | | | 0.87 | 90 |
| macro avg | 0.88 | 0.81 | 0.83 | 90 |
| weighted avg | 0.87 | 0.87 | 0.86 | 90 |

**Figure 1.** Random forest accuracy with default parameters.

For Random Forest Algorithm accuracy depends on data, but also it depends on setting attributes/parameters of algorithm. For this prediction, default parameters are used, and they are represented on Figure 2.

```
{'bootstrap': True,
 'ccp_alpha': 0.0,
 'class_weight': None,
 'criterion': 'gini',
 'max_depth': None,
 'max_features': 'auto',
 'max_leaf_nodes': None,
 'max_samples': None,
 'min_impurity_decrease': 0.0,
 'min_impurity_split': None,
 'min_samples_leaf': 1,
 'min_samples_split': 2,
 'min_weight_fraction_leaf': 0.0,
 'n_estimators': 100,
 'n_jobs': None,
 'oob_score': False,
 'random_state': None,
 'verbose': 0,
 'warm_start': False}
```

**Figure 2.** Random Forest default parameters.

Also, we can see which features from dataset are more important than others. Figure 3 shows that smoking, blood pressure, diabetes and sex, are almost irrelevant but they are also used for prediction.

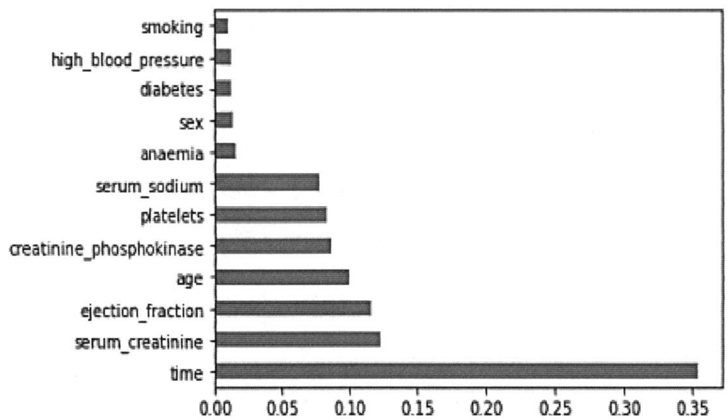


**Figure 3.** Feature importance's.

Random Forest hyperparameter tuning, is done using GridSearchCV [12][22][23][28]. That is process where we obtain the best parameters for Random Forest Algorithm. GridSearchCV is a library function that is a member of sklearn's model_selection package. It helps to loop through predefined hyperparameters and fit model on our training set. So, in the end, we can select the best parameters from the listed hyperparameters [9].

```
grid_forest.best_params_

{'criterion': 'entropy',
 'max_depth': 25,
 'max_features': 'auto',
 'n_estimators': 100}
```

**Figure 4.** Best parameters for Random Forest.

After applying GridSearchCV parameters we have prediction accuracy of 90%.

```
              precision    recall  f1-score   support

           0       0.90      0.97      0.93        62
           1       0.91      0.75      0.82        28

    accuracy                           0.90        90
   macro avg       0.90      0.86      0.88        90
weighted avg       0.90      0.90      0.90        90
```

**Figure 5.** Random Forest accuracy with tuned parameters.

The best score which is achieved with Random Forest Algorithm is 90% accuracy

*2.2. Logistic Regression*

Logistic Regression has ability to show feature importance's too. We can see which features from dataset are more important than others.

```
Feature: 0, Score: 0.69715
Feature: 1, Score: 0.03942
Feature: 2, Score: 0.36148
Feature: 3, Score: 0.11558
Feature: 4, Score: -0.90539
Feature: 5, Score: -0.14069
Feature: 6, Score: -0.10837
Feature: 7, Score: 0.38376
Feature: 8, Score: -0.33606
Feature: 9, Score: -0.42696
Feature: 10, Score: 0.08711
Feature: 11, Score: -1.4599
```

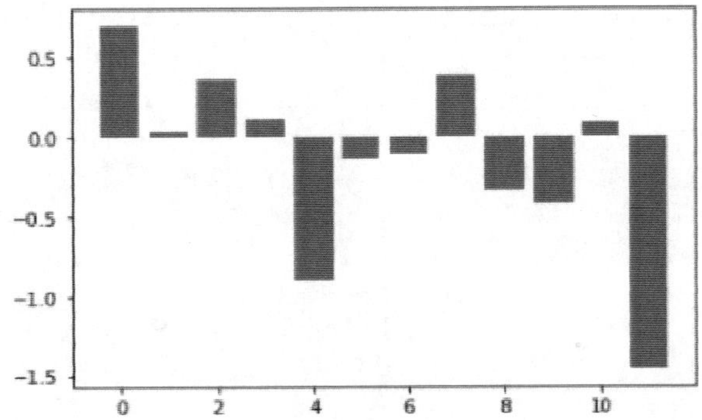


**Figure 6.** Logistic Regression feature importance.

In first phase we tested Logistic Regression with default parameters, shown in the Figure 7.

```
{'C': 1.0,
 'class_weight': None,
 'dual': False,
 'fit_intercept': True,
 'intercept_scaling': 1,
 'l1_ratio': None,
 'max_iter': 100,
 'multi_class': 'auto',
 'n_jobs': None,
 'penalty': 'l2',
 'random_state': None,
 'solver': 'lbfgs',
 'tol': 0.0001,
 'verbose': 0,
 'warm_start': False}
```

**Figure 7.** Logistic Regression default parameters.

After applying Logistic Regression algorithm with default parameters, we achieved accuracy of 81%.

| | precision | recall | f1-score | support |
|---|---|---|---|---|
| 0 | 0.82 | 0.94 | 0.87 | 62 |
| 1 | 0.79 | 0.54 | 0.64 | 28 |
| accuracy | | | 0.81 | 90 |
| macro avg | 0.80 | 0.74 | 0.76 | 90 |
| weighted avg | 0.81 | 0.81 | 0.80 | 90 |

**Figure 8.** Logistic Regression accuracy with default parameters.

For Logistic Regression we also used hyperparameter tuning by GridSearchCV library. After applying GridSearchCV, we found the best parameters for Logistic Regression showed on Figure 9.

```
Best parameters) {'C': 1, 'multi_class': 'ovr', 'penalty': 'l1', 'solver': 'liblinear'}
```

**Figure 9.** Best parameters for Logistic Regression

After applying best parameter for Logistic Regression algorithm, we got prediction accuracy of 82%.

| | precision | recall | f1-score | support |
|---|---|---|---|---|
| 0 | 0.82 | 0.95 | 0.88 | 62 |
| 1 | 0.83 | 0.54 | 0.65 | 28 |
| accuracy | | | 0.82 | 90 |
| macro avg | 0.83 | 0.74 | 0.77 | 90 |
| weighted avg | 0.82 | 0.82 | 0.81 | 90 |

**Figure 10.** Logistic Regression accuracy with tuned parameters.

This can't be considered as significant improvement, but it is worth emphasizing.

## 3. Comparative Analysis

After finishing project with both algorithms, we can compare them by accuracy in prediction, and using AUC (Area Under the Curve) and ROC (Receiver Operator Characteristic Curve) graph. Regarding accuracy in prediction, Random Forest have accuracy of 90%, Logistic Regression 82% with tuned parameters. With default parameters, Random Forest have accuracy 86%, and Logistic Regression have 81% accuracy. The results with tuned parameters are observed below in ROC and AUC graph. ROC graph provides support in deciding the best threshold value of algorithm. ROC graph consists of x-axis which represents False Positive Rate, and y-axis which represents True Positive Rate, and the curve is generated by changing the threshold on the confidence score. AUC and ROC are important evaluation metrics for calculating the performance of any classification model's performance.

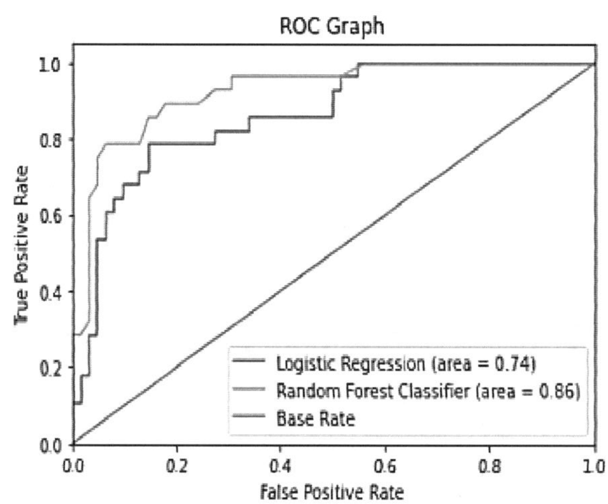


**Figure 11.** ROC graph.

Green diagonal line is used to show when is proportion of correctly classified samples is equal to the proportion of incorrectly classified samples. The area under the ROC curve - AUC is a measure which determine efficiency of a classifier. The AUC makes it easy to compare the ROC curve of one model to another. Higher AUC the better [1].

```
Logistic Regression AUC = 0.74
              precision    recall  f1-score   support

           0       0.82      0.95      0.88        62
           1       0.83      0.54      0.65        28

    accuracy                           0.82        90
   macro avg       0.83      0.74      0.77        90
weighted avg       0.82      0.82      0.81        90


 ---Random Forest Model---
Random Forest AUC = 0.86
              precision    recall  f1-score   support

           0       0.90      0.97      0.93        62
           1       0.91      0.75      0.82        28

    accuracy                           0.90        90
   macro avg       0.90      0.86      0.88        90
weighted avg       0.90      0.90      0.90        90
```

**Figure 12.** AUC for both algorithms.

After the analysis, all the above leads us to a conclusion that the Random Forest is better option for heart failure prediction. Graph (Figure 11) shows that Random Forest line is closer to 1 and have a greater AUC on this dataset.

## 4. Conclusion

The use of Machine Learning in medicine is gaining in importance for the last decade. There are many ways to use machine learning algorithms in medicine and outcome prediction as one of the most used. Random Forest and Logistic Regression algorithms can help predict deaths in heart patients, and the give very good prediction results. But Random Forest Algorithm has better accuracy for default and tuned parameters and have a better AUC result. The main disadvantage of this model is the complexity, in this case, Random Forest Algorithm creates 100 trees in Python sklearn library. To do so, this algorithm requires much computational power and resources. We analysed small dataset which can not affect algorithm performance in noticeable level. To have an even more precise model, it is necessary to increase the amount of data, so not only to get a more reliable model, but also to get more accurate parameters when doing hyperparameter tuning.

## References

[1] Pandey P 2020 Understanding the roc and auc metrics towards data science [cited 2020Nov11] https://towardsdatascience.com/understanding-the-roc-and-auc-curves-a05b68550b69

[2] Li S 2019 Building a logistic regression in python step by step towards data science [cited 2020Dec14] https://towardsdatascience.com/building-a-logistic-regression-in-python-step-by-step-becd4d56c9c8

[3] Donges N 2021 The logistic regression algorithm Experfy Insights [cited 2021May18] https://www.experfy.com/blog/bigdata-cloud/the-logistic-regression-algorithm/

[4] Brownlee J 2020 How to calculate feature importance with python Machine Learning Mastery [cited 2020Dec4] https://machinelearningmastery.com/calculate-feature-importance-with-python

[5] Overfitting in machine learning: What it is and how to prevent it 2020 Elite Data Science [cited 2020Dec5] https://elitedatascience.com/overfitting-in-machine-learning

[6] Kreiger J R 2020 Evaluating a random forest model Analytics Vidhya [cited 2020Dec4] https://medium.com/analytics-vidhya/evaluating-a-random-forest-model-9d165595ad56

[7] Liyanapathirana L 2018 Machine learning workflow on diabetes data Part 02 Towards Data Science [cited 2020Dec2] https://towardsdatascience.com/machine-learning-workflow-on-diabetes-data-part-02-11262b7f7a5c
[8] Fraj M 2017 In depth: parameter tuning for random forest all things AI [cited 2020Nov23] https://medium.com/all-things-ai/in-depth-parameter-tuning-for-random-forest-d67bb7e920d
[9] Borkowski A 2020 Predicting heart failure using machine learning, part 2 Analytics Vidhya [cited 2020Oct30] https://medium.com/analytics-vidhya/predicting-heart-failure-using-machine-learning-part-2-b343471dbde8
[10] How to optimize hyper parameters of a logistic regression model using grid search in python DeZyre [cited 2020Nov2] https://www.dezyre.com/recipes/optimize-hyper-parameters-of-logistic-regression-model-using-grid-search-in-python
[11] Dwivedi R 2020 Guide to hyperparameters tuning using GRIDSEARCHCV and RANDOMIZEDSEARCHCV Analytics India Magazine [cited 2020Dec1] https://analyticsindiamag.com/guide-to-hyperparameters-tuning-using-gridsearchcv-and-randomizedsearchcv
[12] Classification algorithms - random forest Tutorialspoint [cited 2020Dec19] https://www.tutorialspoint.com/machine_learning_with_python/machine_learning_with_python_classification_algorithms_random_forest.htm
[13] Cancer linear regression model tutorial 2017 Dataset by NRIPPNER Data World [cited 2020Oct17] https://data.world/nrippner/cancer-linear-regression-model-tutorial
[14] Stojiljkovic M 2019 Linear regression in python Real python [cited 2020Nov19] https://realpython.com/linear-regression-in-python/#implementing-linear-regression-in-python
[15] Harrison O 2018 Machine learning basics with the K-nearest neighbors algorithm Towards Data Science [cited 2020Dec29] https://towardsdatascience.com/machine-learning-basics-with-the-k-nearest-neighbors-algorithm-6a6e71d01761
[16] Cardiovascular diseases World Health Organization [cited 2020Dec9] https://www.who.int/health-topics/cardiovascular-diseases/#tab=tab_1
[17] Heart and vascular conditions and treatments Heart Failure Statistics [cited 2020Dec30]. https://www.emoryhealthcare.org/heart-vascular/wellness/heart-failure-statistics.html
[18] Probst P, Wright M N and Boulesteix A L 2019 Hyperparameters and tuning strategies for random forest Data Mining and Knowledge Discovery. doi: 9. 10.1002/widm.1301
[19] Babić D 2017 Postupak optimiziranog izbora oglasa na temelju stabala odlučivanja Master thesis (Osijek: Josip Juraj Strossmayer University, Faculty of Electrical Engineering, Computer Science and Information Technology)
[20] Gregorić M 2019 Strojno učenje kao alat za zaključivanje Undergraduate thesis (Zagreb: University of Zagreb, Faculty of Humanities and Social Sciences)
[21] Probst P, Boulesteix A L and Bischl B 2019 Tunability: Importance of hyperparameters of machine learning algorithms, *Journal of Machine Learning Research* **20** pp 1-32
[22] Rijn J, Hutter F 2018 Hyperparameter Importance Across Datasets (NY: Association for Computing Machinery) 24th ACM SIGKDD *International Conference on Knowledge Discovery & Data Mining* pp 2367–2376
[23] Jovanović S, Bajić D, Jovanović M and Milovanović B 2019 Primena tehnika mašinskog učenja u klasifikaciji pacijenata posle infarkta miokarda *18th International Symposium INFOTEH-JAHORINA*
[24] Prasad R, Anjali P, Adil S and Deepa N 2019 Heart disease prediction using logistic regression algorithm using machine learning *International Journal of Engineering and Advanced Technology* **8** 3S
[25] Tanvir A, Assia M, Bhatti S H, Aftab M, Raza M 2017 Survival analysis of heart failure patients: A case study PLoS ONE **12**(7) e0181001. doi:10.1371/journal.pone.0181001
[26] Sidey G J and Sidey G C 2019 Machine learning in medicine: a practical introduction. *BMC Med Res Methodol* **19** 64 https://doi.org/10.1186/s12874-019-0681-4

[27] Selthofer R K and Drenjančević I 2014, Čimbenici rizika za kronično zatajivanje srca, Reviewarticle *Medix* **112**

[28] Schratza P, Muenchowa J, Iturritxab E, Richterc J and Brenninga A 2018 Performance evaluation and hyperparameter tuning of statistical and machine-learning models using spatial data *Ecological Modelling* **406** pp 109-120 arXiv:1803.11266v1

[29] Louppe G 2015 Understanding random forests from theory to practice PhD thesis Machine Learning (stat.ML) arXiv:1407.7502

[30] Hrastić D 2019 Implementacija klasifikacijskog algoritma podizanja gradijenta stablima odluke s primjenom na predikciju ishoda nogometnih utakmica Undergraduate thesis (Zagreb: University of Zagreb, Faculty of Electrical Engineering and Computer Science) https://www.bib.irb.hr/1009326/download/1009326.Dominik_Hrastic_Zavrsni_final.pdf

[31] Tokic L 2017 ROC krivulja – krivulja odnosa specifičnosti I osjetljivosti klasifikatora Master thesis (Zagreb: University of Zagreb, Faculty of Science) https://repozitorij.pmf.unizg.hr/islandora/object/pmf%3A4490/datastream/PDF/view

[32] Chowdary G J, Suganya G and Premalatha M 2020 Effective prediction of cardiovascular disease using cluster of machine learning algorithms *Journal of Critical Reviews* **7** 19 pp 1865-1875. doi:10.31838/jcr.07.19.227

[33] Awan SE, Bennamoun M, Sohel F, Sanfilippo F M and Dwivedi G 2019 Machine learning-based prediction of heart failure readmission or death: implications of choosing the right model and the right metrics *ESC Heart Failure* **6** doi: 10.1002/ehf2.12419.

[34] Latifah F and Slamet I S 2020 Comparison of heart disease classification with logistic regression algorithm and random forest algorithm *AIP Conference Proceedings* **2296** 020021 doi: 10.1063/5.0030579

[35] Chicco D and Jurman G 2020 Machine learning can predict survival of patients with heart failure from serum creatinine and ejection fraction alone *BMC Medical Informatics and Decision Making* **20** 16

IOP Conf. Series: Materials Science and Engineering 1208 (2021) 012040 doi:10.1088/1757-899X/1208/1/012040

# Adaptive car control system based on a predictive model

**Amel Toroman and Samir Vojić**

University of Bihać, Faculty of Technical Engineering, Bosnia and Herzegovina

E-mail: amel.toroman@unbi.ba

**Abstract**. An adaptive control is a control, which by pre-setting the parameters of the controller, enables the control of processes whose parameters are time-varying or are initially uncertain. The possibilities and benefits of adaptive control are versatile and can be best demonstrated by applying the system while driving a car, or maintaining the optimal speed and distance between cars, which is shown in this paper. As the car's weight decreases while driving due to fuel consumption, the control algorithm has to be adapted to the changed driving conditions. Accordingly, an adaptive control system using the Matlab software package, and an adaptive cruise control system (ACC) was created in this paper, which is based on a predictive model. After evaluating the developed model of adaptive car motion control, the output parameters such as speed, acceleration, and distance between the two vehicles were analyzed. In this paper a PID controller is used to reduce oscillations in the system. First, the P controller was used to reduce the rise time of the significant values, then the PI controller improved the rise time, and finally the PID controller achieved overshoot reduction performance without affecting the dynamic response system. The obtained results confirm the justified expectations for the possibility of adaptive car control utilization as one of the possible solutions to the increasing traffic incidents, as well as a measure to improve the reduction of these incidents.

## 1. Introduction

Nowadays, with the increase in traffic demand, there is a need for a flexible way of controlling road traffic. Adaptive control systems are made possible by the emergence of new technologies. With the increase in the number of cars on the road and the increasingly complex transport infrastructure, the need for modernization of the transport system has also increased. Accordingly, new solutions in the field of intelligent transport systems (ITS) have led to new ways of flexible controlling.

Compared to classic cruise control, Adaptive Cruise Control (ACC) contributes to improved comfort and reduced stress for drivers by maintaining tasks at a distance from the Lead Car. The ACC automatically adjusts the travel speed to keep the Lead Car according to the set distance. The system automatically decelerates the car when approaching the Lead Car, if necessary, with a slight activation of the brake, which contributes to and saves fuel consumption [2] and increases economic efficiency [3]. If the Lead Car increases speed, the ACC also increases the car speed, up to the set maximum speed. In case there are no vehicles ahead, the ACC behaves like a classic cruise control. Also, when driving downhill, the ACC brakes maintain the set speed, while a driver only needs to keep the steering wheel under control, and everything else is done by a smart ACC system. Adaptive Cruise Control has proven to be extremely useful in low visibility conditions, such as fog. In this case, the system easily recognizes the Lead Cars, even if a driver has not yet noticed them, and automatically slows down the car.

In order to understand the concept of the adaptive control in this paper, its application in car driving based on a predictive model is presented. Unlike [4] where it is assumed that the Lead Car follows the reference speed set by the driver, and the Ego Car from the ACC system follows the previous vehicle at the desired distance, in this paper the system is implemented using a PID controller was used and a comparison of the effects of P, PI and PID controllers on the dynamic car control system was performed. The simplified dynamics model for vehicle is considered in Longitudinal Vehicle Dynamics and is used in the development of the lower-level controller.

Accordingly, the first part of the paper generally considers adaptive control systems, their characteristics, areas of application, and the model of adaptive speed control and control prediction, as the main elements of an intelligent transport system. In the second part of the paper, an Adaptive Cruise Control system based on a predictive model was implemented using the example of a car control, i.e. maintaining the appropriate speed and distance between cars. In the process of simulating the behavior of this model, which is described by the dynamic behavior of the car, the Simulink tool from the Matlab software package was used. The analysis of the obtained results shows the effect of the ACC system, and the optimal parameters based on which the ACC system maintains a safe distance and appropriate speed between cars, according to their initial speed and position, as well as acceleration of the Ego Car, and how the PID controller affects the behavior of the created system.

## 2. Adaptive control systems

Research on adaptive steering began in the 1950s [1], which was prompted by problems in designing autopilots for aircraft operating at a wide range of speeds and altitudes. Since then, we have strongly developed adaptive control [3], which is applied in: advanced flight control systems for airplanes or spacecraft, robotic manipulators, process control, energy systems, car control and much more. An adaptive control is a systematic approach to an automatic adjustment (setting) of the controller in real time for the maintenance of the desired level of control performance when the parameters of the dynamic process model are unknown and/or change over time [1].

Characteristics of adaptive control are [1]:

- Closed loop control.
- Data on system characteristics are determined during system operation (online).
- Identification of unknown parameters or measurement and calculation of quality criteria.
- Selecting a control strategy and acting on the system by changing the signal, parameter or structure.

The Adaptive Control System is primarily oriented towards eliminating disturbances in the operation of the control system.

Disorders affecting the control system [1]:

- Disorders affecting control variables.
- Disorders affecting the control system performance.

Therefore, a flexible control system is a feedback system whose control variable is the index of performance (IP), and the control monitoring of an IP system with unknown and variable parameters representing precisely the goal of the adaptive control.

The adaptive control system measures a specific IP control system using inputs, states, outputs and known interferences, and based on a comparison of the measured IP and the default set of IPs, the adjustment mechanism changes the adjustable controller parameters and/or generates an auxiliary control to maintain control the IP system of a close default set of IP addresses, as shown in this paper on the example of car control

### *2.1. Adaptive Cruise Control System (ACC)*

Since 1995, when the first cruise control was installed in a car, it has become an almost inevitable in cars, and without it, long journeys would be much more tiring. Nowadays, when traffic is heavier, a conventional cruise control is no longer applicable as it was before. However, instead of becoming obsolete, it has evolved and adapted to new circumstances. For this reason, a cruise control is becoming more adaptable, with the possibility of automatically adjusting the speed in relation to the Lead Car and acceleration when there are no cars ahead. An Adaptive Cruise Control (ACC) is a function of the advanced driver assistance system in the longitudinal dynamics of the car maintaining the desired distance and safe speed from a preceding car [4].

Car speed control is a classic application of the control theory system. Cruise control maintains the speed of a car by adjusting the position of the throttle, which means that sensors are necessary to determine the speed and position of the throttle.

Adaptive Cruise Control in the rear of the car detects the Lead Car and keeps the Lego Car at a safe distance from the Lead Car by controlling its accelerator and brake [5].
When overtaking a slower vehicle, Adaptive Cruise Control uses a radar signal to estimate the distance to it and automatically maintains the distance according to one of three selected settings, such as:

a) Supervision of constant velocity
b) Automatic deceleration
c) Automatic acceleration

The principle of the Adaptive Cruise Control is based on the fact that after setting the desired cruising speed, the system will monitor the traffic in front and automatically adjust the speed to the current situation, while maintaining a safe distance from the Lead Car and those we approach while driving. A driver does not have to brake or depress the accelerator pedal later on. After bypassing a slower car, i.e. crossing the overtaking lane, the car continues to accelerate to the set driving speed.
Reducing traffic congestion and increasing traffic capacity can benefit the environment. Research has shown that ACC systems actually contribute to the environment by reducing fuel consumption and car emissions [6]. Bose and Ioannou [7, 8, 9] suggest that some of the characteristics of ACC-equipped cars (such as accurate car tracking, position error mitigation generated by an armored car during smooth transients, and a smooth response to filtering caused by traffic disturbances) may benefit from air pollution and fuel consumption.

### *2.2. Model Predictive Control (MPC)*

The development of Model Predictive Control (MPC) began in the late 1970s. The Model Predictive Control uses the known dynamics of the process model to calculate a control signal that will achieve the minimum of a given control criterion. Although initially the application was tied exclusively to slow industrial processes, today MPC controllers are used to control a diverse range of systems.

MPC uses a future prediction strategy in order to calculate the input. To ensure that the output of the plant follows the target reference output, the MPC controller uses what is called an optimizer, which comes to the picture by determining the best scenario which achieves the minimum error between the reference and the predicted trajectory. The prediction strategy is based on the use of a plant model, in this case car control by the MPC controller to simulate the car's path in the next P time steps, where P is the prediction horizon which represents the time, the MPC controller looks forward in the future to make the prediction [10].

Thus, the Model Predictive Control (MPC) is an advanced process control method used to control a process satisfying a number of limitations.

## 3. Application of Adaptive Cruise Control System in a car based in the MPC model

Recently, many topics in Intelligent Transportation System (ITS) are researched. One of the topics is Adaptive Cruise Control (ACC). The Adaptive Cruise Control Systems should be designed such that string stability can be guaranteed in addition to that every vehicle in a string of ACC vehicles which use the same control law must track any bounded acceleration and velocity of its preceding vehicle with a bounded distance and velocity error [11]. On the example of maintaining speed between vehicles shows the application of the use of the adaptive control system block in Matlab Simulink and shows the objectives and limitations of managing this block.

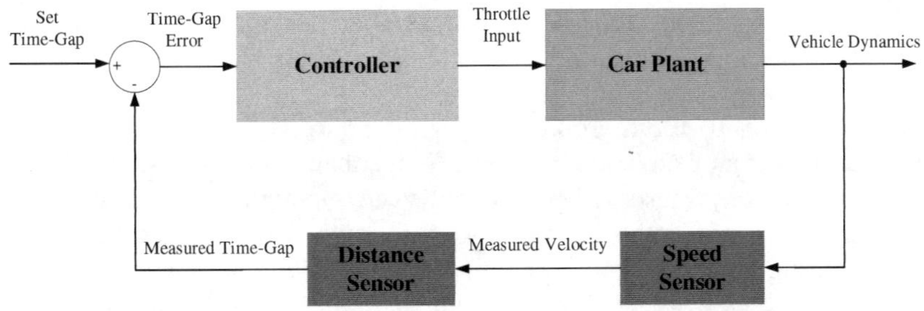


**Figure 1.** Block diagram od Adaptive Cruise Control Systems

Figure 1 shows a block diagram of an Adaptive Cruise Control System. As can be seen the input signal of the regulator is an error that represents the difference between the desired and actual speed, or the distance between the cars. In accordance with this error, the controller will give a gas control signal that adjusts the valve angle. The accelerator pedal opening range changed by the controller will change the engine speed. Cruise control regulates the vehicle speed (v) based on the desired speed and distance, and the actual speed and distance come from the sensor from the feedback branch. The short time-gap gives the driver or system less time to react if the previous vehicle has to slow down very quickly, increasing the possibility of an accidental accident.

A vehicle (Ego Car) equipped with Adaptive Cruise Control (ACC) has a sensor, such as radar, that measures the distance to the preceding vehicle in the same lane (Lead Car), $D_{rel}$. The sensor also measures the relative velocity of the Lead Car, $V_{rel}$. The ACC system operates in the following two modes:

- *Speed control:* The Ego car travels at a driver-set speed.
- *Distance control:* The Ego car maintains a safe distance from the Lead car.

The ACC system decides which mode to use based on real-time radar measurements. For example, if the Lead Car is too close, the ACC system switches from speed control to distance control. Similarly, if the Lead Car is further away, the ACC system switches from distance control to speed control. In other words, the ACC system makes the Ego Car travel at a driver-set speed as long as it maintains a safe distance.

The following rules are used to determine the ACC system operating mode:

- If $D_{rel} \geq D_{safe}$, then speed control mode is active. The control goal is to track the driver-set velocity, $V_{set}$.
- If $D_{rel} < D_{safe}$, then distance control mode is active. The control goal is to maintain the safe distance, $D_{safe}$.

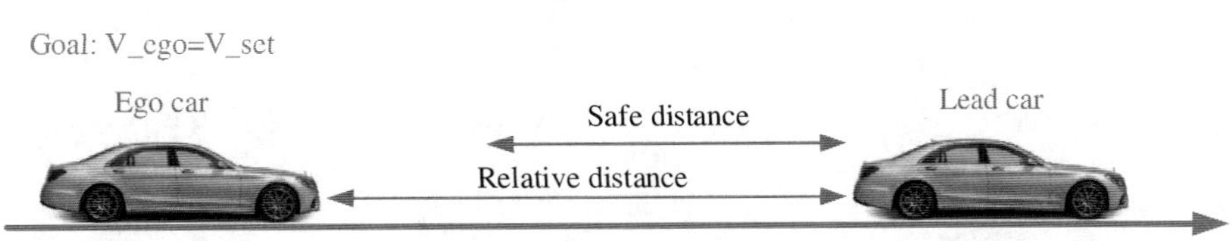


**Figure 2.** Speed control

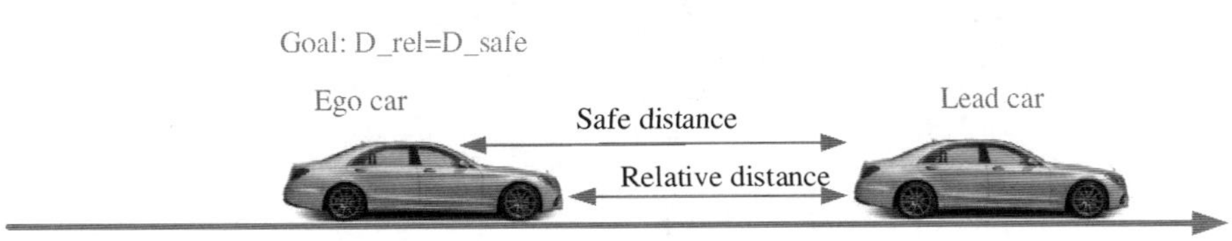


**Figure 3.** Distance control

IOP Conf. Series: Materials Science and Engineering 1208 (2021) 012040 doi:10.1088/1757-899X/1208/1/012040

Therefore, depending on the safe and relative distance between the cars, the Adaptive Control System decides whether to use the appropriate mode, speed control or distance control.

### *3.1. Simulink Model for Lead Car and Ego Car*

Automatic cruise control is an excellent example of a feedback control system found in many modern vehicles. The purpose of the cruise control system is to maintain a constant vehicle speed despite external disturbances, such as changes in wind or road grade. This is accomplished by measuring the vehicle speed, comparing it to the desired or reference speed, and automatically adjusting the throttle according to a control law.

**Figure 4.** Mechanical car control system

Figure 4 shows the mechanical system of vehicle dynamics. A vehicle of mass, m is subjected to a steering force, f. The force f represents the force generated at the road/tire interface. It is assumed that the resistance forces, bv, due to rolling resistance and wind resistance, vary linearly with the vehicle speed, v and act in the opposite direction to the vehicle movement.

Based on the above, we are left with a system of first-order mass silencers. By summing the forces in the x direction and applying Newton's 2nd law, we arrive at the following system Equation (1):

$$m\frac{dv}{dt} + bv(t) = f(t) \tag{1}$$

Taking the Laplace transform of the differential Equation (1) and assuming zero initial conditions, we find the transfer function of the cruise control system to be, Equation (2):

$$ms.V(s) + bV(s) = F(s) \tag{2}$$

$$G_o(s) = \frac{1}{ms + b} \tag{3}$$

The dynamics of control for Lead Car and Ego Car are modeled in Matlab Simulink (Figure 5).

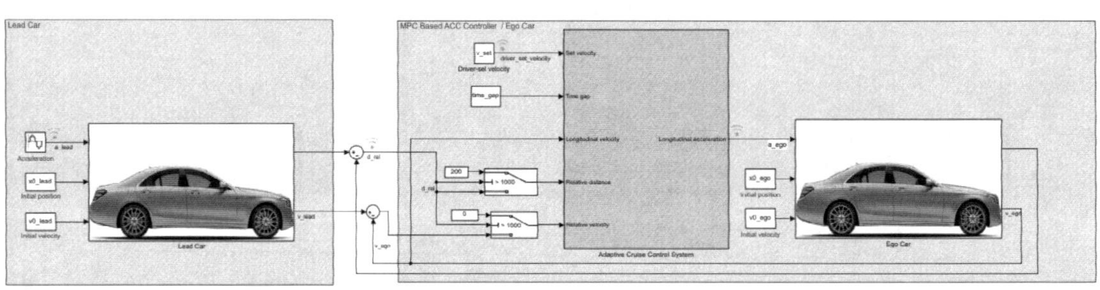

**Figure 5.** Simulink model of Adaptive Car Control System

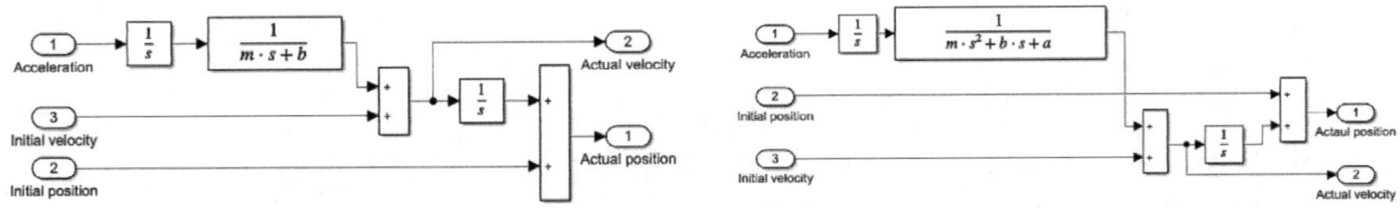


(a) (b)

**Figure 6.** Subsystem of Lead car and Ego car (a) Ego Car (b)

To approximate a realistic driving environment, the acceleration of the Lead Car varies according to a sine wave during the simulation. The Adaptive Cruise Control System (ACC) block outputs an acceleration control signal for the Ego Car, and its complete model is shown in Figure 7.

The analysis of the system operation requires two times: sampling time (Ts) and simulation duration, (T), in seconds. In this case, for the above times it is assumed that Ts = 0.1 [s], and T = 80 [s].

For both, the Ego Car and the Lead Car, whose subsystem is shown in Figure 6, the dynamics between acceleration and velocity are modeled as:

$$G(s) = \frac{1}{s(ms + b)} \tag{4}$$

which approximates the dynamics of the throttle body and vehicle inertia.

By applying the Ziegler Nichols method to equation (3), the parameters of proportional (Kp=1.36), integration (Ki=5.42) and derivative (Kd=0.06) gain were obtained, which were used for the application of P, PI and PID regulators.

Considering Proportional control (Kp), the closed loop transfer function of the cruise control system with a proportional control is obtained (P controller), Equation (5):

$$G_c(s) = \frac{K_p}{ms + (b + K_p)} \tag{5}$$

Considering Proportional (Kp) and integral (Ki) control, the closed loop transfer function of the cruise control system with PI controller is obtained, as in Equation (6):

$$G_c(s) = \frac{K_p s + K_i}{ms^2 + (b + K_p)s + K_i} \tag{6}$$

By considering all three control parameters i.e, proportional (Kp), integral (Ki) and derivative (Kd) control, the closed loop transfer function of the cruise control system with PID controller is obtained, as in Equation (7):

$$G_c(s) = \frac{K_d s^2 + K_p s + K_i}{(m + K_d)s^2 + (b + K_p)s + K_i} \tag{7}$$

As can be seen from Figure 5, the input parameters for the Ego Car are acceleration, initial position (10 m) and initial speed (20 m/s), while the Lead Car for input parameters receives longitudinal acceleration from the ACC system, and initial position (50 m) and the initial speed (25 m/s) it has.

### *3.2. Configuration of Adaptive Cruise Control System*

The ACC system is modelled using the Adaptive Cruise Control System Model in Matlab Simulink. As can be seen from Figure 7, the inputs to the ACC system block are:

- Driver-set velocity: $V_{set}$
- Time gap: $T_{gap}$
- Velocity of the Ego car: $V_{ego}$
- Relative distance to the Lead car (from radar): $D_{rel}$
- Relative velocity to the Lead car (from radar: $V_{rel}$

The output for the ACC system is the acceleration of the Ego Car.
Based on the longitudinal speed of the Ego Car, the relative distance and speed between the Lead and the Ego Car, the ACC generates a longitudinal acceleration for the Ego Car.

The basic logic of an ACC system is, if the relative distance is less than the safe distance, then the primary goal is to slow down and maintain a safe distance. If the relative distance is greater than the safe distance, then the primary goal is to reach driver-set velocity while maintaining a safe distance. These design principles are achieved through the Min and Switch blocks, as can be seen from Figure 7.

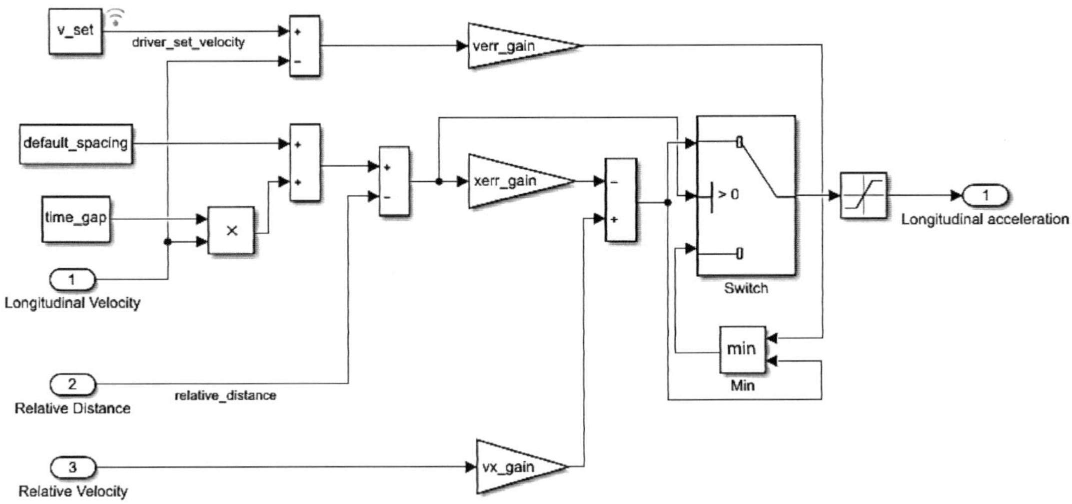


**Figure 7.** Adaptive Cruise Control System Model

The ACC makes the Ego Car travel at a driver-set velocity while maintaining a safe distance from the Lead Car. The safe distance between lead car and ego vehicle is defined as Equation (2), which means that the safe distance between the Lead Car and the Ego car is a function of the Ego Car velocity: $V_{ego}$:

$$D_{safe} = D_{default} + T_{gap} \cdot V_{ego} \tag{8}$$

where:

- $D_{default}$ – the standstill default distance
- $T_{gap}$ – the time gap between the cars
- $V_{ego}$ – Ego car velocity

In this particular case, the set distance between the cars is 10 meters, while the time distance between the cars is 1.4 seconds, the set car speed set by the driver is 30 (m/s).
Considering the physical limitations of the car dynamics, the acceleration is constrained to the range [-3,2] (m/s$^2$), that represents the value of the minimum and maximum acceleration of the Ego Car.

After making the above settings, the default parameters of the Adaptive Cruise Control System block match the parametric simulations, which is shown in Section 3.4.

*3.3. Configuration of Model Predictive Control (MPC)*

Figure 8 shows a Model Predictive Control based Adaptive Cruise Controller, which, as can be seen, improves the performance of relative speed and distance in the ACC system by using Switch blocks. Namely, MPC controller adds the ability to react to more aggressive maneuvers by other vehicles in the environment, in the sense that MPC controller regulates the velocity of the Ego Car while maintaining a strict safe distance constraint. Therefore, the controller can apply more aggressive maneuvers when the environment changes quickly in a similar way to what a human driver would do.

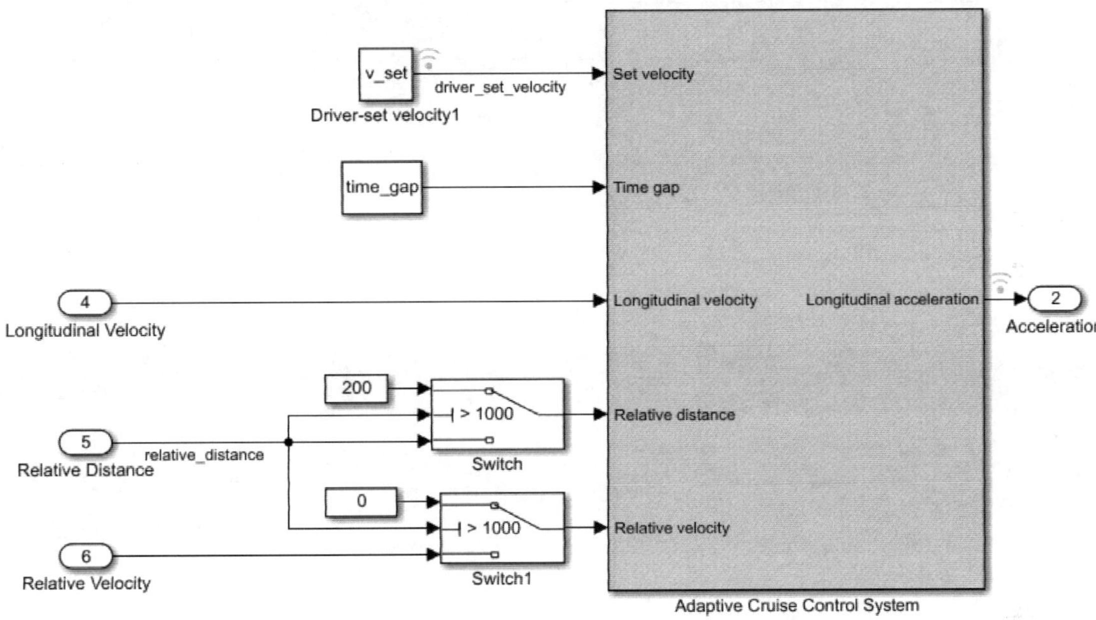


**Figure 8.** Model Predictive Control based Adaptive Cruise Controller

Model predictive control (MPC) is a discrete-time multi-variable control architecture. At each control interval, an MPC controller uses an internal model to predict future plant behavior. Based on this prediction, the controller computes optimal control actions.

*3.4. Simulation analysis*

During the starting of simulation, it is necessary to convert the model to a discrete execution time, assuming that the measured output channels are added to the measured output channel # 2 into integrated white noise, and assuming no disturbance is added to the measured output channel # 1.

By simulating the created model from Figure 5, parameters such as speed, acceleration and distance between cars were observed, and the obtained of simulation results without controller action as well as with P, PI and PID controller action, as can be seen in the figures below.

IOP Conf. Series: Materials Science and Engineering 1208 (2021) 012040 doi:10.1088/1757-899X/1208/1/012040

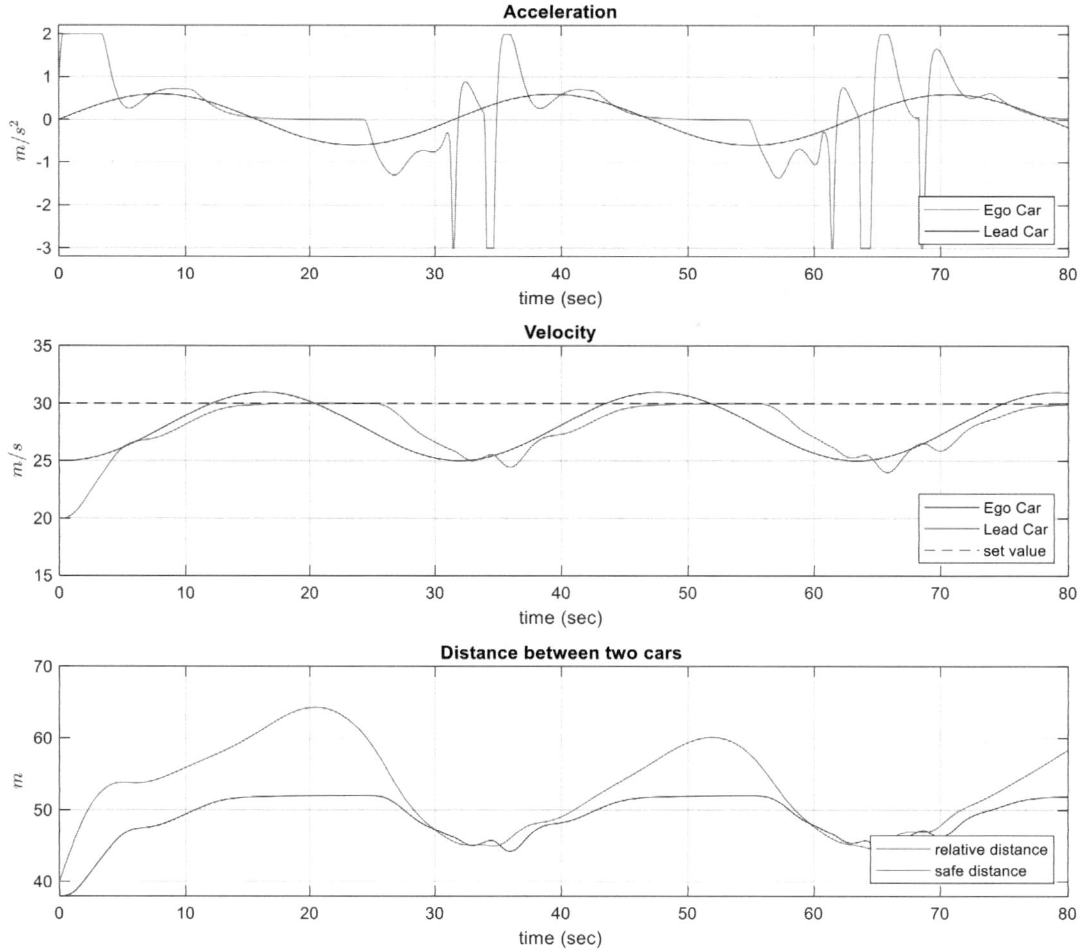


**Figure 9.** Acceleration, Speed, Distance between two cars (without regulator action)

As can be seen from Figure 9, in the first 3 seconds, to reach the driver-set velocity, the Ego Car accelerates at full throttle. From 3 to 13 seconds, the Lead Car accelerates slowly. As a result, to maintain a safe distance to the Lead Car, the Ego Car accelerates with a slower rate.

From 13 to 25 seconds, the Ego Car maintains the driver-set velocity, as shown in Figure 9, the *Velocity* plot. However, as the Lead Car reduces speed, the distance error starts approaching 0 after 20 seconds.

From 25 to 45 seconds, the Lead Car slows down and then accelerates again. The Ego Car maintains a safe distance from the Lead Car by adjusting its speed, as shown in Figure 9, the *Distance between two cars* plots.

From 45 to 56 seconds, the distance error is above 0. Therefore, the Ego Car achieves the driver-set velocity again. After 56 seconds, until 76 seconds, the deceleration/acceleration sequence from the 25 to 45 second interval is repeated.

IOP Conf. Series: Materials Science and Engineering 1208 (2021) 012040 doi:10.1088/1757-899X/1208/1/012040

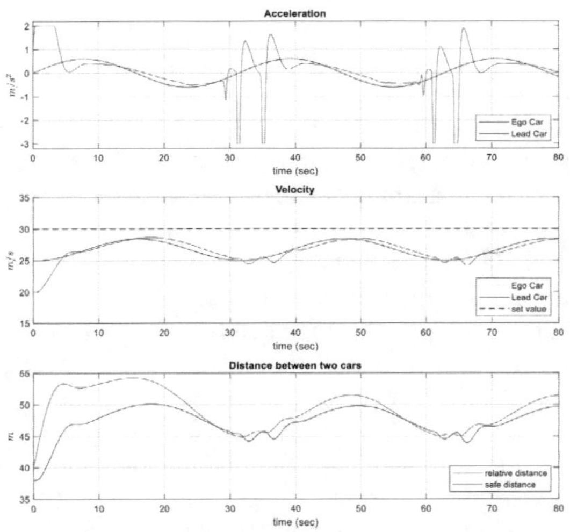


**Figure 10.** Acceleration, Speed, Distance between two cars with the action of the regulator (P controller)

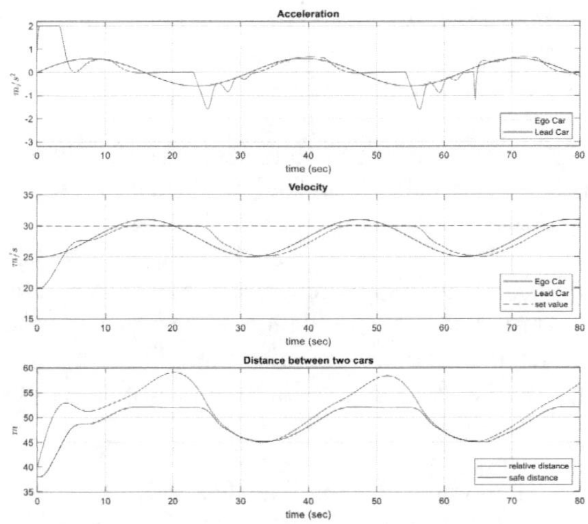


**Figure 11.** Acceleration, Speed, Distance between two cars with the action of the regulator (PI controller)

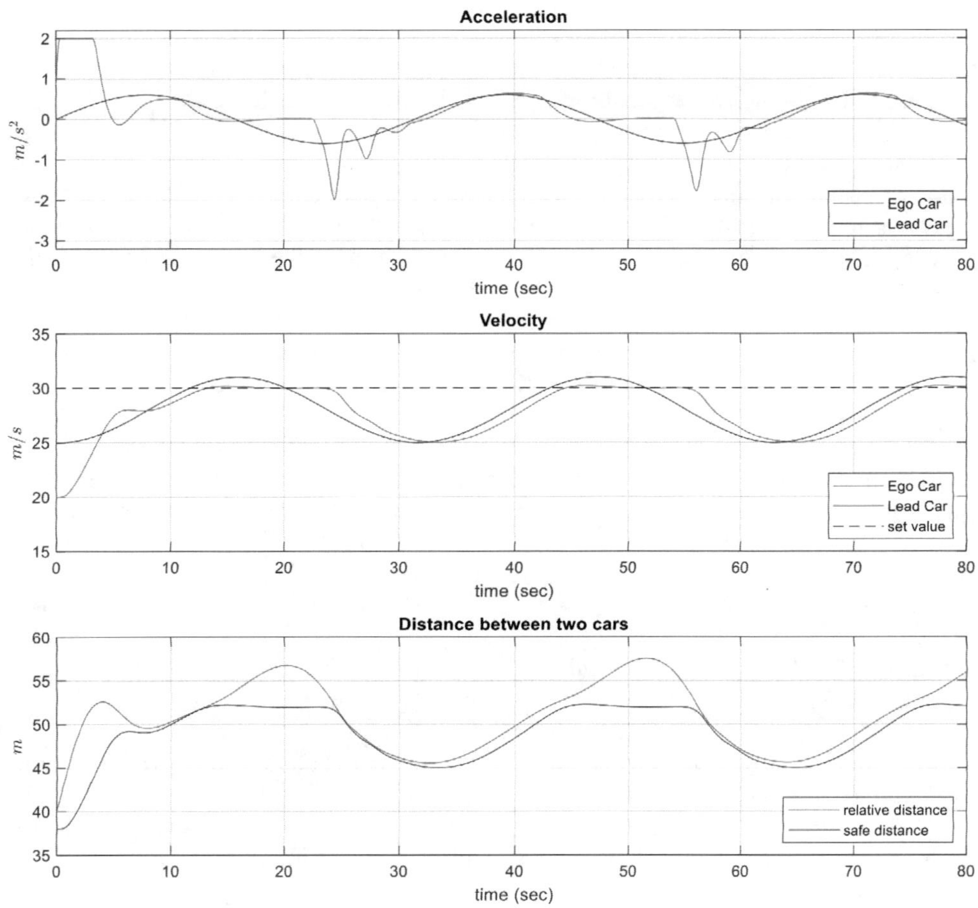


**Figure 12.** Acceleration, Speed, Distance between two cars with the action of the regulator (PID controller)

IOP Conf. Series: Materials Science and Engineering 1208 (2021) 012040 doi:10.1088/1757-899X/1208/1/012040

Based on Equations (5), (6) and (7) applied to the basic model, the result of which is shown in Figure 9, the operation of P, PI and PID controllers is shown, and their comparison is shown in Figure 10, Figure 11, and Figure 12.

As can be seen from Figure 10 by the action of the P controller oscillations occur in the system, which are largely eliminated by the action of the PI controller, and a faster response of the system is obtained, which can be seen from Figure 11. The best effect is achieved by the action of the PID controller, which can be seen from Figure 12 where the oscillations in the system are almost completely eliminated and a rapid response of the system to a sudden change in error is achieved.

Throughout the simulation, the controller ensures that the relative distance between the two cars is greater than the set safe distance. When the actual distance is sufficiently large, then the controller ensures that the Ego Car follows the driver-set velocity.

## 4. Conclusion

The autonomous car consists of several subsystems for the implementation of automotive safety, one of them is the Adaptive Cruise Control System, whose implementation is also shown in this paper on the example of control the appropriate speed and distance between cars, based on the MPC model.

In this paper, despite the relatively small number of Lead Car input parameters (acceleration, initial speed and initial position) and Ego Car output parameters (initial speed, initial position and longitudinal acceleration of the ACC system), the system responds well to changes in all observed driving conditions, especially in the case of PID controller applications. A dynamic low-order model was used to implement the Lead Car, while a higher-order model was used to implement the Ego Car.

The simulation results show that Adaptive Cruise Control based on the MPC model can reduce the number and impact of accidents based on maintaining a safe and relative or actual distance between cars. The driver was shown to ensure that the actual distance between the two cars was greater than the set safe distance. When the actual distance is large enough, then the driver ensures that the Ego Car follows the set speed of the driver, otherwise accelerates, or decelerates according to the difference between the safe and actual distance between the cars. The paper shows that the action of only the P controller on the system control dynamics leads to an increase in oscillations and even car collisions, as shown in Figure 10, Figure 11, and Figure 12 while the action of the PI or PID controllers eliminates system oscillations and increases the speed of reaction.

As the action of the PID controller led to system stabilization, it is assumed that even greater system stabilization would be achieved by using a fuzzy controller. The above indicates the possibility of improving the adaptive car control system presented in this paper, which will leave the next research, the results of which will be compared with the results obtained in this paper.

## References

[1] Verlagic J 2013 *Lecture notes: Introduction to adaptive control* (Faculty of Electrotechnical engineering Sarajevo)

[2] Gao B, Cai K, Qu T, Hu Y and Chen H 2020 Personalized adaptive cruise control based on online driving style recognition technology and model predictive control *IEEE Transactions on Vehicular Technology* **69** 12482–96

[3] Jia Y, Jibrin R and Gorges D 2019 Energy-optimal adaptive cruise control for electric vehicles based on nonlinear model predictive control *2019 IEEE Vehicle Power and Propulsion Conference (VPPC)*

[4] Luu D L, Lupu C and Van Nguyen T 2019 Design and simulation implementation for adaptive cruise control systems of vehicles *2019 22nd International Conference on Control Systems and Computer Science (CSCS)*

[5] Lee M H, Park H G, Lee S H, Yoon K S and Lee K S 2013 An adaptive cruise control system for Autonomous *Vehicles International Journal of Precision Engineering and Manufacturing* **14** 373–80
[6] Xiao L and Gao F 2010 A comprehensive review of the development of Adaptive Cruise Control Systems *Vehicle System Dynamics* **48** 1167–92
[7] Arnab B and Petros I 2000 *Environmental evaluation of intelligent cruise control (ICC) vehicles (IEEEIntelligent Transportation Systems Conference Proceedings Dearborn (MI)).*
[8] Arnab B and Petros I 2001, *Analysis of traffic flow with mixed manual and intelligent cruise control vehicles:Theory and experiments (California PATH Research Report, UCB-ITS-PRR-2001-13)*
[9] Arnab B and Petros I 2003 *Evaluation of the environmental effects of intelligent cruise control vehicles (Transp.Res. Rec. 1774)*, pp. 90–97
[10] Reda A, Bouzid A and Vásárhelyi J 2020 Model Predictive Control for Automated Vehicle Steering *Acta Polytechnica Hungarica* **17** 163–82
[11] Sang-Jin Ko and Ju-Jang Lee 2007 Fuzzy logic based adaptive cruise control with guaranteed string stability *2007 International Conference on Control, Automation and Systems*

# Condition diagnostics of the Stadium "Rođeni" in Mostar

**Jasmina Džino[1,2], Faris Trešnjo[1,2], Merima Šahinagić-Isović[1] and Salko Kulukčija[1,2]**

[1] Civil and Structural Engineering Faculty, University of Mostar, Sjeverni logor bb, Mostar, B&H
[2] Interprojekt d.o.o. Mostar, Maršala Tita 254a, Mostar, B&H

E-mail: jasmina_hadzajlic@outlook.com

**Abstract.** The paper presents a diagnosis of the condition of the stadium FK Velež "Rođeni" in Mostar, which was built in period 2006-2008 and represents an important facility for the citizens of Mostar. The steps in diagnosing the condition of an existing structure are explained, including a collection of existing documentation, an inspection of the structure, test and calculations, and assessment and decision on further action. Drawings of the stadium were given, visual inspection of the stadium was performed and calculations of the current and future layout of the structure. The calculations were performed in software Tower 8 and with respect to European norms EN0, EN1, EN2, EN3 and EN8. A visual inspection revealed a large number of damages to the stadium, which were classified into specific groups. The mechanisms of destruction and classification were performed using EN 1504.

## 1. Introduction

Football club Velež Mostar is a professional football club based in Mostar. The club has a history of being one of the most successful clubs from Bosnia and Herzegovina.

The location of the club stadium is in the northeastern part of the city, in the suburb of Vrapčići. The stadium, at its current stage of development, can accommodate up to 7,000 fans. Lack of funds and the slow process of the country's recovery from the war, as well as the complicated political environment in the city and the country as a whole, have affected the scale and pace of stadium development. Velez returned to the top of Bosnian football after being promoted to the Premier League of Bosnia and Herzegovina in 2006. This return was accompanied by the construction of a large north and a smaller east grandstand. In 2008, a large west stand was built, which was followed by additional works and extensions between 2017 and 2018. In 2019, the roof for the west stand was completed. The construction of a large east stand is in progress [1].

### *1.1. Existing condition*

The stadium is located on the site of Mostar at an altitude of N.V. 76 m. The stadium is oriented east-west with its longer axis. The field is surrounded by a grandstand on the north and west, while on the south side of the field there is a prefabricated grandstand for visiting fans.

The west stand consists of a ground floor, first floor and auditorium, 14 m wide and 70 m long, and is connected by a concrete and metal structure. Metal pillars are continued on concrete pillars, where a canopy is hung on them by the braces. The slope of the canopy is i = 10% (Figure 1.a).

IOP Conf. Series: Materials Science and Engineering 1208 (2021) 012041 doi:10.1088/1757-899X/1208/1/012041

The north grandstand consists of a ground floor and an auditorium and is not covered. The construction consists of three parts that are separated by dilatation 5-7 cm wide. The length of one segment is 35 m and the width is 12.50 m. The total length of the northern stadium, which is the connection between the concrete and masonry structures, is 106.20 m (Figure 1.b).

a) West stand b) North stand

**Figure 1.** Current condition of the west and north stands of the "Rođeni" stadium

*1.2. New designed condition*

In order to achieve stability and durability of the structure in accordance with the importance and tradition of the club, and the relationship with the category of fans with mobility difficulties, a proposal was made to improve the existing condition through redesign of the west and north stands (Figure 2 and Figure 3).

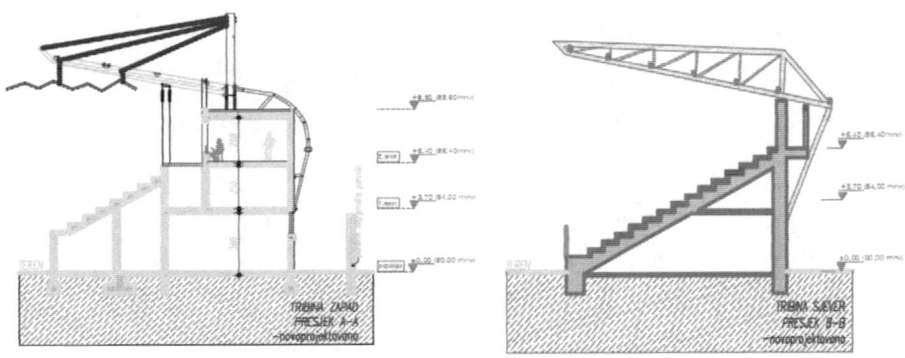


a) West stand b) North stand

**Figure 2.** New designed condition of the west and north stands of the stadium "Rođeni“

Necessary interventions to ensure the stability and durability of the structure or increase the energy efficiency of the building are:

- Installation of the thermal facade on the external facades of the west and north stands,
- Collect the water that remains in the access parts around the stadium with drainage pipes in the tank,
- Installation of mini windmills whose energy would be used to power the display or lighting,
- Landscaping a new garage.

**Figure 3.** Newly designed condition of the west and north stands of the stadium "Rođeni" - 3D view

## 2. Condition diagnostics of the west and north stand

The main goal of diagnostics of the condition of structures is to give answers to the questions in which condition the structure is observed, based on measurements and research. It is necessary to determine the critical elements of the structure, since the overall condition of the building dictates the condition of its weakest part. The diagnostic sequence for existing buildings can be divided into several steps: collecting or reconstructing existing documentation, inspecting the building, testing and calculations, evaluating and deciding on next steps.

Within the condition diagnostics of the west and north stands of the stadium "Rođeni", following the steps of diagnostics of the existing facilities, it was done:
1. Drawings of the west and north stands
2. Overview of stands
3. Static calculation
4. Evaluation and decision on further action

### *2.1. Architectural drawings*

A preliminary review of the west and north stands was approached first. After that, we made the architectural drawings. The geometry of the stands and the cross-sectional were taken. The stands were measured manually, without scaffolding, with the help of a meter, a ribbon and a laser distance meter. The result of the survey are architectural drawings of the existing condition of the building (Figure 4 and Figure 5), which contain:

- situation
- layouts
- cross and longitudinal section
- external and internal facades

IOP Conf. Series: Materials Science and Engineering 1208 (2021) 012041 doi:10.1088/1757-899X/1208/1/012041

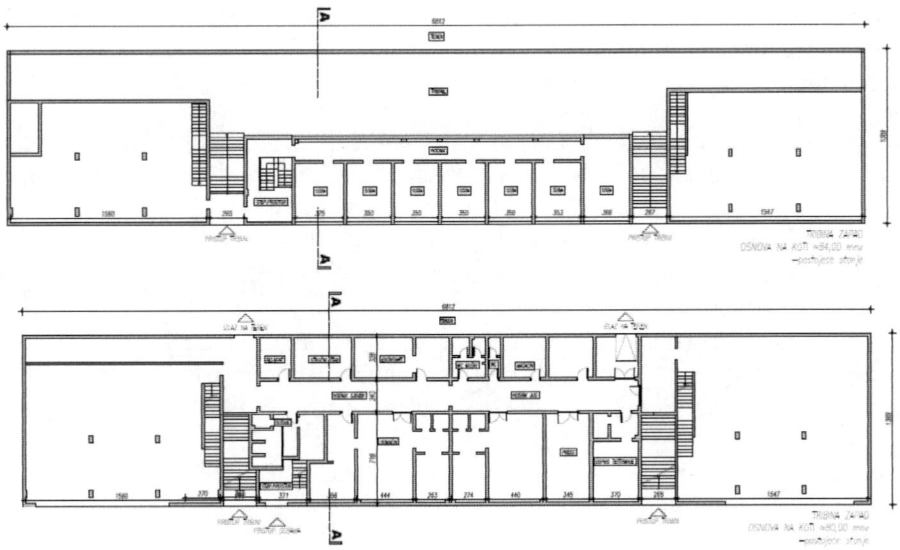

**Figure 4.** Layouts of the west and north stands

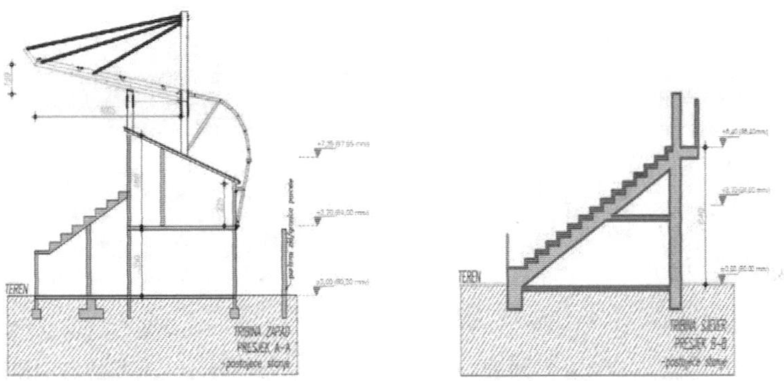

**Figure 5.** Cross-sections of the west and north stands

### *2.2. Damage categorization according to EN 1504*

After that we made the architectural drawings, we started a detailed review of the existing condition, and all identified damages were noticed and inscribed. During the detailed visual inspection of the structure, special attention was paid to: the appearance and differences in the color of the structure surface, the appearance of cracks, their size and arrangement, signs of material degradation on the structure surface, deformations on the structure, and places of leakage or water retention on wet surfaces. A drawing of the existing condition together with damage identification and mapping is the starting point for the next step, the analysis of damage occurrence.

The mechanisms and influences on the deterioration of structures are numerous and very different. The basic question with such damaged structural parts of the structure is can, and for how long, this structural element still performs its (designed) load-bearing task. In summary, the main damage was due to poor design (design phase), poor construction technology and poor material quality (construction phase), subsequent overload of the structure (exploitation phase, but also design), and due to various atmospheric and chemical influences [2].

The categorization of damage on the west and north stands was performed according to the European standard EN 1504 (Figure 6 a and b). The following terms were used to describe the occurrence of damage:

- Damage - physical disruption or change in the condition of the structure or its components due to external actions or influences
- Degradation - a decline in the condition, quality or functionality of a material that most often results in damage
- Deterioration - the process of deterioration or weakening of quality, value, characteristics and other materials and which can result in a progressive reduction of the ability of the structure or its parts to serve its purpose
- Deficiency (structure or components) - Defects occur due to a defect in performance or design or a defect in material and may result in inadequate structural capacity, premature deterioration or aesthetic appearance
- Weathering - any chemical or mechanical process by which a material exposed to atmospheric influences undergoes changes in its character and weakens or loses quality
- Crack - complete or partial cracking of concrete in two or more parts caused by cracking or fracture
- Separation (spall, flake, detachment) of parts of the material due to certain reasons (influence of weather, pressure, etc.)
- Delamination - separation of a certain part of the surface layer but at a certain depth due to a crack below the surface
- Scaling - peeling or intermittent separation of surface pieces of concrete or mortar

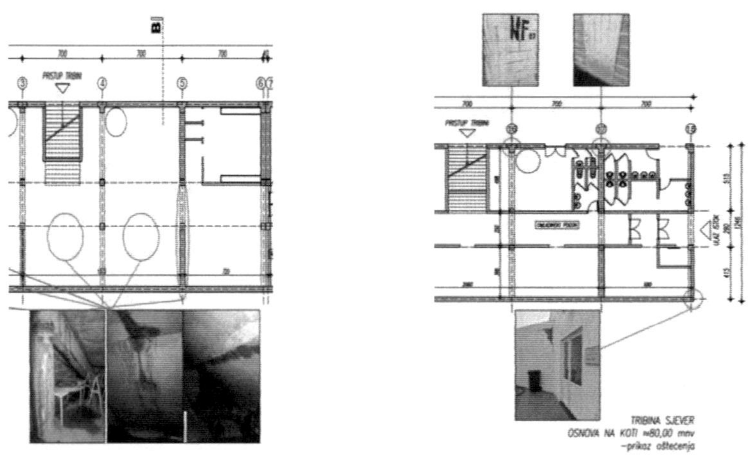

a) West stand b) North stand

**Figure 6.** Drawings of one part of the damages on the west and north stands

*2.3. Static calculation*

Structural calculation refers to the design of critical points on the structure and the calculation of the entire system with realistic parameters, determined through the inspection and testing of the structure. The design model on which the design of the structure is performed must adequately show the behaviour of the structure, the resistance of its parts and the load on the structure in accordance with the actual condition of the existing structure. We used Tower 8 software package in the calculation of the west and north stands, and static analysis of the complete construction of the existing and new designed condition.

Static calculation has proven that the existing and new designed condition satisfies the limit state of bearing capacity and the limit state of usability according to European norms.

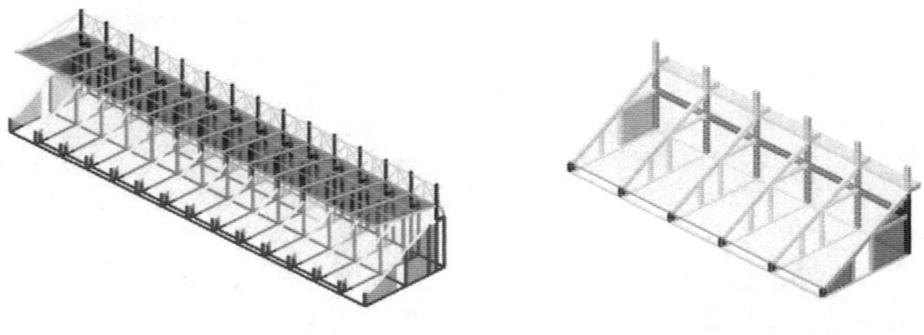

a) West stand b) North stand

**Figure 7.** Static calculation of the existing condition of the north and west stands

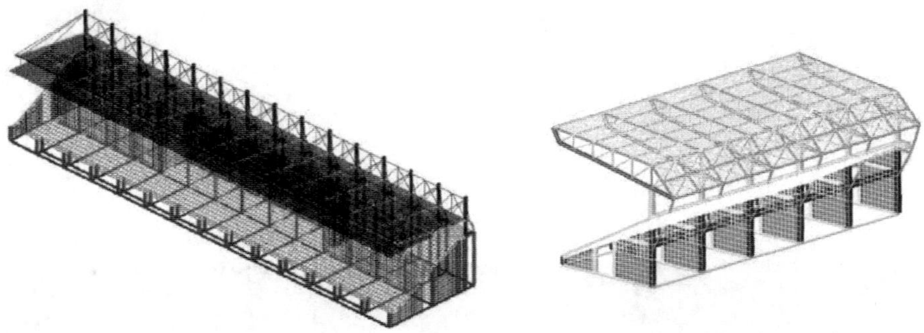

a) West stand b) North stand

**Figure 8.** Static calculation of the new design condition of the north and west stands

*2.4 Proposed reparation measures*

The process of condition diagnostics ends with an answer to the question of what steps should be taken and how to treat structures in the future. Recommendations for short-term and long-term activities should be made.

After the analysis of the occurrence of damage as well as the analysis of individual elements of the structure, we prepared an intervention proposal, based on the findings of the analysis. A damage catalog was made with all the damage and the necessary interventions to repair them. Table 1 shows a part of the damage catalog, with proposals for remedial measures.

For the correct classification of damage and selection of the appropriate method of repair, it is necessary to adjust the diagnosis, design, product selection and repair work to the individual condition of the facility, with the necessary expertise among all people involved in the process. Consequently, only if a systematic assessment clearly identifies the structure of the material, we can select adequate principles and methods of remediation. Also, it is important to assess the further development of the condition of the structure in order to select the optimal time for protection or repair measures. The new European standard EN 1504 provides a systematic approach to the principles and methods of repair and protection of structures, including the selection of appropriate products, and quality control. This approach provides an optimal basis for understanding and designing repairing works.

IOP Conf. Series: Materials Science and Engineering 1208 (2021) 012041 doi:10.1088/1757-899X/1208/1/012041

**Table 1.** Damage catalog and repair methods

| Type of damage | Figure | Detailed description of the damage | Possible cause of damage | Rehabilitation proposal |
|---|---|---|---|---|
| **Crack / deformation (damage)**<br>Separation (separating two or more parts along a certain line or more) | 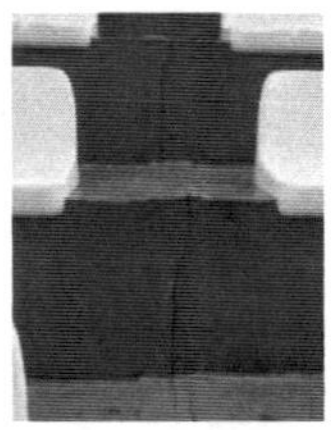 | Separation at the site of dilatation on the north stand.<br><br>It causes further deterioration of water into the expansion layers. | Incorrect installation and execution | -Filling with waterproofing foams<br>-Putting a waterproofing layer with fibers<br>-Applying a new layer of plaster (final layer - facade plaster) |
| **Degradation**<br>Moisture area Water / moisture penetration at the dilatation site |  | Change of the original color of the wall of the west stand. | -Incorrect execution.<br><br>-Unfinished waterproofing.<br><br>-Cracks at the joints of the panels and on the entire panel | -Examine reinforcements bars in the area of water retention<br>-Solve the accumulation of water on the top of the plate.<br>-Waterproofing of the upper part of the structure -Treatment of the structure with anti-mold additives and plastering |
| **Efflorescence**<br>Deposits due to the action of salt |  | Color change and appearance of deposits on the pillar below the north stand. | Appearance of salts | -Treatment with anti-mold additives<br>- Waterproofing on the upper part of the stand<br>-Salt rinsing<br>-Setting up protective layer of mortar |

## 3. Conclusion

This paper presents the diagnostics of the west and north stands of the stadium "Rođeni" and explains the principle of diagnosis which consists of: collecting existing documentation and drawings, visual inspection of the stands, the necessary static calculations and finally evaluations and decisions on further action and maintenance of the stands. Some variants of the new designed solution are given to improve the energy efficiency of the west and north stands.

Static calculations have shown that damage to the stands does not significantly affect the load-bearing capacity of the entire structure. A catalog of all damages has been made with the necessary interventions and guidelines for better and more purposeful maintenance of the stands.

## References

[1] https://fkvelez.ba/
[2] Raupach M and Büttner T 2014 *Concrete Repair to EN 1504*

## Acknowledgments

This paper was published as part of the activities of the Erasmus + project "Promoting academia-industry alliances for R&D through collaborative and open innovation platform" (All4R & D).

IOP Conf. Series: Materials Science and Engineering 1208 (2021) 012042 doi:10.1088/1757-899X/1208/1/012042

# Facade Damage Assessment of the Buildings in Bam, Iran 2003 and Kermanshah, Iran 2017 Earthquakes

**Hamid FARROKH GHATTE**

Civil Engineering Department, Faculty of Engineering, Antalya Bilim University, Antalya, Turkey.

hamid.ghatte@antalya.edu.tr & ghatte@itu.edu.tr

**Abstract.** One of the essential factors in buildings frontage is the continuity of the structural and building envelope parts. In this investigation, a comparison was made between Bam and Kermanshah earthquakes. A strong earthquake (magnitude 6.6) struck the city of Bam in southeast Iran on 26 December 2003, and similarly, another strong earthquake struck the city of Kermanshah (magnitude 7.3) in Iran on 12 November 2017. Damage in the facades of the buildings was a clear contributor to the overall building damage. This paper presents the damage assessment of the different facade systems from multi-story buildings in Bam and Kermanshah, Iran. The survey covers the buildings greater than three stories in height, excluding most unreinforced masonry facades. As far as a building can have more than one facade system, any facade systems are evaluated individually. Observation of facade damage is discussed and is presented in terms of its performance level.

## 1. Introduction

On Friday, December 26, 2003, at 1:56:56 GMT (5:26:56 local time), a powerful earthquake struck Bam, a city located in southeastern Iran. The magnitude is 6.6 by USGS1, killing more than 40,000 and injuring more than 25,000 people in Bam Township(Fig 1a). The depth of this earthquake is estimated to be about 7 km. The epicenter of this earthquake is located at N29.2 E58.4 coordinates, as reported by the Institute of Geophysics of Tehran University. The construction quality in this area of Iran is generally poor, particularly in small towns and villages. The buildings in these regions are highly vulnerable even to moderate earthquakes, and most of them completely collapse when subjected to these excitations [5].

On Sunday, November 12, 2017, at 21:48 local time, a powerful earthquake struck Kermanshah, a city located northwest of Iran. The magnitude is 7.3 by USGS1, killing more than 630 and injuring more than 8100 people in Kermanshah Township (Fig 1b). The depth of this earthquake is estimated to be about 19 km. It should be noted that the seismic zones are classified into four categories (Low, Moderate, High, and very high-risk categories), and Bam is in high fisk, and Kermanshah is in a very high-risk area. These earthquakes occurred due to the stresses generated by the motion of the Arabian plate northward against the Eurasian plate (Fig 2). Deformation of the earth's crust in response to the plate motion takes place in a broad zone that spans the entire width of Iran and extends into

Turkmenistan. Earthquakes occur due to both reverse faulting and strike-slip faulting within the zone of deformation [1-6].

However, the seismic performance of structural elements had been widely investigated in the literature, and a lack of information regarding the damage assessments of facades performance is one of the main topics in the construction industry [7 -9]. Moreover, the central part of the previous studies focused on the structural behavior and lateral load systems. Limited published studies are dealing with damage assessments of facades and the connection of the structural elements.

This research aims to investigate the damage assessments of different types of facades during seismic actions. The facades and structural response throughout the earthquake were evaluated based on Bam and Kermanshah earthquakes. For this purpose, currently used different types of facades for buildings in Iran were considered for better understanding the behavior of the facades during the earthquake.

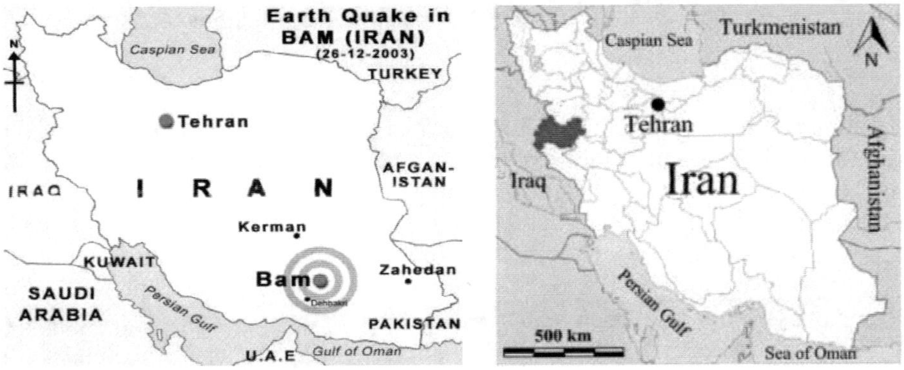


(a) Bam 2003 (b) Kermanshah 2017

**Figure 1.** The geographic location of Bam and Kermanshah on the map.

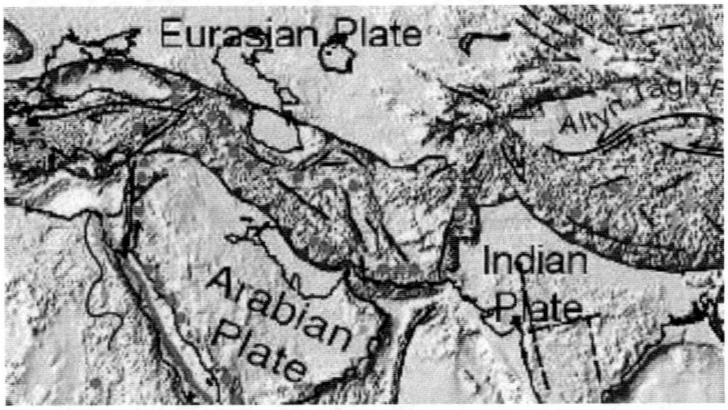


**Figure 2.** Plate tectonic map of Iran region.

## 2. Facad technology in Iran

Facade systems can be classified into the following types; composite facade, glass facade, bayramix facade claddings, Stone facade, Ceramics facade, and Brick facade. The simplest way to differentiate between the two types is to construct in dry and wet conditions, while all of them are attached externally to the primary structure. Types of facades used in the existing buildings in Iran are illustrated in Fig 3.

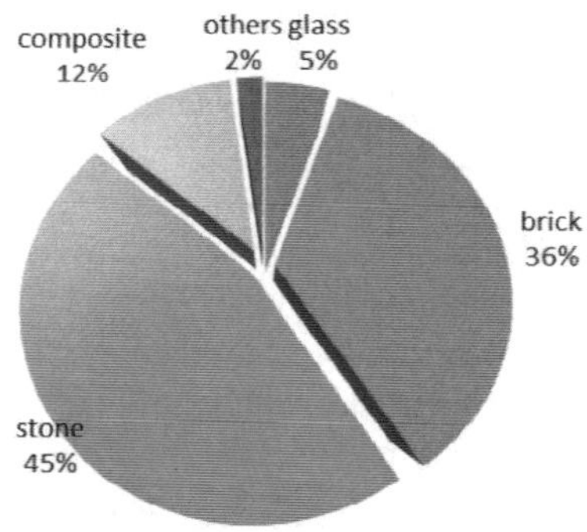


**Figure 3.** Types of building's facade in area of study.

### *2.1. Composite facade*

Composites are unique types of materials that are formed by two or more substances, so the combined substance has more strength and durability than the individual ingredients. Today, composites are typically used for exterior building facades. The composites consist of two layers of aluminum and a core of plastic or mineral fillers located between these two layers. The properties of composites are generally better than any of their ingredients. It is noteworthy that various ingredients improve performance. Although composites are weak in terms of flexibility, however, we can mention the lightweight, ease of assembling, and maintenance as their advantages. Composite facades used in Bam buildings in Iran are illustrated in Fig 4.

Bam 2003

**Figure 4.** Composite facades failures during the Bam 2003.

### *2.2. Glass facade*

Glass facades have been used since the modern architecture era as the buildings cover. They cause lots of problems for residents as the result of their low thickness as well as lack of thermal resistance. Glass facades used in Bam and Kermanshah buildings in Iran are illustrated in Fig 5.

(a) Bam 2003 (b) Kermanshah 2017

**Figure 5.** Glass facades failures during the Bam 2003 and Kermanshah 2017 earthquakes.

### *2.3. Bayramix facade*

Bayramix is a decorative cover that is utilized as an interior or exterior facade. This product is applicable on painted walls, stone, cement walls, wood, cardboard, plastic, metal, and glass. A Turkish construction group used Bayramix for the first time in 1993, and it was presented to the world. The most recent product is in the form of mortar, and it is a mixture of granite grains and marble with different colors, numerous grading, natural and artificial resins, and other additives. Bayramix is not applicable on greasy, dusty, rugged, wet surfaces or in stormy or dusty weather. Although this type of facade has some advantages; however, it has not been welcomed by Iranian engineers and construction contractors because this type of facade is inconsistent with their cultural context and tendencies. So it will lose its beauty and effect in a short time. Bayramix facades used in Bam and Kermanshah buildings in Iran are illustrated in Fig 6.

(a) Bam 2003 (b) Kermanshah 2017

**Figure 6.** Bayramix facades failures during the Bam 2003 and Kermanshah 2017 earthquakes.

### 2.4. Stone facade

Two groups of natural stones are executable in the facade; limestones and igneous stones. Limestone is not usually durable due to the existence of strands. On the other side, low water absorption and poor adhesion are the main igneous stones problems. Moreover, the facade is significant as it is exposed to weather conditions. The stones used for the facade should have particular features to sustain against these natural factors. Stone facades used in Bam and Kermanshah buildings in Iran are illustrated in Fig 7.

(a) Bam 2003 (b) Kermanshah 2017

**Figure 7**. Stone facades failures during the Bam 2003 and Kermanshah 2017 earthquakes.

### 2.5. Brick facade

The brick frontage is the other type of frontage, and it is divided into two groups; clay bricks and chili bricks. Brick facades have some benefits, such as beauty, variety, and color stability. In addition, it plays the role of cryogenic and thermal insulation because of the low heat transfer coefficient. As a result, brick building facades prevent wasting heating and cooling energy. According to their color and material, brick facades cause reflection of the heat in summer and absorb solar energy in winter. Brick facades used in Bam and Kermanshah buildings in Iran are illustrated in Fig 8.

(a) Bam 2003 (b) Kermanshah 2017

**Figure 8**. Brick facades failures during the Bam 2003 and Kermanshah 2017 earthquakes.

## 3. General behavior of the systems

The old facades like brick facades and stone that employ the wet material for pasting the facades could not present an acceptable performance during the seismic actions, even during the service period. In the case of seismic categories, facade damage belongs to limited damage during the seismic actions. The surveying of the different facades showed the weaknesses of traditional types like brick and stone facades. The new methods like glass and composite that can be connected to the structural elements had better performance. To produce accurate results that benefit the construction industry, the performance of the facades should be evaluated in detail. Therefore, the industry needs to use s proper system that can have a minimal displacement or facade system that can accommodate a certain level of displacements during the earthquake. However, the composite is primarily typical in business and commercial buildings because they want to be different and make the building more exciting, or a city sign, stone, or brick facades are typically used in residential buildings to protect the privacy and save energy and money. Among the factors for selecting the facades, price and general view of the facades are more critical than others parameters. Nevertheless, the parameters of the seismic performance should evaluate for any of the existing facades. In the case of fixing methods after the earthquake, traditional methods are more complex than new technics since wet paste materials have been used.

## 4. Conclusion

According to the reasons mentioned above, the conclusions are summarized below:

- It is noticeable that the seismic behavior of the facades of the building should be one of the main factors in selecting the type of facade.
- The traditional and old methods of facades like brick facades and stone could not present an acceptable performance during the seismic actions in Bam 2003 and Kermanshah 2017, and generally, the damages cannot be relatively easy to fix.
- Earthquake damage to facade systems undoubtedly poses a significant threat to life. The economic implication from facade damage is also significant due to business downtime and repair costs.
- Although facade damage belongs to limited damage during the seismic actions, surveying the different types of facades showed the weaknesses of traditional types like brick and stone facades.
- In fixing schema after the seismic actions, traditional methods are more complex since wet paste materials have been used and repair of new technics is relatively easy.
- In order to reduce damage to facade systems in the future, both technical and political issues need to be addressed. Improvements are required to better understand the behavior of many facade systems and whether the methods used to isolate them are satisfactory. Design guidelines are required for both designers and installers of facade systems. Communicating common errors that should be avoided is also essential.

## References

[1] Høseggen R, Wachenfeldt B J and Hanssen S O 2008 Building simulation as an assisting tool in decision making *Energy and Buildings* **40** 821–7
[2] Shakib H, Ahmadizadeh M and Rashtian M 2002 Evaluation of beha-Concrete vior of buildings in Changureh-Avaj earthquake stricken region *12th International Symposium on Earthquake Engineering Roorkee*
[3] USGS U S Geological Survey Preliminary earthquake report Magnitude 6.6 southeasten Iran 2003 December 26 *National earthquake Information Center World data Center for Seismology*
[4] EERI Newsletter.Earthquake Engineering Research Institute 2011 Learning from Earthquakes-The M 6.3 *Christchurch*
[5] Applied Technology Council 1989 Procedures for Postearthquake Safety Evaluation of Buildings & Addendum, *ATC-20.*

[6] Baird A, Palermo A, Pampanin S, Riccio P and Tasligedik A S 2011 Focusing on reducing the earthquake damage to facade systems *Bulletin of the New Zealand Society for Earthquake Engineering* **44** 108–20
[7] Farrokh Ghatte H 2020 External steel ties and CFRP jacketing effects on seismic performance and failure mechanisms of substandard rectangular RC Columns *Composite Structures* **248** 112542
[8] Farrokh Ghatte H 2020 Failure mechanisms and cracking performance of T-shaped SCC beam-column connections at top floor: Test results and Fe *Modeling Structures* **28** 1009–18
Farrokh Ghatte H, Comert M, Demir C, Akbaba M and Ilki A 2019 Seismic retrofit of full-scale substandard extended rectangular RC columns through CFRP jacketing: Test results and design recommendations *Journal of Composites for Construction* **23** 04018071

# Numerical modelling of the bearing capacity of a beam cross-section

**Besim Demirović [1], Rašid Hadžović [2] and Nedim Osmić [1]**

[1] Faculty of Mining Geology and Civil Engineering, University of Tuzla, Bosnia and Herzegovina
[2] Faculty of Civil Engineering, University of "Džemal Bijedić" in Mostar, Bosnia and Herzegovina

E-mail: besim_demirovic@yahoo.com

**Abstract.** The paper presents a procedure for numerical modelling of the rod cross-section bearing capacity. Equilibrium between cross sectional forces and cross-sectional stresses is determined by iterative procedures. According to the described procedure, the load-bearing capacity of the cross-section is determined according to the isotropic linear and nonlinear behavior of the material, for homogeneous and inhomogeneous cross-sections. The nonlinear behavior of the material reduces the stiffness of the cross section of the rod EA and EI, with a significant increase in the deformation values ε and κ. The applicability of the calculation and analysis of obtained results is presented using numerical examples.

## 1. Introduction

During exploitation, building structures should be constructed to safely receive and transfer the projected loads to the base. It is necessary to ensure that the stresses and deformations of the structure remain below the limit state of bearing capacity, limit state of usability and limit state of durability. The aim of the calculation, according to the limit state method, is to achieve an appropriate probability that the structure will remain reliable for use during exploitation [8]. The limit state can be reached by: loss of balance of the system or one structural element, transformation of the structure into a mechanism, failure and excessive deformation of the critical cross-section and failure due to material fatigue. It is necessary to determine the load-bearing capacity of the cross section of the rod and required coefficient of safety against fracture for the purpose of ensuring the limit state of the cross section fracture [2]. The load-bearing capacity of the cross section of the rod is the basis for the analysis of the material non-linearity of the system. It depends on the material behavior, the cross-sectional geometry and the ratio of the cross-sectional forces that are acting at the centroid of the cross section. The load-bearing capacity of the cross section is ensured if all its parts are within the permissible stresses and strains. The load-bearing capacity of the cross section of the rod is determined by the elastic and plastic area of behavior of the material. Some parts of the cross-section under the action of cross-sectional forces can be located in the elastic area, while others can be located in the plastic area. In this case, it is the elastic-plastic bearing capacity of the rod cross section. The stress and strain area of the rod cross section depends on the adopted material behavior diagram, or more of them. The behavior of the material under load is described by a stress-strain relation diagram. The dependence of normal stress and longitudinal deformation $\sigma=\sigma(\varepsilon)$, and shear stress and shear deformation $\tau=\tau(\gamma)$ is complex. The stress-strain ratio depends on the type of material and the stresses. Figure 1 presents the behavior diagrams of materials in structures [7].

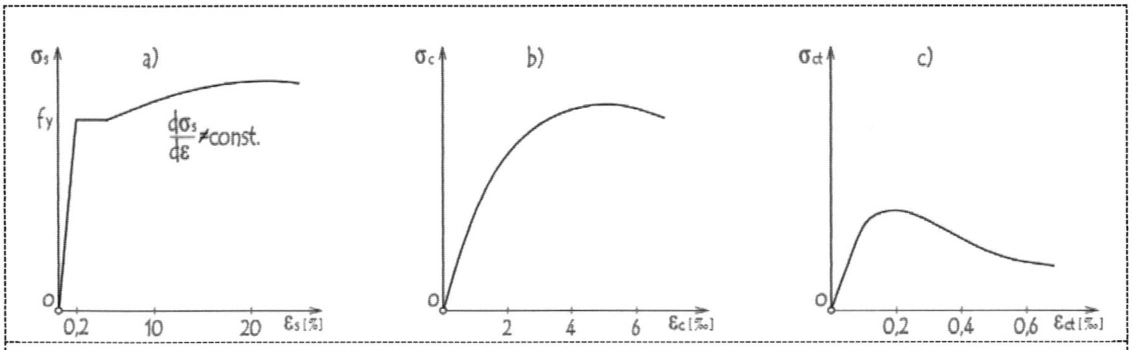


**Figure 1.** a) Behavior diagram of steel, b) stress-strain ratio of concrete under pressure, c) stress-strain tension ratio of concrete.

Figure 1.a presents a behavior diagram characteristic of homogeneous isotropic materials such as steel. At lower stresses, the stress-strain relation is linear. The stress becomes constant and independent of deformation after reaching the yield strength $f_y$. An increase of load makes the material to harden therefore the stress-strain relation is nonlinear. In the case of in-homogeneous isotropic materials such as concrete (Figure 1.b), initially at low compressive stress the relation of stress-strain is linear. The function of the $\sigma_c$-$\varepsilon_c$ ratio with an increase of compressive stress on concrete is nonlinear. In the case of concrete, in addition to the concrete class, the duration of the load is significantly affected by the stress-strain ratio [6]. Stress-strain behavior diagrams are conveniently idealized [1]. During idealizing material behavior diagrams, we should strive to be as simple as possible and as close as possible to the actual material behavior under load. The manner of idealizing the diagram depends on its application in certain areas of construction.

## 2. Analysis of load capacity of homogeneous cross section rod

The behavior of the material in the form of a bilinear diagram describes the actual behavior of the material. The real behavior of the material is approximated by an ideal elastic-plastic diagram of the material behavior in the bilinear form, Figure 2.d. In the diagram, we distinguish between a linear stress-strain ratio in the range of 0-$\varepsilon_y$, and a linear relation in the range of $\varepsilon_y$-$\varepsilon_u$. The secant modulus of elasticity of the material $E_s$ is valid up to the yield strength of the material $\varepsilon_y$. After the material softening, the tangent modulus of the material $E_t$ occurs [5].

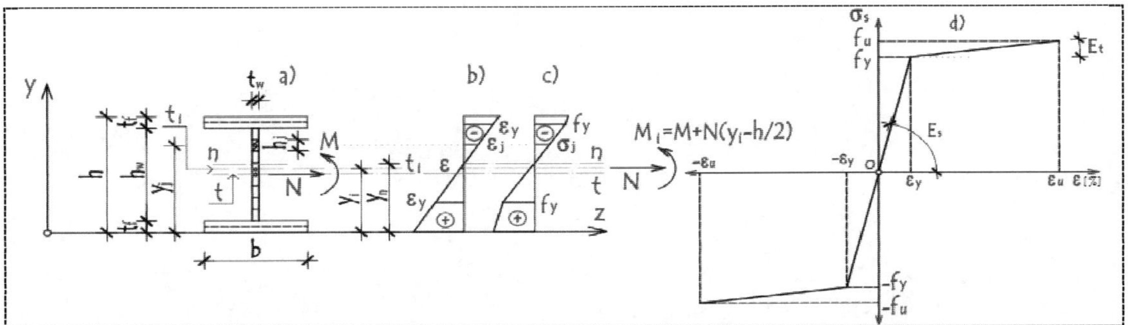

**Figure 2.** a) Layered cross section I, b) cross section deformations, c) stresses, d) bilinear diagram of material behaviour.

The cross section is divided into equal layers of rib panels and flanges, thickness $h_j$. Each layer corresponds to a deformation $\varepsilon_j$ and a normal stress $\sigma_j$. The stress of individual layers in the elastic or plastic area depends on the deformation. Due to the different behavior of the material in the tension and compressed zone, the centroid of the cross section shifts, therefore the problem must be solved iteratively. In the initial iterative step, the secant modulus of elasticity of the material $E_s$ is adopted for the selected combination of section forces at the centroid of the cross section. On the basis of this, the deformation and curvature of the cross section are calculated in the form:

$$\varepsilon^{(0)} = \frac{N}{E_s A_i}, \quad \kappa^{(0)} = \frac{M}{E_s \mathrm{I}_i} \tag{1}$$

$E_sA_i$ - axial stiffness of the cross section of the rod; $E_sI_i$ - bending stiffness of the rod cross section. Deformations of layers in the initial iteration are determined by the expression:

$$\varepsilon_j^{(0)} = \varepsilon + \kappa \cdot \left(y_i - y_j\right) \tag{2}$$

The deformation of the j-th layer in the initial iteration corresponds to the stress in the first iteration, according to the stress diagram:

$$\begin{aligned} \sigma_j^{(1)} &= E_s \cdot \varepsilon_j^{(0)} & & \left(0 \le \varepsilon \le \varepsilon_y\right) \\ \sigma_j^{(1)} &= f_y + E_t \cdot \left(\varepsilon_j^{(0)} - \varepsilon_y\right) & & \left(\varepsilon_y \le \varepsilon \le \varepsilon_u\right) \end{aligned} \tag{3}$$

An initial iterative step is sufficient if all the cross-sectional layers are in an elastic stress state. The iterative procedure continues in the case of plasticization of one or more layers. We determine the cross-sectional forces of the first iteration by integrating the stresses by layers:

$$N^{(1)} = \sum_j \sigma_j^{(1)} \cdot A_i, \qquad M_i^{(1)} = \sum_j \sigma_j^{(1)} \cdot A_i \left(y_i - y_j\right) \tag{4}$$

In each iterative step, it is determined the stiffness of the cross section and the position of the new centroid of the cross-section. In the n-th iteration we get:

$$E^{(n)}A_i^{(n)} = \sum_j E_j^{(n)} \cdot A_j$$

$$E^{(n)}\mathrm{I}_i^{(n)} = \sum_j E_j^{(n)} \cdot A_j \left(y_j - y_i\right)^2 \quad \begin{bmatrix} \left(0 \le \varepsilon \le \varepsilon_y\right) \Rightarrow E_j^{(n)} = E_s \\ \left(\varepsilon_y \le \varepsilon \le \varepsilon_u\right) \Rightarrow E_j^{(n)} = E_t \end{bmatrix} \tag{5}$$

$$y_i^{(n)} = \frac{\sum_j E_j^{(n)} \cdot A_j \cdot y_j}{E^{(n)}A_i^{(n)}}$$

The increment of deformations and curvature of the cross-section of the shape corresponds to the new cross-sectional characteristics:

$$\Delta\varepsilon^{(n)} = \frac{\Delta N^{(n)}}{E^{(n)}A_i^{(n)}}, \quad \Delta\kappa^{(n)} = \frac{\Delta M_i^{(n)}}{E^{(n)}\mathrm{I}_i^{(n)}} \tag{6}$$

the increment of normal force and bending moment is determined by expression (7):

$$\Delta N^{(n)} = N - N^{(n-1)}, \quad \Delta M_i^{(n)} = M_i - M_i^{(n-1)} \tag{7}$$

The deformation quantities in the n-th iterative step are:

$$\begin{aligned} &\varepsilon^{(n)} = \varepsilon^{(n-1)} + \Delta\varepsilon^{(n)}, \quad \kappa^{(n)} = \kappa^{(n-1)} + \Delta\kappa^{(n)} \\ &\varepsilon_j^{(n)} = \varepsilon^{(n)} + \kappa^{(n)} \cdot \left(y_i - y_j\right) \end{aligned} \tag{8}$$

The stresses of the cross-sectional layers in the n+1 step are:

$$\begin{aligned} &\sigma_j^{(n+1)} = E_s \cdot \varepsilon_j^{(n)} && \left(0 \le \varepsilon \le \varepsilon_y\right) \\ &\sigma_j^{(n+1)} = f_y + E_t \cdot \left(\varepsilon_j^{(n)} - \varepsilon_y\right) && \left(\varepsilon_y \le \varepsilon \le \varepsilon_u\right) \end{aligned} \tag{9}$$

The iterative procedure is repeated until the loading rate of forces and deformation becomes a small value.

## 3. Load capacity of inhomogeneous rod cross section

An analysis of nonlinear load-bearing capacity has been presented on the example of reinforced concrete section. In determining the relation of the M-k and N-ε cross-section, the following assumptions are adopted: a) the cross-sections remain straight and perpendicular to the axis of the girder before and after deformation; b) the concrete does not participate in the tensile of the load-bearing capacity; c) no slipping between concrete and reinforcement; the influence of transverse forces on bearing capacity is neglected; e) the participation of forks in bending capacity is neglected [3]. The cross section is loaded by the bending moment around the z axis and the normal force, which in interaction causes normal cross-sectional stresses. The concrete behavior diagram was adopted in the form of polynomials, while reinforcement is adopted as a bilinear stress-strain dependence, without taking into account the material safety coefficients (Figure 3.e, and 3.f). Normal stresses occur at each point of the section, depending on the deformation of the section when the combination of bending moment and normal force is combined. The solution of the cross-section load-bearing capacity problem is sought iteratively for the adopted load increment. The deformations of the section in the zero iterative step are described by the deformation $\varepsilon^{(0)}$ and the curve $\kappa^{(0)}$ which are determined at the centroid of the cross-section according to the expressions:

$$\varepsilon^{(0)} = \frac{N}{EA_i}, \quad \kappa^{(0)} = \frac{M_i}{EI_i} \tag{10}$$

$EA_i$ - axial stiffness of the rod cross section; $EI_i$ - bending stiffness of the rod cross section. The described cross-sectional stiffness are determined on the basis of the initial modulus of elasticity of the material and the surface area of the entire cross-section of concrete and reinforcement. The stiffness of the cross section of the rod changes and the deformation size of the rod increases during use of nonlinear stress-strain ratios. In the n-th iterative step, the increment $\Delta\varepsilon$ and $\Delta\kappa$ is in the form of:

$$\Delta\varepsilon^{(n)} = \frac{\Delta N^{(n)}}{E_t^{(n)} \cdot A_i^{(n)}}, \quad \Delta\kappa^{(n)} = \frac{\Delta M_i^{(n)}}{E_t^{(n)} \cdot I_i^{(n)}} \tag{11}$$

$E_t^{(n)} \cdot A_i^{(n)}$ - reduced axial stiffness of the rod cross section in the n-th iterative step; $E_t^{(n)} \cdot I_i^{(n)}$ - reduced bending stiffness of the rod in the n-th iterative step. The increment of cross-sectional forces and bending moments is determined by the expression:

$$\Delta N^{(n)} = N - N^{(n-1)}, \quad \Delta M_i^{(n)} = M_i - M_i^{(n-1)} \tag{12}$$

The deformation sizes of the rod are in the form of:

$$\varepsilon^{(n)} = \varepsilon^{(n-1)} + \Delta\varepsilon^{(n)}, \quad \kappa^{(n)} = \kappa^{(n-1)} + \Delta\kappa^{(n)} \tag{13}$$

In order to calculate the ideal surface $A_i$ and the ideal moment of inertia $I_i$ of the cross section, the pressed part of the concrete and the longitudinal reinforcement are taken. The tangent modulus $E_t$ of the material changes along the height of the cross section depending on $\sigma$ and $\varepsilon$ when the stress changes. In this case, the strain quantities $\varepsilon$ and $\kappa$ in expression (13) are defined through the functional relations of stresses and strains. The calculation of deformation quantities with complex functions can be simplified by dividing the cross section into layers. Each j-th cross-sectional layer at its centroid of the cross section corresponds to the stress, strain and tangent modulus of the material.

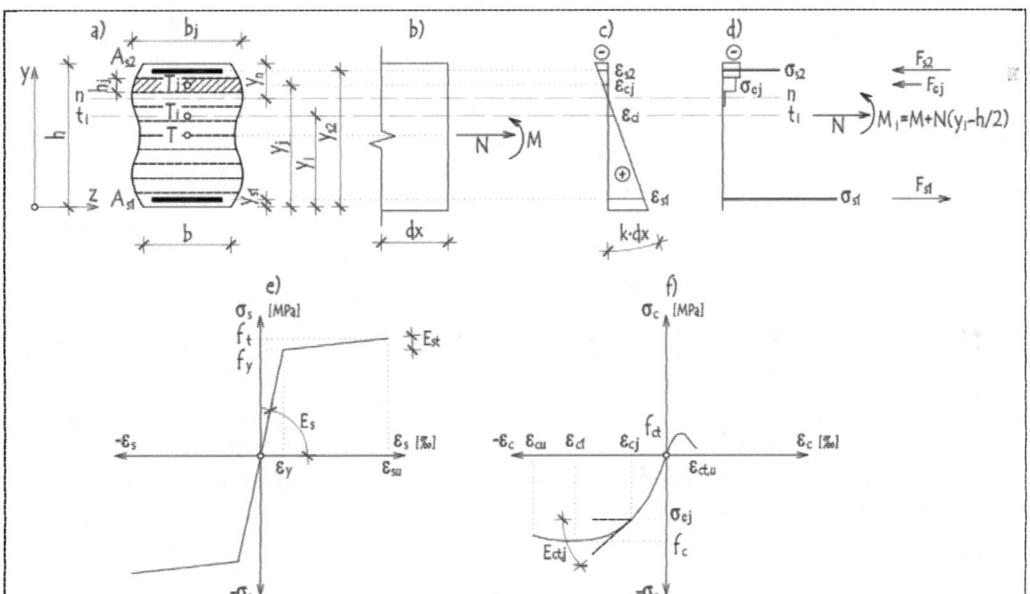


**Figure 3.** a) cross section; b) an elementary part of a rod length dx; c) distribution of deformations; d) stresses and cross-sectional forces, e) relation diagram of $\sigma_s$-$\varepsilon_s$ reinforcement; f) diagram of the relation of $\sigma_c$-$\varepsilon_c$ concrete.

Static and geometric values of the cross section of the rod are determined in relation to the position of the ideal of the centroid cross section $T_i$. The cross-sectional stiffness of the rod $E_tA_i$ and $E_tI_i$ is determined by the layers in relation to the ideal centroid of the cross-section according to the expressions:

$$E_t A_i = \sum_j E_{ct,j} \cdot A_j + E_{st} \cdot A_{s1} + E_{st} \cdot A_{s2}$$
$$E_t \mathrm{I}_i = \sum_j E_{ct,j} \cdot A_j \left(y_j - y_i\right)^2 + E_{st} \cdot A_{s1}\left(y_i - y_{s1}\right)^2 + E_{st} \cdot A_{s2}\left(y_{s2} - y_i\right)^2 \tag{14}$$

On the basis of the deformation quantities ε and κ, the distribution of deformations along the height of the cross section is in the form of:

$$\varepsilon_{cj} = \varepsilon_{ci} + \kappa \cdot \left(y_i - y_j\right), \quad \varepsilon_s = \varepsilon_{ci} + \kappa \cdot \left(y_i - y_s\right) \tag{15}$$

Sectional plasticization is based on one-axial material properties. The deformations of the cross-sectional layers from the stress-strain relation diagram determines the stress $\sigma_{cj}$ of the concrete layers and $\sigma_s$ of the reinforcement On the basis of the obtained stress of concrete and reinforcement, the forces $F_{cj}$ of concrete layers and $F_s$ reinforcement are determined. The bending moments $M_i$ and the force N, which are not in equilibrium with the given cross-sectional forces, are obtained from the equilibrium of all forces in the direction of the x-axis and the sum of all moments around the z-axis in the centroid $T_i$ of the cross section. Due to the increase in stress of the concrete layers and reinforcement, there is a change in the static and geometric characteristics of the cross section. In the next iterative step, for the new position of the centroid cross section and static characteristics of the cross section, the increment of the deformation quantities Δε and Δκ is determined, and the distribution of deformations along the height of the cross section. The deformation of loading rate is determined according to expression (11) due to the difference between the given cross-sectional forces and the forces in the (n-1) iterative step. The course of the iterative process of balancing forces and cross-sectional deformations is shown in Figure 4 [3].

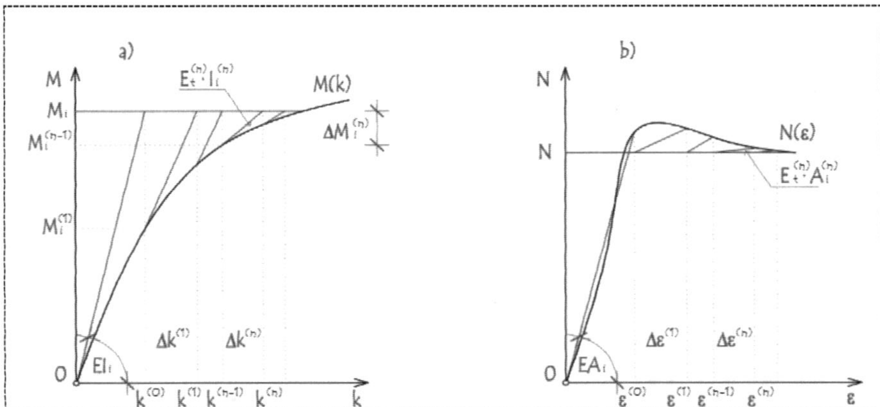


**Figure 4.** a) The course of the iterative procedure of the M-κ relation; b) the course of the iterative procedure of N-ε relation.

## 4. Numerical examples

### *4.1. Example 1*

In the first numerical example, the bearing capacity of a homogeneous cross-section of a HEA profile made of S 235 JR steel ($f_y$ =235 MPa, $F_u$=360 MPa) is analysed. Load capacity is determined by the interaction of bending moments and normal forces for different $N/N_{pl}$ ratios. Reference values of bending moments at the yield strength $M_y$, plasticity moments $M_{pl}$ with and without the action of normal force according to expressions (16) are used [2, 4] for the purpose of comparing the results:

$$M_y = f_y \cdot W_{el}, \quad M_{pl} = f_y \cdot W_{pl} \qquad [N = 0$$

$$M_y = \left( f_y - \left| \frac{N}{A} \right| \right) W_{el}, \quad M_{pl,N} = 1{,}2 \cdot M_{pl} \left( 1 - \frac{N}{N_{pl}} \right), \quad N_{pl} = f_y \cdot A \qquad \begin{bmatrix} N \neq 0 \\ 0{,}15 N_{pl} \leq N \leq N_{pl} \end{bmatrix} \tag{16}$$

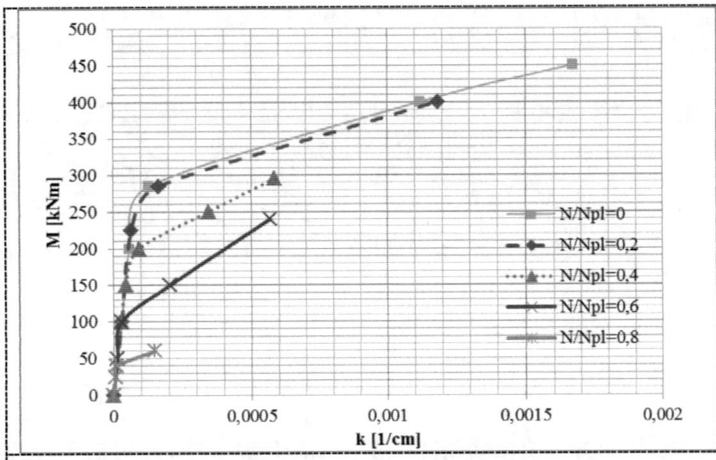


**Figure 5.** Bending-curvature ratio of HEA 300 profile.

The calculation results are given in Table 1.

**Table 1.** Results of the calculation of the cross-sectional forces of the HEA profile.

| | | | | Literature [2] | | | | Numerical calculation | |
|---|---|---|---|---|---|---|---|---|---|
| Bar | $N/N_{pl}$ | A [cm²] | $W_{el}$ [cm³] | $W_{pl}$ [cm³] | $N_{pl}$ [kN] | $M_y$ [kNm] | $M_{pl}$ [kNm] | $M_y$ [kNm] | $M_{pl}$ [kNm] |
| HEA 200 | 0 | 53.8 | 389 | 430 | 1264.30 | 91.42 | 101.05 | 89.03 | 142.80 |
| | 0.2 | | | | | 73.13 | 97.01 | 70.20 | 129.05 |
| | 0.4 | | | | | 54.85 | 72.76 | 51.50 | 91.30 |
| | 0.6 | | | | | 36.57 | 48.50 | 32.70 | 78.81 |
| | 0.8 | | | | | 18.28 | 24.25 | 13.96 | 17.35 |
| | 1 | | | | | 0.00 | 0.00 | 0.00 | 4.70 |
| HEA 300 | 0 | 113 | 1260 | 1384 | 2655.50 | 296.10 | 325.24 | 286.70 | 458.00 |
| | 0.2 | | | | | 236.88 | 312.23 | 225.70 | 413.60 |
| | 0.4 | | | | | 177.66 | 234.17 | 164.80 | 296.00 |
| | 0.6 | | | | | 118.44 | 156.12 | 103.70 | 250.80 |
| | 0.8 | | | | | 59.22 | 78.06 | 42.80 | 52.80 |
| | 1 | | | | | 0.00 | 0.00 | 0.00 | 18.00 |
| HEA 400 | 0 | 159 | 2310 | 2560 | 3736.50 | 542.85 | 601.60 | 533.45 | 859.60 |
| | 0.2 | | | | | 434.28 | 577.54 | 422.35 | 780.50 |
| | 0.4 | | | | | 325.71 | 433.15 | 311.30 | 601.00 |
| | 0.6 | | | | | 217.14 | 288.77 | 200.20 | 477.50 |
| | 0.8 | | | | | 108.57 | 144.38 | 89.10 | 95.00 |
| | 1 | | | | | 0.00 | 0.00 | 0.00 | 25.00 |

### *4.2. Example 2*

In the second example, the load-bearing capacity of the reinforced concrete section is determined. The cross section is made of concrete class C25/30 ($E_c$=3x$10^7$ kN/$m^2$), reinforced with bars 4 Φ16 B 500 B ($E_s$=2x$10^8$ kN/$m^2$). The stress-strain relation for reinforcement was adopted in the form of a bilinear diagram ($f_y$ =500 MPa, $F_u$=550 MPa), and concrete in the form of a third degree polynomial ($f_{ck}$=25 MPa). Cross-section failure occurs when a stress limit occurs in a single concrete cross-sectional layer or in reinforcement. The stiffness of the cross section of the bending rod is minimal at the fracture boundary of the cross section, Figure 6.b. The load-bearing capacity of the cross-section is defined by the bending-curvature ratio, Figure 6.a. The change in static and geometric cross-sectional sizes of the rod is shown in Tables 2 and 3.

**Table 2.** Results of the change in cross-sectional stiffness depending on M and N = 0 kN.

| Beam (N =0 kN) | | | | | | |
|---|---|---|---|---|---|---|
| **M [kNm]** | **0** | **50** | **100** | **150** | **175** | **190** |
| **EI[kNm²]** | 84813.17 | 23146.31 | 22932.17 | 22678.11 | 2324.28 | 1376.95 |
| **κ [1/m]** | 0 | 0.002186 | 0.004434 | 0.006771 | 0.016738 | 0.062313 |

**Table 3.** Results of the change in cross-sectional stiffness as a function of M and N=100 kN.

| Beam (N =100 kN) | | | | | | |
|---|---|---|---|---|---|---|
| **M [kNm]** | **0** | **50** | **100** | **150** | **160** | **170** |
| **EI[kNm²]** | 84813.17 | 23284.50 | 23086.35 | 22730.36 | 2359.62 | 1664.49 |
| **κ [1/m]** | 0 | 0.001516 | 0.003739 | 0.006077 | 0.012655 | 0.051915 |

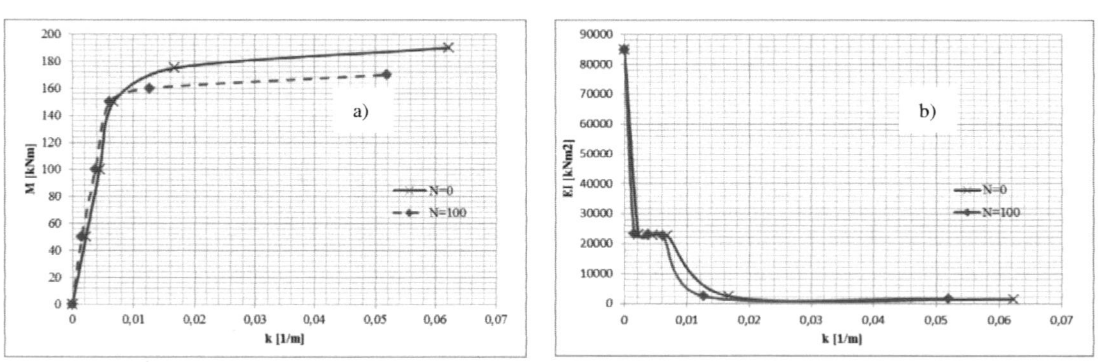


**Figure 6.** a) Bending moment-curvature ratio of cross-section b/h = 25/50 cm, b) change of stiffness of EI concrete cross-section.

## 5. Conclusion

It is shown in the paper that the bearing capacity of the cross section can be determined numerically, with linear and nonlinear behavior of the material. The bending stiffness up to the yield strength of the material $f_y$ is constant in the rod made of a homogeneous material. The cross-sectional stiffness decreases after reaching the yield point. During a further increase in bending moments and cross-sectional forces, the curvature of the cross-section has a multi-stage increment until the section break. The action of normal force on the cross section reduces the bending moment at the yield point of the material. Analysis of the inhomogeneous reinforced concrete cross-section concluded that after the appearance of the first cracks in the tension zone, the stiffness of the EI cross-section decreases significantly. Thereafter, the cross-sectional stiffness does not change significantly, until the stress in the reinforcement reaches the yield strength $f_y$. There is a decrease in stiffness and an increase in the curvature of the cross section after the yield strength. Analysis of the load-bearing capacity of the cross-section under the nonlinear behavior of the material showed that the cross-sections have a significant plastic reserve in the load-bearing capacity. The consequence of the cross-sectional stress in the nonlinear range of material behavior are deformations of the cross-sections of the rods and the bearing system.

**References**

[1] Alfirević I and group of authors 1996 The Engineering Manual 1 *Školska knjiga Zagreb* pp 587-588

[2] Androić B, Dujmović D and Džeba I 1994 Steel constructions 1 *Civil Engineering Institute of Croatia* pp 182-197

[3] Demirović B and Osmić N 2019 Numerical analysis of reinforced concrete beam in two-dimensional form *Electronic Journal of the Faculty of Civil Engineering Osijek-e-GFOS* **18** 36-47 https://doi.org/10.13167/2019.18.

[4] Demirović B, Požegić Z and Memić M 2015 Numerical modelling of material nonlinearity of the rod *10th International Scientific Conference RIM Dubrovnik* **10** 415-420.

[5] HRN EN 1993-1-1 Eurocode 3 2008 Design of steel structures – Part 1-1: General rules and rules for buildings (EN 1993-1-1:2005+AC:2006) *HZN*

[6] Hasanović V 2010 Calculation of reinforced concrete structures according to EC2 and DIN 1045-1 *Faculty of Civil Engineering of the University "Džemal Bijedić" in Mostar* pp 55-65

[7] Mihanović A, Marović P and Dvornik J 1993 Nonlinear calculations of reinforced concrete structures *Croatian Society of Structural Engineers* pp 2-77

[8] Tomičić I 1996 Concrete structure *Croatian Society of Structural Engineers* pp 143-148

IOP Conf. Series: Materials Science and Engineering 1208 (2021) 012044 doi:10.1088/1757-899X/1208/1/012044

# Assessment of masonry structure "Radnički dom" in Mostar

**Amra Šarančić-Logo, Marko Ćećez, Merima Šahinagić-Isović**

Dzemal Bijedic University of Mostar

E-mail: marko.cecez@unmo.ba

**Abstract**. The paper presents the assessment of the building "Radnički dom" (Workers' Home) in Mostar, which was built in the Austro-Hungarian period, and represents one of the buildings of cultural and historical significance, located in the area of the historic urban core of the city. The paper explains the steps in assessing the condition of the existing structure, which include the collection of existing documentation, structural inspections, tests and calculations, and the assessment and decision on further action. The paper presents the drafts and gives descriptions of the performed visual inspection and the performed static calculation of the existing structure. At the end of the paper, recommendations are given for the rehabilitation and consolidation of the walls of the building: classical methods (injection and grouting) as well as modern methods (carbon strips). The paper points out the complexity of the procedure for the restoration of cultural and historical heritage buildings, the need for valid expertise of the condition and causes of building degradation, the importance of designing details of new structural elements and their corresponding and adequate connections with the original structure of the object.

## 1. Introduction

The historic urban core of the city of Mostar abounds in buildings from different historical periods, from the pre-Ottoman, Ottoman to Austro-Hungarian periods. All these objects have experienced, for various reasons, a significant degradation of the basic constructive elements. More recently, there has been an increase in tourism as the leading tertiary activity in this region. The area of the historic urban core, recognized as a cultural monument in the middle of the 20th century, has been revalorized in the last 15 years from the point of view of cultural and historical heritage protection, and appropriate protection measures have been defined by national heritage protection institution. Part of the area of the historic core was inscribed on the UNESCO World Heritage List in 2005 (https://whc.unesco.org/). From 1998 to 2004, the renovation of buildings in the historic urban core took an active part. The conservation of a number of buildings from all historical periods continues today, but with a different approach, more precisely with the different attitude towards heritage.

Challenges in the conservation of cultural and historical heritage (conservation meaning general approach and not particular technique) are numerous, such as the choice of original, identical or modern but appropriate materials, preservation, as much as possible of the traditional way of construction, and similar conservation-constructive challenges and aspects of performing the necessary works. In addition to the above, there are a number of requirements and restrictions, especially in the process of adapting buildings to new purposes and requirements for use. Namely, it is very important that during the intervention, in the process of planning and execution, there is no damage or loss of those characteristics

IOP Conf. Series: Materials Science and Engineering 1208 (2021) 012044 doi:10.1088/1757-899X/1208/1/012044

of the object that give it importance or make its values, affect the authenticity, integrity and / or originality.

In the current practice of working on objects of cultural and historical heritage and in historical cores, certain principles and principles of action have crystallized. Some of the relevant ones for this paper can be formulated as follows [3]:

- not to "invent" or "falsify" the historical features of the object / environment,
- not to copy the heritage of other environments and cultures,
- ensure a clear legibility of the intervention, i.e. without conspicuous emphasis on the new as well as without changing the authentic and original,
- ensure readability and presentation of the historical stratification of buildings,
- not to adapt the objects of cultural and historical heritage to new contents, but to adapt the content to the given space and construction,
- use, as far as possible, materials identical to or identical to those found on the building.

In the area of the historic urban core of Mostar there are a large number of buildings of cultural-historical and / or ambient values, in which the basic elements of the structure are largely damaged by war destruction or degradation during a long period of lack of maintenance and / or exposure to atmospheric influences. The approach to conservation may be similar for several buildings; although there are common elements in all buildings, the solution to each problem is individual and depends on a number of specific details [1]. Solving these problems considers following steps: determination the characteristics of the basic materials of the building, and the possibility of applying new materials, in the spirit of the original construction of the building [3]. In this way, in compliance with all conservation guidelines, we conserve or restore buildings as much as possible on the original condition. Buildings become not only new polyvalent spaces in the city center, but also make a significant contribution to the appearance of streets in the historic urban core. Proper approach to conservation directly contributes to the preservation and presentation of the values and significance of the historic urban core of the city, the rehabilitation of buildings as well as ambient values of the streets and, ultimately, life in them.

**Figure 1.** West facade of the "Radnički dom" building

The procedures and tests of structures themselves require a multidisciplinary approach, observation of the building in the construction, urban, conservation and architectural aspects [2]. In a number of examples in the area of Mostar, the approach was one-sided, so we have witnessed how during the "renovation" some of buildings lost their originality, values, and some became inconsistent with the environment, threatening the monumental features of the historic core.

The basic principle in working on objects of cultural and historical heritage is to respect the existing or original and authentic condition, which we determine through authentic documentation [1]. Respect refers primarily to design, construction and construction details, and materialization. This can be quite difficult, be-cause for most buildings from the 19. century or older, generally do not exist or it is difficult to find the original designs and photographs, and the buildings are of-ten more or less changed over time. In conservation practice, the application of the principle of analogy in some cases is a solution for buildings that have been significantly changed or damaged over time, and there is no authentic documentation on their original appearance. The application of the principles must be based on detailed research and analysis and clearly explained in the project documentation. During conservation works, it is necessary to use materials of the same or identical origin as much as possible, and to return the original material back to the building if the decomposition procedure was necessary. Of course, the application of new materials and construction techniques is also possible. The degree of application of new materials as well as the type of the same directly depends on the degree of authenticity and integrity, i.e. the values of the object on which it operates. For buildings such as the Radnički dom building, since its basic value lies in the exterior design and streetscape, new types of mortars and grouting methods can be used to consolidate the walls and inject agents to create an anti-moisture barrier. When forming the slabs, foam concrete or granular slab structure with gypsum boards or similar can be used (drywall). On the other hand, however, it is necessary to preserve as much as possible the elements of the building that characterize the facades, especially the street one. Profiles (in stone or mortar), due to the exceptional degradation of the elements over time, need to be upgraded with elements of similar composition, and the joint should be reinforced with graphite or other sticks. During the procedure, it is necessary to fully preserve the design of the profile and its position on the facade. Plaster, as one of the key parts of the appearance of the building from the Austro-Hungarian period, must be examined in detail with regard to mechanical and chemical characteristics, as well as color. Testing is necessary in order to be able to determine the properties of the material from which the new plaster will be made or to repair the old ones in places, and to define the color of the object. It is also necessary, if new hydro-phobic coatings and new binders are used, to determine their compatibility according to the chemical composition of the stone as the basic raw material from which the walls are built.

Only if it is not possible to repair the damaged element with this or similar repair, the stone or part of the facade can be replaced with new materials and elements.

The aim of this paper is to present the assessment of the building Radnički dom (Workers' Home) in Mostar in Maršala Tita Street (Figure 1). The building is today a neglected and abandoned building, of which only the outer structure of the walls remains visible. The ground floor of the building was walled up for security reasons, so it was not possible to see the entire interior of the building during investigation work. Therefore, the work is based on the basic methods and principles of facility rehabilitation.

## 2. Construction assessment

The main goal of structural assessment is to provide answers to questions about the condition of the observed structure on the basis of measurements and tests. It is necessary to determine the critical elements of the structure since the overall condition of the building dictates the condition of its weakest part. The assessment sequence for existing buildings can be divided into several steps [2, 5]:

- collecting or reconstructing existing documentation,
- inspecting the building,
- testing and calculations,
- evaluating and deciding on next steps.

The first historical sources and documents about the building, which was later named Radnički dom, date from 1890. The first preserved written documents about the Radnički dom building, that is, the original drawings of the building designed by the famous builder Miroslav Losse dates from September 1927. A historical review shows that this building has served several purposes throughout history and over time, various reconstructions, finishing and adaptations have been made on it (Figure 2).

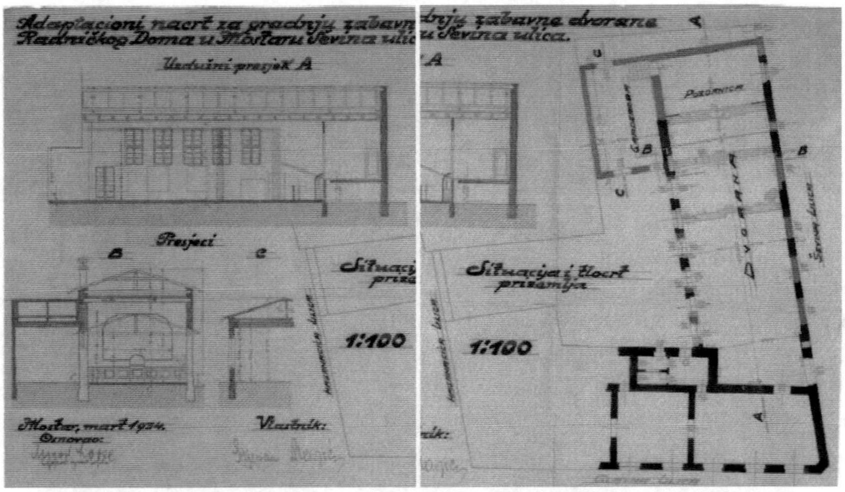


**Figure 2.** The first blueprints of the "Radnički dom" building

During the war in Bosnia and Herzegovina, the building suffered extensive damage. In the Radnički dom building, on July 19, 2012, a fire occurred (Figure 3), during which the interior of the building was additionally destroyed, on which ad-equate rehabilitation and adaptation had not been done for years. The fire caused additional damage to the building. The roof of the building burned in the fire, and great material damage was done on the first floor of the building.

**Figure 3.** Fire in the "Radnički dom" building in 2012

The Radnički dom building has a simple, rectangular floor plan, with approximate dimensions of 24.50m x 9.40m. The building is oriented with a longer façade to the west. Height is two floors (ground floor and first floor). The facility is currently not used. All visible exterior facades of the building have plastered facades. Major damage to the plaster is visible on the facades.

The base of the ground floor of the building is lined with carved stone. All openings on the facades are framed by stone chambers. All the carpentry of the building is visibly damaged, and the openings of the ground floor are walled up with concrete blocks. Due to the inaccessibility of the interior space of the slabs, this condition of the interior has not been examined in detail.

### *2.1. Damage analysis*

After making the architectural drawings of the current condition of the building, a detailed review of the existing condition has started, and all identified damages were identified and positioned on the drawings.

During the detailed visual inspection of the structure, special attention was paid to: appearance and differences in the color of the structure surface, cracks, their size and arrangement, signs of material degradation on the structure surface, deformations on the structure, and places of leakage or water retention, i.e. on wet surfaces. Going out on the field, the following was established (Figure 4):

- greater damages to the plaster are visible on both facades,
- all the carpentry is visibly damaged, and the openings on the ground floor are walled up with concrete blocks,
- there is no roof and floor structures,
- local cracks and decay of plaster are visible all over the facade,
- damage to the cantilever part is visible on the parts of the cornice,
- inspection of the yard does not show traces of cracks or differential subsidence of the building,
- there are mostly local damages on the windowsills.

**Figure 4.** North (left) and south (right) facade

### *2.2. Testing methods*

As part of the construction assessment, the next step is testing and calculations. Within the testing methods non-destructive and destructive testing can be used and should be carried out on the site of the building or in the laboratory [2]. Some common non-destructive tests that can be performed on these types of buildings are: rebound hammer, pull-out, ultrasonic, thermography, radiography etc. [6]. Since the building is in an extremely dilapidated condition, a decision was made not to test the material, and the assessment of further action was made based on previous experiences on similar structures [5].

The calculation model on which the calculation of the existing structure is per-formed should most adequately show the behavior of the structure, the resistance of its parts and the loading on the structure in accordance with the actual condition of the existing structure. The available documentation on the Radnički dom building did not contain data for the design of the structure.

Structural calculation refers to the design of critical points on the structure and the calculation of the entire system with realistic parameters, determined through the review and characteristics of the basic material. Based on the above, it was necessary to make a calculation of the roof structure, slabs and stairs in the program Tower 6 (Figure 5).

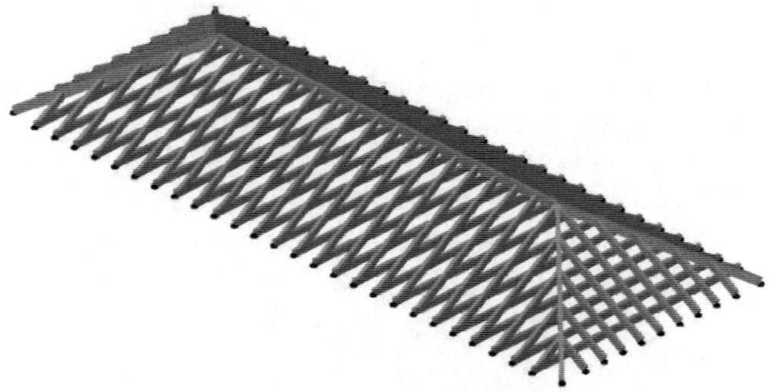

**Figure 5.** Model of the roof

### *2.3. Assessment and decision on further action*

The course of construction assessment ends with an answer to the question of what steps should be taken and how to deal with structures in the future. Recommendations for short-term and long-term activities should be made. After the analysis of the occurrence of damage, as well as the analysis of individual elements of the structure through the calculation of the existing structure, the proposal of intervention was prepared:

- Reconstruction of the slabs between the ground floor and the first floor
- Roof construction reconstruction
- Reconstruction and extension of stairs
- Conservation and consolidation of walls

*2.3.1. Reconstruction of the slabs between the ground floor and the first floor.* The new construction is designed as a series of IPE 360 steel girders that carry a span of approx. 860 cm and are spaced apart with a span of approx. 140 cm. A 10 cm thick reinforced concrete slab is made between the girders. The girders are not calculated as composite, but the coupling is performed to achieve a quality fit of the plate. The girders are made in the minimum class of steel S 275 J0, and the reinforced concrete slab of concrete C25/30 (Figure 6).

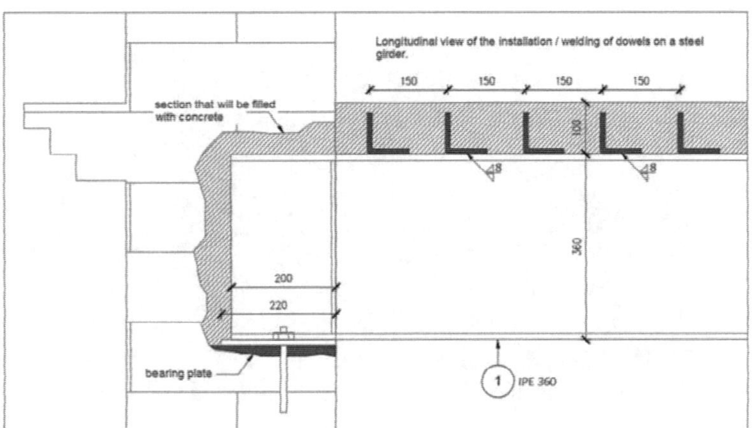


**Figure 6.** Detail of the bracket resting on the wall

*2.3.2. Roof construction reconstruction.* The roof is designed as a gabled roof (gabled with sloping roofs). The rafters are made on an axial grid of approximately 89 cm and are made in class C30 wood. The dimensions of the rafters are 14cm x 26cm. The rafters of the east-west orientation rest on horizontal beams that rest on the wall (16cm x 10cm). The rafters of the north-south orientation rest directly on the wall. The masonry is anchored in a new reinforced concrete horizontal circle, which is performed along the perimeter of the entire building. Horizontal beams have a dimension of 14cm x 26cm and are made of wood class C30. All other wooden beams are also made of C30 class wood (Figure 7).

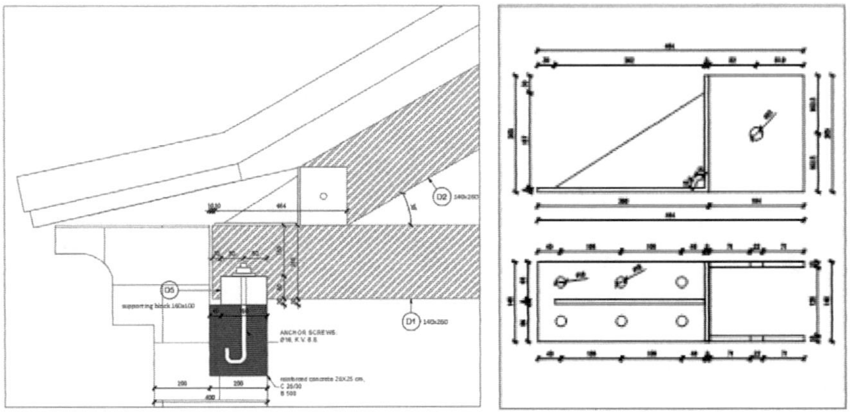

**Figure 7.** Detail of the connection between the roof rafter and the lower girder

*2.3.3. Conservation and consolidation of walls.* Which type of strengthening method will be applied depends on type of the structure, quality of the material of the existing structure, availability of adequate equipment and workers' skills and knowledge, and above all its seismic resistance [7]. Dilapidated, weakened or insufficient quality walls are strengthened depending on the nature of the damage. The basic remedial intervention to strengthen the walls is the appropriate treatment of joints. When the binder in the joints is missing or extremely dilapidated, the restoration of the joints is a much more effective remedial intervention than it seems at first glance, both in terms of ensuring the durability of the wall, and in a static sense. The work should be done with quality and appropriate materials. This requires that the joints be cleaned to a depth of 2 to 3 cm and a new binder installed [3]. It should be a lime mortar possibly with the addition of white cement. Mortar is installed in joints with mason's spoons, but there are also aids by which the mortar is pressed with a pressure nozzle. Such equipment ensures better filling of especially narrow joints (Figure 8).

**Figure 8.** Mortar injection under pressure

The next commonly used procedure for repairing and strengthening walls is 'grouting' [3]. It can also repair cracks and the entire structure of the wall structure. These can sometimes be separate jobs (cracks - wall structure), and some-times it can be done at the same time. Injection is most often performed through a series of injection wells arranged in the so-called chess scheme. The number of wells is calculated approximately - one well per 1 -2 m$^2$ of wall. The expected consumption of the injection mixture ranges from about 3 - 6% of the wall volume [3]. Weakened corners of the walls of their collisions, etc. we can effectively strengthen it by installing rod anchors. Rod anchors are usually made of ribbed reinforcement (stainless steel reinforcement should be used, i.e. ordinary reinforcement should be adequately corrosively protected). These are usually bars with a diameter of 14 to a maximum of 20 mm, and a length of about 1-2 m (rarely longer with this type of construction). Anchors are installed in previously drilled wells (diameter about 20 - 30 mm, length about 20 cm more than the length of the steel anchor). Drilling, especially of longer lengths, also requires appropriate drills - drilling accessories. After the anchor is installed, the well is filled with injection mixtures, from the bottom up. For this purpose, a suitable injection tube should be attached to the steel rod. When the damage to the stone wall is pronounced and where it is not possible to apply the described method, the only remaining solution is rebuilding, i.e., to dismantle (carefully decompose) the wall and rebuild it (re-composition) using the same stone. It must be borne in mind that a piece of new stone will also be needed. When rebuilding, the masonry style of the original decomposed wall should be respected [3].

The goal of each method, traditional and modern, is to make the body of the wall compact (grouting and pointing), and to allow additional materials (mortars, shotcrete or carbon strips) to give the wall in critical sections elements that can with-stand tensile stresses. More modern methods of remediation are based on new materials, primarily carbon strips and biaxial carbon panels (panels that carry in two directions) [5].

The existing building is built of stone walls. There is plaster on the outside of the building. If larger crack systems are detected during the removal of the plaster, one of the ways to fix the walls can be with carbon strips according to Figure 9.

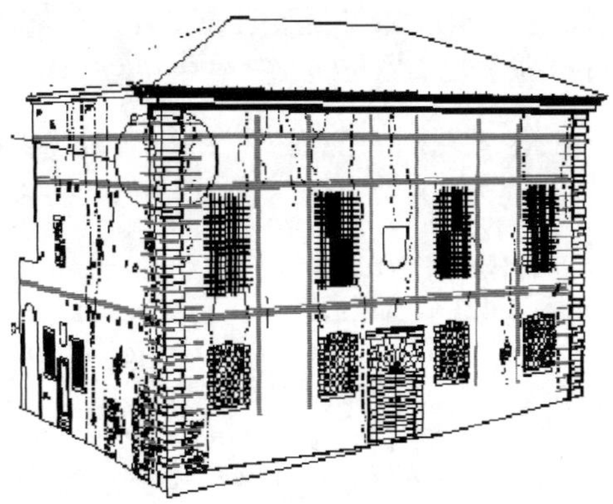

**Figure 9.** Strengthening with carbon stripes (www.akroterij.hr)

Fiber-reinforced polymers (FRP) are composite materials, developed during the 1940s for military purposes and aerospace applications, and are now used to rehabilitate and upgrade existing buildings in the form of reinforcements. The application of these fibers in the reconstruction of old stone buildings was due to the conclusion that a large number of damages to the buildings occurred due to tensile stresses in the wall that can not be overcome by classical methods (mortars and grouting) [8]. In such cases, we resort to the method of strengthening the structure with a fiber grid (Figure 9).

## 3. Conclusion

This paper presents the procedure of construction assessment, rehabilitation and partial reconstruction that can be applied to buildings that have certain values and degrees of authenticity and integrity in the historic urban core of the city of Mostar. These procedures are carried out for the purpose of conservation and adaptation to the new purpose of these facilities. On the example of the building of the Workers' Home (Radnički dom) in Maršala Tita Street, a building from the Austro-Hungarian period of Mostar architecture, the principle of assessment consisting of: collection of existing documentation and drawings, visual inspections, tests and static calculations and finally evaluations and decisions and maintenance.

Methods for consolidating and strengthening stone masonry are presented, depending on the causes of degradation and / or cracking of the wall itself. Classical methods (joint treatment - grouting, grouting, wall reinforcement by plastering) are presented, as well as newer methods such as carbon strips as anti-seismic elements in wall reinforcement.

The paper aims to point out the complexity of the procedure in the conservation of historic buildings, the need for valid examination of the condition and causes of degradation of the structure, and the importance of designing details of new structural elements and their fit with the original structure of the building.

## References

[1] Radić J 2010 *Trajnost konstrukcija* (Zagreb: University of Zagreb)

[2] Zlomušica E, Šahinagić-Isović M, Ademović N 2020 *Elementi održivosti okolinskih infrastrukturnih sistema* (Mostar: Faculty of Civil Engineering, Dzemal Bijedic University of Mostar)

[3] Živković Z 2015 *Tradicijska kamena kuća dalmatinskog zaleđa* (Zagreb: Ministarstvo turizma Republike Hrvatske Ministarstvo vanjskih i europskih poslova Republike Hrvatske)

[4] Španić M, Hadzima-Nyarko M., Morić D 2012 Strengthening of historical buildings with composite polymers *Electronic journal of the Faculty of Civil Engineering Osijek* e-GFOS **3** **5** 74-85

[5] Šahinagić-Isović M and Ćećez M 2017 Reconstruction of the municipal court, national monument building in Mostar *Proceedings of the 1st International Conference on Construction Materials for Sustainable Future Zadar* 718 – 723

[6] Glavaš H, Hadzima-Nyarko M, Haničar Buljan I, Barić T 2019 Locating Hidden Elements in Walls of Cultural Heritage Buildings by Using Infrared Thermography *Buildings* **9** **2** 32 doi:10.3390/buildings9020032

[7] Ademović N and Hadzima-Nyarko M 2018 Seismic Vulnerability, Damage and Strengthening of Masonry Structures in the Balkans with a Focus on Bosnia and Herzegovina *Proceedings of 16th European Conference on Earthquake Engineering Solun* 1-12.

[8] Šahinagić-Isović M and Ćećez M 2015 The application of fiber reinforced materials for cultural-historical heritage buildings *3rd International Conference The Importance of Place, Sustainability and heritage in a world of change Sarajevo B&H* 92.

**Institute of Physics**
Dirac House, Temple Back
Bristol BS1 6BE UK

ISSN: 1757-8981
ISBN 978-1-7138-4667-3

CURRAN ASSOCIATES INC.
proceedings.com